STOCHASTIC DYNAMICS OF MARINE STRUCTURES

This book is meant to serve as a text for students and a reference for professionals on the basic theory and methods used for stochastic modeling and analysis of marine structures subjected to environmental loads. The first part of the book provides a detailed introduction to the basic dynamic analysis of structures, which serves as a foundation for later chapters on stochastic response analysis. This includes an extensive chapter on the finite element method. A careful introduction to stochastic modeling is provided, which includes the concepts of stochastic process, variance spectrum, random environmental processes, response spectrum, response statistics, and short- and long-term extreme value models. The second part of the book offers detailed discussions of limit state design approaches, fatigue design methods, equations of motion for dynamic structures, and numerical solution techniques. The final chapter highlights methods for prediction of extreme values from measured data or data obtained by Monte Carlo simulation.

Arvid Naess has been a professor of structural engineering since 1987 and a professor of mathematical statistics since 2001 at the Norwegian University of Science and Technology. He works on a wide range of problems related to stochastic dynamics of structures and structural safety and reliability. Professor Naess has published more than 250 scientific papers and lectured at conferences and universities worldwide. He is the associate editor of several international engineering journals. Professor Naess is a recipient of the Alfred M. Freudenthal Medal from ASCE and is an elected Fellow of the ASME and ASCE.

Torgeir Moan has been a professor of marine structures at the Norwegian University of Science and Technology since 1977. He was the first (adjunct) Keppel professor at the National University of Singapore (2002–2007) and Director of the Centre for Ships and Ocean Structures (CeSOS), a Norwegian Centre of Excellence since 2002. Professor Moan's work focuses on structural analysis and design of marine structures, with an emphasis on structural risk and reliability analysis, as well as probabilistic analysis of wave- and wind-induced stochastic dynamic load effects. He has published more than 450 refereed journal and conference papers. He is an editor of the *Journal of Marine Structures* and serves on the boards of several other journals. Professor Moan received various awards for his research, including the Statoil and the ASME James W. Rice Awards. He is an elected Foreign Member of the Royal Academy of Engineering in the UK and Fellow of ASCE, IABSE, and The Offshore Energy Center Hall of Fame.

Stochastic Dynamics of Marine Structures

Arvid Naess
Norwegian University of Science and Technology

Torgeir Moan
Norwegian University of Science and Technology

CAMBRIDGE
UNIVERSITY PRESS

CAMBRIDGE
UNIVERSITY PRESS

University Printing House, Cambridge CB2 8BS, United Kingdom

One Liberty Plaza, 20th Floor, New York, NY 10006, USA

477 Williamstown Road, Port Melbourne, VIC 3207, Australia

4843/24, 2nd Floor, Ansari Road, Daryaganj, Delhi - 110002, India

79 Anson Road, #06-04/06, Singapore 079906

Cambridge University Press is part of the University of Cambridge.

It furthers the University's mission by disseminating knowledge in the pursuit of education, learning and research at the highest international levels of excellence.

www.cambridge.org
Information on this title: www.cambridge.org/9780521881555

First published 2013

A catalogue record for this publication is available from the British Library

Library of Congress Cataloging in Publication data
Naess, Arvid.
Stochastic dynamics of marine structures / Naess, Norwegian University of Science and Technology, Moan, Norwegian University of Science and Technology.
pages cm.
Includes bibliographical references and index.
ISBN 978-0-521-88155-5 (hardback)
1. Offshore structures. 2. Structural dynamics. 3. Stochastic processes – Congresses. I. Moan, Torgeir. II. Title.
TC1665.N34 2012
627′.980151923–dc23 2012024936

ISBN 978-0-521-88155-5 Hardback

Additional resources for this publication at www.cambridge.org/naess

Contents

Preface

This textbook provides the material for both basic and intermediate modern courses in dynamic analysis of ships and offshore structures. The word "modern" is used to signify that both deterministic and stochastic dynamics are covered. Because the main goal is to provide an introduction to dynamic analysis, the basic elements are described in some detail. A consequence of this is that the majority of the book deals with structures or structural elements that can be modeled or reduced to a single-degree-of-freedom (SDOF) system. However, realizing that multi-degree-of-freedom (MDOF) systems are unavoidable in many practical situations the engineer is likely to meet, and, consequently, that the basic principles for their analysis should be understood, a chapter on linear MDOF systems is included. This is also done to clearly demonstrate the principle of modal decomposition whereby an MDOF system is reduced to a set of uncoupled SDOF systems.

Broadly speaking, a dynamic analysis is carried out in two different ways according to how the loading is specified. If the time-variant loading is given in such a way that we may consider it to be exactly known as a function of time, the same will apply to the response. In such a case, the dynamic analysis is called deterministic. This is in contrast to a stochastic analysis, where the loading is specified using probabilistic concepts. This implies that the corresponding displacements and tensions can only be described in the same way. Even if naturally occurring loading to which a structure is subjected, such as wind and waves, can be claimed to be deterministic, its specification in terms of fundamental physical laws will remain beyond reach for any foreseeable length of time. For such types of loading, a stochastic description has proven to be exceedingly useful.

The first part of the book (Chapters 2–4) describes fundamental aspects of a deterministic dynamic analysis, with emphasis on simple but important dynamic problems relevant for marine structures. The second part (Chapters 5–16) provides a rather extensive introduction to stochastic dynamics of marine structures. Even though the book is focused on marine structures, with a suitable selection of material, it may also serve as a textbook for a more general course in the deterministic and stochastic dynamics of structures.

This book is supported by a Web site (www.cambridge.org/naess) containing numerous problems, many related to ships and offshore structures, that will make it useful not only for students, but also for professional engineers.

Many colleagues and friends offered their comments and suggestions for improving the book manuscript. In particular, the authors want to thank Professors K. M. Mathisen, E. Hjorth-Hansen, G. Moe, L. V. S. Sagrilo, B. J. Leira, S. Haver, N. Saha, and Z. Gao. The help rendered by Huirong Jia, Biao Su, and Mahmoud Etemaddar is also greatly appreciated.

Figures and illustrations are an important ingredient in a book like this. Most of these were done by Ole Erik Brandrud Naess and Gaute Halvorsen. Their expert assistance is much appreciated.

The authors gratefully acknowledge the support of the Norwegian Research Council through the Centre for Ships and Ocean Structures (CeSOS) at the Norwegian University of Science and Technology. The first author would also like to thank the Norwegian Non-Fiction Writers and Translators Association (NFF) for its support.

1 Preliminaries

1.1 Introduction

This book provides the reader with the basic procedures for performing a dynamic analysis of marine structures subjected to environmental stochastic load processes such as wind and ocean waves. The dynamics of rigid bodies and flexible structures are considered. As opposed to the static analysis of structures, the dynamic analysis of a structure concerns itself with time-variant external forces, inertia, damping, reaction forces, and the corresponding responses. Dynamic behavior differs from static behavior. Consider, for instance, the cantilevered tower shown in Fig. 1.1a.

Figures 1.1(b) and 1.1(d) display the bending moment in the tower with a deck (i.e., mass) on the top when the hydrodynamic forces from a long wave are considered as static and dynamic, respectively. Obviously, the dynamic case is relevant for a certain time variation and an instant in time. The difference is due to the inertia forces (on the deck Q_i and along the tower q_i).

Clearly, dynamic behavior is more complex than static, and its calculation is also more demanding. This becomes even more clear by considering a buoyant, rigid, articulated tower such as the one shown in Fig. 1.2. The static response is such that the tower rotates in such a way that the net effect of buoyancy and gravity balances the external forces (expressed by zero total moment with respect to the pin joint). If the external forces are considered as time variant, the external force at any time instant is balanced by the mentioned reaction forces as well as by inertia and damping forces. In practical situations (when resonance is not an issue), the inertia forces would typically dominate.

The loading on the structure is then described in terms of a load vector where one or more of the quantities, size, direction, or position, varies with time. As a consequence, the structural response to the dynamic loading that is, the resulting displacements, internal forces etc. will also be time varying or dynamic.

Broadly speaking, a dynamic analysis is performed in two different ways according to how the loading is specified. If the time-variant loading is given in such a way that we may consider it to be exactly known as a function of time, the same will apply to the response. In such a case, the dynamic analysis is called deterministic. This is in contrast to a stochastic analysis, where the loading is specified using probabilistic concepts. When this is done, the corresponding displacements and tensions can only

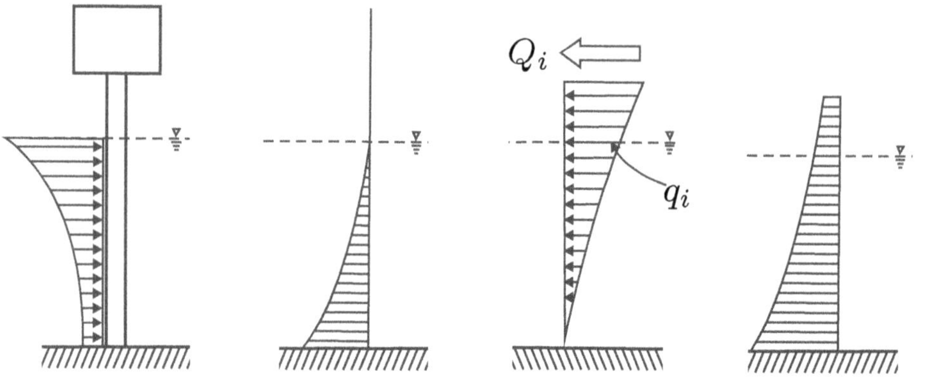

(a) Tower and external load due to a long wave (b) Static bending moment (c) Inertia reaction forces (d) Dynamic bending moment

Figure 1.1. Schematic of wave-induced static versus instantaneous dynamic forces and moments in a bottom-fixed cantilevered tower.

be described similarly. Even if, in principle, the naturally occurring loading that a structure is subjected to, such as wind and waves, can be claimed to be deterministic if the laws of physics and the initial conditions were known, their exact description and analysis within such a framework are believed to remain beyond reach for any foreseeable length of time. Fortunately, the complexity of these natural phenomena is such that they can be accurately modeled as random processes. This opens the way for practical analysis and prediction of the response of structures subjected to environmental loads.

The response analysis is performed as part of the design verification of structures, according to serviceability and safety requirements. Serviceability requirements relate to the function of the structure and the possible equipment it is carrying, while safety requirements refer to avoidance of ultimate or fatigue failure.

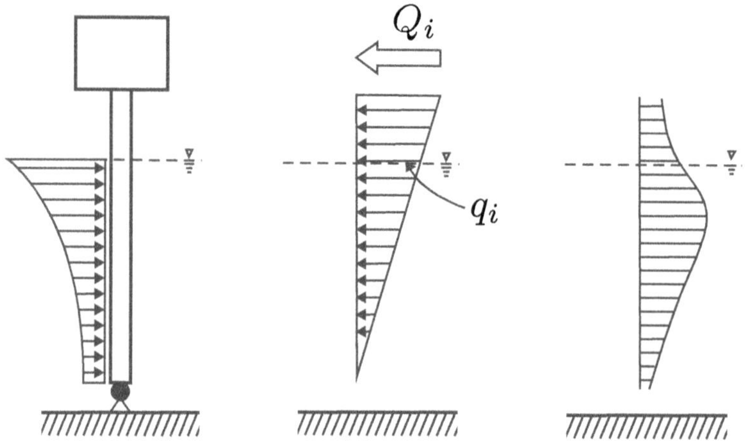

(a) Tower and external load due to a long wave (b) Inertia reaction forces (c) Dynamic bending moment

Figure 1.2. Schematic of instantaneous wave-induced forces and moments in an articulated tower.

The corresponding verification of ultimate or fatigue limit state criteria requires the extreme structural response and the whole response history, respectively.

1.2 Equations of Motion

The goal of a dynamic analysis of structures is to calculate the time history of the displacements, internal forces, or stresses at specific places in the structure. Before a dynamic analysis can be performed, most real structures would normally have to be represented by an idealized model.

Dynamic models fall into two basic categories, namely, continuous and discrete models. The number of variables – normally, displacement components – that need to be applied to describe the behavior is termed *the number of degrees of freedom* (nDOF). According to this definition, a continuous model represents an infinite DOF system, while a discrete model represents a finite DOF system.

Once the computational model of the structure is defined, the proper formulation of the equations of motion creates the mathematical model of the structure. A continuous model leads to partial differential equation(s), while a discrete model leads to ordinary differential equation(s).

When the mathematical model has been formulated, the next step deals with the solution of the differential equation(s) to predict the response of the real structure. Finally, the last step is the verification and confirmation of the results. This can be achieved by comparison of the computed results with dynamic tests or the responses of the models arrived at in different ways.

For plane motions of a rigid body in the reference (x, y)-plane, Newton's second law yields to the following equations:

$$m\ddot{u}_{cx} = \sum P_x(t), \tag{1.1}$$

$$m\ddot{u}_{cy} = \sum P_y(t), \tag{1.2}$$

$$J_c\ddot{\phi} = \sum M_c(t). \tag{1.3}$$

Herein, $\ddot{u}_{cx}$ and $\ddot{u}_{cy}$ denote the acceleration components of the center of mass C of the body along the x and the y axes; $\sum P_x(t)$ and $\sum P_y(t)$ are the corresponding resultant forces and $\ddot{\phi}$ is the angular acceleration; J_c represents the mass moment of inertia of the body with respect to an orthogonal axis through C, the center of mass; and $\sum M_c(t)$ is the sum of all moments acting on the body with respect to the axis through C.

Sometimes, especially when dealing with multiple degrees-of-freedom mass points with complex kinematic features, it is convenient to use Newton's second law recast in terms of Lagrange equations with reference to work and energy.

In analogy with the condition for static equilibrium, it is possible to introduce a corresponding condition for dynamic equilibrium by defining an auxiliary (vector) force, $\mathbf{f}_i(t) = -(m\ddot{u}_{cx}, m\ddot{u}_{cy}, J_c\ddot{\phi})^T$, which is often referred to as the inertia force. By rewriting Eqs. (1.1)–(1.3), letting $\mathbf{f}(t) = (\sum P_x(t), \sum P_y(t), \sum M_c(t))^T$, it is then obtained that

$$\mathbf{f}(t) + \mathbf{f}_i(t) = \mathbf{0}. \tag{1.4}$$

This rephrasing of Newton's second law of motion is usually called the *principle of d'Alembert*. Expressed in words, it may be stated as follows:

> *The condition for dynamic equilibrium is that the total force is in equilibrium with the inertia force.*

It is emphasized that the extension of the equilibrium concept we obtained by the introduction of the concept of dynamic equilibrium is synonymous with compliance with Newton's second law. This reformulation has resulted in a general condition of equilibrium, which is completely analogous to the condition of static equilibrium, and it contains this condition as a special case. When the acceleration is known (or is considered known), one can therefore recast the dynamic problem into the framework of a static problem by including the inertia force.

Although the treatment of a system of (rigid) mass points mainly requires the use of Newton's second law or derived energy formulations, the principles of continuum mechanics are needed to handle deformed bodies. They include equilibrium, kinematic compatibility and constitutive (stress–strain) relations, and may be used to formulate differential or energy expressions. In particular, the principle of virtual work provides a versatile tool for finite element discretizations of deformable bodies.

In this book, only the structural mechanics for bars, truss works, beams, and frames are described, while the continuum mechanics of plane stress, plate bending, shell, and solids are treated only in principle.

The principle of virtual work is briefly discussed. This principle is expressed here as an integrated form of d'Alembert's principle. Verbally expressed, it assumes the following form.

If the (generalized) forces in Eq. (1.4) are considered as functions of position, the principle of virtual work can be expressed as

$$\int_V \delta\mathbf{u}^T\mathbf{f}\,dV + \int_V \delta\mathbf{u}^T\mathbf{f}_i\,dV = 0, \tag{1.5}$$

where $\delta\mathbf{u}$ is a virtual displacement vector and V is the extension pr volume of the system. The force $\mathbf{f}$ is then understood as the external forces minus possible forces in the structure from damping and elastic resistance to displacement.

1.3 Stochastic Models

After the publication of the classical paper "On the motions of ships in confused seas," by St. Denis and Pierson (1953), where stochastic concepts were used for modeling the waves and the wave forces on a ship subjected to the ocean environment, it was generally recognized that this approach represented a rational and suitable way to account for the "unpredictability" of the hydrodynamic loading on a ship during an ocean voyage. During the more than 50 years since then, the stochastic modeling of environmental forces and the responses of the structures subjected to them has reached a fairly mature state. Today, stochastic modeling of the environmental loads and the induced responses is a routine procedure when designing ships and other ocean structures, even if the degree of sophistication varies depending on the type of structure and design provisions. In particular, taking account of the random character of wave and other environmental loads is important when the loads induce inertia and damping reaction forces. In this light, the main aspects of stochastic modeling of the environmental processes and loads, as well as the induced responses of structures, should be an important part of any textbook on the dynamics of marine structures. In this book, we therefore make an effort to discuss these aspects in some detail.

1.4 Organization of the Book

The first part of the book (Chapters 2–4) describes fundamental issues of a deterministic dynamic analysis with emphasis on simple, but important, vibration problems. The second part (Chapters 5–16) provides a fairly extensive introduction to stochastic dynamic analysis of marine structures, as well as applications of such analyses in the design process.

It is inevitable that a book of this type also to some extent reflects the authors' "world view." This is particularly apparent in the choice of a number of advanced topics that the authors have been particularly involved in.

2 Dynamics of Single-Degree-of-Freedom Linear Systems

2.1 Introduction

This chapter deals with vibrations of structures that can be represented as a single-degree-of-freedom (SDOF) system. This means that the oscillatory response can be completely described by one displacement variable. This may seem like a gross oversimplification for structures of engineering interest that leads to a theory of little practical significance. However, the theory of vibrations for systems of an SDOF is crucial for understanding the vibration response of more complex structures. Frequently, it is also the case that one may investigate the vibration response characteristics of apparently complex structures by directly applying the theory of vibrations of SDOF systems. This is demonstrated in Chapter 3 on multi-degrees-of-freedom (MDOF) structures.

The word "vibration" used in this chapter should be interpreted as meaning oscillatory response in a fairly general sense, e.g., as applied to marine structures.

2.2 Harmonic Oscillator – Free Vibrations

Free vibrations or oscillations occur when there are no external forces imposed on the structure, e.g., after an initial displacement and release. Two different situations are discussed: translational oscillations and rotational oscillations.

2.2.1 Motions of Marine Structures

Because the main focus of this book is the motion response of marine structures, it is expedient to define the terms commonly used to describe the rigid-body motions of floating structures. This is most easily done by referring to Fig. 2.1. For a shiplike structure, it is common practice to place the x-axis along the beam of the ship (for the body-fixed coordinate system), and call the corresponding translational motion for surge. The y-axis is placed in the water plane (or parallel to it). Whether the z-axis points into the water (downward) or out of the water (upward) may vary. In this book, the z-axis is invariably positive upward. Thus, there are three translatory motions: surge (x-axis), sway (y-axis), and heave (z-axis). Similarly, there are three rotational motions: roll (x-axis), pitch (y-axis), and yaw (z-axis).

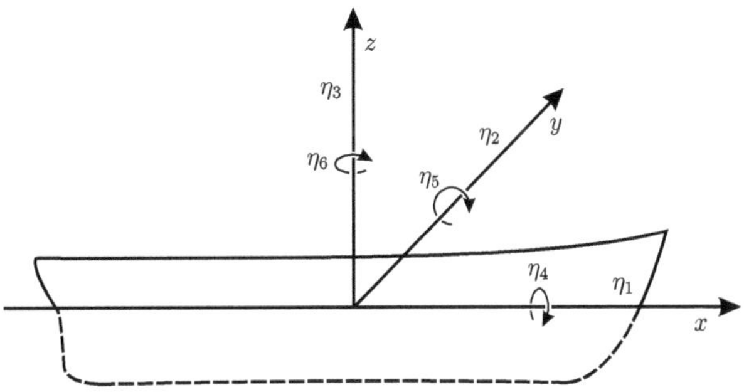

Figure 2.1. Definition of the motion response modes of a marine structure. η_1 = surge, η_2 = sway, η_3 = heave, η_4 = roll, η_5 = pitch, η_6 = yaw (SNAME, 1988).

For many offshore platform structures, there is no obvious "x-axis" in the same way as for a ship, and the placement of a local coordinate system is therefore to a larger extent arbitrary. In such cases, the x-axis is often placed along, or close to, the main wave direction.

2.2.2 Translational Oscillations

Figure 2.2 displays a schematic of an undamped (without friction) vibration system of an SDOF. The spring k is assumed to be without mass. It is also assumed to comply with Hooke's law; that is, it is assumed to be linearly elastic. The displacement u is considered positive to the right of the equilibrium point; that is, $u = 0$ at the equilibrium point. When the mass m is displaced a distance u, the spring k will impose a force $-ku$ on m, where the minus sign indicates that the force is directed against the displacement. Invoking Newton's second law then gives the relation

$$-ku = m\ddot{u} \tag{2.1}$$

or

$$m\ddot{u} + ku = 0 \tag{2.2}$$

To simplify language and notation, we let the letter m denote both the physical mass itself and its size in kilograms (kg). Similarly for the letter k, which denotes both a linearly elastic spring and the corresponding spring constant in Hooke's law ([k] = N/m).

Equation (2.2) can be rewritten as

$$\ddot{u} + \omega_e^2 u = 0 \tag{2.3}$$

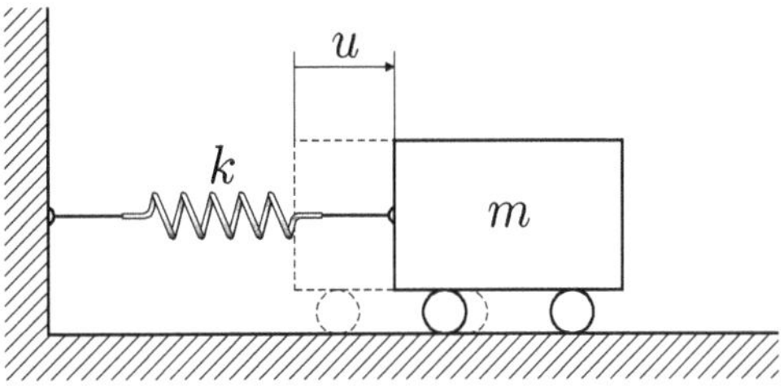

Figure 2.2. Principle sketch of the vibration system of an SDOF.

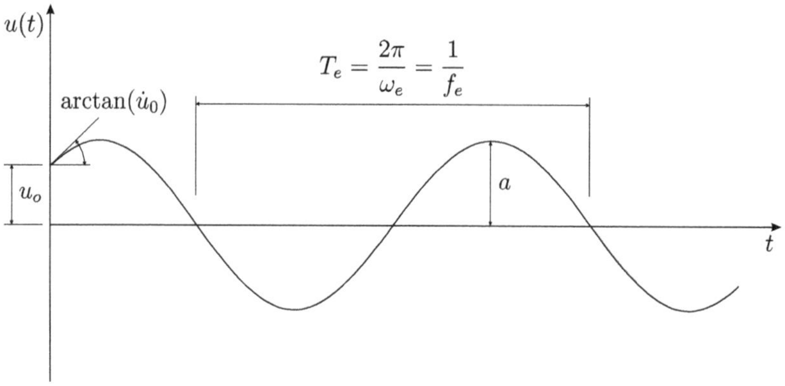

Figure 2.3. Undamped free oscillation response of amplitude a and period T_e.

by introducing the angular frequency ω_e defined by

$$\omega_e^2 = \frac{k}{m} \quad ([\omega_e] = \text{rad/s}) \tag{2.4}$$

ω_e is referred to as the natural frequency, or the eigenfrequency, of the oscillatory system.

One can easily verify that the harmonic functions $\cos\omega_e t$ and $\sin\omega_e t$ both satisfy Eq. (2.3). The general solution can therefore be written as

$$u(t) = a_1 \cos\omega_e t + a_2 \sin\omega_e t, \tag{2.5}$$

where a_1 and a_2 are two constants. The solution $u(t)$ represents an oscillation response with constant angular frequency ω_e, hence the name harmonic oscillator. Because the differential Eq. (2.3) is of second order, two constants are required to describe the general solution. If position and velocity at a particular point in time are given, for instance, $u(0) = u_0$ and $\dot{u}(0) = v_0$, then the solution will be uniquely determined as follows,

$$u(t) = u_0 \cos\omega_e t + \frac{v_0}{\omega_e} \sin\omega_e t. \tag{2.6}$$

This harmonic oscillation is illustrated in Fig. 2.3. It is seen from Eqs. (2.5) and (2.6) that $u(t) = u(t + 2\pi/\omega_e)$, while $u(t) \neq u(t + t')$ when $0 < t' < 2\pi/\omega_e$. This leads to the expression for the natural period of the system:

$$T_e = \frac{2\pi}{\omega_e} = 2\pi\sqrt{\frac{m}{k}}. \tag{2.7}$$

The natural frequency f_e ($[f_e] = \text{s}^{-1}$ or Hz) is therefore

$$f_e = \frac{1}{T_e} = \frac{\omega_e}{2\pi}. \tag{2.8}$$

Equation (2.5) may be rewritten as

$$u(t) = \sqrt{a_1^2 + a_2^2}\left(\frac{a_1}{\sqrt{a_1^2 + a_2^2}} \cos\omega_e t + \frac{a_2}{\sqrt{a_1^2 + a_2^2}} \sin\omega_e t\right). \tag{2.9}$$

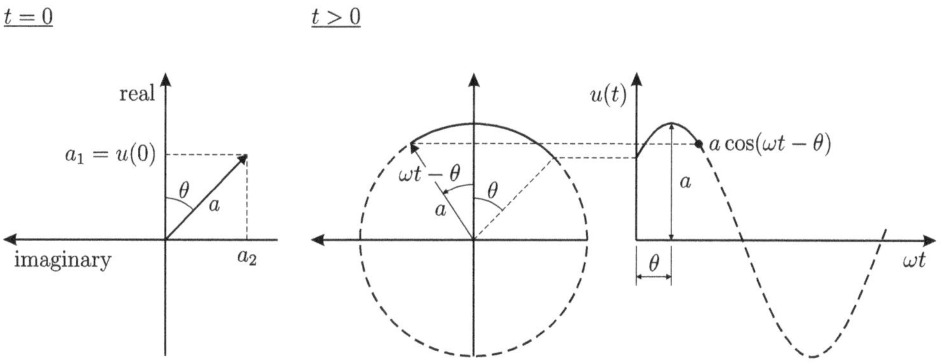

Figure 2.4. Argand diagram for the oscillation response shown in Fig. 2.3.

Because

$$\left(\frac{a_1}{\sqrt{a_1^2 + a_2^2}}\right)^2 + \left(\frac{a_2}{\sqrt{a_1^2 + a_2^2}}\right)^2 = 1, \tag{2.10}$$

there must exist an angle θ $(0 \leq \theta < 2\pi)$ such that $\cos\theta = a_1/\sqrt{a_1^2 + a_2^2}$ and $\sin\theta = a_2/\sqrt{a_1^2 + a_2^2}$. This implies that

$$u(t) = \sqrt{a_1^2 + a_2^2}\,(\cos\omega_e t\,\cos\theta + \sin\omega_e t\,\sin\theta) = a\,\cos(\omega_e t - \theta), \tag{2.11}$$

where the amplitude $a = \sqrt{a_1^2 + a_2^2}$ and the phase angle θ are determined by the initial conditions at $t = 0$. Because $\tan\theta = a_2/a_1$, θ can also be calculated from the relation $\theta = \arctan(a_2/a_1)$. There are two solutions for θ in the interval $[0, 2\pi]$, and the sign of $u(0) = u_0$ determines the correct one. When $u(0)$ is positive, the phase angle will be in the first or fourth quadrant. If $u(0)$ is negative, the phase angle will be in the second or third quadrant.

The oscillation response can be represented in the complex plane by using an Argand diagram, where $u(t)$ is the real part of the complex number $\mathbf{a}$ ($|\mathbf{a}| = a$), which rotates with constant angular frequency ω_e, see Fig. 2.4.

2.2.3 Example – Amplitude and Phase of a Free Oscillation

Assume a vibration response determined by Eq. (2.2) with $m = 10$ kg and $k = 400$ N/m. What will be the resulting amplitude and phase angle for the free oscillation when $u_0 = 0.1$ m and $v_0 = -0.2$ m/s?

The natural frequency of the system is $\omega_e = \sqrt{k/m} = 2$ rad/s. From Eqs. (2.2) and (2.11), it follows that the amplitude $a = \sqrt{u_0^2 + (v_0/\omega_e)^2} = \sqrt{0.1^2 + 0.1^2} = 0.14$ m. The phase angle $\theta = \arctan(v_0/(\omega_e u_0)) = \arctan(-1) = 7\pi/4$, corresponding to a positive displacement and negative velocity.

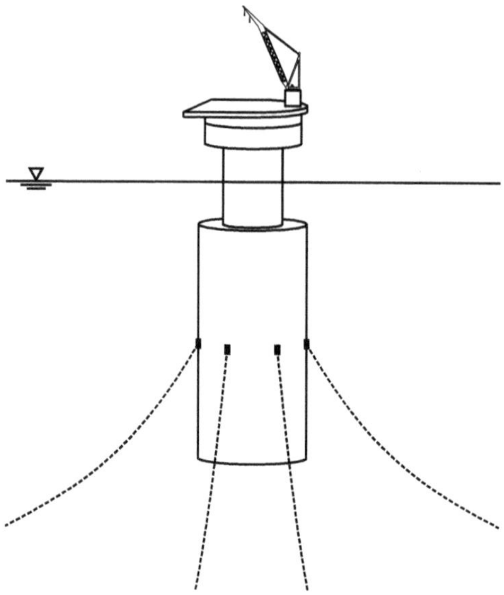

Figure 2.5. A coarse sketch of the Brent Spar.

2.2.4 Example – Heave Oscillations of a Spar Buoy

In the mid-1970s, the oil company Shell installed an oil storage and offloading spar buoy at the Brent offshore oil field in the North Sea at a water depth of 140 meters, see Fig. 2.5. The Brent Spar, or Brent E, was a conventional cylindrical buoy moored to the sea floor by six anchors. The cylindrical tank was 137 meters high with a diameter of 29 meters and a displacement of 66000 metric tons, which corresponds to a draft of 109 meters. The spar buoy was made up of oil storage tanks at the bottom, buoyancy tanks toward the middle and a topside containing the offloading and other equipment. Decommissioned in the mid-1990s, the Brent Spar was in operation for about 20 years. The original plans for decommissioning, which were basically to sink it in deep water, were fiercely attacked by Greenpeace and other environmental organizations, and these plans were eventually abandoned. Accounts of the "battle" of the Brent Spar can be found on the Internet.

The deep draft of the spar buoy makes it nonresponsive to normal sea states, and the heave response is hardly noticeable under ordinary operating conditions. This is connected with the long natural period of the heave oscillations of the spar, which can be determined by calculating the restoring force produced by a vertical displacement of the spar. The restoring force originates mainly from a change in the buoyancy effect. The mooring lines will contribute only marginally to the total restoring force in the heave direction. Hence, a vertical displacement of x meters will produce a restoring force $f_r(x)$ approximately equal to

$$f_r(x) = \rho g \pi R^2 x, \tag{a}$$

where $\rho = 1{,}025$ kg/m^3 = density of sea water, $g = 9.81$ m/s^2, and $R = 14.5$ m = radius of cylindrical tank. It follows that the spar buoy offers a restoring force that is a linear function of the displacement x; that is, $f_r(x) = Kx$. It is found that $K = 6.6 \cdot 10^6$ N/m for the Brent Spar.

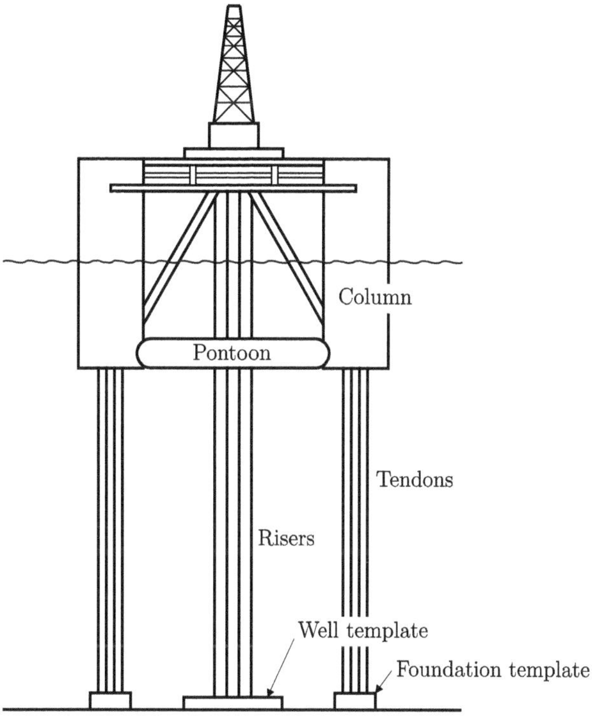

Figure 2.6. A sketch of a TLP structure.

To calculate the natural period in heave, it is necessary to determine the total effective mass, which will be somewhat larger than $6.6 \cdot 10^7$ kg given by the displacement alone. The reason for this adjustment is due to the fact that when an object is accelerated through water, a certain amount of water also has to be accelerated, thereby increasing the apparent inertia of the object. This increase of inertia can be interpreted as an added mass. In the specific case of this spar buoy, it is estimated to be about 10% of the displacement, giving a total mass of $M_h = 7.26 \cdot 10^7$ kg. The natural period in heave T_h of the spar buoy is now given as

$$T_h = 2\pi\sqrt{\frac{M_h}{K}} \approx 21\,\text{s}. \qquad \text{(b)}$$

2.2.5 Example – Heave and Surge Oscillations of a TLP

Figure 2.6 shows a sketch of a tension leg platform (TLP), which is a floating platform concept almost exclusively used for production purposes offshore, and thus represents a permanent installation. The name derives from the way this kind of structure is moored. From each leg of the platform there are vertical tendons or tethers, usually steel pipes, extending all the way down to an anchor template at the sea bottom. During installation of the tendons, the platform is ballasted down to a preassigned draft. Once the tendons are installed, ballast is then removed to create a desired tension in the tendons. This tension is made sufficiently high to avoid snap loads in the tendons during the passage of very high waves. A characteristic feature of the TLP is a very high resistance toward vertical motions of the platform, while there is low resistance toward horizontal motions.

The first TLP was installed at the Hutton Field at a water depth of 148 meters in the UK sector of the North Sea in 1984 by Conoco (it was removed in 2001). This was followed in 1988 by the Jolliet TLP at a water depth of 535 meters in the Gulf of Mexico. In the Norwegian North Sea, the first TLP was the Snorre platform, which was installed in 1992 at a water depth of 310 meters by Saga Petroleum (later merged with Statoil). Two years later the Heidrun TLP was in place at a location with a 350-meter water depth. Although all TLP hulls had previously been steel structures, the Heidrun platform had a concrete hull. The TLP concept continues to be a viable concept for deep water production facilities.

If we consider the heave motions of the TLP, the tendons will work as one big linear spring with a stiffness k_h determined by the cross-sectional area of steel in each tendon A_t, the modulus of elasticity of the steel E, and the length of the tendons L_t according to the formula

$$k_h = \frac{n_t\, A_t\, E}{L_t}. \tag{a}$$

In this formula, a possible contribution from the production risers was neglected. If relevant, such a contribution should be included. With a total mass of M_h, including added mass in heave (cf. Example 7.5.1), the natural period in heave T_h will be

$$T_h = 2\pi \sqrt{\frac{M_h\, L_t}{n_t\, A_t\, E}}. \tag{b}$$

The steel tendons are typically designed so that the TLP has T_h in the range 2 to 4 seconds. The reason for this will become apparent later in this chapter.

The restoring force $f_r(x)$ resisting a small horizontal displacement x of the platform is given approximately by

$$f_r(x) = n_t\, T_t\, \frac{x}{L_t}, \tag{c}$$

where $n_t =$ number of tendons, $T_t =$ the tension in each tendon (assumed equal), and $L_t =$ the length of each tendon. Note that if the risers are in tension, they may in fact act as virtual tendons in this context and their contribution should be included. The preceding equation is also based on the assumption that the tendons are neutrally buoyant. For a case with constant tendon weight in water w per unit length, the horizontal restoring force becomes approximately (Demirbilek, 1989)

$$f_r(x) = \left(n_t\, T_t - \frac{w\, L_t}{2}\right) \frac{x}{L_t}. \tag{d}$$

For the case of neutrally buoyant tendons, the natural period of the surge oscillations is given by

$$T_s = 2\pi \sqrt{\frac{M_s\, L_t}{n_t\, T_t}}, \tag{e}$$

where $M_s =$ total mass of the TLP, including added mass in surge (cf. Example 7.5.1). With $M_s = 2 \cdot 10^7$ kg (20.000 tons), $n_t\, T_t = 10^7$ N, and $L_t = 200$ m, it

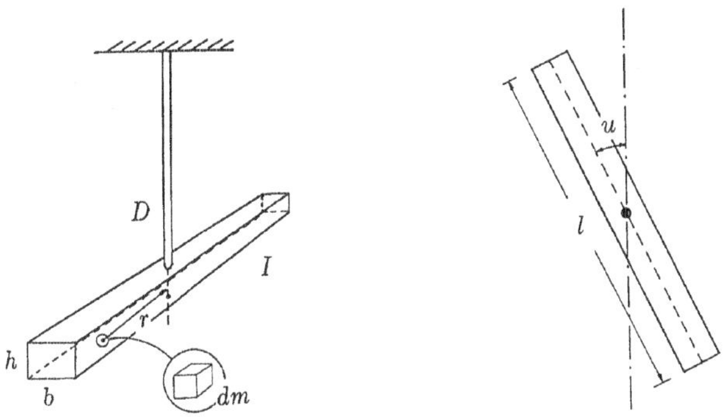

Figure 2.7. Stiff beam with mass moment of inertia I suspended in a torsion bar with torsional stiffness D.

is obtained that $T_s \approx 120$ s, which is a typical natural period for the surge and sway motions of a TLP.

2.2.6 Rotational Oscillations

Figure 2.7 shows a beam suspended horizontally by a bar assumed to be without internal friction, but with a linear rotational or torsional stiffness D. That is, at an angular displacement u the bar will impose on the beam a moment equal to $-Du$, when $u = 0$ defines the position at rest. By invoking d'Alembert's principle (moment equilibrium), it is obtained that

$$D\,u + \int_{beam} a_t\, r\, dm = 0, \tag{2.12}$$

where $u = u(t)$ denotes the angular response relative to the position of rest, r is the distance from the mass element dm to the axis of rotation, and a_t is tangential acceleration; that is, $a_t = r\,\ddot{u}$. This leads to the equation

$$I\,\ddot{u} + D\,u = 0, \tag{2.13}$$

where

$$I = \int_{beam} r^2\, dm \tag{2.14}$$

denotes the mass moment of inertia of the beam. For a homogeneous beam of mass m, length l, and cross section $b \times h$ suspended at the midpoint, it can be shown that $I = m(b^2 + l^2)/12$.

As for the case of translational vibration response, the solution to Eq. (2.13) will assume the form

$$u(t) = c_1 \cos \omega_e t + c_2 \sin \omega_e t \tag{2.15}$$

where the angular frequency ω_e is now given by the equation

$$\omega_e^2 = \frac{D}{I}. \tag{2.16}$$

The constants c_1 and c_2 are again determined by the initial conditions.

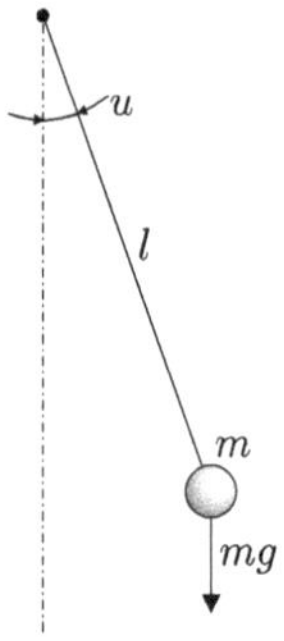

Figure 2.8. Ideal pendulum of mass m and length l suspended in the gravity field. The acceleration of gravity is denoted by g.

2.2.7 Example – Ideal Pendulum

Figure 2.8 shows a sketch of an ideal pendulum of mass m and length l suspended in a gravity field with gravitational acceleration g. Ideal here means a mass point connected to a point of rotation with a weightless bar with no friction against rotation. The moment m_D of the gravity force with respect to the point of rotation is given as

$$m_D = -m g l \sin u \quad (g = 9.81\,\text{m/s}^2). \tag{a}$$

The mass moment of inertia for the pendulum becomes $I = ml^2$, leading to the following criterion for dynamic equilibrium:

$$ml^2 \ddot{u} + m g l \sin u = 0. \tag{b}$$

This equation has no simple solution; however, for small motions, $\sin u \approx u$ so that an approximate solution can be found by assuming that $u(t)$ is of the form specified by Eq. (2.14). The natural frequency ω_e of the pendulum is then given by

$$\omega_e = \sqrt{\frac{mgl}{ml^2}} = \sqrt{\frac{g}{l}}, \tag{c}$$

provided small values of u. ω_e therefore depends only on the length l of the pendulum (for the same value of g). To a certain extent, this is also true for larger values of u. It can be shown that $\tilde{T}_e/T_e \approx 1 + u_0^2/16$ for $u_0 = \max|u(t)| \leq \pi/2$ where $T_e = 2\pi\sqrt{l/g}$ and $\tilde{T}_e$ denotes the exact natural period. This gives $\tilde{T}_e/T_e \approx 1.02$ for $u_0 = \pi/6$ ($= 30$ degrees) and $\tilde{T}_e/T_e \approx 1.18$ for $u_0 = \pi/2$

2.2.8 Example – Tilting Oscillations of an ALP

Articulated loading platforms (ALPs) have been in use on the Norwegian Continental Shelf for offshore oil loading operations for more than two decades. Dynamically, this type of platform has a certain similarity to an inverted pendulum. In this example, a simplified model of this structure will be considered. Specifically, it is assumed that the load-carrying part of the structure consists of a steel truss tower mounted on a frictionless universal joint at the sea floor, as illustrated in Fig. 2.9. Apart from the steel truss structure, the main elements of the considered ALP system that determine its dynamic behavior are the main deck structure carrying the offloading equipment, etc., and a buoyancy tank.

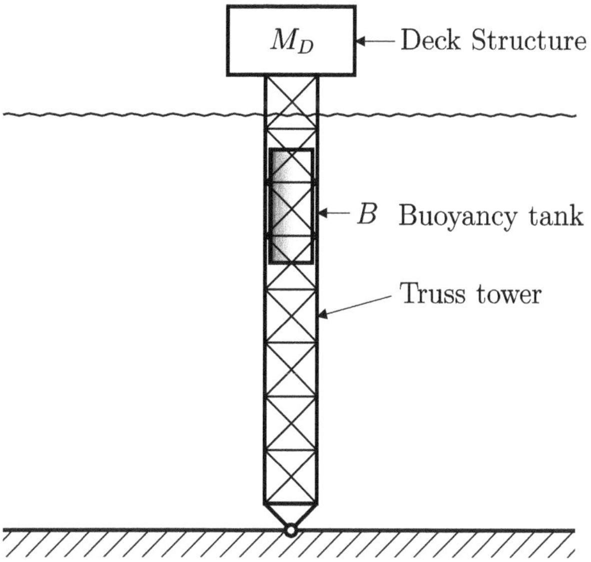

Figure 2.9. Sketch of an ALP structure.

To further simplify the calculations, it is assumed that the steel truss structure can be considered without mass and volume. When the ALP is tilted at an angle θ, the righting moment can be expressed as

$$\left[(B\rho - M_B)gl_B - M_Dgl_D\right]\theta, \tag{a}$$

where B = the buoyancy provided by the buoyancy tank, M_B = the mass of the buoyancy tank; and M_D = the mass of the deck structure and the offloading equipment mounted on it, whose center of gravity is at a distance l_D from the articulated joint. It is assumed that the center of buoyancy and the center of gravity of the buoyancy tank coincide at distance l_B from the articulated joint. Thus, the buoyancy tank must be large enough to provide a righting moment that returns the ALP to an upright position when it is released from a tilted position.

A first estimate of the mass moment of inertia I of the ALP for rotations around the articulated joint is obtained by assuming that masses are point masses. It is then obtained that

$$I = M_Dl_D^2 + (M_B + M_{BH})l_B^2, \tag{b}$$

where M_{BH} denotes the hydrodynamic mass, or added mass, of the buoyancy tank.

An approximate formula for the natural frequency of the tilting (or pitching) oscillations of the ALP then becomes

$$\omega_e = \sqrt{\frac{g((B\rho - M_B)l_B - M_Dl_D)}{(M_B + M_{BH})l_B^2 + M_Dl_D^2}}. \tag{c}$$

In Chapter 3 we show how to calculate the natural frequency of the tilting oscillations of the ALP more accurately by accounting for the distributed mass of the steel truss structure.

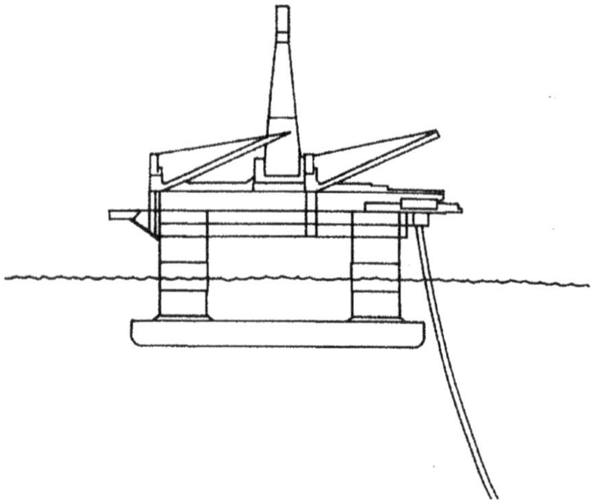

Figure 2.10. Sketch of a semisubmersible drilling rig.

2.2.9 Example – Pitch and Roll Oscillations of a Semisubmersible

A sketch of a typical semisubmersible floating platform used for drilling operations offshore is shown in Fig. 2.10. It is often referred to as a semisubmersible mobile offshore drilling unit. Semisubmersibles are usually kept in position by a catenary spread mooring arrangement, which is sometimes augmented by a thruster-based automatic positioning system.

Let us derive expressions for the natural periods of roll and pitch for such a platform. When the platform is tilted a small angle θ_r in roll, the resulting righting moment M_r is proportional to the tilting angle. Neglecting the influence of the mooring cables, this is expressed as $M_r = -\rho g V \overline{GM}_r\, \theta_r$, where the parameter $\overline{GM}_r$ is defined by this relation. $\overline{GM}_r$ is referred to as the metacentric height for roll (Faltinsen, 1990). Analogously, for pitch, it is obtained that $M_p = -\rho g V \overline{GM}_p\, \theta_p$.

Hence, the expression for the natural period in roll becomes

$$T_r = 2\pi \sqrt{\frac{I_r}{\rho g V \overline{GM}_r}}, \qquad \text{(a)}$$

where I_r denotes the total mass moment of inertia in roll, including added mass effects.

Similarly, the natural period in pitch is given as

$$T_p = 2\pi \sqrt{\frac{I_p}{\rho g V \overline{GM}_p}}, \qquad \text{(b)}$$

where I_p denotes the total mass moment of inertia in pitch, including added mass effects. Note that entirely similar formulas apply to any free-floating structure, e.g., a ship.

Typically, for a semisubmersible, T_r assumes a value in the range of 30 to 50 seconds, while T_p has a value ranging from 20 to 40 seconds.

2.2.10 Example – Yaw Oscillations of a TLP

Returning to the TLP of Example 2.2.5, it is now assumed for simplicity that all tendons have the same distance R_y from the axis of rotation when the TLP is subjected to a pure turning moment. For a rotation by a small angle u, the tangential displacement of the tendon fairleads in the platform legs will be equal to $R_y\,u$. Hence, the total restoring moment $D_y(u)$ becomes

$$D_y(u) = R_y\, n_t\, T_t \frac{R_y\, u}{L_t} = R_y^2 \frac{n_t\, T_t}{L_t} u, \tag{a}$$

assuming neutrally buoyant tendons, see Example 2.2.5. Denoting the total mass moment of inertia in yaw by I_y, including added mass effects, the natural period in yaw, T_y, becomes

$$T_y = \frac{2\pi}{R_y}\sqrt{\frac{I_y\, L_t}{n_t\, T_t}}. \tag{b}$$

2.3 Free Damped Oscillations

It is apparent from the solution Eq. (2.11) that an oscillation response described by Eq. (2.2) will continue with the same amplitude forever when it is initiated. In real structures, there will always be a certain loss of kinetic energy due to, e.g., heat generation or wave propagation. This transfer of kinetic energy to other forms of energy is expressed through the notion of damping.

A common and much used model for damping is that of linear (viscous) damping. Such a model is often represented as a dashpot damper, where the damping force is proportional and oppositely directed to the velocity. That is, the damping force = $-c\dot{u}$, where c is a (positive) constant. This damping model is especially advantageous with respect to calculations because of the direct proportionality to the velocity. Fortunately, it is also sufficiently accurate for many applications, particularly when the damping force is small relative to other forces.

One may envisage that the linear, viscous damping will apply to the damping force on a perforated plate that is moved slowly through a fluid with internal friction, like water or oil, such that the flow pattern is smooth and steady. By increasing the speed of the plate through the fluid, eventually a transition in the flow pattern from smooth to chaotic (turbulent) will start to appear. When this happens, the linear model for the damping force starts to break down. To represent the damping under such circumstances, the usual model is to assume that the damping force is proportional to the square of the velocity.

Of other damping mechanisms it may be worth mentioning dry friction or Coulomb damping, which is typically due to friction between sliding dry surfaces, e.g., in bolted or riveted structures. Material damping and hysteretic damping are due to internal friction and hysteresis when a material is deformed, which may be caused by material defects and internal cracking.

Figure 2.11 depicts an SDOF vibration system with linear damping. When there are no external forces, the equation of motion for the oscillator becomes $-c\,\dot{u} - ku = m\,\ddot{u}$, or

$$m\ddot{u} + c\,\dot{u} + k\,u = 0, \tag{2.17}$$

To solve this problem, the mass moment of inertia I of the door with respect to the axis of the hinge pins is needed. On the basis of the given information, it is obtained that

$$I = \int_0^{0.80} \frac{36}{0.80} r^2 \, dr = 36 \frac{0.80^2}{3} = 7.68 \, \mathrm{kgm}^2.$$

The thickness of the door was neglected. Because the total mass of the door is given, the height of the door does not enter into the calculation. It is now obtained that $\omega_e = \sqrt{10/7.68} = 1.141$ rad/s. The critical damping is then $c_{cr} = 2 \cdot 36 \cdot 1.141 = 82.152$ kgm^2/s.

If the door is released from 90 degrees, how long does it take before the door is one degree from being closed?

The solution to this problem is found by using Eq. (2.23) with initial conditions $u_0 = \pi/2$ and $v_0 = 0$. The value of t satisfies the equation $u(t) = 0.5\pi(1 + 1.141t)\, e^{-1.141t} = 0.0175$ (rad), which corresponds to one degree. By using an iterative solution method, it is found that $t = 5.71$ s.

Supercritical Damping, $\xi > 1$

When $\xi > 1$, the expression under the root sign becomes positive, leading to two unequal real solutions to Eq. (2.19), which are both negative. This gives the displacement response

$$u(t) = e^{-\xi\omega_e t} \left(c_1 \, e^{\varpi t} + c_2 \, e^{-\varpi t} \right), \tag{2.25}$$

where

$$\varpi = \omega_e \sqrt{\xi^2 - 1} < \omega_e \, \xi. \tag{2.26}$$

The response $u(t)$ becomes a strongly damped motion toward the position of rest $u = 0$. For marine structures, (critical or) supercritical damping is usually not met in practice. How such a motion develops in time is illustrated in Fig. 2.12.

Subcritical Damping, $\xi < 1$

When the damping ratio $\xi < 1$, the damping is referred to as subcritical. This is by far the most common case, and the roots of Eq. (2.20) can be written as

$$\lambda_{1,2} = -\omega_e \, \xi \pm \mathrm{i}\omega_e \sqrt{1 - \xi^2}. \tag{2.27}$$

Here, $\mathrm{i} = \sqrt{-1}$, the imaginary unit, and the roots become complex numbers in this case. It is convenient to introduce what is referred to as the *damped natural frequency* ω_d, which is given as

$$\omega_d = \omega_e \sqrt{1 - \xi^2}. \tag{2.28}$$

The expression "damped natural frequency" is a simplified, perhaps somewhat unfortunate, but very widespread notation for the "natural frequency in a system with (linear) damping."

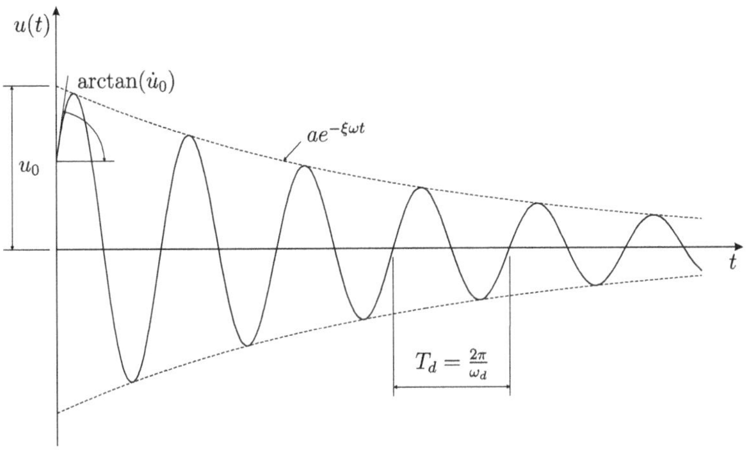

Figure 2.14. Example of a subcritically damped vibration response.

Inserting the roots into the expression for the trial solution, $u(t) = e^{\lambda t}$, leads to the solution

$$u(t) = e^{-\xi\omega_e t}\left(c_1\, e^{i\omega_d t} + c_2\, e^{-i\omega_d t}\right). \tag{2.29}$$

By exploiting Euler's identity $e^{ix} = \cos x + i\,\sin x$, Eq. (2.29) can be rewritten as

$$u(t) = e^{-\xi\omega_e t}\left(a_1\,\cos\omega_d t + a_2\,\sin\omega_d t\right). \tag{2.30}$$

For given initial conditions, a_1 and a_2 can be determined. In the same way that Eq. (2.5) was rewritten as Eq. (2.11), Eq. (2.30) can be expressed as

$$u(t) = e^{-\xi\omega_e t}\,a\,\cos(\omega_d t - \theta). \tag{2.31}$$

Figure (2.14) shows a damped vibration response as determined by Eq. (2.30) or (2.31).

By comparing Eqs. (2.31) and (2.11), there are two aspects of the solutions that should be noted. As opposed to Eq. (2.11), the vibration response given by Eq. (2.31) is characterized by an amplitude that is decreasing with time. This reflects the fact that damping manifests itself by extracting energy from the vibration response and dissipates that energy, usually as heat. Another distinction between Eqs. (2.31) and (2.11) is that the free vibrations in a system with damping do not take place at the natural frequency ω_e, but at the damped natural frequency ω_d given by Eq. (2.28), which directly depends on the damping. However, the difference between the two frequencies are in most cases very small. For example, a damping ratio $\xi = 0.05$ gives $\omega_d = 0.9987\,\omega_e$. This difference is hardly measurable in practice. Values of the damping ratio for welded steel structures could be 0.02–0.04, while for bolted or riveted structures, it could be 0.05–0.07. These statements apply first and foremost to structures of some size and complexity. For simple and homogeneous structures the damping level can, in fact, be significantly lower than this.

The ratio between the amplitude values at times t_i and $t_i + nT_d$, where $T_d = 2\pi/\omega_d$ is given by

$$\frac{u_i}{u_{i+n}} = \frac{e^{-\xi\omega_e t_i}\, a}{e^{-\xi\omega_e(t_i+nT_d)}\, a} = e^{\xi\omega_e n T_d}. \tag{2.32}$$

This equation is the basis for determining the damping in a structure when a free vibration response has been recorded. By comparing the amplitudes at a time separation of T_d, the logarithmic decrement Λ is defined as follows:

$$\Lambda = \ln\left(\frac{u_i}{u_{i+1}}\right). \tag{2.33}$$

The connection between Λ and ξ is brought to the fore by substituting from Eq. (2.32), which leads to

$$\Lambda = \xi\, \omega_e\, T_d = 2\pi \frac{\xi}{\sqrt{1-\xi^2}}. \tag{2.34}$$

Because ξ is usually a small number, Λ is approximately given by the relation

$$\Lambda \approx 2\,\pi\,\xi. \tag{2.35}$$

When performing a practical assessment of the damping from a recorded vibration response, it would usually be necessary to modify the expression for the logarithmic decrement by invoking Eqs. (2.32) and (2.34), which gives

$$\Lambda = \frac{1}{n}\ln\left(\frac{u_i}{u_{i+n}}\right). \tag{2.36}$$

From Eq. (2.35), it is therefore obtained that

$$\xi \approx \frac{1}{2\pi n}\ln\left(\frac{u_i}{u_{i+n}}\right). \tag{2.37}$$

This equation allows for a more accurate assessment of the damping by choosing a suitable value for n.

In addition to those already mentioned, other quantities for characterizing the damping in a structure were introduced. Here, only one such quantity is noted, viz. the damping coefficient δ, which is defined by

$$\delta = \xi\,\omega_e. \tag{2.38}$$

Misunderstandings often occur because sufficient care is not exercised in specifying which damping parameter is being used, and as stated, there are several other variants of damping parameters than those discussed here. Note that while ξ and Λ are dimensionless, δ has dimension s^{-1}.

2.3.2 Example – Logarithmic Decrement

Assume that the recording of the free vibration response shows that it has stabilized after the first three periods and that it then proceeds as damped free oscillations. By measurements on the recorded response signal it was found that $u_3 = 35$ mm and $u_{10} = 17$ mm. Equation (2.36) gives an average logarithmic

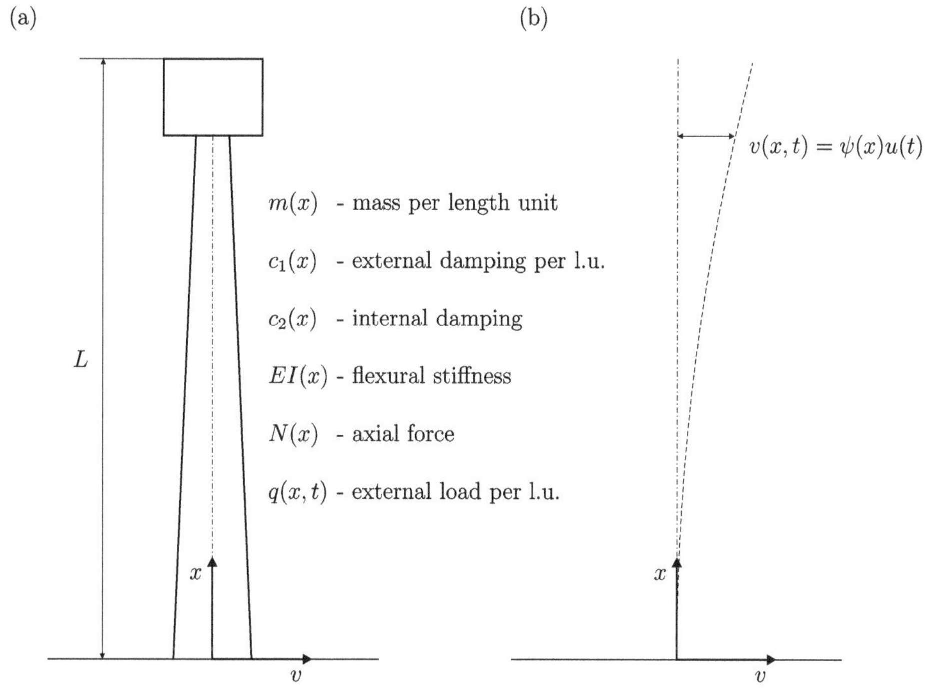

Figure 2.15. Sketch of a vibrating tower.

decrement within this amplitude range:

$$\Lambda = \frac{1}{7}\ln(35/17) = 0.103 \quad (10.3\%),$$

and the damping ratio ξ is from Eq. (2.35) to be $\xi \approx 0.0164$ (1.64%).

2.3.3 Example – Vibrating Tower

It is apparent that a vibration problem with only an SDOF is a special and idealized case. In real structures, it may be necessary to use an overwhelming number of degrees of freedom if, for instance, the goal is to perform detailed and accurate calculations of how an impact is transmitted throughout the structure as stress waves. For other types of problems, the required degrees of freedom may be much more modest for the same structure. As an example of this, we focus on the vibration response of a tower structure as illustrated in Fig. 2.15. The structure could represent an offshore concrete gravity platform for oil production in deep water.

Note that the aim of presenting this example is to illustrate the important concept of a generalized degree of freedom. Thus, the details of the derivations in this example may be skipped at this point; the main message to take away is that a complex vibration problem can be hugely simplified when approximate answers are sufficient, typically in preliminary dynamic analyses.

To simplify the problem, it is presupposed that the oscillations of the tower take place in the "pageplane." Let us assume that on the basis of experience we know that such a tower will oscillate as a cantilever beam with one dominating response shape. That is, it is assumed that the displacement $v(x, t)$, cf. Fig. 2.15, can be written as

$$v(x, t) = \psi(x)\, u(t). \tag{a}$$

Here, the shape function $\psi(x)$ enters together with a quantity $u(t)$ representing the displacement response as a function of time of a suitable reference point of the structure, for instance, the top of the tower. The quantity $u(t)$ in this context is often referred to as a generalized degree of freedom, or simply generalized DOF. Figure 2.15 also shows a snapshot of the tower axis when the response displacement is $v(x, t) = \psi(x)\, u(t)$. The shape function is given here in a symbolic form, but it must be chosen and specified before any calculations can be done.

The subsequent considerations are based on a number of assumptions that are not always satisfied. To make the example easier to write out, and thereby more transparent, the true mass distribution is simplified to a mass distribution along the tower axis, that is, an idealized mass string. The displacement responses orthogonal to the tower axis are assumed so small that we may neglect any motions and accelerations along the tower axis, that is, in the vertical direction.

It should also be mentioned that the example is somewhat elaborate to cover a few secondary goals as well. It is taken into consideration that the tower is subjected to gravity forces and that this results in the presence of axial forces in the tower. Damping is introduced in two different ways, and both are conceivable in cases where the system lacks identifiable dampers of the type shown in Fig. 2.11.

To derive the equation of motion for the tower, the principle of virtual work is used. The derivation of the various contributions is shown directly in the following sketches, but a few explanations are required:

- A virtual displacement is chosen that is similar with the real displacement; that is,

$$\delta v(x, t) = \psi(x)\, \delta u(t). \tag{b}$$

- *External* damping is introduced by imagining that the tower is supported by a row of linear, viscous dampers characterized by a damping coefficient $c_1(x)$ per unit of length of the damping array.
- *Internal* damping is introduced by complementing the linearly elastic relation between moment and curvature with an additional term:

$$M(x, t) = -EI(x)\, v''(x, t) - c_2(x)\, \frac{\partial v''(x, t)}{\partial t}, \tag{c}$$

where $v''(x, t) = d^2 v(x, t)/dx^2$.

The following illustrating vignettes are imagined to represent an infinitesimal part of the tower. Forces that do not contribute to the virtual work for the chosen virtual displacement pattern, for example shear forces, are not shown. For identification of the various contributions, cf. Fig. 2.15.

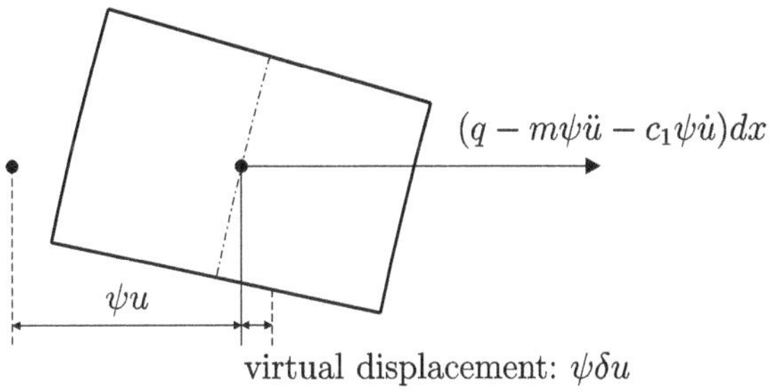

Figure 2.16. Virtual work of external forces, inertia forces and external damping.

External Forces, Inertia Forces and External Damping

The virtual work associated with these contributions becomes (Fig. 2.16)

$$\begin{aligned}\delta W_1 &= \int_0^L [q(x,t) - m(x)\psi(x)\ddot{u} - c_1(x)\psi(x)\dot{u}]\psi(x)dx\delta u \\ &= \left(\int_0^L q(x,t)\psi(x)dx - \int_0^L m(x)[\psi(x)]^2 dx\ddot{u} - \int_0^L c_1(x)[\psi(x)]^2 dx\dot{u}\right)\delta u. \qquad \text{(d)}\end{aligned}$$

Elastic Forces and Internal Damping

Eqs. (a) and (c) show that the virtual work in this case adds up to (Fig. 2.17)

$$\begin{aligned}\delta W_2 &= \int_0^L M(x,t)\,\psi''(x)\,dx\,\delta u \\ &= \left(-\int_0^L EI(x)\,[\psi''(x)]^2\,dx\,u - \int_0^L c_2(x)\,[\psi''(x)]^2\,dx\,\dot{u}\right)\delta u. \qquad \text{(e)}\end{aligned}$$

Effect of Axial Forces

The virtual work associated with the axial forces becomes (Fig. 2.18)

$$\delta W_3 = \int_0^L N(x)\,[\psi'(x)]^2\,dx\,u\,\delta u. \qquad \text{(f)}$$

According to Eq. (1.3), $\delta W_1 + \delta W_2 + \delta W_3 = 0$. Because the terms δW_j for $j = 1, 2, 3$ contain δu as a common factor, it can be deleted, and the equation can be expressed as

$$\hat{m}\ddot{u} + \hat{c}\,\dot{u} + [\hat{k}_e - \hat{k}_g]\,u = \hat{f}, \qquad \text{(g)}$$

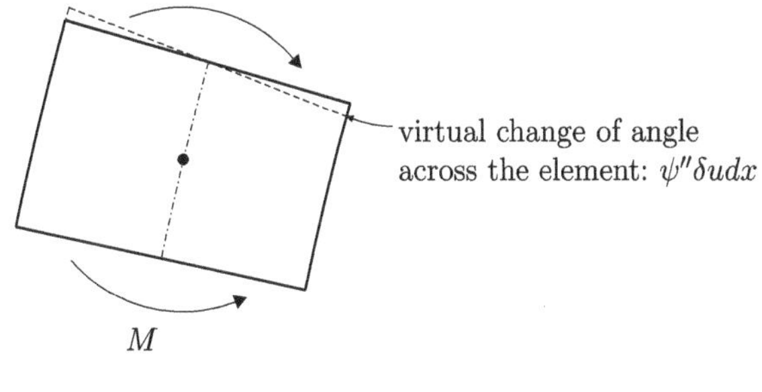

Figure 2.17. Virtual work of elastic forces and internal damping.

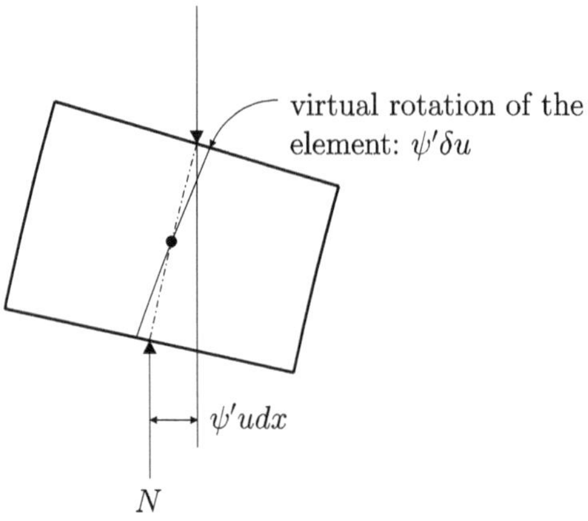

Figure 2.18. Virtual work of axial forces

where

$$\hat{m} = \int_0^L m(x)\,[\psi(x)]^2\,dx \quad = \text{(generalized mass)}, \tag{h}$$

$$\hat{c} = \int_0^L \{c_1(x)\,[\psi(x)]^2 + c_2(x)\,[\psi''(x)]^2\}\,dx \quad = \text{(generalized damping)}, \tag{i}$$

$$\hat{k}_e = \int_0^L EI(x)\,[\psi''(x)]^2\,dx \quad = \text{(generalized stiffness)}, \tag{j}$$

$$\hat{k}_g = \int_0^L N(x)\,[\psi'(x)]^2\,dx \quad = \text{(generalized geometrical stiffness)}, \tag{k}$$

$$\hat{f} = \int_0^L q(x,t)\,\psi(x)\,dx \quad = \text{(generalized load)}. \tag{l}$$

Equation (g) without external forces, can be written as

$$\hat{m}\,\ddot{u} + \hat{c}\,\dot{u} + \hat{k}\,u = 0, \tag{m}$$

where $\hat{k} = \hat{k}_e - \hat{k}_g$. This is an equation of motion of the exact same form as Eq. (2.17) for the free vibrations of an SDOF damped oscillator. Hence, this simple equation of motion can be used to study important aspects of the vibration response of a structure that at the outset appeared to be much more complicated.

Even if the derivations in this example are difficult to follow on a first reading, it is still recommended to spend a few minutes to study Eqs. (g)–(l) to see how the various distributed quantities are integrated. This eventually leads to the simple equation of motion (Eq. (g)) for an SDOF vibrating system.

Instead of the prefix "generalized" in Eqs. (h)–(l), one could just as well have used the word "modal," referring to a rather general approach called modal analysis for studying the vibration response of extended structures. This topic is discussed in more detail later in the book.

Because compression forces will be the usual loading in the axial direction for an example structure of the kind considered here, we consider the axial force $N(t)$ positive when in compression. Equation (g) shows that compression

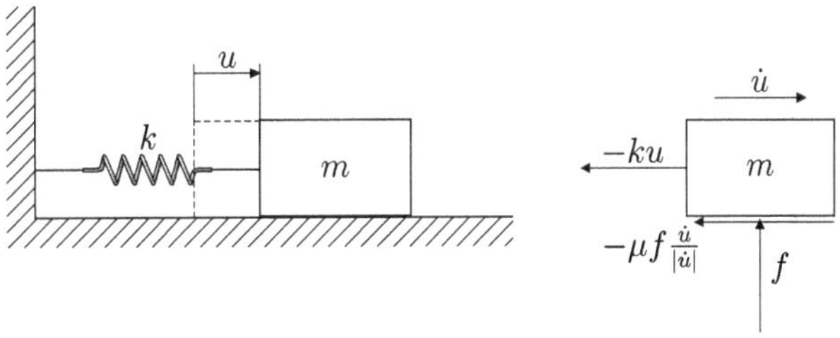

Figure 2.19. Vibration system with Coulomb damping

forces in the tower will lead to increased displacement response and reduce the effective stiffness. In practice, such an effect is usually negligible. It is included here to illustrate the phenomenon.

Equations (h)–(l) contain as a special case the discrete case with point masses, point forces, and dampers acting at discrete points. The integrals are then replaced by sums. The derivation of Eqs. (h)–(l) are quite general for structures under bending, and it is in no way limited to the tower structure shown in Fig. 2.15, which was primarily used to catch the eye and serve as a motivation. To extend the analysis to several degrees of freedom is, in fact, not very difficult, and Eq. (g) then becomes a matrix equation with a possibly time-varying column matrix on the right-hand side (rhs).

The degree of realism when it comes to the constants and the forces in Eq. (g), depends on how well the modal shape function $\psi(x)$ was estimated. If this function gives a realistic response shape leading to a fairly accurate curvature shape, Eq. (c) will provide a good moment function, otherwise not. For example, $\psi(x) = (x/L)^2$ would lead to a dismal moment function.

The shape function $\psi(x)$ applied in this example to the whole structure. In other cases, it may be desirable to impose a clear division into elements and nodal points along the structure, in which case several shape functions may be needed, for example those that underlie the element stiffness matrices that the reader may have encountered in other courses.

2.3.4 Example – Coulomb Damping

The advantage of using the linear (viscous) damping model is the possibility to achieve simple mathematical solutions. At the same time, this model is sufficiently accurate for many applications. Despite this, let us have a look at a damping model that deviates strongly from the linear one. If the damping mechanism is mainly due to dry friction, the damping force will be proportional to the contact force and oppositely directed to the relative velocity. This can be expressed as ($\dot{u} \neq 0$)

$$f_c = -\mu f \frac{\dot{u}}{|\dot{u}|}, \tag{a}$$

where μ = friction coefficient, f = contact force, and $\dot{u}$ = velocity. This damping model is usually known as Coulomb damping, and it may be appropriate for modelling damping in mechanically joined structural elements.

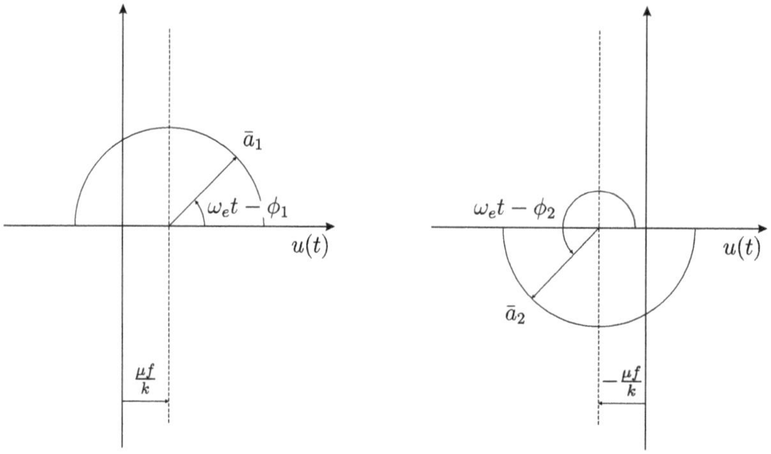

Figure 2.20. Argand diagram for each half period.

As an example, we consider the system schematically shown in Fig. 2.19. The equation of motion in this case becomes

$$m\ddot{u} + \mu f \frac{\dot{u}}{|\dot{u}|} + ku = 0. \tag{b}$$

To solve the equation of motion (b), it is necessary to consider the two cases $\dot{u} < 0$ and $\dot{u} > 0$ separately:

$\dot{u} < 0$

In this case, Eq. (b) becomes

$$m\ddot{u} - \mu f + ku = 0 \quad \text{or} \quad m\ddot{u} + k\left(u - \frac{\mu f}{k}\right) = 0. \tag{c}$$

Because $\mu f/k =$ constant, the new variable $v(t) = u(t) - \mu f/k$ may be introduced, which allows us to rewrite Eq. (c) as

$$\ddot{v} + \omega_e^2 v = 0, \tag{d}$$

where $\omega_e^2 = k/m$. This equation has the solution $v(t) = a_1 \cos(\omega_e t - \phi_1)$; that is,

$$u(t) = a_1 \cos(\omega_e t - \phi_1) + \frac{\mu f}{k}. \tag{e}$$

$\dot{u} > 0$

Analogously, the solution can be written as

$$u(t) = a_2 \cos(\omega_e t - \phi_2) - \frac{\mu f}{k}. \tag{f}$$

Figure 2.20 shows the Argand diagrams for the two parts of the solution.

To determine the integration constants a_1, a_2, ϕ_1 and ϕ_2, it is assumed that the initial conditions are $u(0) = u_1 > \mu f/k$ and $\dot{u}(0) = 0$, which will result in a motion toward the left; that is, $\dot{u} < 0$. This leads to the equations

$$u_1 = a_1 \cos\phi_1 + \frac{\mu f}{k} \tag{g}$$

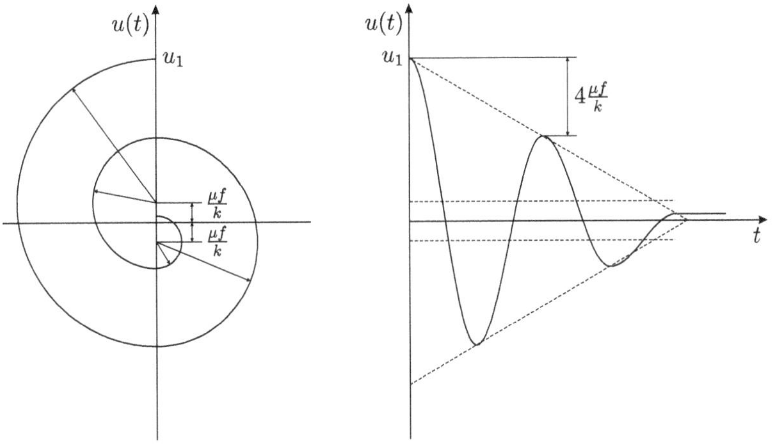

Figure 2.21. Example of the free vibration of an oscillator with Coulomb damping.

and

$$0 = \omega_e\, a_1 \sin \phi_1, \tag{h}$$

which gives $\phi = 0$, $a_1 = u_1 - \mu f/k$, and

$$u(t) = \left(u_1 - \frac{\mu f}{k}\right) \cos \omega_e t + \frac{\mu f}{k}, \quad \text{for } 0 \le \omega_e t \le \pi, \tag{i}$$

because $\dot{u} < 0$ for $0 < \omega_e t < \pi$. The initial conditions for the next half period become $\dot{u}(\pi/\omega_e) = 0$ and $u(\pi/\omega_e = -u_1 + 2\mu f/k$, which is found from (i) at $t = \pi/\omega_e$. This gives the equations

$$-u_1 + 2\frac{\mu f}{k} = a_2 \cos(\pi - \phi_2) - \frac{\mu f}{k} = -a_2 \cos \phi_2 - \frac{\mu f}{k} \tag{j}$$

and

$$0 = -\omega_e\, a_2 \sin(\pi - \phi_2) = -\omega_e\, a_2 \sin \phi_2, \tag{k}$$

which imply that $\phi_2 = 0$, $a_2 = u_1 - 3\mu f/k$, and

$$u(t) = (u_1 - 3\frac{\mu f}{k}) \cos \omega_e t - \frac{\mu f}{k}, \quad \pi \le \omega_e t \le 2\pi. \tag{l}$$

Figure 2.21 shows the graph of (i) and (l), and the vibration response takes place around the two mean positions $u = \pm \mu f/k$. Also, the amplitudes are reduced by $2\mu f/k$ over each half period. For the nth half period, the initial value $u_n = u\big((n-1)\pi/\omega_e\big)$ is given as follows:

$$u_n = u_1 - 2(n-1)\frac{\mu f}{k} \quad \text{for} \quad n = 1, 3, 5, \ldots, \tag{m}$$

and

$$u_n = -u_1 + 2(n-1)\frac{\mu f}{k} \quad \text{for} \quad n = 2, 4, 6, \ldots. \tag{n}$$

When the final value for an oscillation is less than $2\mu f/k$ in absolute value, it stops.

2.4 Forced Vibrations by Harmonic Excitation

Some examples of forced vibrations are as follows:

- A TV tower or high-rise building will experience vibration response due to wind forces.
- A moored, semisubmersible oil production platform will exhibit oscillatory response due to wind and wave forces.
- A structure on an elastic bed for earthquake protection will oscillate due to the forces transmitted from a vibrating foundation caused by an earthquake.

Let us first study the simplest and most fundamental type of excitation. It is assumed that the system in Fig. 2.11 is excited by a harmonic force $f(t) = f_0 \cos \omega t$. The condition for dynamic equilibrium is in this case

$$m\ddot{u} + c\dot{u} + ku = f_0 \cos \omega t, \tag{2.39}$$

where $u = u(t)$ denotes the displacement response from the position of rest ($u = 0$), when there is no external force. The general solution to Eq. (2.39) consists of the sum of a particular solution, that is, any function of time that satisfies Eq. (2.39), and the homogeneous solution Eq. (2.30) or (2.31), cf. Kreyszig (1993), as we assume that $c < c_{cr}$. To find a particular solution $u_p(t)$, according to Kreyszig (1993), we may assume that

$$u_p(t) = b_1 \cos \omega t + b_2 \sin \omega t, \tag{2.40}$$

with two constants b_1 and b_2. By substitution into Eq. (2.39) and sorting the terms, the following equations are obtained:

$$-m\omega^2 b_1 + c\,\omega\, b_2 + k b_1 = f_0, \tag{2.41}$$

$$-m\omega^2 b_2 - c\,\omega\, b_1 + k b_2 = 0. \tag{2.42}$$

Solving these equations with respect to b_1 and b_2 gives

$$b_1 = \frac{f_0}{k} \cdot \frac{1-\beta^2}{(1-\beta^2)^2 + (2\xi\beta)^2} \tag{2.43}$$

and

$$b_2 = \frac{f_0}{k} \cdot \frac{2\xi\beta}{(1-\beta^2)^2 + (2\xi\beta)^2}, \tag{2.44}$$

where the *frequency ratio* β is introduced:

$$\beta = \frac{\omega}{\omega_e}. \tag{2.45}$$

A particular solution therefore becomes

$$u_p(t) = \frac{f_0}{k} \cdot \frac{1}{(1-\beta^2)^2 + (2\xi\beta)^2} \left((1-\beta^2)^2 \cos \omega t + 2\xi\beta \sin \omega t\right). \tag{2.46}$$

Because of the exponential term in Eq. (2.30), the homogeneous solution (often referred to as the *transient term*) will appear as an initial disturbance, which will die out eventually. Hence, the particular solution (the steady state or nontransient term) will, after a while, dominate. The transient behaviour of the dynamic system is then

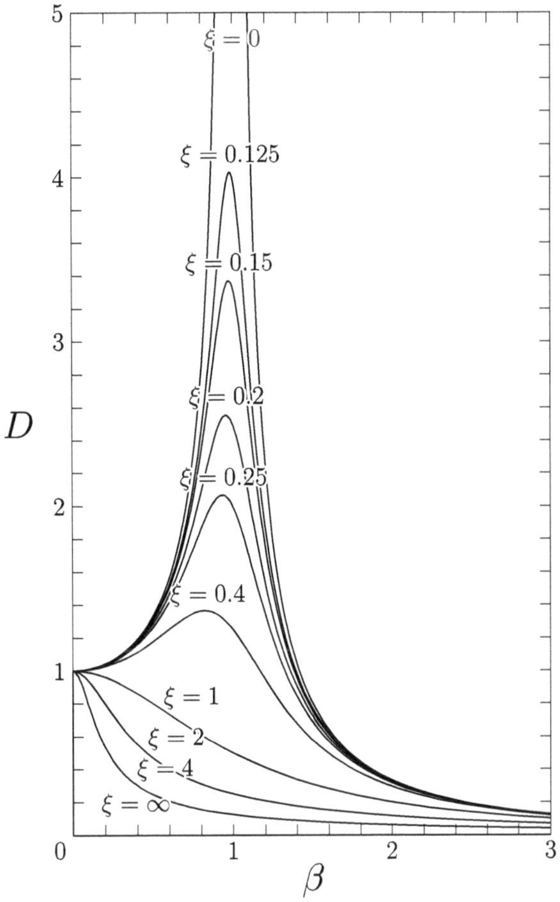

Figure 2.22. Dynamic amplification factor D as a function of the frequency ratio β.

determined by the homogeneous solution together with the initial conditions. Analogously to the equivalence of Eq. (2.5) and Eq. (2.11), the harmonic (nontransient) response can be written as

$$u_p(t) = b\cos(\omega t - \phi), \tag{2.47}$$

where

$$b = \sqrt{b_1^2 + b_2^2} = \frac{f_0}{k} \cdot \frac{1}{\left((1-\beta^2)^2 + (2\xi\beta)^2\right)^{1/2}}, \tag{2.48}$$

$$\phi = \arctan(b_2/b_1) = \arctan\left(2\xi\beta/(1-\beta^2)\right). \tag{2.49}$$

The system's (nontransient) response to a static load, f_0, is clearly f_0/k. This solution is also obtained by putting $\omega = 0$, that is, $\beta = 0$, in Eqs. (2.47)–(2.49). The ratio between the steady-state dynamic and the corresponding static displacement response is

$$D = \left|\frac{u_p(t)_{max}}{f_0/k}\right| = \frac{kb}{f_0} = \frac{1}{\left((1-\beta^2)^2 + (2\xi\beta)^2\right)^{1/2}}. \tag{2.50}$$

D is often called the *dynamic amplification factor*. Figure 2.22 shows D as a function of the frequency ratio β for several values of the damping ratio ξ. The corresponding diagram for the phase angle is shown in Fig. 2.23. It may be noted that that the

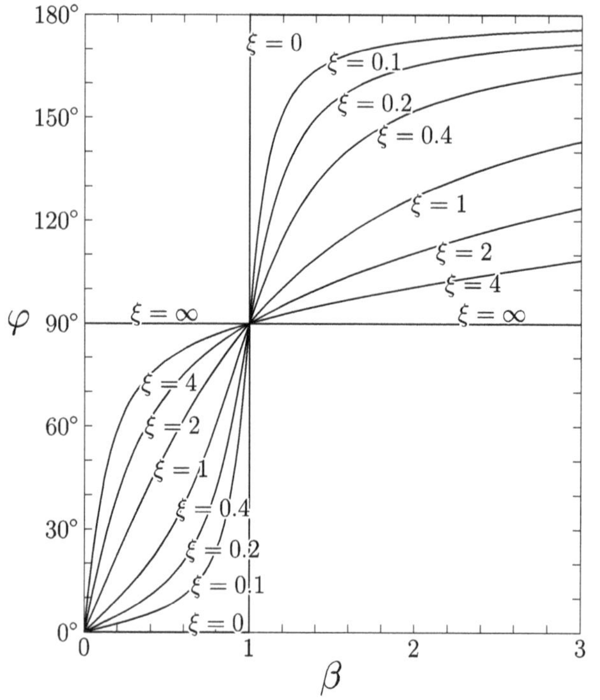

Figure 2.23. Phase angle ϕ as a function of the frequency ratio β.

phase angle ϕ has values in the range from 0 to 180 degrees, which means that the displacement response $u(t)$ is always lagging the force $f(t)$ (notice the minus sign in front of the phase angle in Eq. (2.47)). The diagram for D also includes results for critical and supercritical damping. A large part of Fig. 2.22 has limited practical value for us, but it is included to illustrate the transition from supercritical to subcritical damping. Figure 9.5 shows D, or rather $D/k = |H(\omega)|$ ($H(\omega)$ is explained later in this chapter) as a function of the frequency ratio for damping values down to $\xi = 0.01$.

From Fig. 2.22 (and in particular Fig. 9.5 it is seen that the displacement amplitude for lightly damped systems increases strongly when the frequency of the imposed forcing function is approaching ω_e ($\beta \to 1$). For excitation frequencies $\omega > \omega_e$, that is, $\beta > 1$, the response will decrease rapidly with increasing values of β, and for $\beta > 1.41$, the response amplitude is smaller than for $\beta = 0$ irrespective of the value of ξ. This is easily seen by making a note of the fact that the numerator in the expression for D is larger than 1 when $\beta^2 > 2(1 - 2\xi^2)$, and therefore $D < 1$ when $\beta^2 > 2 > 2(1 - 2\xi^2)$. The practical interpretation of this is that the system does not manage to react fully to the rapidly varying external force.

D assumes its maximal value when $d(\beta) = (1 - \beta^2)^2 + (2\xi\beta)^2$ assumes its minimal value. The minimum point β_r is determined from the equation $d'(\beta) = 0$. It is found that

$$\beta_r = \sqrt{1 - 2\xi^2}\,, \quad \xi < \frac{1}{\sqrt{2}}, \tag{2.51}$$

$$\beta_r = 0\,, \quad \xi \geq \frac{1}{\sqrt{2}}. \tag{2.52}$$

The dynamic amplification factor D_{max} that corresponds to β_r is given by

$$D_{max} = \frac{1}{2\xi\sqrt{1-\xi^2}}, \quad \left(\xi < \frac{1}{\sqrt{2}}\right). \tag{2.53}$$

The value β_r corresponds to resonance between the external force and the dynamic system. It is therefore natural to define the resonance frequency:

$$\omega_r = \omega_e\sqrt{1-2\xi^2}, \quad \left(\xi < \frac{1}{\sqrt{2}}\right). \tag{2.54}$$

Comparing Eqs. (2.28) and (2.54) leads to the relations

$$\omega_r < \omega_d < \omega_e, \quad \left(\xi < \frac{1}{\sqrt{2}}\right). \tag{2.55}$$

For weakly damped systems, $1 - 2\xi^2 \approx 1$, such that $\omega_r \approx \omega_d \approx \omega_e$. Hence, the following simple approximate formula is obtained:

$$D_{max} \approx \frac{1}{2\xi}, \quad \left(\text{or } \frac{\pi}{\Lambda}\right). \tag{2.56}$$

For example, at 2% of critical damping $D_{max} \approx 25$. This means that even small oscillating forces may lead to large responses and perhaps damage if the forcing frequency is close to the resonance frequency.

For the analysis of structural response to various forcing frequencies, it is expedient to introduce the complex valued function $H(\omega) = |H(\omega)|\, e^{-i\phi}$. $|H(\omega)|$ and ϕ are determined by writing the nontransient vibration response as

$$u(t) = |H(\omega)|\, f_0 \cos(\omega t - \phi). \tag{2.57}$$

It is seen that the factor $|H(\omega)|$ gives the amplification of the amplitude, while ϕ gives the phase shift. If, for example, $|H(\omega)| = 0.001$ for a particular frequency ω, then a force amplitude of 100 N will give rise to a displacement amplitude of 0.1 meters at this frequency.

From Eqs. (2.47) and (2.48), it follows that

$$|H(\omega)| = \frac{1}{k\sqrt{(1-\beta^2)^2 + (2\xi\beta)^2}}. \tag{2.58}$$

$H(\omega)$ is called the *transfer function* or the *frequency response function* between the external forcing and the response. The name *mechanical transfer function* is also used for this function. It characterizes the system response to a given harmonic excitation. $|H(\omega)|$ is proportional to the amplification factor D; that is, $H(\omega)$ contains all relevant information about dynamic amplifications and phase shift, which is determined by the angle ϕ.

To get an impression of how the transient part of the vibration response will behave, we look at a situation with harmonic external forcing with amplitude f_0 and circular frequency $\omega = \omega_r \approx \omega_e$ (that is, $\xi \ll 1$) which is imposed on a system at rest at $t = 0$. The full solution $u(t)$ consists of a sum of the homogeneous solution Eq. (2.31) and the particular solution Eq. (2.47). The constants a and θ are determined by the

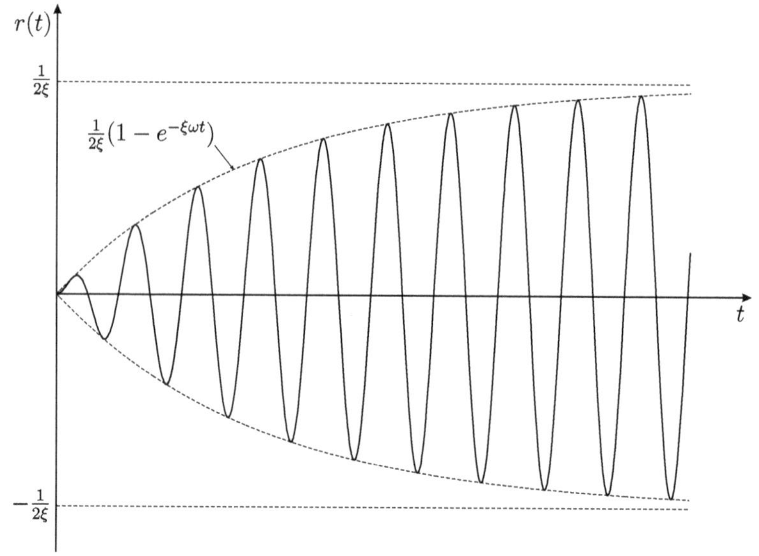

Figure 2.24. Example of starting from rest of the vibration response of a damped system.

initial conditions $u(0) = \dot{u}(0) = 0$ (the system is at rest at $t = 0$). This leads to the solution

$$u(t) \approx \frac{f_0}{k} D(1 - e^{-\xi\omega t}) \sin \omega t. \tag{2.59}$$

Let us have look at the ratio

$$r(t) = \frac{u(t)}{f_0/k} \approx \frac{1}{2\xi}(1 - e^{-\xi\omega t}) \sin \omega t, \tag{2.60}$$

because $D \approx 1/(2\xi)$. Figure 2.24 shows $r(t)$ as a function of t. It is seen that the response is gradually building up to an oscillatory response with amplitude $f_0/(2\xi k)$. It also follows that with zero damping, the response will grow beyond any limit. The shape of $r(t)$ for $\xi = 0$ can be derived from Eq. (2.60) by invoking l'Hôpital's rule:

$$r(t) = \frac{1}{2}\omega t \sin \omega t \quad \text{for } \xi = 0. \tag{2.61}$$

The vibration amplitude grows without limit as a linear function of time. Such a situation can, of course, never occur in practice.

2.4.1 Example – Harmonic Force

As an example of a harmonic forcing function on a damped system, we consider an idealized model of a machine of total mass M, including a rotating part of mass m, see Fig. 2.25. The mass center of the rotating part has a distance e from the axis of rotation, and the rotation takes place at a constant angular frequency ω. The machine is restricted to move vertically by the suspension arrangement.

The vertical component f of the centrifugal force in the rotor is given as

$$f = m e \omega^2 \sin \omega t. \tag{a}$$

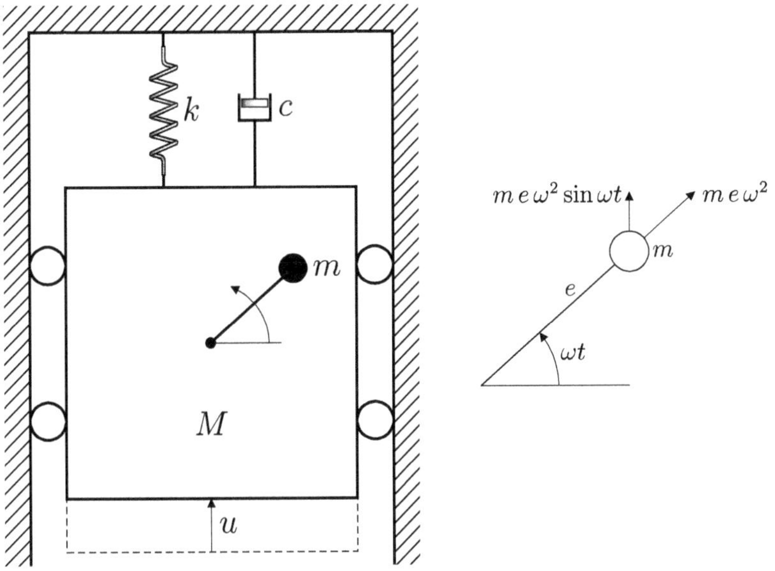

Figure 2.25. Machine with a rotating part.

The equation of dynamic equilibrium in the vertical direction becomes

$$-M\ddot{u} - c\,\dot{u} - k\,u + m\,e\,\omega^2 \sin\omega t = 0 \tag{b}$$

or

$$M\ddot{u} + c\,\dot{u} + k\,u = m\,e\,\omega^2 \sin\omega t = m\,e\,\omega^2 \cos(\omega t - \pi/2). \tag{c}$$

This equation has the same form as Eq. (2.39). The transient part of the solution will be damped out, and we are only interested in the steady-state solution, which was found previously in Eq. (2.47). It is obtained that

$$u_p(t) = b\,\cos(\omega t - \pi/2 - \phi) = b\,\sin(\omega t - \phi). \tag{d}$$

For the system, the following relations apply:

$$\omega_d = \omega_e\sqrt{1-\xi^2}\,, \quad \omega_e = \sqrt{k/M}\,, \quad \xi = c/(2M\omega_e). \tag{e}$$

By introducing the frequency ratio $\beta = \omega/\omega_e$ and replacing f_0 by $me\omega^2$, the expression for the response amplitude is obtained from Eq. (2.48):

$$b = \frac{m\,e}{M}\cdot\frac{\beta^2}{\sqrt{(1-\beta^2)^2+(2\xi\beta)^2}}. \tag{f}$$

The phase angle ϕ is given by Eq. (2.49). The reaction force f_r on the foundation is determined by the equation $f_r = k\,u + c\,\dot{u}$, leading to the expression

$$\begin{aligned} f_r &= kb\sin(\omega t - \phi) + \omega cb\cos(\omega t - \phi) \\ &= b\sqrt{k^2+(c\omega)^2}\cos(\omega t - \phi - \psi) \\ &= bk\sqrt{1+(2\xi\beta)^2}\cos(\omega t - \phi - \psi), \end{aligned} \tag{g}$$

where ψ is a properly chosen phase angle.

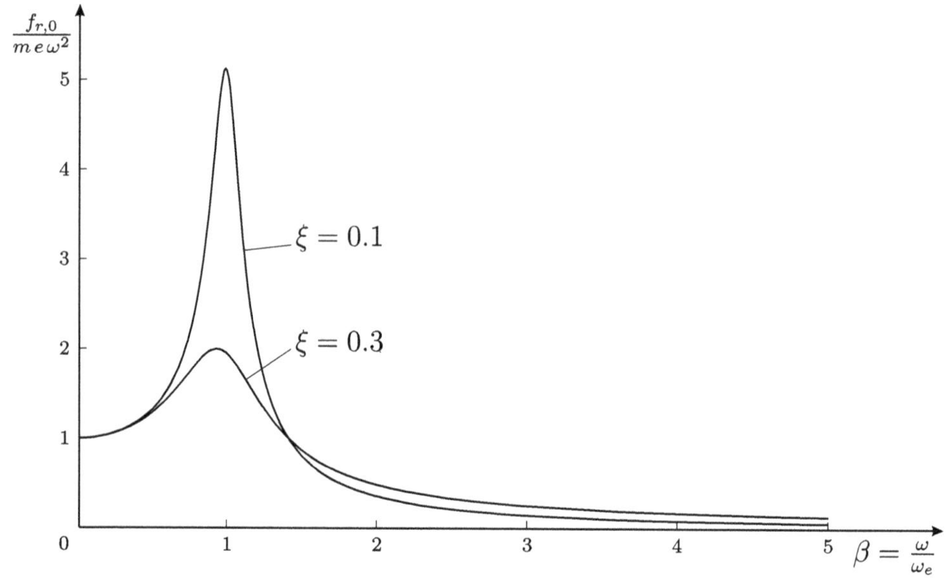

Figure 2.26. Transmission ratio as a function of β for two damping ratios.

It is seen from the expression for b that the amplitude grows strongly when the frequency ratio tends to 1 and resonance occurs. When starting a machine, it may be necessary to pass through a resonance frequency. To avoid violent shaking, it is advantageous for the starting phase to be as brief as possible, such that the critical frequency range is quickly passed. The amplitude can also be reduced by introducing more damping because the maximum value of the amplitude b is inversely proportional to 2ξ. However, it is important to note that larger damping will lead to a larger reaction force at higher frequency ratios. This can be seen by substituting the expression for b into the expression for the amplitude $f_{r,0}$ of the reaction force f_r:

$$f_{r,0} = m e \omega^2 \frac{\sqrt{1 + (2\xi\beta)^2}}{\sqrt{(1 - \beta^2)^2 + (2\xi\beta)^2}}. \tag{h}$$

The first factor $me\omega^2$ represents a characteristic property of the machine, and it is often difficult to change that. The damping and the stiffness of the suspension enters through ξ and β. Alternative arrangements for the suspension are assessed by calculating the ratio of the two square roots, which is often referred to as the *transmission ratio*. It is a direct expression for the ratio between the amplitude of the imposed force and the amplitude of the force being transmitted to the foundation. The transmission ratio is shown as a function of β for two values of the damping ratio in Fig. 2.26. For example, at $\beta = 3$, $\xi = 0.3$ leads to almost twice the force amplitude compared to $\xi = 0.1$. Low damping and low stiffness provide good isolation of the dynamic system from the foundation at high values of β. However, this must be compared with the disadvantage of resonance during startup. The ideal arrangement would be a damper that was only active during startup and slowdown.

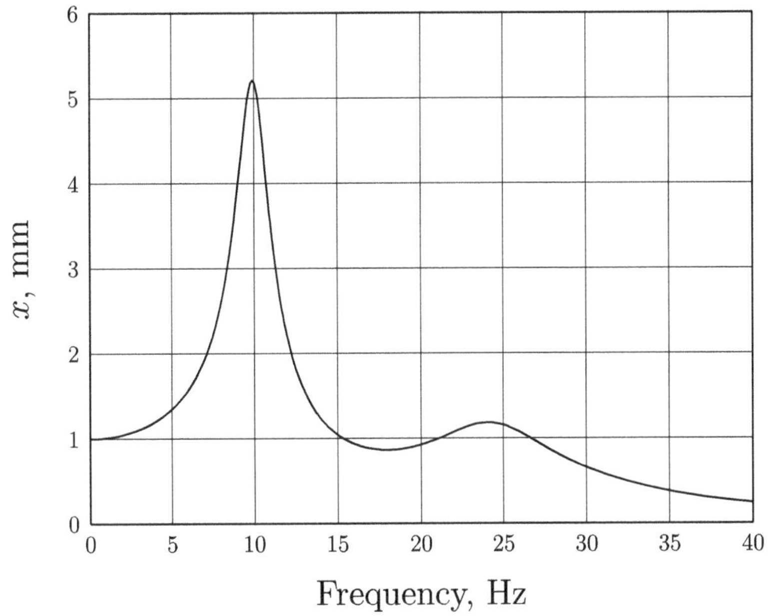

Figure 2.27. Displacement amplitude as a function of frequency for harmonic excitation at constant force amplitude.

We close this section by describing a method for estimating the damping ratio from the behavior of the dynamic amplification factor D, or equivalently, from the behavior of $|H(\omega)|$. It is assumed that the damping is small, for example, $\xi < 0.1$. The method consists of determining the two frequencies ω_+ and ω_- ($\omega_- < \omega_+$) such that $|H(\omega_\pm)|^2 = |H(\omega_r)|^2/2 \approx |H(\omega_e)|^2/2$. Clearly, $\omega_- < \omega_e < \omega_+$. To determine $\omega_\pm$, the following equation must be solved ($\beta_\pm = \omega_\pm/\omega_e$)

$$\frac{1}{(1-\beta_\pm^2)^2 + 4\xi^2\beta_\pm^2} = \frac{1}{8\xi^2}.$$

The solution is $\beta_\pm \approx 1 \pm \xi$. Hence, we find that the half-value width $\omega_+ - \omega_- \approx 2\xi\omega_e$. That is, the width of $|H(\omega)|^2$ at half the peak value provides a direct measure of the damping ratio, because the frequency at the peak value gives an estimate of the natural frequency. If the frequency is given in Hz and denoted by f, the corresponding result becomes $f_+ - f_- \approx 2\xi f_e$.

2.4.2 Example – Damping Ratio from Half-Value Width

A graphic representation of the amplitude x as a function of frequency for the displacement response of a linear dynamic system to harmonic excitation with constant amplitude is shown in Fig. 2.27. Let us use this information to estimate the damping ratio ξ.

On the basis of the given information, we know that the ordinate values are proportional to the corresponding values of the dynamic amplification factor. From Fig. 2.27, the half-value width, which occurs at $x_{max}/\sqrt{2}$, is approximately 2 Hz, and the natural frequency is close to 10 Hz. This gives $\xi \approx 2/(2 \cdot 10) = 0.1$.

2.5 Forced Vibration – Complex Analysis

In the previous section, the analysis of the vibration system was carried out entirely by using real harmonic functions, that is, cosine and sine functions. The goal of this section is to show that it is usually much more convenient to use complex quantities when analyzing linear vibration systems. In practice, this means that we replace $a\cos\phi$ by $ae^{\mathrm{i}\phi}$. The justification for this can be shown as follows.

By introducing the differential operator

$$L[\cdot] = m\frac{d^2}{dt^2} + c\frac{d}{dt} + k, \tag{2.62}$$

Eq. (2.39) can be written as

$$L[u] = f_0 \cos \omega t. \tag{2.63}$$

Instead of finding the vibration response $u(t)$ as in the previous section, we make the following observations. From Euler's formula, $e^{\mathrm{i}\phi} = \cos\phi + \mathrm{i}\sin\phi$, it is obtained that $f_0 \cos\omega t = \Re f_0 e^{\mathrm{i}\omega t}$, where $\Re z$ denotes the real part of the complex number z. Let us assume that $w(t)$ satisfies the equation

$$L[w(t)] = f_0 e^{\mathrm{i}\omega t}. \tag{2.64}$$

$w(t)$ will be a complex function; that is, $w(t) = u(t) + \mathrm{i}v(t)$, where $u(t)$ and $v(t)$ are real functions, and $u(t) = \Re w(t)$. Because m, c, and k are real constants, it must be true that $\Re L[w(t)] = L[u(t)]$. This means that if $w(t)$ satisfies Eq. (2.64), then $u(t) = \Re w(t)$ must satisfy Eq. (2.63). That is, Eq. (2.63) can be solved by first solving Eq. (2.64) and then taking the real part of the obtained solution. Let us assume a trial solution $w(t) = ae^{\mathrm{i}\omega t}$ of Eq. (2.64), where a may be a complex number. This leads to the equation

$$\left(-\omega^2 m + \mathrm{i}\omega c + k\right) ae^{\mathrm{i}\omega t} = f_0 e^{\mathrm{i}\omega t}. \tag{2.65}$$

Solving for a gives

$$a = \frac{f_0}{-\omega^2 m + \mathrm{i}\omega c + k} = \frac{f_0}{k} \cdot \frac{1}{1 - \beta^2 + 2\mathrm{i}\xi\beta}. \tag{2.66}$$

A complex number $z = x + \mathrm{i}y$ can be written as $z = \sqrt{x^2 + y^2}\,\exp\{\mathrm{i}\arctan(y/x)\}$ ($\exp\{x\} = e^x$). This implies that

$$a = \frac{f_0}{k} \cdot \frac{1}{\sqrt{(1-\beta^2)^2 + (2\xi\beta)^2}} \exp\{-\mathrm{i}\phi\}, \tag{2.67}$$

where

$$\phi = \arctan\left(\frac{2\xi\beta}{1-\beta^2}\right). \tag{2.68}$$

The following solution is thereby obtained

$$w(t) = \frac{f_0}{k} D \exp\{\mathrm{i}(\omega t - \phi)\}, \tag{2.69}$$

which leads to a particular solution of Eq. (2.63) of the form

$$u(t) = \Re w(t) = \frac{f_0}{k} D \cos(\omega t - \phi), \tag{2.70}$$

and this solution agrees with Eq. (2.47).

In Section 2.4, the complex transfer function $H(\omega)$ for the transfer between the force $f(t)$ and the response $u(t)$ was mentioned, and $|H(\omega)|$ was identified. By going back to Eqs. (2.66) and (2.67), we may also identify $H(\omega)$ directly by using $a = H(\omega)\, f_0$ by definition; that is,

$$H(\omega) = \frac{1}{k} \cdot \frac{1}{1-\beta^2+2\mathrm{i}\xi\beta} = \frac{1}{k} \cdot \frac{1}{\sqrt{(1-\beta^2)^2+(2\xi\beta)^2}} \exp\{-\mathrm{i}\phi\}. \tag{2.71}$$

By comparing Eqs. (2.58) and (2.71), it is seen that $H(\omega) = |H(\omega)|\, e^{-\mathrm{i}\phi}$, in agreement with the preceding definition.

The reason why it is easier to use $e^{\mathrm{i}\phi}$ than $\cos\phi$ and $\sin\phi$ is due to the following two properties of $e^{\mathrm{i}\phi}$:

- $\frac{d}{dt} e^{\mathrm{i}\phi} = \mathrm{i}\frac{d\phi}{dt}\, e^{\mathrm{i}\phi}$; that is, the $e^{\mathrm{i}\phi}$ factor does not change.
- $e^{\mathrm{i}\phi_1} \cdot e^{\mathrm{i}\phi_2} = e^{\mathrm{i}(\phi_1+\phi_2)} = e^{\mathrm{i}\phi_3}$; that is, the product of two factors of type $e^{\mathrm{i}\phi}$ is again of the same kind.

By going back to the previous derivations, it can be verified how these two properties in particular were used. This opportunity to apply an auxiliary complex analysis is convenient, and is used repeatedly in this book.

So far, we have only defined a transfer function $H(\omega)$ between $f(t)$ and $u(t)$, where $L[u] = f$, see Eq. (2.62). $H(\omega)$ was determined by the fact that $u(t) = H(\omega)e^{\mathrm{i}\omega t}$ if $f(t) = e^{\mathrm{i}\omega t}$. This is an example of a more general situation. Let us assume that the connection between two time-variant functions $x(t)$ and $y(t)$ is given as $L_1[y] = L_2[x]$, where $L_1[\cdot]$ and $L_2[\cdot]$ denotes two linear operations. That $L_j[\cdot]$, $j = 1, 2$, is linear means that $L[a_1\, f_1(t) + a_2\, f_2(t)] = a_1\, L[\,f_1(t)] + a_2\, L[\,f_2(t)]$ for arbitrary constants a_1 and a_2 and functions $f_1(t)$ and $f_2(t)$. The relation between $x(t)$ and $y(t)$ can be expressed by a (linear) transfer function $H_{xy}(\omega)$ if $L_1[H_{xy}(\omega)\, e^{\mathrm{i}\omega t}] = L_2[e^{\mathrm{i}\omega t}]$. That is, if there exists a function $H_{xy}(\omega)$ such that $y(t) = H_{xy}(\omega)\, e^{\mathrm{i}\omega t}$ satisfies the equation $L_1[y] = L_2[x]$ for every $x(t) = e^{\mathrm{i}\omega t}$, then the relation between $x(t)$ and $y(t)$ is determined by a (linear) transfer function, which is $H_{xy}(\omega)$. It is important to note that $H_{xy}(\omega)$ is a function of ω only, and independent of t. This is often emphasized by calling the linear relation time invariant. The opposite is then time variant. An illustration of both cases is given in the next example.

2.5.1 Example – Transfer Function

The following linear equation is an example of a linear and time-invariant relationship between $x(t)$ and $y(t)$:

$$m\,\ddot{y} + c\,\dot{y} + k\,y = q\,\ddot{x}, \tag{a}$$

where m, c, and k are positive constants and q is a real constant. By putting $x(t) = e^{\mathrm{i}\omega t}$ and assuming that $y(t) = ae^{\mathrm{i}\omega t}$, it is obtained by substituting into Eq. (a) that

$$a = \frac{-\omega^2\, q}{-\omega^2\, m + \mathrm{i}\omega\, c + k}. \tag{b}$$

Because the rhs of Eq. (b) depends only on ω, the relationship between $x(t)$ and $y(t)$ is described by a (linear) transfer function $H_{xy}(\omega)$, which equals the rhs of Eq. (b).

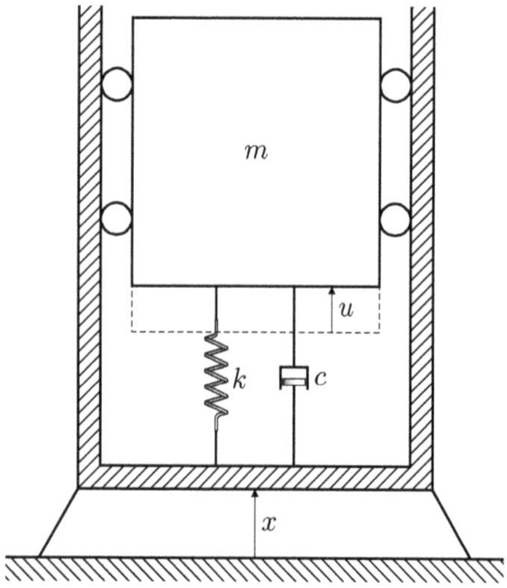

Figure 2.28. Principle sketch of a structure on a vibrating foundation.

However, if, for example,

$$\dot{y} + \frac{t}{\sqrt{1+t^2}}\, y = x, \tag{c}$$

then the relationship between $x(t)$ and $y(t)$ is still linear, but it is now time-variant. This is easily seen by inserting $x(t) = e^{\mathrm{i}\omega t}$ and $y(t) = a e^{\mathrm{i}\omega t}$ into Eq. (c). Under the premise that a does not depend on t, it follows immediately that

$$a = \left(\mathrm{i}\,\omega + \frac{t}{\sqrt{1+t^2}}\right)^{-1}, \tag{d}$$

which clearly depends on both ω and t, and thereby violates the premise about a. Therefore, a transfer function $H_{xy}(\omega)$ does not exist in this case.

2.5.2 Example – Structure on a Vibrating Foundation

The vibration system in Fig. 2.28 is excited by a vibrating foundation. This vibration is assumed to be given as $x(t) = x_0 \cos \omega t$. Forces and displacements are considered as positive upward. The restoring force for a relative displacement $u - x$ is given as $-k(u - x)$, that is, u and x is chosen such that $u - x = 0$ in the position of rest. Similarly, the damping force becomes $-c(\dot{u} - \dot{x})$. The condition for dynamic equilibrium can then be written as

$$m\ddot{u} + c\,(\dot{u} - \dot{x}) + k(u - x) = 0 \tag{a}$$

or

$$m\ddot{u} + c\,\dot{u} + k\,u = c\,\dot{x} + k\,x. \tag{b}$$

It is worth noting that by introducing the relative displacement $y = u - x$, Eq. (a) may be rewritten in the form

$$m\,\ddot{y} + c\,\dot{y} + k\,y = -m\ddot{x}, \tag{c}$$

cf. Eq. (a) in the previous example.

In the same manner as previously, it is seen that if $w(t)$ is a solution of Eq. (b) with $x(t)$ replaced by $x_0 e^{i\omega t}$, then $u(t) = \Re\{w(t)\}$ is a solution of (b) with $x(t) = x_0 \cos \omega t$ (x_0 is a real constant). We now assume a trial solution of the form $w(t) = be^{i\omega t}$. This leads to the equation

$$(-\omega^2 m + i\omega c + k)be^{i\omega t} = (i\omega c + k)x_0 e^{i\omega t}. \tag{d}$$

This equation determines b as

$$b = \frac{i\omega c + k}{-\omega^2 m + i\omega c + k} x_0. \tag{e}$$

A particular solution of Eq. (b) with $x(t) = x_0 \cos \omega t$ then becomes

$$u(t) = \Re\left\{\frac{i\omega c + k}{-\omega^2 m + i\omega c + k} x_0 e^{i\omega t}\right\} = x_0 \Re\left\{\frac{1 + 2i\xi\beta}{1 - \beta^2 + 2i\xi\beta} e^{i\omega t}\right\}. \tag{f}$$

We may write $1 + 2i\xi\beta = \sqrt{1 + (2\xi\beta)^2} \exp\{i\phi_1\}$, where $\phi_1 = \arctan(2\xi\beta)$. Thereby, we obtain

$$\begin{aligned} u(t) &= x_0 \Re\left\{\sqrt{\frac{1 + (2\xi\beta)^2}{(1 - \beta^2)^2 + (2\xi\beta)^2}} e^{i(\omega t + \phi_1 - \phi_2)}\right\} \\ &= x_0 \sqrt{\frac{1 + (2\xi\beta)^2}{(1 - \beta^2)^2 + (2\xi\beta)^2}} \cos(\omega t + \phi_1 - \phi_2), \end{aligned} \tag{g}$$

where ϕ_2 is given by Eq. (2.68). It is recommended that the reader does the same derivation of $u(t)$ without using complex functions.

From Eq. (b), it is seen that the relationship between $x(t)$ and $u(t)$ is a special case of the general linear relationship described previously. If $H_{xu}(\omega)$ is defined as the rhs of Eq. (e) for $x_0 = 1$, it has just been shown that $H_{xu}(\omega)\, e^{i\omega t}$ satisfies Eq. (b) where $x(t)$ is replaced by $e^{i\omega t}$. $H_{xu}(\omega)$, which is a function of ω only, is therefore the transfer function for the relationship between $x(t)$ and $u(t)$. The transfer function $H_{xy}(\omega)$ between $x(t)$ and relative displacement $y(t)$ can easily be found by using the results from Example 2.5.1.

2.5.3 Example – Vibrating Beam

A cantilever beam is fixed to a foundation as shown in Fig. 2.29(a). The goal is to investigate the vibrations (in the page plane) of the beam end due to vibrations in the foundation. It is assumed that this vibration system can be described as a linear $\hat{m}$-$\hat{c}$-$\hat{k}$ oscillator by using the method described in Example 2.3.3. Figure 2.29(b) shows a principle sketch of the resulting vibration model.

By observing free vibrations of the beam end, it is found that the vibration amplitude is halved after 10 oscillations. What will be the maximum vibration amplitude when the foundation is assumed to oscillate harmonically with amplitude 1 mm?

The logarithmic decrement can be calculated, and it assumes the value $\Lambda = 0.1 \ln 2 = 0.0693$. This gives a relative damping $\xi = 0.011$. By comparing

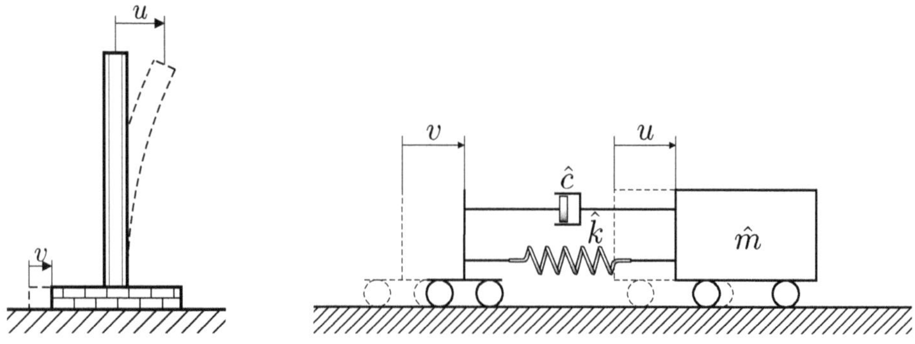

(a) Physical model (b) Vibration model

Figure 2.29. Vibrating beam.

the vibration model from Fig. 2.29(b) with the vibration system in the previous example, it is seen that the transfer function is formally the same for the two cases. The amplitude amplification $|H(\omega)|$ therefore becomes

$$|H(\omega)| = \sqrt{\frac{1 + (0.022\beta)^2}{(1 - \beta^2)^2 + (0.022\beta)^2}}.$$

It follows that the maximum amplitude amplification will take place for $\beta \approx 1$ and is close to $|H(\omega_e)| = 1/(2\xi) = 1/0.022 = 45.5$. The beam end will consequently oscillate with a maximum amplitude of approximately 45.5 mm when the excitation period is close to the natural period.

2.6 Forced Vibrations by Periodic Excitation

In some practical situations, a structure may be excited by an approximately periodic forcing function, which cannot be expressed as a simple harmonic function of time. Such forces may be due to mechanical production equipment, or, for example, propeller forces on ship structures. Figure 2.30 displays an example of such a force function. It is known (Kreyszig, 1993) that such a periodic function can be expanded into a Fourier series under fairly general conditions, which can usually be assumed in practice. If $f(t)$ denotes a periodic forcing function with period T_p, we may therefore write

$$f(t) = a_0 + \sum_{n=1}^{\infty} \{a_n \cos \omega_n t + b_n \sin \omega_n t\}, \tag{2.72}$$

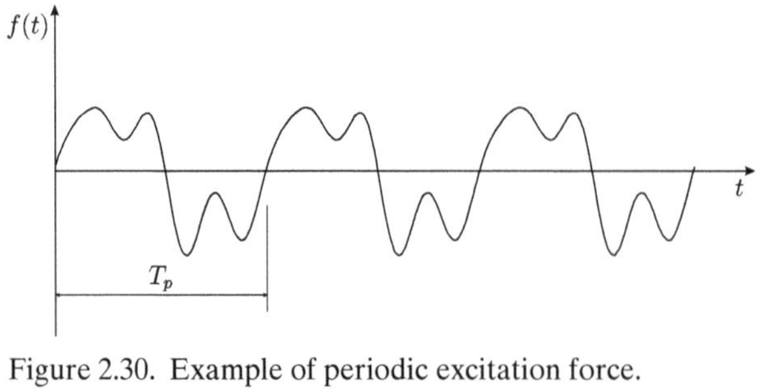

Figure 2.30. Example of periodic excitation force.

where

$$\omega_n = n\,\frac{2\pi}{T_p} = n\,\omega_p\,, \quad n = 1, 2, \ldots, \tag{2.73}$$

$$a_0 = \frac{1}{T_p}\int_{-T_p/2}^{T_p/2} f(t)\,dt = \overline{f} \quad \text{(the mean force)}, \tag{2.74}$$

$$a_n = \frac{2}{T_p}\int_{-T_p/2}^{T_p/2} f(t)\,\cos\omega_n t\,dt, \tag{2.75}$$

$$b_n = \frac{2}{T_p}\int_{-T_p/2}^{T_p/2} f(t)\,\sin\omega_n t\,dt. \tag{2.76}$$

The integration limits are chosen to be $-T_p/2$ and $T_p/2$. Because the integrands are all periodic functions with period T_p, one could just as well have chosen t' and $t' + T_p$ as integration limits, where t' is arbitrary. In particular, one may choose 0 and T_p, which is also a common choice.

The forcing function is now expressed as a sum of harmonic components. To achieve simplicity of notation, Eq. (2.72) is written in the form ($f_0(t) = \overline{f}$):

$$f(t) = \overline{f} + \sum_{n=1}^{\infty} f_n(t) = \sum_{n=0}^{\infty} f_n(t), \tag{2.77}$$

where $f_n(t) = a_n \cos\omega_n t + b_n \sin\omega_n t$ denotes the harmonic component of frequency ω_n. In a manner analogous to the derivation of Eq. (2.11) from Eq. (2.5), we may write $f_n(t) = f_n \cos(\omega_n t - \psi_n)$. In Section 2.3 (and 2.4), we find the solution to the vibration problem

$$m\ddot{u} + c\dot{u} + ku = f_n \cos(\omega_n t - \psi_n), \tag{2.78}$$

except that $\psi_n = 0$ was assumed. The solution with $\psi_n \neq 0$ is then obtained by replacing ϕ_n with $\phi_n + \psi_n$ in the old solutions. A particular solution $u_n(t)$ of Eq. (2.78) for $n = 1, 2\ldots$ is therefore

$$u_n(t) = |H(\omega_n)|\, f_n \cos(\omega_n t - \psi_n - \phi_n), \tag{2.79}$$

where $H(\omega)$ is given by Eq. (2.71).

Because the dynamic model given by the left-hand side of Eq. (2.78) is linear, it is realized that

$$u(t) = \sum_{n=0}^{\infty} u_n(t), \tag{2.80}$$

where $u_0(t) = \overline{f}/k$ satisfies the equation

$$m\ddot{u} + c\dot{u} + ku = f(t). \tag{2.81}$$

$u(t)$ is also a periodic function of period T_p. Note how the Fourier coefficients of $u(t)$ are determined by the Fourier coefficients of $f(t)$ and the transfer function $H(\omega)$.

It is instructive to carry out the same calculation using complex functions as described in Section 2.5. We then write $f(t) = \sum_{n=0}^{\infty} \Re\{f_n \exp[\mathrm{i}(\omega_n t - \psi_n)]\} = \Re\{\sum_{n=0}^{\infty} f_n \exp[\mathrm{i}(\omega_n t - \psi_n)]\}$. A solution $w_n(t)$ of Eq. (2.81) with $f_n \exp[\mathrm{i}(\omega_n t - \psi_n)] = f_n \exp[-\mathrm{i}\psi_n)] \cdot \exp[\mathrm{i}\omega_n t]$ as $f(t)$ becomes, according to Eqs. (2.64)–(2.66) and Eq. (2.71),

$$w_n(t) = H(\omega_n)\, f_n\, e^{-\mathrm{i}\psi_n}\, e^{\mathrm{i}\omega_n t}\,, \quad n = 0, 1, 2, \ldots \tag{2.82}$$

In the same manner as previously, $w(t) = \sum_{n=0}^{\infty} w_n(t)$ satisfies Eq. (2.81). The desired solution $u(t)$ is therefore $u(t) = \Re\{w(t)\}$; that is,

$$u(t) = \sum_{n=0}^{\infty} \Re\{H(\omega_n)\, f_n\, e^{-i\psi_n}\, e^{i\omega_n t}\} = \sum_{n=0}^{\infty} |H(\omega_n)|\, f_n \cos(\omega_n t - \psi_n - \phi_n), \qquad (2.83)$$

where $H(\omega_n) = |H(\omega_n)|\, e^{-i\phi_n}$. This is in agreement with Eqs. (2.79) and (2.80).

2.6.1 Example – Periodic Excitation

Figure 2.31a shows a principle sketch of a linear vibration system excited by a rotating cam-wheel that is driving a follower pin, which is connected to the system via a linear spring. The cam-wheel is rotating at 30 revolutions per minute and gives the follower pin a periodic motion $y(t)$, as shown in Fig. 2.31(b).

To calculate the expression for the resulting displacement response $x(t)$, we first determine the Fourier series Eq. (2.72) of the motion of the follower pin. The period is $T_p = 60/30 = 2$ s. The Fourier coefficients become

$$a_0 = \frac{1}{2}\int_0^2 \frac{t}{2}\, dt, \qquad \text{(a)}$$

$$a_n = \int_0^2 \frac{t}{2} \cos \pi n t\, dt, \qquad \text{(b)}$$

$$b_n = \int_0^2 \frac{t}{2} \sin \pi n t\, dt. \qquad \text{(c)}$$

By partial integration, it is found that $a_0 = 0.5$, $a_n = 0$, and $b_n = -1/(\pi n)$, $n = 1, 2, \ldots$. It leads to the Fourier series

$$y(t) = \frac{1}{2} - \sum_{n=1}^{\infty} \frac{1}{\pi n} \sin \pi n t. \qquad \text{(d)}$$

Our next task is to calculate the displacement response $x(t)$ when it is specified that $m = 100$ kg, $c = 8$ kg/s and $k = 200$ N/m. This can now be done by calculating the Fourier series of $x(t)$, and for this we need the transfer function between $y(t)$ and $x(t)$. The requirement of dynamic equilibrium leads to the equation

$$m\ddot{x} + c\dot{x} + kx + k(x - y) = 0. \qquad \text{(e)}$$

By substituting $y(t) = e^{i\omega t}$ and $x(t) = H_{yx}(\omega)\, e^{i\omega t}$ into (e), it is found that

$$H_{yx}(\omega) = \frac{k}{-\omega^2 m + i\omega c + 2k} = \frac{1}{2(1 - \beta^2 + 2i\xi\beta)}. \qquad \text{(f)}$$

Before we proceed, it is expedient to make some observations. The natural frequency $\omega_e = 2$ rad/s, and the damping ratio $\xi = 0.02$. Because $\omega_n = n\pi$, $n = 1, 2, \ldots$, it follows that $1 - \beta_n^2$ ($\beta_n = \omega_n/\omega_e$) is much larger than $2\xi\beta_n$ in absolute value. This means that the damping can be neglected in the transfer function for the frequencies that are relevant in the Fourier series. This implies

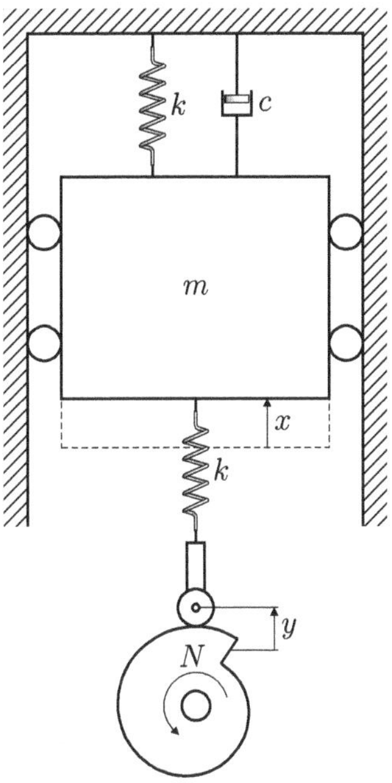

(a) Vibration model

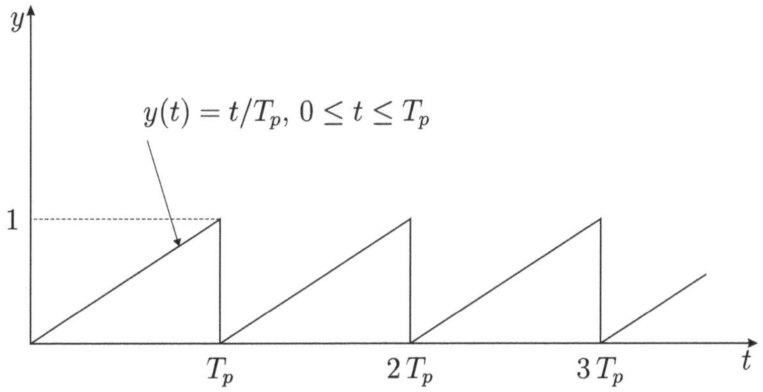

(b) Motion of cam-wheel follower pin

Figure 2.31. Linear vibration system driven by a cam-wheel mechanism.

that the transfer function can be considered as real, and we obtain that

$$x(t) \approx \frac{1}{4} + \sum_{n=1}^{\infty} \frac{1}{2\pi n\left[(\pi n/2)^2 - 1\right]} \sin \pi n t. \tag{g}$$

Note that when the transfer function is real, the Fourier coefficients are multiplied by the transfer function as it is. It is not necessary to take the absolute value and calculate the associated phase shift separately. The sign of the transfer function automatically determines the phase shift. When it is positive, the phase shift is 0 degrees. When it is negative, the phase shift is 180 degrees.

2.7 Forced Vibrations by Arbitrary Excitation

In this section, the case of excitation forces that are not periodic is considered. A nonperiodic function can be viewed as a limit of periodic functions where the period $T_p \to \infty$. In this limiting case the summation in Eqs. (2.77) and (2.80) must be replaced by integration. The desired relations are derived by using heuristic arguments; a rigorous mathematical derivation is found in the specialist literature.

The starting point is an excitation force $f(t)$ that is not periodic. An associated periodic function is defined as follows:

$$f_T(t) = \begin{cases} f(t), & -T/2 \le t \le T/2, \\ \text{periodic of period } T. \end{cases} \tag{2.84}$$

Clearly, $f_T(t) \to f(t)$ when $T \to \infty$ for every t. Because $f_T(t)$ is periodic, we saw in the previous section that

$$f_T(t) = \sum_{n=0}^{\infty} f_n \cos(\omega_n t - \psi_n), \tag{2.85}$$

where f_n, ω_n, and ψ_n depends on T. For the derivations in this section, it is expedient to rewrite Eq. (2.85). By using the formula $\cos x = (e^{ix} + e^{-ix})/2$, and introducing the constants $c_n = f_n e^{-i\psi_n}/2$, $c_{-n} = f_n e^{i\psi_n}/2$, for $n = 1, 2, \ldots$, Eq. (2.85) can be written in the form

$$f_T(t) = \sum_{n=-\infty}^{\infty} c_n\, e^{i\omega_n t}. \tag{2.86}$$

A function $F_T(\omega_n)$ is now defined by the equation

$$F_T(\omega_n)\, \Delta\omega = c_n, \tag{2.87}$$

where $\Delta\omega = 2\pi/T$; that is,

$$f_T(t) = \sum_{n=-\infty}^{\infty} F_T(\omega_n)\, e^{i\omega_n t}\, \Delta\omega. \tag{2.88}$$

When $T \to \infty$, then $\Delta\omega \to 0$, and we assume that the sum Eq. (2.88) will converge to the integral $\int_{-\infty}^{\infty} F(\omega)\, e^{i\omega t}\, d\omega$, where $F_T(\omega) \to F(\omega)$ when $T \to \infty$. This leads to the formula

$$f(t) = \int_{-\infty}^{\infty} F(\omega)\, e^{i\omega t}\, d\omega. \tag{2.89}$$

To the periodic excitation $f_T(t)$, there is associated a periodic response $u_T(t)$, which can be determined by using Eq. (2.83). By a transition entirely similar to that from Eq. (2.85) to (2.89), it can be shown that

$$u(t) = \int_{-\infty}^{\infty} H(\omega)\, F(\omega)\, e^{i\omega t}\, d\omega, \tag{2.90}$$

where $u(t) = \lim u_T(t)$, $T \to \infty$. Note that the imaginary part of the integrand in Eqs. (2.89) and (2.90) integrates to zero because the final result must be a real number for each value of the time t (after all, the integral expresses a force or a displacement response).

To verify that this leads to the desired solution, we have to show that $u(t)$ given by Eq. (2.90) satisfies the equation $L[u] = m\ddot{u} + c\dot{u} + ku = f(t)$, where $f(t)$ is given by Eq. (2.89). From Section 2.5, we know that $L[H(\omega)e^{\mathrm{i}\omega t}] = e^{\mathrm{i}\omega t}$ (by definition). Let us assume that derivation with respect to t can be interchanged with the integration in Eq. (2.90). This leads to $L[u(t)] = \int_{-\infty}^{\infty} F(\omega)\, L[H(\omega)\, e^{\mathrm{i}\omega t}]\, d\omega = \int_{-\infty}^{\infty} F(\omega)\, e^{\mathrm{i}\omega t}\, d\omega = f(t)$, which confirms $u(t)$ as a solution.

How the function $F(\omega)$ is given can be derived as follows. From Eq. (2.86), it is found that

$$c_n = \frac{1}{T}\int_{-T/2}^{T/2} f_T(t)\, e^{-\mathrm{i}\omega_n t}\, dt = \Delta\omega \frac{1}{2\pi}\int_{-T/2}^{T/2} f(t)\, e^{-\mathrm{i}\omega_n t}\, dt. \tag{2.91}$$

Comparison with Eq. (2.87) gives the equation

$$F_T(\omega_n) = \frac{1}{2\pi}\int_{-T/2}^{T/2} f(t)\, e^{-\mathrm{i}\omega_n t}\, dt, \tag{2.92}$$

which leads to

$$F(\omega) = \frac{1}{2\pi}\int_{-\infty}^{\infty} f(t)\, e^{-\mathrm{i}\omega t}\, dt, \tag{2.93}$$

provided the integral exists. Equations (2.89) and (2.93) establish $f(t)$ and $F(\omega)$ as a Fourier transform pair. In the same manner, it is seen that $u(t)$ and $H(\omega)F(\omega)$ also form a Fourier transform pair.

2.8 Impulse Response Function and Duhamel Integral

We have seen that the transfer function $H(\omega)$ provides a complete description of the response properties of the vibration system in that the response $u(t)$ is uniquely determined by Eq. (2.90) because the excitation $f(t)$ determines $F(\omega)$ according to Eq. (2.93). Since $H(\omega)$ is a function of the frequency ω, $H(\omega)$ is said to provide a description of the vibration system along the frequency axis, or in the frequency domain. The impulse response function, to be discussed in the following, provides a corresponding description along the time axis, or in the time domain.

Impulse Load

Assume that $f(t)$ in Eq. (2.81) is an impulse load of short duration (sometimes referred to as a force impact) at time $t = \tau$; that is, we assume that $f(t) = 0$, except in a short time interval $\Delta\tau$ after $t = \tau$ such that

$$\int_{-\infty}^{\infty} f(t)\, dt = \int_{\tau}^{\tau+\Delta\tau} f(t)\, dt = I, \tag{2.94}$$

where I represents an impulse or force impact ([I] = Ns).

As $f(t) = 0$ up to $t = \tau$, the response $u(t) = 0$ for $t \le \tau$. Similarly, $f(t) = 0$ for $t > \tau' = \tau + \Delta\tau$; that is, $u(t)$ is equal to the homogeneous solution of Eq. (2.30) for $t > \tau'$ because we have assumed $\xi < 1$. Hence,

$$u(t) = e^{-\xi\omega_e(t-\tau')}\left(a_1 \cos\omega_d(t-\tau') + a_2 \sin\omega_d(t-\tau')\right), \tag{2.95}$$

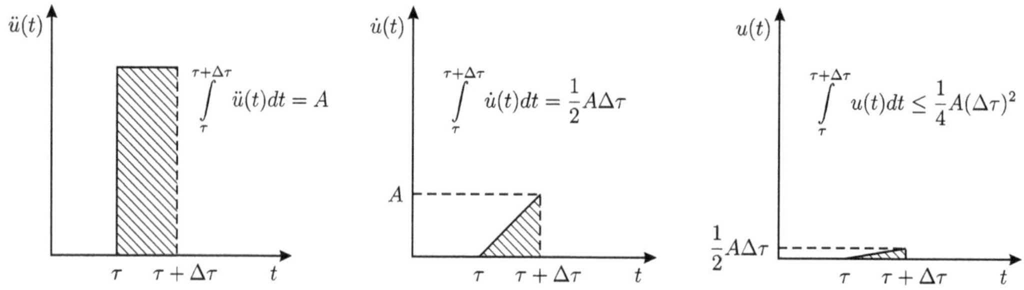

Figure 2.32. Illustration of the relative size of the terms in Eq. (2.96).

where a_1 and a_2 are determined by the initial conditions $u(\tau')$ and $\dot{u}(\tau')$. To derive these, Eq. (2.81) is integrated over the impact time $(\tau, \tau + \Delta\tau)$, which leads to the equation

$$m \int_\tau^{\tau+\Delta\tau} \ddot{u}(t)\, dt + c \int_\tau^{\tau+\Delta\tau} \dot{u}(t)\, dt + k \int_\tau^{\tau+\Delta\tau} u(t)\, dt = I. \tag{2.96}$$

When the impact time is sufficiently short, the two last integrals on the left hand side of Eq. (2.96) become negligible compared with the first. Figure 2.32 shows why it is the integral of $\ddot{u}(t)$ that will give the dominating contribution to the left hand side of Eq. (2.96). It is seen that for the case shown in Fig. 2.32, the integrals of $\dot{u}(t)$ and $u(t)$ can be made as small as we want compared with the integral of $\ddot{u}(t)$, if only the impact time is small enough. By assuming this situation, it is found that

$$m \int_\tau^{\tau+\Delta\tau} \ddot{u}(t)\, dt = I \tag{2.97}$$

and

$$\int_\tau^{\tau+\Delta\tau} \dot{u}(t)\, dt = 0. \tag{2.98}$$

Because the system is at rest at $t = \tau$, $\dot{u}(\tau) = u(\tau) = 0$, which leads to the equations

$$m\dot{u}(\tau + \Delta\tau) = m \int_\tau^{\tau+\Delta\tau} \ddot{u}(t)\, dt = I \tag{2.99}$$

and

$$u(\tau + \Delta\tau) = \int_\tau^{\tau+\Delta\tau} \dot{u}(t)\, dt = 0. \tag{2.100}$$

$u(\tau + \Delta\tau) = u(\tau') = 0$ leads to $a_1 = 0$ in Eq. (2.95). $\dot{u}(\tau + \Delta\tau) = \dot{u}(\tau') = I/m$ gives $a_2 = I/(\omega_d m)$. Hence, it is obtained that $u(t) = 0, t \le \tau$, and

$$u(t) = \frac{I}{\omega_d m} e^{-\xi\omega_e(t-\tau')} \sin \omega_d(t - \tau'), \quad t \ge \tau', \tag{2.101}$$

is the response to an impulse I of short duration $\Delta\tau$ during the time interval $\tau \le t \le \tau'$. The solution during the time interval itself is left unspecified.

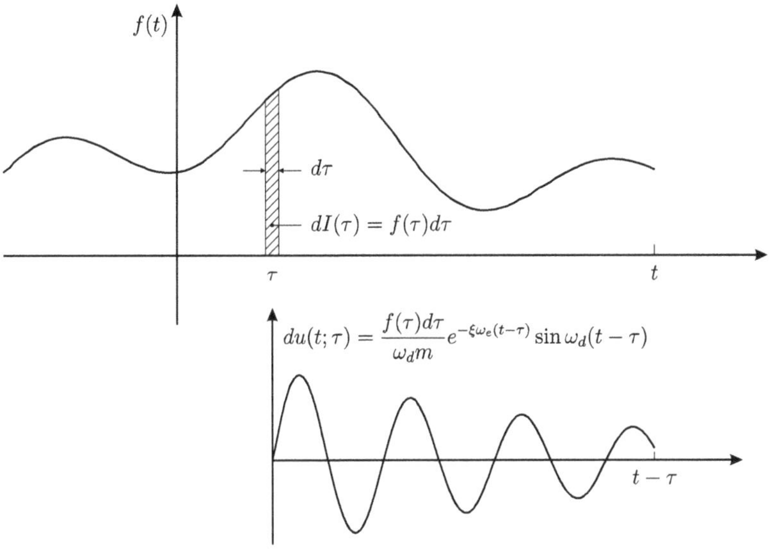

Figure 2.33. Response $du(t;\tau)$ from the impulse load $dI(\tau)$.

We now demonstrate that the response to a general excitation $f(t)$ can be built up using Eq. (2.101). To this end, the following function is usually introduced:

$$h(t) = \begin{cases} \frac{1}{\omega_d m} e^{-\xi\omega_e t} \sin \omega_d t, & t \geq 0, \\ 0, \quad t < 0. \end{cases} \tag{2.102}$$

This gives $u(t) = Ih(t-\tau)$ in the limit when $\Delta\tau \to 0$, that is, for a force impact of infinitesimal (infinitely short) duration. $h(t)$ is called the impulse response function because it describes the response to a unit impulse (at $t = 0$). Note in particular that $h(t) = 0$ for $t < 0$. This expresses the principle of causality: no effect before the cause.

General Load

Assume a general load time history $f(t)$ on the rhs of Eq. (2.81). The effect of $f(t)$ on the vibration system at an arbitrary point in time $t = \tau$ can be considered as the effect of an infinitesimal impulse load (Fig. 2.33)

$$dI(\tau) = f(\tau)\, d\tau, \tag{2.103}$$

According to Eqs. (2.101) and (2.102), the corresponding response at time $t > \tau$, denoted by $du(t;\tau)$, is given by

$$du(t;\tau) = f(\tau)\, d\tau\, h(t-\tau). \tag{2.104}$$

The expression for the total response $u(t)$ at time t due to the whole load time history before t is obtained by summing (integrating) all response contributions from $\tau = -\infty$ to $\tau = t$. That is,

$$u(t) = \int_{-\infty}^{t} f(\tau)\, h(t-\tau)\, d\tau. \tag{2.105}$$

Because $h(t-\tau)=0$ for $t<\tau$, Eq. (2.105) can be rewritten as

$$u(t)=\int_{-\infty}^{\infty} f(\tau)\,h(t-\tau)\,d\tau. \tag{2.106}$$

In Eq. (2.106), we may introduce the new integration variable $s=t-\tau$, which expresses the time since the impulse load was exerted. This gives the equation

$$u(t)=\int_{-\infty}^{\infty} h(s)\,f(t-s)\,ds. \tag{2.107}$$

Substituting from Eq. (2.102), it is obtained that

$$u(t)=\frac{1}{\omega_d m}\int_{0}^{\infty} e^{-\xi\omega_e s}\sin(\omega_d s)\,f(t-s)\,ds. \tag{2.108}$$

These integral expressions for the response $u(t)$ are often referred to as Duhamel integrals.

We have now shown that the complete solution of the equation of motion Eq. (2.81) with a load time history $f(t)$ is given by Eq. (2.108). This solution requires, however, that the whole time history of the load is known, that is, from $t=-\infty$. If we know the values of $u(t)$ and $\dot{u}(t)$ at a specific point in time, $t=t_0$ say, then only the time history of $f(t)$ for $t\ge t_0$ needs to be known. In this case, the complete solution of Eq. (2.81) for $t\ge t_0$ can then be written in the form

$$u(t)=e^{-\xi\omega_e(t-t_0)}\Big(a\cos\omega_d(t-t_0)+b\sin\omega_d(t-t_0)\Big)+\int_{t_0}^{t} h(t-\tau)f(\tau)d\tau. \tag{2.109}$$

where the constants a and b are determined by the values of $u(t_0)$ and $\dot{u}(t_0)$. That this is indeed the complete solution can be seen in several ways. Here, we first show that

$$u(t)=\int_{t_0}^{t} h(t-\tau)\,f(\tau)\,d\tau \tag{2.110}$$

is a particular solution of Eq. (2.81) for all $t\ge t_0$. Then, as seen previously in this chapter, Eq. (2.109) provides us with the complete solution.

We know that Eq. (2.110) for $t\ge t_0$ determines the complete solution of Eq. (2.81) for the case when $f(t)=0$ for all $t<t_0$. This implies that $u_p(t)$ given by Eq. (2.110) satisfies Eq. (2.81) for all $t\ge t_0$, irrespective of which values $f(t)$ assumes for $t<t_0$, because these values are not used in the expression for $u_p(t)$. Of course, in this last case, $u_p(t)$ no longer represents the complete solution. This, however, is of no concern now as we only need to ensure that $u_p(t)$ given by Eq. (2.110) satisfies Eq. (2.81). Hence, we have shown that $u_p(t)$ given by Eq. (2.110) is a particular solution of Eq. (2.81) for every $t\ge t_0$. From Eq. (2.110) it follows that both $u_p(t_0)=0$ and $\dot{u}_p(t_0)=0$. The constants a and b are therefore determined solely by $u(t_0)$ and $\dot{u}(t_0)$.

Relationship Between $h(t)$ and $H(\omega)$

In Section 2.5, if we put $f(t)=e^{i\omega t}$, a particular solution of Eq. (2.81) was then given by $u(t)=H(\omega)e^{i\omega t}$. Together with Eq. (2.107), this gives the relation

$$H(\omega)e^{i\omega t}=\int_{-\infty}^{\infty} h(s)\,e^{i\omega(t-s)}\,ds. \tag{2.111}$$

Deleting equal terms leads to

$$H(\omega) = \int_{-\infty}^{\infty} h(s)\, e^{-i\omega s}\, ds. \tag{2.112}$$

Conversely, it can be shown that

$$h(t) = \frac{1}{2\pi} \int_{-\infty}^{\infty} H(\omega)\, e^{i\omega t}\, d\omega. \tag{2.113}$$

The two last equations show that $h(t)$ and $H(\omega)$ constitute a Fourier transform pair, but there is an important difference from before. By comparing Eqs. (2.112) and (2.113) with Eqs. (2.89) and (2.93), it is seen that the factor $1/(2\pi)$ has switched place.

2.8.1 Example – Suddenly Applied Force

Assume that the vibration system described by Eq. (2.81) is excited by a force time history that can be modeled as

$$f(t) = \begin{cases} f_0\, e^{-at}, & t \geq 0\ (a \geq 0), \\ 0, & t < 0. \end{cases} \tag{a}$$

Using Eq. (2.108) to calculate the response, it is found that

$$\begin{aligned} u(t) &= \frac{1}{\omega_d m} \int_0^t e^{-\xi\omega_e s} \sin(\omega_d s)\, f_0\, e^{-a(t-s)}\, ds \\ &= \frac{f_0\, e^{-at}}{\omega_d m} \int_0^t e^{(a-\xi\omega_e)s} \sin(\omega_d s)\, ds \end{aligned} \tag{b}$$

Carrying out the integration leads to the result

$$\begin{aligned} u(t) &= \frac{f_0\, e^{-at}}{\omega_d m} \left[\frac{e^{(a-\xi\omega_e)s}}{(a-\xi\omega_e)^2 + \omega_d^2} \{(a-\xi\omega_e)\sin(\omega_d s) - \omega_d \cos(\omega_d s)\} \right]_{s=0}^{t} \\ &= \frac{f_0}{\omega_d m[(a-\xi\omega_e)^2 + \omega_d^2]} \{[(a-\xi\omega_e)\sin(\omega_d t) - \omega_d \cos(\omega_d t)]e^{-\xi\omega_e t} + \omega_d e^{-at}\}. \end{aligned} \tag{c}$$

In this case, the Duhamel integral provided a solution in a direct manner without (explicitly) having to involve the differential equation.

2.9 Maximum Response to Various Force Time Histories

In some cases, it is of particular interest to calculate the maximum response of a dynamic structure to a given force time history. We now assume that the structural response can be modeled by using a standard $m - c - k$ oscillator with low damping. The calculations are based on the Duhamel integral and some of the derivations of the previous section.

Suddenly Applied Constant Force of Long Duration

An excitation force $f(t)$ of this kind can be represented as $f(t) = 0$ for $t < 0$, and $f(t) = f_0 = \text{constant}$ for $t \geq 0$, where t represents the time that has elapsed since the force was applied. The corresponding displacement response $u(t)$ can be calculated by using Eq. (2.108), which leads to the expression

$$u(t) = \frac{f_0}{\omega_d m} \int_0^t e^{-\xi\omega_e s} \sin(\omega_d s)\, ds. \tag{2.114}$$

The expression for the rhs of this equation can be obtained from Eq. (c) of Example 2.13 by setting $a = 0$, which gives the result

$$u(t) = \frac{f_0}{k} \{1 - [\cos(\omega_d t) + \frac{\xi}{\sqrt{1-\xi^2}} \sin(\omega_d t)]\, e^{-\xi\omega_e t}\}, \tag{2.115}$$

where f_0/k, as we have seen, expresses the static displacement response.

The maximum response u_{max} occurs at the time $t \approx T_d/2$ because the damping is assumed to be small, and it is given as

$$u_{max} \approx \frac{f_0}{k}\left(1 + e^{-\xi\pi}\right). \tag{2.116}$$

When $\xi = 0.05$, then $u_{max} \approx 1.85\, f_0/k$. The classical result $u_{max} = 2\, f_0/k$ is obtained when $\xi = 0$. This result, which says that the maximum dynamic response to a suddenly applied load of long duration is twice as large as the corresponding static response, may serve as a simple rule of thumb for the case of low damping. We show in the following that this result may also apply to situations with more general force time histories.

2.9.1 Example – Torsional Rotation of a Suspension Bridge

Irvine (1988) reports on an event where a large crowd of people were gathered along one of the side rails of a suspension bridge to watch a marine flotilla that was approaching the bridge. When the flotilla passed under the bridge, the crowd quickly moved over to the other side of the bridge. As a result, the rotational moment on the bridge deck changed rapidly, which lead to a motion response of the deck that frightened many people to the extent that they fled the bridge in panic.

Let us investigate this effect by using what we have learned. Assume that the crowd, when it is first gathered at the side rail exerts a rotational moment m_t on the bridge deck. When the crowd "suddenly" moves over to the other side of the bridge deck, it leads to a change of sign of the rotational moment. If the moment m_t gave a static angle of rotation equal to θ, then a sudden change from m_t to $-m_t$ would give rise to a rotation response that could reach almost 4θ, that is, from θ to -3θ. The resulting oscillations would lead to many large response amplitudes ($\approx \pm 3\theta$) before the damping eventually kills off the motions.

Suddenly Applied Constant Force of Limited Duration

If the force is applied for only a limited period of time t_*, then the force time history is given as $f(t) = 0$ for $t < 0$ and $t > t_*$, and $f(t) = f_0 = \text{constant}$ for $0 \le t \le t_*$.

Before we do any calculations for this specific case, we can immediately say that if $t_* \ge T_d/2$, that is, the force is applied for a time at least as long as half the natural period, then the maximum response is given by Eq. (2.116). This is clearly so because the present force time history is identical to the one giving rise to Eq. (2.116) for

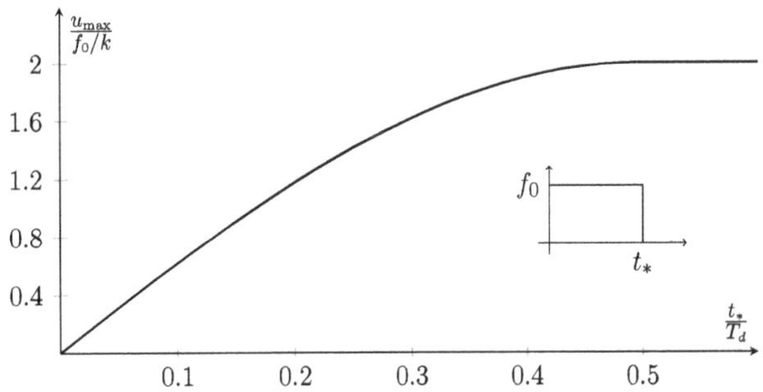

Figure 2.34. Maximum response to a suddenly applied constant force of limited duration.

all times up to the time of maximum response, and the fact that the force suddenly disappears at a later time cannot lead to a response that exceeds u_{max} as given by Eq. (2.116).

It only remains to establish the maximum response when $t_* < T_d/2$. We saw that for small damping, a reasonably accurate estimate of the maximum response is obtained by neglecting the damping. To simplify the calculations, we therefore assume that $\xi = 0$. The response for $0 \leq t \leq t_*$ is then given by Eq. (2.115) with $\xi = 0$. This gives

$$u(t) = \frac{f_0}{k}\left\{1 - \cos(\omega_d t)\right\}, \quad 0 \leq t \leq t_*. \tag{2.117}$$

For $t > t_*$, the response is given by the homogeneous solution because the force is now zero; that is,

$$u(t) = a \cos \omega_d(t - t_*) + b \sin \omega_d(t - t_*), \tag{2.118}$$

where the constants a and b are determined by $u(t_*)$ and $\dot{u}(t_*)$. This gives $a = f_0/k(1 - \cos \omega_d t_*)$ and $b = f_0/k \sin \omega_d t_*$. Note that $\dot{u}(t)$ is positive in the entire time interval $0 \leq t \leq t_*$ because $t_* < T_d/2$. This implies that the response grows in this time interval and that the maximum response therefore must occur after the time t_*.

Because the response after $t = t_*$ is a harmonic oscillation, the maximum response must equal the amplitude of this oscillation, that is,

$$\begin{aligned} u_{max} &= \sqrt{a^2 + b^2} = \frac{f_0}{k}\sqrt{(1 - \cos \omega_d t_*^2) + (\sin \omega_d t_*)^2} \\ &= \frac{\sqrt{2} f_0}{k}\sqrt{1 - \cos \omega_d t_*} = \frac{2 f_0}{k} \sin \frac{\omega_d t_*}{2} = \frac{2 f_0}{k} \sin \frac{\pi t_*}{T_d}. \end{aligned} \tag{2.119}$$

Note that our previous result, $u_{max} = 2 f_0/k$, is obtained when $t_* = T_d/2$. The result is shown graphically in Fig. 2.34.

The amplitude of the homogeneous solution as it is given by Eq. (2.118) is valid whatever the value of t_*. This shows that the maximum response for a suddenly applied constant force of finite duration cannot exceed the corresponding maximum response for a suddenly applied constant force of the same size and of infinite duration. Even if this result was derived for the idealized case of no damping, it should apply even better to systems with positive damping because the response amplitudes then decrease with time.

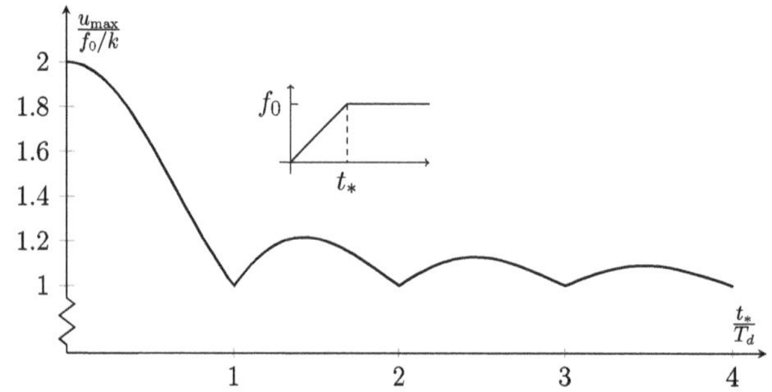

Figure 2.35. Maximum response to a constant force with a finite rise time.

Other Simple Force Time Histories

The maximum response for three simple force time histories for the case of no damping are shown in Figs. 2.35 to 2.37. They are based on material from Biggs (1964).

Figure 2.35 shows $u_{max}/(f_0/k)$ when the force increases steadily toward a constant level. It is interesting to observe from Fig. 2.35 that when t_* is small (e.g., $t_* < T_d/4$), then the reduction in the maximum response is insignificant compared to the situation with a suddenly applied force. This means that if the time it takes before the applied force reaches its peak value is less than 25% of the natural period, then, in practice, one may assume that the applied force takes on its peak value immediately. In such cases, one may also expect that the exact behavior of the force time history prior to the peak value is of little concern. It is a peculiarity of this system that the maximum response equals the static response f_0/k when the rise time t_* is an integer multiple of the natural period T_d. The fact that the maximum response approaches the static value when the rise time becomes long should surprise nobody.

Figure 2.36 illustrates the situation with a suddenly applied force that decreases linearly toward zero. It is natural to compare this case with the situation illustrated in Fig. 2.34, which is the case of a suddenly applied force of finite duration. The triangular force time history will always lead to lower maximum response than the corresponding rectangular one for the simple reason that the total impulse is smaller. When the duration increases, the maximum response to the triangular force time history will approach twice the static value, but only for values of t_* that are much larger than in Fig. 2.34.

Figure 2.37 shows the maximum response when the force time history looks like a sawtooth. In this case, the maximum response is largest when the duration of the applied force is equal to the natural period. Then the maximum response is approximately 50% larger than the static response f_0/k. When the duration increases, the maximum response will, as could be expected, approach the static value.

Impulse Load

A characteristic feature of the maximum response to load time histories of limited duration, as it can be gleaned from Figs. 2.34 to 2.37, is that the duration relative to the natural period to a large extent determines the response. In the following, we consider the situation where the load duration is short compared to the natural period. This will typically be the case for loads caused by explosions.

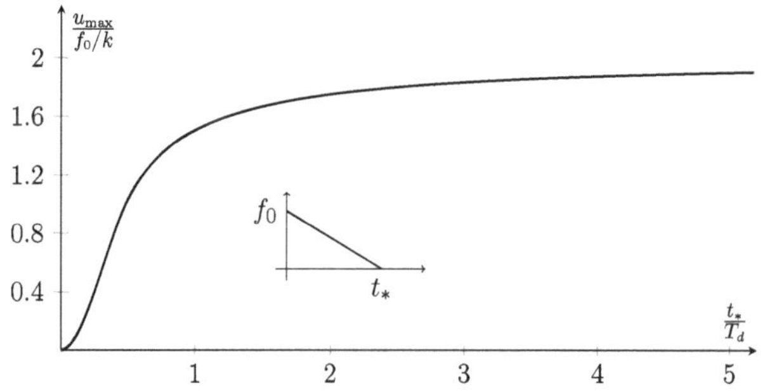

Figure 2.36. Maximum response to a suddenly applied triangular force time history.

The result we need is found in the section on the Duhamel integral. Equation (2.101) provides us with the response to a load of short duration of impulse I. An upper bound on the maximal response is therefore

$$u_{max} = \frac{I}{\omega_d m}. \tag{2.120}$$

Hence, when the duration is short, it is the impulse that is important, not the size of the load or its time history.

2.9.2 Example – Response to Collision Load

An offshore monotower is located rather close to a busy ship route, and the authorities have requested that the effect on the tower of a ship colliding with the deck structure be investigated. A sketch of the situation is shown in Fig. 2.38. The purpose of this investigation is to determine whether the tower will incur more than serious local damage after such a collision, that is, if the tower itself will survive.

As a reference case, a ship of mass 10,000 tons and speed 5 m/s is assumed. After the collision, it is assumed that the ship is stuck in the deck structure. With reference to Example 2.3.3, a set of generalized parameters for the dominant vibration mode of the tower after the collision was found. These refer to the displacement response of an assumed collision point in the tower. The relevant quantities are equivalent mass $\hat{M} = 10{,}000$ tons and natural period $T_e = 2$ s.

On the basis of this information, the task is now to estimate a maximum displacement response after the collision with the provision that we may use linear elastic analysis. That is, we neglect the energy lost in the collision.

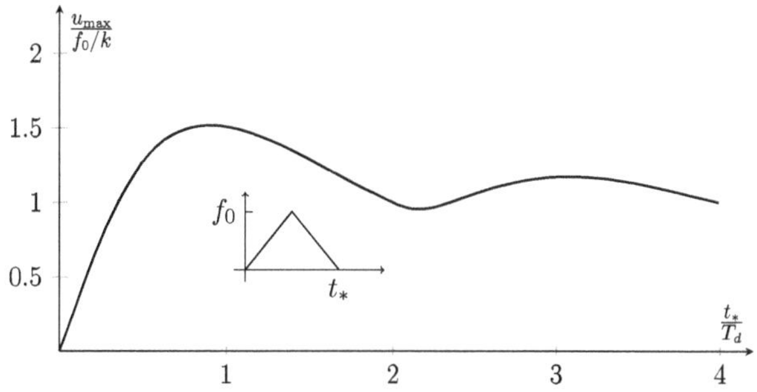

Figure 2.37. Maximum response to a symmetric triangular force time history.

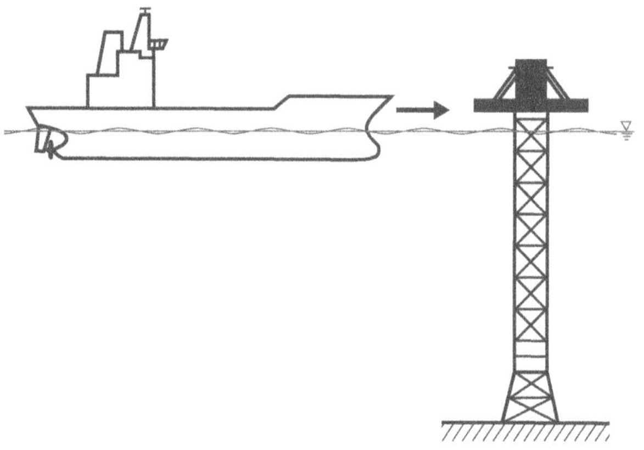

Figure 2.38. Sketch of ship on collision course with the deck structure of an offshore tower.

As the ship follows the tower after the collision, we may, according to the principle of conservation of impulse, assume that the equivalent tower mass plus that of the ship, denoted by $\hat{M}_t$, receives an impulse equal to that of the ship immediately prior to impact, that is, $I = 5 \cdot 10^7$ kgm/s. According to Eq. (2.120), this gives a maximum response of the reference point given by

$$u_{max} \approx \frac{I T_e}{2\pi \hat{M}_t} = 1.06 \text{ m}.$$

On the basis of this result, one can, in principle, calculate the maximum stresses in the tower to verify that they do not exceed allowed values.

3 Dynamics of Multi-Degree-of-Freedom Linear Systems

3.1 Introduction

In Chapter 2, the dynamic equation of motion was established for an SDOF system, and the solutions to different deterministic load conditions were described. It was also shown that the SDOF system could be used to model important features of rigid-body and flexible motions of various structures. The SDOF can be translational or rotational motion. The number of degrees of freedom is equal to the number of variables found necessary to describe the dynamic behavior of a structure with the required accuracy. This chapter deals with formulation and solution of dynamic problems which include both rigid-body motions and structural deformations, and are based on multiple degrees of freedom.

In Example 2.3.3, the concept of a generalized DOF is introduced, and it is illustrated how this could be used to effectively reduce a large number of DOFs to just one generalized DOF by combining this with the appropriate shape function. Thus, the instantaneous position of all mass points in the system were specified without ambiguity and without redundancy. The simplest example of this is a mass point moving in a plane, which is described by two coordinates. Of course, the concept of generalized coordinates is not needed in this case. However, a rigid disc moving in a plane needs three generalized coordinates. A rigid body moving in the paper plane, see Fig. 3.1, may be described by $u_G(t)$, $v_G(t)$, and $\phi_G(t)$ of the center of gravity. Clearly, using the three quantities with reference to one of the ends of the body also represents a (unique) set of generalized coordinates. If considered as a rigid body, the articulated loading tower in Example 2.2.8 can be described by a single generalized coordinate (DOF) when a plane motion is considered. A continuous body may be considered to have infinitely many mass points linked together. However, elastic bodies may be described by a finite (often a small) number of generalized coordinates. In this case, the coordinate, q_i, is such that $q_i w_i(x)$ represents a deformation pattern of a one-dimensional structure, e.g., a beam. By summing these contributions, the total deformation is obtained.

Although the dynamic equations of motion can be established by equilibrium considerations for rigid-body motions, a proper description of structural behavior requires consideration of equilibrium, kinematic compatibility, and constitutive relations when the structural deformations are to be accounted for. The governing

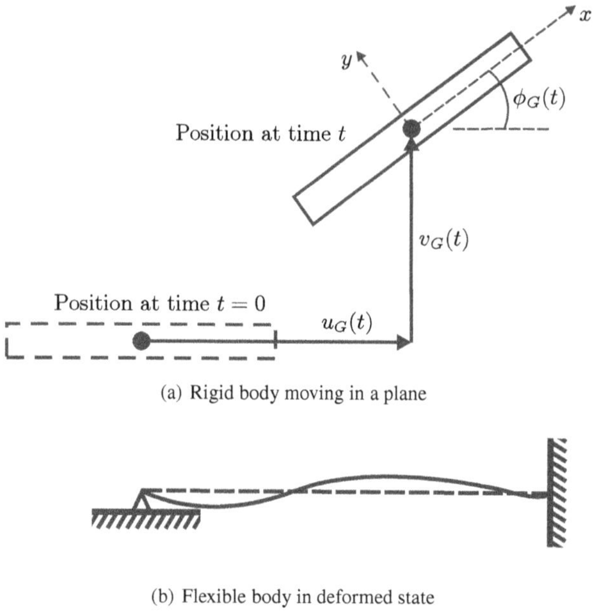

Figure 3.1. Generalized and other coordinates.

dynamic equations of motion can then be obtained by using the principle of virtual displacements.

The dynamic behavior of flexible structures may be determined in different ways. The main approaches include using some kind of discretization or solving the continuous dynamics by solving partial differential equations. Flexible structures may be discretized in different manners. Consider, for instance, a beam under bending. In principle, it is a continuous structure with infinitely many degrees of freedom. However, it may be represented in a rough way by lumping the mass in the midpoint. Then the dynamic behavior is described by the lateral displacement of the mass point (an SDOF). Another matter is that the axial displacement requires another degree of freedom. Moreover, an SDOF is an approximate representation and the accuracy of results obtained depends on the properties of loading in time and space. As subsequently discussed, a more accurate approach can be obtained by discretizing the structure in space by the FE method and then establishing the dynamic equations in time. Yet another method is to use normal vibration modes (Chopra, 2001) or some other assumed modes of vibration.

Structures are supported on soils or by water, and interact with those media. For this reason the dynamic properties of surrounding soil or water are also briefly touched on.

In Section 3.2 the modeling and analysis of some simple discrete systems under deterministic loads are described. Section 3.3 deals with the solution of some simple continuous structures, by exact solution of the governing differential equation where relevant, or by the use of various approximate techniques. Chapter 4 is devoted to finite element discretization of multi-degree-of-freedom, continuous systems.

The purpose of this chapter is partly to provide insight into establishing dynamic models relevant for marine structures and solving the resulting equations. It is also to illustrate characteristic features of the dynamic behavior.

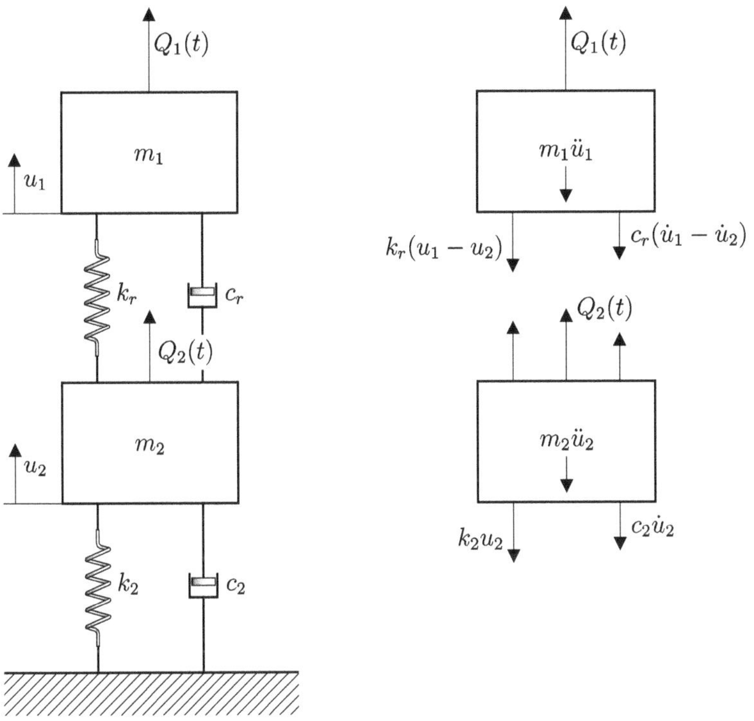

(a) 2DOF model (b) Free body diagram

Figure 3.2. Two degrees of freedom system.

3.2 Discrete Systems

3.2.1 Discrete Systems of Rigid Bodies

The formulation and solution of a simple example of a two-degrees-of-freedom system shown in Fig. 3.2 is first considered. The system consists of two masses connected by a spring and a viscous damper.

In this case, the dynamic equilibrium equation for the two masses yield two equations. According to d'Alembert's principle, the two dynamic equations of motion for this system are (Fig. 3.2)

$$m_1\,\ddot{u}_1 + c_r(\dot{u}_1 - \dot{u}_2) + k_r(u_1 - u_2) = Q_1(t) \tag{3.1}$$

and

$$m_2\ddot{u}_2 + c_r(\dot{u}_2 - \dot{u}_1) + k_r(u_2 - u_1) + c_2\dot{u}_2 + k_2u_2 = Q_2(t). \tag{3.2}$$

Because there are two unknowns, $u_1(t)$ and $u_2(t)$, this is a two-degrees-of-freedom system.

The equations of motion may be formulated in matrix notation as

$$\begin{bmatrix} m_1 & 0 \\ 0 & m_2 \end{bmatrix}\ddot{\mathbf{u}} + \begin{bmatrix} c_r & -c_r \\ -c_r & c_r + c_2 \end{bmatrix}\dot{\mathbf{u}} + \begin{bmatrix} k_r & -k_r \\ -k_r & k_r + k_2 \end{bmatrix}\mathbf{u} = \begin{bmatrix} Q_1(t) \\ Q_2(t) \end{bmatrix}, \tag{3.3}$$

or

$$\mathbf{M}\ddot{\mathbf{u}} + \mathbf{C}\dot{\mathbf{u}} + \mathbf{K}\mathbf{u} = \mathbf{Q}(t), \tag{3.4}$$

where the displacement vector $\mathbf{u}$ is

$$\mathbf{u} = \begin{pmatrix} u_1 \\ u_2 \end{pmatrix}.$$

The natural frequencies are found by assuming a harmonic oscillation $\mathbf{u} = \mathbf{u}_0 \sin \omega t$ and neglecting the damping term $\mathbf{C}\dot{\mathbf{u}}$ and $\mathbf{Q}(t)$, that is,

$$(-\omega^2 \mathbf{M} + \mathbf{K})\mathbf{u}_0 = \mathbf{0}. \tag{3.5}$$

The natural frequencies ω can then be found from

$$\det\left(-\omega^2\mathbf{M} + \mathbf{K}\right) = \det \begin{bmatrix} k_r - \omega^2 m_1 & -k_r \\ -k_r & k_r + k_2 - \omega^2 m_2 \end{bmatrix} = 0, \tag{3.6}$$

and the eigenvectors $\mathbf{u} = \boldsymbol{\phi}_i$ can be determined from Eq. (3.5) when ω is known.

The solution to Eq. (3.4) can be conveniently found by modal superposition. The coupled differential equations are then transformed into two uncoupled equations, which can be solved separately according to the methods for SDOF systems in Chapter 2.

In this case, the general solution to Eq. (3.4) can then be written as

$$\mathbf{u} = \boldsymbol{\phi}_1 Y_1(t) + \boldsymbol{\phi}_2 Y_2(t). \tag{3.7}$$

This modeling and solution method can be generalized for systems with many degrees of freedom.

3.2.2 Other Examples

Consider now the system in Fig. 3.3, which shows a floating platform equipped with a buoy that can move along a guide beam. The relative motion is transformed into mechanical and then electrical power; that is, it is performing as a wave energy converter. It is assumed for simplicity that the power take-off is proportional with the relative velocity between the platform and buoy. By taking the buoy and the floating platform as bodies 1 and 2, the model in Fig. 3.2 needs to be extended by restoring and damping forces relative to the seabed, which would introduce downward forces $k_1 u_1$ and $c_1 \dot{u}_1$, respectively, on body 1, for positive u_1 and $\dot{u}_1$ upward. The equations for dynamic equilibrium can then be written as Eq. (3.4), with the matrix $\mathbf{M}$ and $\mathbf{Q}(t)$ as in Eq. (3.3) and with

$$\mathbf{C} = \begin{bmatrix} c_1 + c_r & -c_r \\ -c_r & c_2 + c_r \end{bmatrix}, \quad \mathbf{K} = \begin{bmatrix} k_1 + k_r & -k_r \\ -k_r & k_2 + k_r \end{bmatrix}. \tag{3.8}$$

The power take-off is $P = c_r(\dot{u}_1 - \dot{u}_2)^2$.

In Chapter 2 the equations of motion for heave, pitch, and surge of a TLP were considered separately. If, for instance, surge and heave are assumed to be uncoupled, the equations of motion may be written as Eq. (3.4), with diagonal matrices $\mathbf{M}$, $\mathbf{C}$ and $\mathbf{K}$.

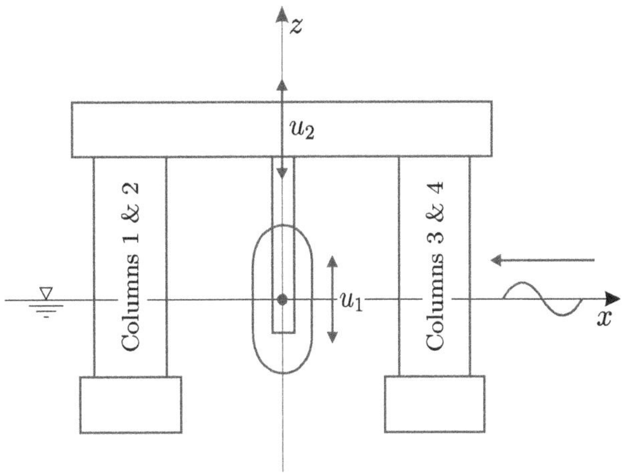

Figure 3.3. Sketch of a two-body wave energy converter, based on relative heave motion.

The first equation represents surge, while the second one expresses vertical balance of forces. Coupling means that, for instance, a surge motion introduces inertia, damping, or stiffness (restoring forces) relating to the vertical dynamic equilibrium, i.e., the second equation in Eq. (3.4) adapted to this case. And, indeed, there is a slight coupling between these modes because a surge motion implies a vertical motion (the TLP is an inverted pendulum). The heave displacement u_2 caused by a surge u_1 is approximately $0.5(u_1/l)u_1$, where l is the length of the tether. This is a nonlinear effect.

Next, consider the possible coupling between heave and pitch. If the structure possesses a vertical symmetry plane, a pitch motion produces upward and downward vertical forces of equal magnitude, and hence, the resultant vertical force becomes zero, implying no coupling. For a body of arbitrary shape, there will be coupling between heave and pitch.

In Chapter 2, tower-type structures were modeled by a single degree of freedom. Now, consider a tower with a horizontal (u) and rotational (θ) degree of freedom, as indicated in Fig. 3.4a. By assuming the tower to be rigid but the soil to be flexible, the motion is defined as shown in Fig. 3.4a. The excitation force is chosen as $q(z,t) = q_0(z/h_1)\cdot f(t)$. The mass m (including added mass) is assumed to be constant per unit length. The forces acting are shown in Fig. 3.4b. By establishing the two equations for dynamic equilibrium for horizontal forces and moment, respectively, the final equations may be written as shown in Eq. (3.9).

$$\begin{bmatrix} m_{11} & m_{12} \\ m_{21} & m_{22} \end{bmatrix} \ddot{\mathbf{u}} + \begin{bmatrix} c_{11} & c_{12} \\ c_{21} & c_{22} \end{bmatrix} \dot{\mathbf{u}} + \begin{bmatrix} k_{11} & k_{12} \\ k_{21} & k_{22} \end{bmatrix} \mathbf{u} = \begin{bmatrix} Q_1(t) \\ Q_2(t) \end{bmatrix}, \tag{3.9}$$

where

$$\mathbf{u} = \begin{bmatrix} u_1 \\ u_2 \end{bmatrix} = \begin{bmatrix} u \\ \theta \end{bmatrix}.$$

Here, the (generalized) mass $m_{11} = M + mh$, $m_{12} = m_{21} = Mh + mh^2/2$, $m_{22} = Mh^2 + mh^3/3$, and stiffness $k_{11} = k_u$, $k_{22} = k_\theta + bh_1^2/2 - (Mgh + mgh^2/2)$, $k_{12} = k_{21} = 0$. The parenthesis of the k_{22} term represents the so-called $P - \Delta$ effect; i.e., the moment caused by gravitational forces when the tower gets a finite rotation.

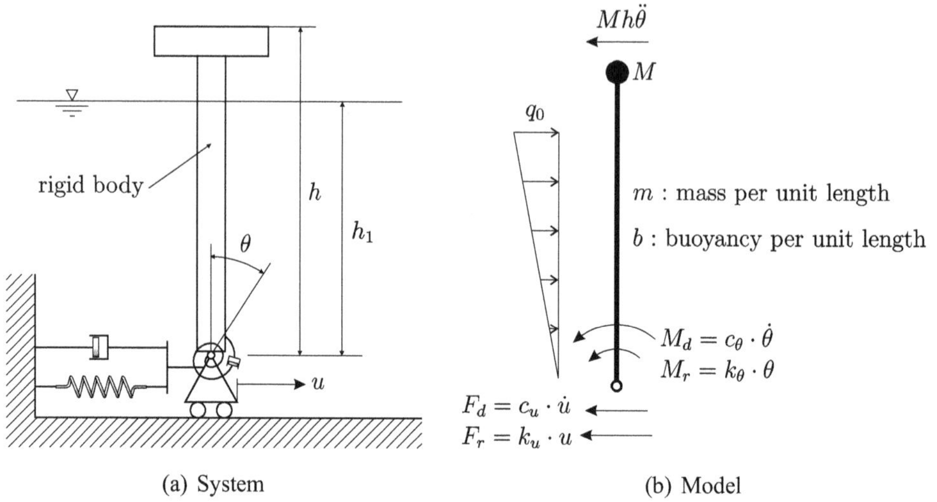

Figure 3.4. Articulated tower, supported on the seabed, with two degrees of freedom. The tower has a uniform buoyancy and mass as well as concentrated mass at the top.

Because the moment acts in the same direction as the rotation, this represents a negative stiffness. The damping coefficients also generally depend on the soil and water, but they would have to be determined experimentally or from field data.

Another example is shown in Fig. 3.5, where Fig. 3.5(a) shows a lifting operation with a crane vessel. The lifted object has a mass, including added mass, of M. To control the (vertical) motions or tensions during such a lifting operation, a heave compensator is often applied. Figure 3.5(b) shows a model of a passive heave compensator (Nielsen, 2007). u_0 represents the heave motion of the floating crane structure. u_{hc} is the motion of the (small) mass (m_{hc}) of the heave compensator, while k_{hc} and c_{hc} represent the stiffness and damping of the (passive) heave compensator, respectively. The stiffness is represented by the gas accumulator, while the damping is primarily due to viscous losses in the hydraulic system. k_w represents the stiffness

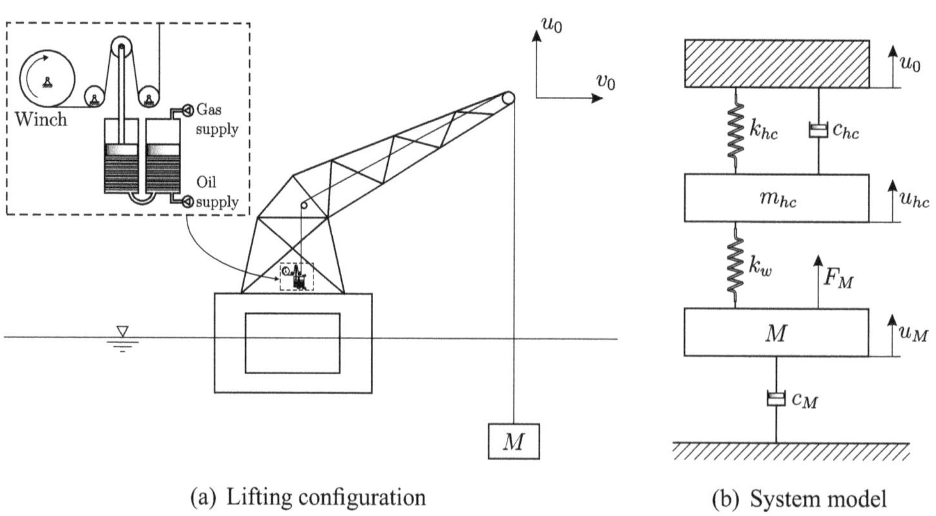

Figure 3.5. Heave compensation of lifting operation.

of the wire between the compensator and the hoisted, submerged object, with a total mass M and a vertical displacement of u_M. c_M is the viscous damping acting on the object. The motion compensator ensures that the motion u_M of M is less than the excitation motion, u_0. The equilibrium equations for the mass points m_{hc} and M can then be established follows:

$$m_{hc}\ddot{u}_{hc} + c_{hc}(\dot{u}_{hc} - \dot{u}_0) + k_{hc}(u_{hc} - u_0) + k_w(u_{hc} - u_M) = 0 \tag{3.10}$$

and

$$M\ddot{u}_M + c_M\dot{u}_M + k_w(u_M - u_{hc}) = F_M(t), \tag{3.11}$$

where $F_M(t)$ is the vertical force acting on M, positive upward. If the top of the motion compensator is given a forced displacement of u_0, the corresponding force is given by

$$F_0(t) = c_{hc}\dot{u}_0 + k_{hc}u_0. \tag{3.12}$$

If the mass m_{hc} is neglected and the stiffness k_w is assumed to be large, the SDOF system can be solved for harmonic excitation with a frequency ω to give

$$\left|\frac{u_M}{u_0}\right|^2 = \frac{1 + (\omega/\omega_0)^2 c_{hc}^2/(k_{hc}M)}{\left[1 - (\omega/\omega_0)^2\right]^2 + (\omega/\omega_0)^2(c_{hc} + c_M)^2/(k_{hc}M)}, \tag{3.13}$$

where $\omega_0^2 \approx k_{hc}/M$ is the approximate undamped natural frequency. For frequency ratios less than 1, the normalized response approaches 1, while at resonance it is

$$\left|\frac{u_M}{u_0}\right|^2 = \frac{k_{hc}M + c_{hc}^2}{(c_{hc} + c_M)^2}, \tag{3.14}$$

which achieves a minimum for $c_{hc} = k_{hc}M/c_M$ and yields

$$\left|\frac{u_M}{u_0}\right|^2 = \frac{k_{hc}M + c_{hc}^2}{(c_{hc} + c_M)^2} = \left[\frac{\omega_0 M}{\sqrt{\omega_0^2 M^2 + c_M^2}}\right]^2. \tag{3.15}$$

For frequency ratios ω/ω_0 larger than 1, the following expression applies asymptotically:

$$\left|\frac{u_M}{u_0}\right|^2 = (\omega_0/\omega)^4 + c_{hc}^2/(\omega M)^2. \tag{3.16}$$

This expression shows that the efficiency of the heave compensator increases with a small natural frequency (ω_0) and damping (c_{hc}) of the heave compensator. In general, Eqs. (3.10) and (3.11) may be solved in terms of u_{hc} and u_M if the displacement u_0 (as a forced motion) and the external force F_M on the mass M are given.

3.2.3 Vibrating Bars and Strings

Vibrating Bar

The bar in Fig. 3.6 is subjected to axial vibrations. The bar in Fig. 3.6a has a constant mass per unit length and an axial stiffness EA/l, and it is discretized in two or three mass points by simple lumping of the distributed mass, m. The mass assigned to the

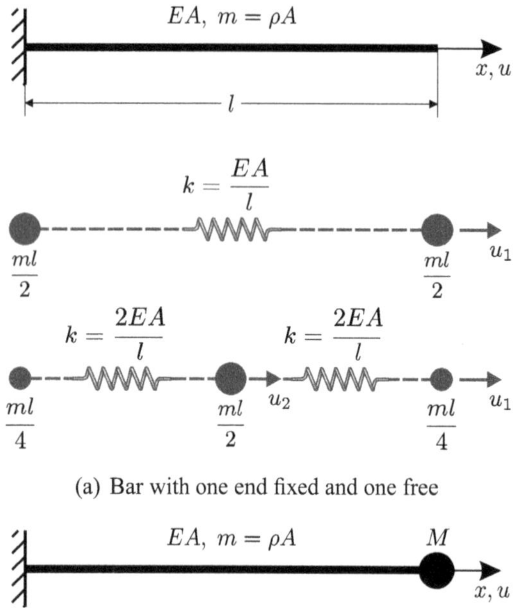

(a) Bar with one end fixed and one free

(b) Bar with one end fixed and one free with a concentrated mass attached to it

Figure 3.6. Bar with axial vibrations. Two discretizations are shown, involving 1DOF (SDOF) and 2DOF after introduction of the boundary condition at $x = 0$.

left end of the bar will not be vibrating due to the fixed-end boundary condition. The elasticity in the bar provides the stiffness represented by springs between the mass points and the support. It is hence seen that the model with three discrete masses corresponds to the system in Fig. 3.2 when damping and external forces are neglected. For the system described by an SDOF u_1, the natural frequency is

$$\omega = \sqrt{(EA/l)/(ml/2)} \approx (1.414/l)\sqrt{E/\rho}. \tag{3.17}$$

The second discretization in Fig. 3.6a corresponds to a system with two degrees of freedom. With reference to Eq. (3.9), the (generalized) mass is $m_{11} = ml/4, m_{12} = m_{21} = 0, m_{22} = ml/2$, and stiffness $k_{11} = 2EA/l, k_{22} = 4EA/l, k_{12} = k_{21} = -2EA/l$. The natural frequencies for this approximate model can then be determined by the determinant method, Eq. (3.6), for a 2DOF system. It is obtained that $\omega_1 = (1.53/l)\sqrt{E/\rho}$ and $\omega_2 = (3.7/l)\sqrt{E/\rho}$, respectively.

These results may be compared with the exact solution for the first two (lowest) eigenfrequencies, which are $\omega_1 = (\pi/2l)\sqrt{E/\rho} \approx (1.57/l)\sqrt{E/\rho}$ and $\omega_2 = (3\pi/2l)\sqrt{E/\rho} \approx (4.71/l)\sqrt{E/\rho}$ (see Section 3.3.2).

The eigenvectors $\boldsymbol{\phi}_i$ for the two modes of the approximate solution of this problem can then be determined from Eq. (3.5) as

$$\boldsymbol{\phi}_1 = [1, \sqrt{2}/2]^T; \quad \boldsymbol{\phi}_2 = [1, -\sqrt{2}/2]^T. \tag{3.18}$$

It is seen that the approximation with two mass points yields relatively accurate natural frequencies, even for the second one. However, no approximation of higher-order natural frequencies are obtained by this model. The model may also be easily applied to consider a case that includes a concentrated mass at the free end, as shown in Fig. 3.6.

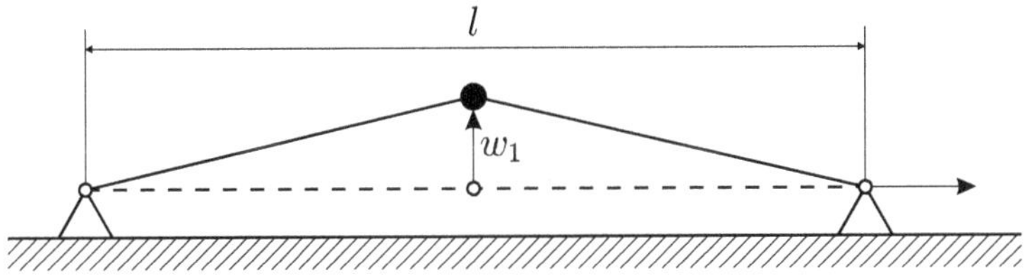

(a) Discretization by one mass point

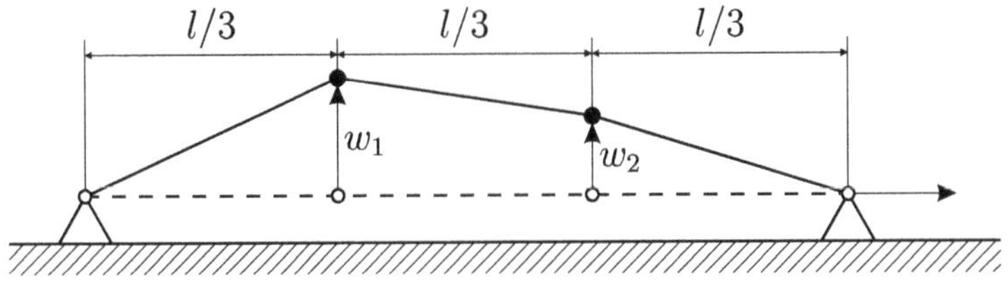

(b) Discretization by two mass points

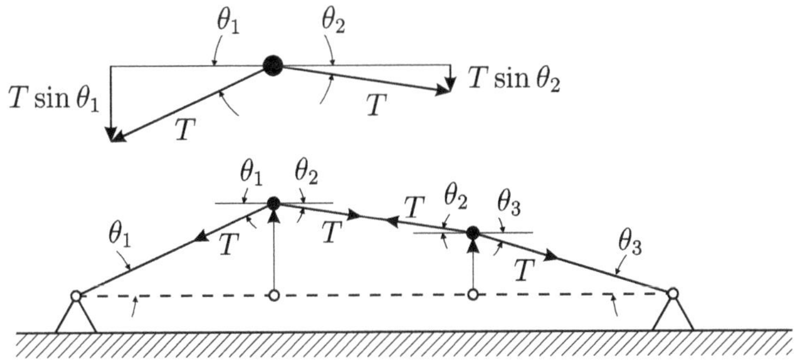

(c) Force diagram for the case with two mass points

Figure 3.7. Vibrating string discretized by one and two point masses.

Vibrating String

Consider the string with an axial (prestressing) force T in Fig. 3.7. For small displacements w_1 and w_2, the axial forces are assumed constant and equal to T. Consider the deformation mode in Fig. 3.7.

Assume that the mass is concentrated in two points: $M = ml/3$. Then the following dynamic equilibrium equations are obtained for the two mass points:

Mass 1

$$M\ddot{w}_1 = -T\sin\theta_1 - T\sin\theta_2, \tag{3.19}$$

where $\sin\theta_1 = \dfrac{w_1}{l/3}$; $\sin\theta_2 = \dfrac{w_1 - w_2}{l/3}$. Hence,

$$M\ddot{w}_1 + T\left(\frac{w_1}{l/3} + \frac{w_1 - w_2}{l/3}\right) = 0; \tag{3.20}$$

that is,

$$M\ddot{w}_1 + \frac{2T}{l/3}w_1 - \frac{T}{l/3}w_2 = 0. \tag{3.21}$$

Mass 2

$$M\ddot{w}_2 = T\sin\theta_2 - T\sin\theta_3, \tag{3.22}$$

where $\sin\theta_3 = \dfrac{w_2}{l/3}$; that is,

$$M\ddot{w}_2 - \frac{T}{l/3}w_1 + \frac{2T}{l/3}w_2 = 0. \tag{3.23}$$

The resulting equations may be written as

$$\begin{bmatrix} ml/9 & 0 \\ 0 & ml/9 \end{bmatrix}\begin{bmatrix} \ddot{w}_1 \\ \ddot{w}_2 \end{bmatrix} + \begin{bmatrix} 2T/l & -T/l \\ -T/l & 2T/l \end{bmatrix}\begin{bmatrix} w_1 \\ w_2 \end{bmatrix} = 0. \tag{3.24}$$

Assuming that

$$\begin{bmatrix} w_1 \\ w_2 \end{bmatrix} = \begin{bmatrix} A_1 \\ A_2 \end{bmatrix}\sin\omega t,$$

and solving for the natural frequencies, it is found that

$$\omega_1 = \frac{3}{l}\sqrt{\frac{T}{m}} \quad \text{and} \quad \omega_2 = \frac{3\sqrt{3}}{l}\sqrt{\frac{T}{m}} \approx \frac{5.2}{l}\sqrt{\frac{T}{m}}. \tag{3.25}$$

The corresponding eigenvectors are

$$\begin{bmatrix} A_1 \\ A_2 \end{bmatrix} = \begin{bmatrix} 1 \\ 1 \end{bmatrix} \quad \text{and} \quad \begin{bmatrix} A_1 \\ A_2 \end{bmatrix} = \begin{bmatrix} 1 \\ -1 \end{bmatrix}. \tag{3.26}$$

The lowest eigenfrequency for the two-mass discretization is $\omega = 3\sqrt{T/(ml^2)}$. Similarly, it is found that the eigenfrequency is $\omega = \sqrt{8T/(ml^2)}$ when the mass is represented by a single mass, $0.5ml$ (Fig. 3.7). These results can be compared to the exact solution for a vibrating string of a length l and with uniform mass m and pretension T, which is $\omega_{n,exact} = n\pi\sqrt{T/(ml^2)}$. It is seen that a discretization with both one and two mass points gives a relatively accurate estimate of the first natural frequency.

3.3 Beams Under Axial and Lateral Loads

3.3.1 Basic Principles of Structural Mechanics

General Laws Used to Solve Structural Problem

Structural behavior is generally solved by applying the following principles:

- Balance of forces of all parts of the structure (in terms of stresses and internal forces)
- Compatibility in the material (in terms of displacements and strains)
- Stress-strain relationship (i.e., Hooke's law for a linearly elastic material)

An exact solution to a structural problem means that these principles are satisfied in all infinitesimal parts of the structure, that is, point by point. The differential equations for the one-dimensional problems (bar, beam) can be readily solved for many practical problems. However, plane stress and plate bending problems encountered in practice may be difficult to solve.

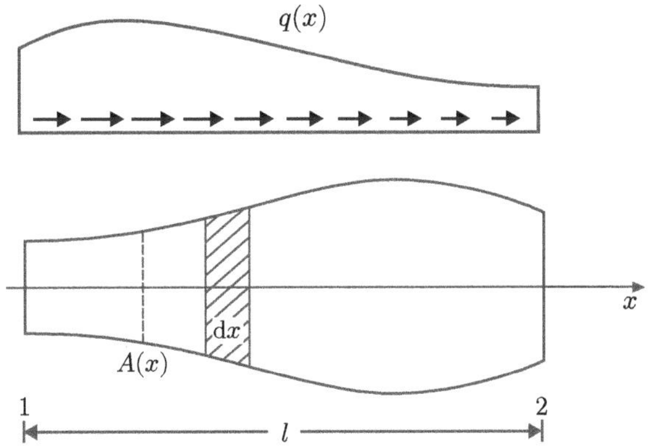

Figure 3.8. Bar with a varying cross section area, A subjected to an axial load, $q(x)$.

The FE method represents a numerical method to obtain an approximate solution to two- and three-dimensional structural mechanics problems. The general idea is that the error in approximation is decreasing when the number of elements that are applied is increasing. The concept of how to establish the approximate solution by the FE method is outlined in Chapter 4. In this case, the three basic laws are applied to cast the governing equation on an integral form rather than a differential form. This approach is shown in the next section.

In this section, the basic principles for mechanical analysis of bar and beam problems are reviewed, and the differential and variational form of the governing equations are established. In the next section, dynamic features are addressed.

It is noted that for a straight member subjected to axial and lateral loads that cause bending and shear forces in the member and exhibit small displacements, these two modes are uncoupled and can be treated separately and superimposed.

Differential Formulation for a Bar Element

The differential and energy formulation of a structural problem are commonly referred to as the strong and weak formulations of the problem, respectively. The difference between the two is demonstrated in the following for a truss element with varying cross section, see Fig. 3.8.

By looking at a cross-sectional disk of length dx of the truss element, the horizontal equilibrium of the disk is expressed in Fig. 3.9:

$$dP + q dx = 0, \tag{3.27}$$

which may be written in terms of stresses:

$$\frac{\partial}{\partial x}(A\sigma) + q = 0. \tag{3.28}$$

Introducing Hooke's law for uniaxial stress,

$$\sigma_x = E \cdot \varepsilon_x, \tag{3.29}$$

and by assuming compatibility, a relationship between displacement and strain is obtained. By looking at a small fiber of the material, the state of deformation appears as shown in Fig. 3.9.

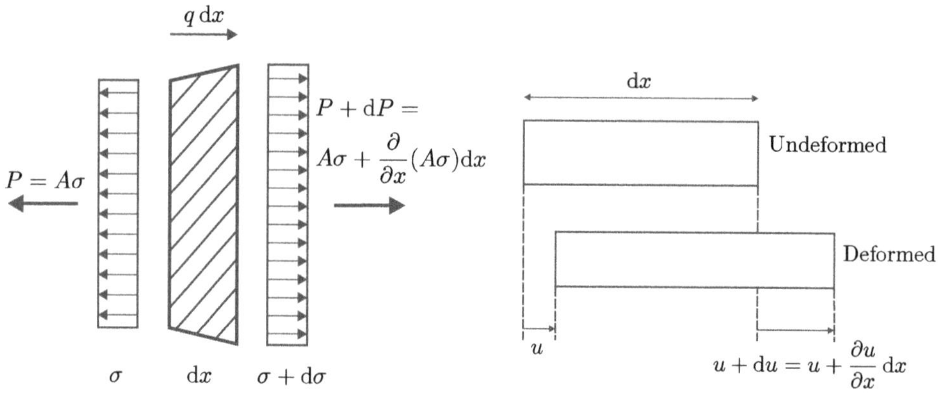

(a) Stresses (b) Deformation in terms of displacement

Figure 3.9. State of stress and deformation of a bar. Note that P is positive in tension.

The strain can be expressed in terms of the axial displacement $u(x)$ as

$$\varepsilon_x = \frac{\Delta \ell}{\ell} = \frac{u + du - u}{dx} = \frac{\partial u}{\partial x}. \tag{3.30}$$

By combining the preceding equations, and keeping in mind that the cross-sectional area $A(x)$ is varying along the element, the following equation is obtained:

$$\frac{\partial}{\partial x}(EA\frac{\partial u}{\partial x}) + q = 0, \tag{3.31}$$

which is the differential equation for the bar or truss element. This is called the *strong form* of the problem.

Virtual Work Formulation for a Bar Element

In a general case, it may not be possible to obtain a solution to the differential equation. An alternative is to assume a function for the displacement, $u = u(x)$, which is believed to be a good approximation of the exact solution.

If the approximate solution is entered into the differential equation, the left-hand side of the equation have a value that is dependent on the position along the element axis. The residue may vanish for some values of x, but it will not be zero in general unless we have the exact solution:

$$\frac{\partial}{\partial x}(EA\frac{\partial u}{\partial x}) + q = r(x) \neq 0. \tag{3.32}$$

The value found from the expression, $r(x)$, is regarded as a residual force.

The selected approximation is presumed to be good if the residue is small. There are several schemes to obtain small residues. The most obvious is to demand the residue to be zero for some prescribed points within the element or to minimize the square of the residue. The scheme that is discussed here is the Galerkin method.

The Galerkin method requires a weighted average of the residuals to be zero over the element:

$$\int_{\ell} r(x)\, w(x)\, dx = 0, \tag{3.33}$$

where the weight function $w(x)$ is usually selected to be an independent virtual displacement pattern, $\tilde{u}(x)$, of the same form as the proposed approximate solution. The proposed criteria will, hence, read

$$\int_{\ell} r(x)\tilde{u}(x)\,dx = 0. \tag{3.34}$$

The integrand, $r(x)\tilde{u}(x)$, expresses the work performed by the residual force along the element. In other words, the Galerkin method demands that the work performed by $r(x)$ summed up over the element should be zero.

Substituting for the residual force, $r(x)$, gives

$$\int_{\ell} \left[\frac{\partial}{\partial x}(EA\frac{\partial u}{\partial x}) + q\right]\tilde{u}dx = 0. \tag{3.35}$$

This is the *weak form* formulation for the problem and is sometimes referred to as an integrated equilibrium of the structure.

The next step is to transform the previous equation to a form that is convenient in the general case. This is done by partial integration of the equation.

The integration by parts rule is derived by differentiating the product of two terms u and v, $(uv)' = u'v + uv'$. Reordering the terms and integrating each of them gives

$$\int u'v = uv - \int uv'. \tag{3.36}$$

This rule is applied to the first term in the bracket of the weak form of the problem:

$$\int_0^{\ell} \frac{\partial}{\partial x}(EA\frac{\partial u}{\partial x})\tilde{u} = EA\frac{\partial u}{\partial x}\tilde{u}\Big|_0^{\ell} - \int_0^{\ell} EA\frac{\partial u}{\partial x}\frac{\partial}{\partial x}(\tilde{u})\,dx. \tag{3.37}$$

The first term on the rhs can be simplified to

$$EA\frac{\partial u}{\partial x}\tilde{u}\Big|_0^{l} = EA\varepsilon_x\tilde{u}|_0^l = (EA\varepsilon_x\tilde{u})_{x=l} - (EA\varepsilon_x\tilde{u})_{x=0}. \tag{3.38}$$

Realizing that $AE\varepsilon = A\sigma = S$ is axial tension in the cross section and that

$$\frac{\partial}{\partial x}(\tilde{u}) = \tilde{\varepsilon}_x \text{ and } EA\frac{\partial u}{\partial x} = A\sigma_x \tag{3.39}$$

gives

$$\int_0^{\ell} \frac{\partial}{\partial x}(EA\frac{\partial u}{\partial x})\tilde{u}dx = (S\tilde{u})_{x=\ell} + (-S\tilde{u})_{x=0} - \int_0^{\ell} A\sigma\tilde{\varepsilon}_x\,dx. \tag{3.40}$$

The terms involving $S\tilde{u}$ express the virtual work performed by the end forces. A negative sign at $x = 0$ is needed because the force and displacement are defined in opposite directions.

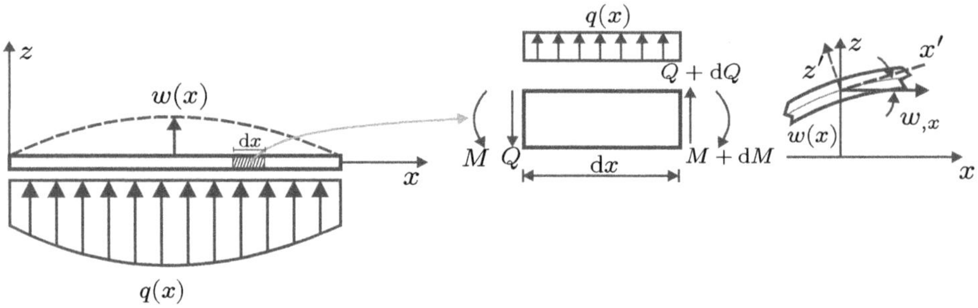

Figure 3.10. Beam displacement and forces.

Substituting into the weak form formulation and ordering the terms gives

$$\underbrace{\int_0^{\ell} A\sigma\tilde{\varepsilon}_x \, dx}_{\substack{\text{Internal} \\ \text{virtual work}}} = \underbrace{(-S\tilde{u})_{x=0} + (S\tilde{u})_{x=\ell} + \int_0^{\ell} q\tilde{u} \, dx}_{\substack{\text{External} \\ \text{virtual work}}} \tag{3.41}$$

Note that the virtual strain energy per unit volume is $\sigma\tilde{\varepsilon}$ and not $\frac{1}{2}\sigma\tilde{\varepsilon}$. The reason for this is that the stress σ is constant when the strain is acting.

An interpretation of the equation is that by keeping all internal stresses and external loads fixed, the internal and external work performed by the stresses and forces when they are combined with a compatible virtual displacement (weighting function) should be equal. This is the principle of virtual displacements (PVD).

In this case, the principle of virtual displacements is derived by the Galerkin weighted residual method which gives a physical interpretation of the PVD applied to approximate displacement functions (zero average work done by the residue).

Differential Equation for the Bending of a Beam Under Lateral Loads

Equilibrium is fulfilled by considering the equilibrium of an arbitrary "slice" of the beam, with length dx, see Fig. 3.10.

Hence, equilibrium demands

$$\frac{\partial Q}{\partial x} = -q \text{ (vertical)}, \quad Q = \frac{\partial M}{\partial x} \text{ (moment)}. \tag{3.42}$$

Substituting the second term into the first gives

$$\frac{\partial^2 M}{\partial x^2} = -q. \tag{3.43}$$

Compatibility in the structure means that all adjacent cross sections will have the same displacement and deformation, and that the material itself will be continuous as it deforms. No cracks will occur, and the strain will be finite. The compatibility of a beam-bending problem is ensured by describing the displacement by a continuous lateral displacement w and ensuring proper boundary conditions.

By assuming the cross-sections to be plane and perpendicular to the beam axis at all times (Navier's hypothesis), the displacement in the (axial) x direction will be

$$u = -z \cdot w_{,x}, \tag{3.44}$$

where z denotes the distance to the neutral axis (in the paper plane). The notation $w,_x = \partial w/\partial x$ has also been introduced, and similarly, $w,_{xx} = \partial^2 w/\partial x^2$. The strain in the axial direction is

$$\varepsilon_x = \frac{\partial u}{\partial x} = -z \cdot w,_{xx} . \tag{3.45}$$

According to Hooke's law, the stress σ_x is then

$$\sigma_x = E\varepsilon_x = -E \cdot z \cdot w,_{xx} . \tag{3.46}$$

The bending moment is

$$M = \int_A z\sigma_x \, dA \quad (A = \text{cross-sectional area}). \tag{3.47}$$

This gives

$$M = -EI\frac{\partial^2 w}{\partial x^2}, \tag{3.48}$$

where $I = \int_A z^2 dA$.

Navier's hypothesis is an approximation that does not ensure that all small particles along the cross-section are in equilibrium, but the error is small for slender beams.

Equations (3.43) and (3.48) combined give

$$\frac{\partial^4 w}{\partial x^4} = \frac{q}{EI}, \tag{3.49}$$

which is a differential equation relating the deflection $w(x)$ and the load $q(x)$. This is commonly referred to as the differential equation for the beam.

The boundary conditions for beams are

- Clamped end: $w = 0$, $w,_x = 0$
- Simply supported end: $w = 0$, $M = -EIw,_{xx} = 0$
- Free end: $M = 0$, $Q = \dfrac{\partial M}{\partial x} = 0$

Modification of Beam Theory

The differential equation derived previously applies for a slender beam. If the beam has a height/length ratio that is larger than, e.g., one fifth, shear deformation, and in the case of dynamics, the rotational inertia effect may be of importance. It should be noted that the length in this case is not the span length, but the wavelength of the relevant deformation pattern of the beam, which depends on the loading.

Another issue is the effect of (large) axial forces on the lateral stiffness. In the extreme case of axial compressive loads, the beam will buckle, implying that both the effective bending stiffness and the natural frequency are zero.

Before these issues are addressed, the virtual work formulation for the slender beam under static lateral loads is established.

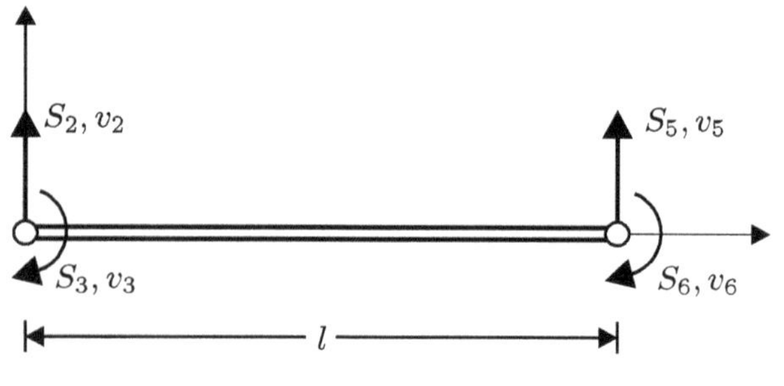

Figure 3.11. Nodal displacements and forces for a beam element.

Virtual Work for Beam Under Lateral Loads

The weak formulation of the beam can be established in an analogous manner as for the beam with axial force. The starting point is then

$$\int_{\ell} [EI\frac{\partial^4 w}{\partial x^4} - q]\tilde{w} dx = 0. \quad (3.50)$$

Integrating by parts, the following result is obtained:

$$\int_0^{\ell} EI\frac{\partial^4 w}{\partial x^4}\tilde{w} dx = EI[\frac{\partial^3 w}{\partial x^3}\tilde{w}]\Big|_0^{\ell} - EI[\frac{\partial^2 w}{\partial x^2}(\frac{\partial \tilde{w}}{\partial x})]\Big|_0^{\ell} + \int_0^{\ell} EI\frac{\partial^2 w}{\partial x^2}(\frac{\partial^2 \tilde{w}}{\partial x^2})dx. \quad (3.51)$$

Realizing that

$$-EI\frac{\partial^2 w}{\partial x^2} = M, \quad (3.52)$$

$$(\frac{\partial^2 \tilde{w}}{\partial x^2}) = \tilde{\kappa}, \quad (3.53)$$

$$EI\frac{\partial^3 w}{\partial x^3} = -\frac{\partial M}{\partial x} = -Q, \quad (3.54)$$

it is obtained that

$$\int_0^{\ell} EI\frac{\partial^4 w}{\partial x^4}\tilde{w} dx = (-Q\tilde{w})_{x=\ell} + (Q\tilde{w})_{x=0} + (M\tilde{w},_x)_{x=\ell} + (-M\tilde{w},_x)_{x=0} + \int_0^{\ell} EI\kappa\tilde{\kappa} dx, \quad (3.55)$$

where the curvature is $\kappa = \partial^2 w/\partial x^2$.

It is noted that the internal virtual work can also be expressed as follows:

$$\int_V \tilde{\varepsilon}_x \cdot \sigma_x dV = \int_{\ell}\int_A \tilde{\varepsilon}_x \cdot E \cdot \varepsilon_x dA dx = \int_{\ell} E(\int_A z^2 dA)(\tilde{\kappa})\kappa dx = \int_{\ell} \tilde{\kappa}(EI)\kappa \cdot dx, \quad (3.56)$$

using the relation $\varepsilon_x = -z \cdot w,_{xx} = -z \cdot \kappa$. Equation (3.55) may be written more concisely by using the definitions of w, $w,_x$, M and Q and the nodal displacements, v_i, and stress resultants, S_i, in Fig. 3.11.

The following relations can then be established:

$$\begin{aligned} v_2 &= w(0), & v_5 &= w(\ell), \\ v_3 &= -w,_x(0), & v_6 &= -w,_x(\ell), \\ S_2 &= -Q(0), & S_5 &= Q(\ell), \\ S_3 &= -M(0), & S_6 &= M(\ell). \end{aligned} \tag{3.57}$$

By introducing a pattern of virtual displacements, $\tilde{w} = \tilde{w}(x)$, with corresponding nodal displacements $\tilde{\mathbf{v}}'$; substituting Eq. (3.55) into the weak formulation Eq. (3.50) and using the relations in Eqs. (3.57), we get

$$\begin{aligned} \underbrace{\int_0^\ell \tilde{\kappa}(EI)\kappa \, dx}_{\text{Internal virtual work}} &= \tilde{v}_2 S_2 + \tilde{v}_3 S_3 + \tilde{v}_5 S_5 + \tilde{v}_6 S_6 + \int_0^\ell \tilde{w} q \, dx \\ &= \underbrace{(\tilde{\mathbf{v}}')^T \mathbf{S}' + \int_0^\ell q \tilde{w} \, dx}_{\text{External virtual work}} \, , \end{aligned} \tag{3.58}$$

where

$$\mathbf{v}'^T = [v_2, v_3, v_5, v_6], \tag{3.59}$$

$$\mathbf{S}'^T = [S_2, S_3, S_5, S_6]. \tag{3.60}$$

Equation (3.58) is the equation for virtual work (virtual displacements) for a beam.

The expression on the left-hand side of Eq. (3.58) may be written with the constant (EI) on any place among $\tilde{\kappa}$ and κ, or even on the outside of the integral. The location used is convenient when this expression is generalized (e.g., to plates).

Differential and Virtual Work Formulation for a Beam with Lateral Loads and an Axial Force

Reference is made to Fig. 3.10. If the beam is subjected to an axial compression force N, all relationships presented previously for the beam remain the same except that the moment equation (3.42) becomes

$$Q = \frac{\partial M}{\partial x} - N\frac{\partial w}{\partial x}. \tag{3.61}$$

By assuming that the force N varies along the beam, the resulting differential equation then becomes

$$\frac{\partial^2}{\partial x^2}\left(EI\frac{\partial^2 w}{\partial x^2}\right) + \frac{\partial}{\partial x}(N\frac{\partial w}{\partial x}) = q(x). \tag{3.62}$$

The sign convention applied here is that the axial force N is positive when it is compressive.

Equation (3.62) may be solved exactly, and the solution used to express an exact stiffness relationship for the beam with axial force, by using so-called Livesley or stability functions (Livesley, 1983). However, this method cannot be extended to two-dimensional problems. For this reason, the stiffness relationship is established by using the principle of virtual work, which can serve as a basis for establishing a numerical (FE) method, as shown in Section 4.5.

The virtual work equation may be obtained by applying the same procedure used in connection with Eqs. (3.50)–(3.55) to the differential equation (Eq. 3.49).

The additional term to be considered is

$$\int_0^\ell N w_{,xx} \tilde{w}\, dx = [N w_{,x} \tilde{w}]_0^\ell - \int_0^\ell N w_{,x}\, \tilde{w}_{,x}\, dx. \tag{3.63}$$

The first term vanishes on a clamped and simply supported end. The resulting equation involves the internal virtual work, which equals

$$\int_0^\ell EI\, \tilde{\kappa}\, \kappa\, dx - \int_0^\ell N w_{,x}\, \tilde{w}_{,x}\, dx, \tag{3.64}$$

where the last term is a correction.

Differential Formulation for a High Beam with Shear Deformation – Shear Beam

Consider a beam that deforms laterally only due to shear stress. The constitutive relation for the shear angle Y_{zx} and the shear stress T_{zx} is

$$Y_{zx} = \frac{\partial w_s}{\partial x} = \frac{T_{zx}}{G} = \frac{Q}{GA_s}, \tag{3.65}$$

where w_s is the shear deflection, and the average shear area A_s is such that $Q/A_s = T_{zx}$ and G is the shear modulus. Combining this with the equilibrium equation (Eq. (3.42)), the following governing equation results:

$$\frac{\partial}{\partial x}\left[GA_s \frac{\partial w_s}{\partial x}\right] = -q. \tag{3.66}$$

This differential equation is of similar type as that for the bar.

For a cantilever beam with a point load P at the tip $x = l$, the deflection is given by $w_s(x) = Px/(GA_s)$.

Differential Formulation of Timoshenko Beam Theory – with Account of Shear Deformation

The equilibrium equations are as shown previously, cf. Eq. (3.42). For a high beam, the hypothesis is that a line normal to the neutral axis in undeformed state is not normal after deformation. This implies that the displacement of a point not on the neutral axis is not governed by the slope, $w_{,x}$, of the neutral axis as in Kirchhoff's theory. Rather, its motion depends on rotations $\theta = \theta(x)$ of lines that were normal to the neutral axis of the undeformed beam (Fig. 3.12). Thus, with θ being small angles of rotation, the displacement of a point not on the neutral axis is given as

$$u(x) = z \cdot \theta(x). \tag{3.67}$$

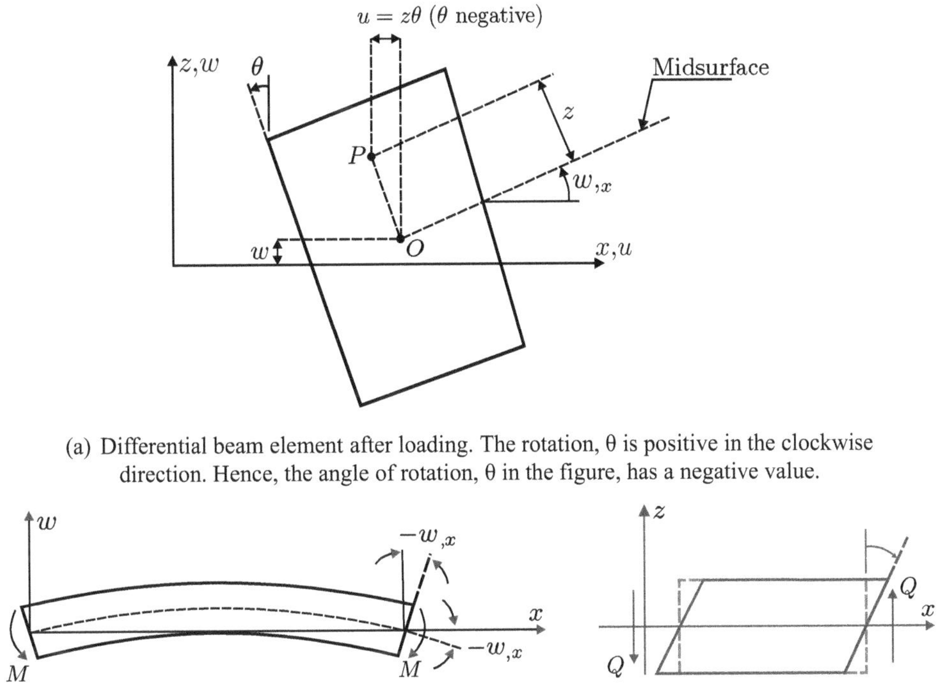

(a) Differential beam element after loading. The rotation, θ is positive in the clockwise direction. Hence, the angle of rotation, θ in the figure, has a negative value.

(b) Bending deformation (c) Shear deformation

Figure 3.12. Bending and shear deformations. Differential beam element after loading, analogous to Fig. 3.10, but with transverse shear deformation allowed ($w,_x + \theta \neq 0$).

Hence, the following expressions for strains in terms of displacements apply:

$$\varepsilon_x = u,_x = z \cdot \frac{\partial \theta}{\partial x}, \tag{3.68}$$

$$\varepsilon_z = 0 \qquad \text{(as for slender beams)}, \tag{3.69}$$

$$\gamma_{zx} = w,_x + u,_z = w,_x + \theta. \tag{3.70}$$

The finite element model will be based on assumed w and θ. If $w,_x + \theta$ is different from zero, shear deformation Y_{xz} in the xz-plane will be accounted for. It is noted that in the slender beam (thin plate) theory, $\gamma_{zx} \equiv 0$, and the expressions for strains specialize into those of the slender beam.

The stress – strain relationships are

$$\sigma_x = E\varepsilon_x = z\,E\,\frac{\partial \theta}{\partial x} \tag{3.71}$$

$$T_{zx}/G = \gamma_{zx} = \gamma \qquad \text{(as Eq. (3.65))}, \tag{3.72}$$

where $G = E/(2(1+\nu))$ for a homogeneous material. G is the shear modulus of elasticity, E is the modulus of elasticity, and ν is the Poisson's ratio.

By combining Eqs. (3.47), (3.71), Eqs. (3.65) and (3.70), it follows that

$$M = \int_A z \cdot \sigma_x dA = EI\theta_{,x} \tag{3.73}$$

$$Q = GA_s\gamma_{zx} = GA_s(w_{,x} + \theta). \tag{3.74}$$

Inserting Eqs. (3.73) and (3.74) into the the equilibrium equations, Eq. (3.42), yields the following set of differential equations:

$$GA_s(w_{,x} + \theta)_{,x} = -q, \tag{3.75}$$

$$-(EI\theta_{,x})_{,x} + GA_s(w_{,x} + \theta) = 0, \tag{3.76}$$

which specializes into Eq. (3.49) when $w_{,x} + \theta = 0$.

It is important to recognize that both $w(x)$ and $\theta(x)$ are such that they fulfill Eqs. (3.75) and (3.76), as well as the following boundary conditions:

$$\begin{aligned} &\text{- Fixed end: } w = \theta = 0 \\ &\text{- Simply supported end: } w = \theta_{,x} = 0 \\ &\text{- Free end: } GA_s(w_{,x} + \theta) = EI\theta_{,x} = 0 \end{aligned} \tag{3.77}$$

In particular, it is noted that the essential boundary condition $\theta = 0$ for a fixed-end support implies that $w_{,x}(x) = \gamma_{zx} \neq 0$. This fact should be observed if Eqs. (3.75) and (3.76) are reformulated in terms of $w(x)$.

Consider now a simple example in terms of a cantilever beam, fixed at $x = 0$ and free with a lateral point load P at $x = l$. The boundary conditions are given by Eq. (3.77), except that $GA_s(w_{,x} + \theta) = P$ at the free end. In this case, the solution can be obtained by integrating Eq. (3.76) and fulfilling the boundary conditions.

Because $\gamma_{zx} = w_{,x} + \theta = P/(GA_s)$ is known to be constant, the final result is obtained as $w(x) = P(3lx^2 - x^3)/(6EI) + Px/(GA_s)$.

The effect of shear deformation is particularly significant when there is limited shear area, A_s, as compared to the total cross-section area, or when the shear modulus of the shear area is small. A typical example of the latter case is a sandwich structure composed of a fiber reinforced panel (FRP) skin and a lightweight, soft core with a much lower stiffness than the FRP.

3.3.2 Differential Equation for Dynamic Behavior

Bar

The differential equation for a bar with static axial loading $q(x)$ is given by Eq. (3.31). If the structure experiences sufficiently large motions, inertia and damping have an effect and may be included, according to d'Alembert's principle, by replacing the loading $q(x)$ in Eq. (3.31) by

$$q(x,t) = q_{ext}(x,t) - m(x)\ddot{u} - c(x)\dot{u}, \tag{3.78}$$

where $q_{ext}(x,t)$ is due to external dynamic loads, and the other terms represent inertia and damping due to continuously distributed mass and viscous damping. The distributed damping may be due to internal damping in the bar, as well as external sources of damping.

By introducing this expression in Eq. (3.31), the dynamic equilibrium equation of a bar is obtained as follows:

$$\frac{\partial}{\partial x}(EA\frac{\partial u}{\partial x}) - c(x)\dot{u} - m(x)\ddot{u} + q_{ext}(x,t) = 0. \tag{3.79}$$

In addition, the boundary conditions have to be considered. Representative conditions would be

- Fixed end: $u = 0$.
- Free end: Axial force $P = A\sigma_x = EA\varepsilon_x = EAu_{,x} = 0$.
- Free end with a concentrated mass M: $P = AEu_{,x} = -M\ddot{u}$.

The solution is illustrated for the case with a constant EA, no damping and external loading, i.e., the free vibration problem. This is done by assuming a solution of the form,

$$u(x,t) = \psi(x)Y(t), \tag{3.80}$$

based on the principle of separation of variables. It is found that

$$Y(t) = A\sin\omega_0 t + B\cos\omega_0 t, \tag{3.81}$$

$$\psi(x) = C_1\sin\lambda x + C_2\cos\lambda x, \tag{3.82}$$

where $\lambda = \omega_0\sqrt{\rho/E}$ and the constants A, B, C_1 and C_2 are determined by the boundary conditions. This leads to the determination of λ and hence ω_0, which is the natural frequency of vibration.

For the case of a bar fixed at one end and free at the other, the natural frequencies are found to be: $\omega_{0,n} = (n - 1/2)(\pi/l)\sqrt{E/\rho}$ with the following natural modes of vibration: $\psi_n(x) = C_n \cdot \sin[(n - 1/2)\pi(x/l)]$, where C_n is an undetermined constant to be chosen.

Consider, then, a wire fixed at one end and with a concentrated mass M attached to the free end and otherwise having a uniform stiffness and mass, m. By the introduction of the boundary conditions into the general solution, Eqs. (3.80)–(3.82), the governing equation that determines the eigenfrequencies is

$$\frac{ml}{M} = \lambda l \cdot \tan(\lambda l). \tag{3.83}$$

The natural frequencies $\omega_0 = \lambda\sqrt{E/\rho}$. In the special case when the distributed mass m is neglected, the natural frequency specializes into $\omega = \sqrt{EA/(Ml)}$. If M is zero, the solution specializes into the exact solution $\omega_{0,n} = (n - 0.5)(\pi/l)\sqrt{E/\rho}$.

Slender Beam

The differential equation of a slender beam is given by Eq. (3.49) or (3.62) when the lateral loading is static. If the structure experience sufficiently large motions, inertia and damping have an effect, and may be included according to d'Alembert's principle, by replacing the lateral loading $q(x)$ with

$$q(x,t) = q_{ext}(x,t) - m(x)\ddot{w} - c(x)\dot{w} - k(x)w, \tag{3.84}$$

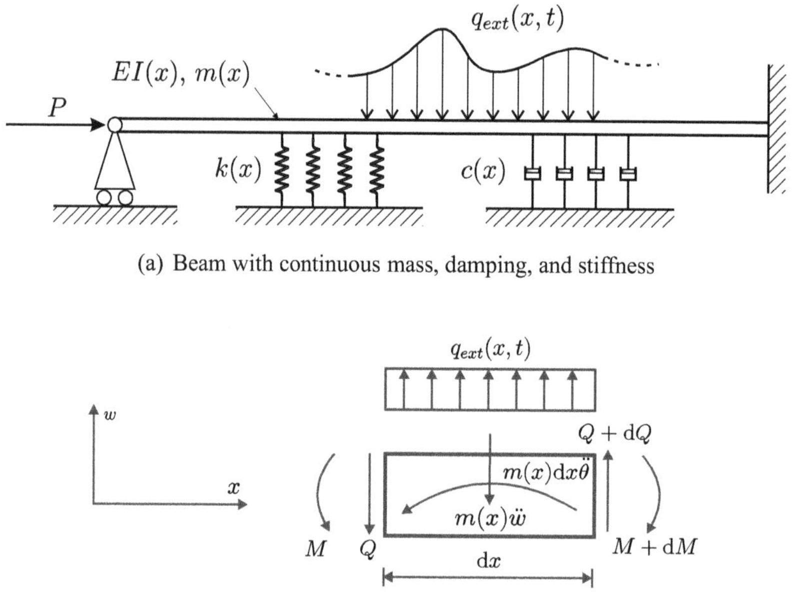

(a) Beam with continuous mass, damping, and stiffness

(b) Elastic and inertia forces acting on a beam element of length dx

Figure 3.13. Dynamic equilibrium equation.

where $q_{ext}(x, t)$ is due to external loads, and the other terms represent inertia, damping, and stiffness forces due to continuous distributed mass, viscous damper and spring stiffness, as shown in Fig. 3.13a. The distributed damping may be due to internal damping in the beam, as well as external sources of damping such as when the beam floats on water or rests on a soil. The distributed stiffness is due to external stiffness provided by the buoyancy and soil support. Two inertia terms are given in Fig. 3.13. One is associated with the lateral motion of the beam, and the other is due to the rotational inertia of the cross section. The latter term is only of interest when then the height/length ratio of the beam is small. Hence, this feature is associated with problems where the shear deformation is also of some importance and is treated in connection with the Timoshenko beam theory.

By introducing Eq. (3.84) into Eq. (3.49) or (3.62), the following equation results:

$$\frac{\partial^2}{\partial x^2}\left(EI\frac{\partial^2 w}{\partial x^2}\right)+\frac{\partial}{\partial x}(P\frac{\partial w}{\partial x})+k(x)w+c(x)\dot{w}+m(x)\ddot{w}=q_{ext}(x,t). \tag{3.85}$$

In addition, the boundary conditions need to be considered. Representative conditions are

- Simply supported end: with $w = w,_{xx} = 0$.
- Fixed end: with $w = w,_{x} = 0$.
- Free end: with $w,_{xx} = w,_{xxx} = 0$.

It is convenient to solve Eq. (3.85) step by step, starting with the case with $k = c = q_{ext} = 0$ corresponding to an undamped free vibration. This equation can be solved for the case that P is constant force along the beam by assuming a solution of the form: $w(x, t) = \psi(x) \cdot \sin(\omega t)$.

Table 3.1. *Eigenmodes and eigenfrequencies of a slender beam with a uniform bending stiffness and mass distribution for different boundary conditions*

Beam boundary condition	Mode No.	Mode shape, $\psi(x)$	Natural frequency, ω (rad/s)
Simply supported	n	$\sin(n\pi/l)$	$(n\pi/l)^2\sqrt{EI/m}$
Fixed ends	1		$(4.73/l)^2\sqrt{EI/m}$*
Free ends	3 (1.flex.)		$(4.73/l)^2\sqrt{EI/m}$*
Cantilever beam	1	$-a\sinh\alpha + b\cosh\alpha +$ $a\sin\alpha - b\cos\alpha$ $a = 0.7341$; $b = 1$; $\alpha = 1.8751x/l$	$(1.875/l)^2\sqrt{EI/m}$*
	2		$(4.69/l)^2\sqrt{EI/m}$*

* Approximate solution.

For a simply supported slender beam with a length l, the following function satisfies the differential equation and boundary conditions, $\psi(x) = \sin(n\pi x/l)$, and the natural frequencies are

$$\omega^2 = \frac{EI(n\pi/l)^4 - P(n\pi/l)^2 + k}{m}. \tag{3.86}$$

This equation shows that an axial compressive load, P, reduces the frequency. If $k = 0$, the frequency ω is zero when $P = EI(n\pi/l)^2$, which corresponds to the Euler buckling load.

Moreover, when the terms with the bending stiffness and elastic foundation are neglected, Eq. (3.85) represents the governing equation for the vibrating string – and, hence, Eq. (3.86) represents its natural frequency. It shows that the natural frequency of a vibrating string with a constant tensile load $T = -P$ and no bending stiffness is $\omega = n\pi\sqrt{T/ml^2}$, which is the exact value mentioned previously.

If $P = k = c = q_{ext} = 0$, i.e., retaining the bending term and mass terms, Eq. (3.85) has the general solution

$$\psi(x) = A\cos ax + B\sin ax + C\cosh ax + D\sinh ax, \tag{3.87}$$

where $a^4 = m\omega_i^2/(EI)$. The four constants A, B, C, D are determined by the two boundary conditions at each end of the beam. The resulting equations are homogenous and require the determinant of the coefficients to be equal to zero.

This gives an equation to determine the unknown, a. For a simply supported beam (with $w = w_{,xx} = 0$ at both ends), the exact solution is $a = r\pi/l$.

For a cantilevered beam, the unknown a is given by

$$\cosh al \ \cos al = -1. \tag{3.88}$$

When both ends are clamped, the unknown a is given by

$$\cosh al \ \cos al = +1. \tag{3.89}$$

Actually, the a for a beam with both ends free needs to satisfy the same equation as for the clamped beam. It is noted that this equation applies to the flexible modes. The free-free beam also has to have zero frequencies corresponding to the rigid body motions. The corresponding frequencies, ω_r, are given in Table 3.1.

Other solutions for the first flexible mode of vibration, $\psi(x)$, and the corresponding eigenfrequency, are shown in Table 3.1.

It is noted that the boundary condition has decreasing influence on the eigenmode form and eigenfrequency for increasing mode number, n, i.e., toward the results for the simply supported beam. However, at the same time the effect of shear deformation and rotational inertia becomes increasingly important to include for increasing n, as discussed subsequently.

The general solution of the dynamic equation of motion Eq. (3.85) can be obtained by so-called modal superposition. If the different mode shapes $\psi_n(x)$ and natural frequencies ω_n are obtained, the general solution may be written as follows:

$$w(x,t) = \sum_n \psi_n(x) \cdot Y_n(t). \tag{3.90}$$

This method is explained in more detail in Section 3.3.10.

Timoshenko Beam

If the inertia effects due to lateral and rotational motion are included in the equilibrium equations for a Timoshenko beam, the following equations result:

$$GA_s(w_{,x} + \theta)_{,x} = -q + m\ddot{w}, \tag{3.91}$$

$$-(EI\theta_{,x})_{,x} + GA_s(w_{,x} + \theta) = -mr^2\ddot{\theta}. \tag{3.92}$$

The latter inertia contribution for a unit beam length is obtained due to the rotational acceleration as follows:

$$m_x = \ddot{\theta} \int_A \rho z^2 dA = \ddot{\theta}\rho I = \ddot{\theta}\rho A r^2 = \ddot{\theta} m r^2 \tag{3.93}$$

where $r = \sqrt{I/A}$. Also, let $A_s = A/k$. m_x is acting in the clockwise direction.

In the free vibration problem ($q = 0$), the resulting equation in terms of $w(x,t)$ may be written as

$$EI\frac{\partial^4 w}{\partial x^4} - \rho I\left(\frac{EA}{GA_s} + 1\right)\frac{\partial^4 w}{\partial x^2 \partial t^2} + \rho A\frac{\partial^2 w}{\partial t^2} + \rho^2 I\frac{A}{GA_s}\frac{\partial^4 w}{\partial t^4} = 0. \tag{3.94}$$

Again, the boundary conditions in Eq. (3.77) – in terms of w and θ – should be observed when solving the problem. For the simply supported beam, the function $w(x,t) = \sin(n\pi x/l)\sin\omega t$ satisfies the boundary condition $w = 0$, but only approximately the zero moment condition. This fact influences the calculated lower modes. The smaller the shear and rotational inertia effects, the better the boundary conditions are satisfied. Moreover, this function satisfies Eq. (3.94). The natural frequencies are then given by

$$\left[n\frac{\pi}{l}\right]^4 - \left[\frac{\rho}{E}\left(\frac{Ek}{G} + 1\right)\left(n\frac{\pi}{l}\right)^2 + \frac{\rho A}{EI}\right]\omega_n^2 + \frac{\rho^2 k}{EG}\omega_n^4 = 0 \tag{3.95}$$

If bending deformation (b) together with translational (t) and rotational (r) inertia are considered, and shear deformation is neglected, the eigenfrequency is obtained by neglecting the terms that contain the factor k. The eigenfrequency, $\omega_{n(b,t-r)}$, is then written as

$$\omega_{n(b,t-r)} = \omega_{n(b,t)}\sqrt{\frac{1}{1+(I/A)(n\pi/l)^2}} = \omega_{n(b)}\sqrt{\frac{1}{1+(n\pi r/l)^2}}\,. \tag{3.96}$$

For a (relatively short) beam with $h/l = 0.3$, i.e., $r/l \sim 0.1$–0.15, the effect of rotational inertia on the first eigenfrequency is a 5% to 10% reduction.

If bending (b) and shear (s) deformation together with translational inertia are included, while rotational inertia is neglected, the eigenfrequency may be written as

$$\omega_{n(b-s,t)} = \omega_{n(b,t)}\sqrt{\frac{1}{1+[(n\pi/l)^2(EI)/(GA_s)]}}. \tag{3.97}$$

Other Cases

Often, more general problems are encountered in practice. For instance, concentrated masses, dampers, or spring stiffnesses may be connected to the beams. Moreover, complex spatial systems of beams (frames) and possibly other types of structural components may be dealt with. The FE technique provides a basis for dealing with such problems. This technique is described in Chapter 4. Before addressing FE modeling, other discretization approaches are considered as an introduction to the FE method.

3.3.3 Approximate Solution of Dynamic Response Based on Discretization

Continuous structures, such as beams, may be discretized in various ways. In this section, the focus is on representing the stiffness and mass that normally constitute the dominant dynamic properties. A beam with bending deformations (i.e., no shear deformation) and lateral inertia effects (i.e., no rotational inertia effects) is considered. Excitation forces and damping reaction forces can be modeled afterward.

Approximations by Discrete Masses

Figure 3.14 shows a simply supported beam with continuous stiffness and mass. The simplest model is obtained by representing the uniform mass with a single mass located in the midpoint. If the vertical motion of this mass is considered as a generalized coordinate for the beam, the resulting SDOF has a stiffness $k = P/\delta$, where δ is the displacement due to a vertical force P acting at the mass point. The mass is denoted ηml. The factor η is less than 1 because a given displacement (and acceleration) of the beam will be less that those of the mass point. With a stiffness $k = 48EI/l^3$, the natural frequency becomes $\omega = \sqrt{48EI/(\eta ml^4)}$. It is seen that the exact solution for the first natural frequency of the simply supported beam, mentioned previously is obtained with an η of about 0.5. However, this model with a single mass point cannot represent the other modes of vibration.

By discretizing the beam with more mass points, a better model is obtained. However, to represent the stiffness corresponding to the 4 DOF for the lower model in Fig. 3.14a, a FE discretization of the structural behavior is most practical. Such models are pursued in Chapter 4.

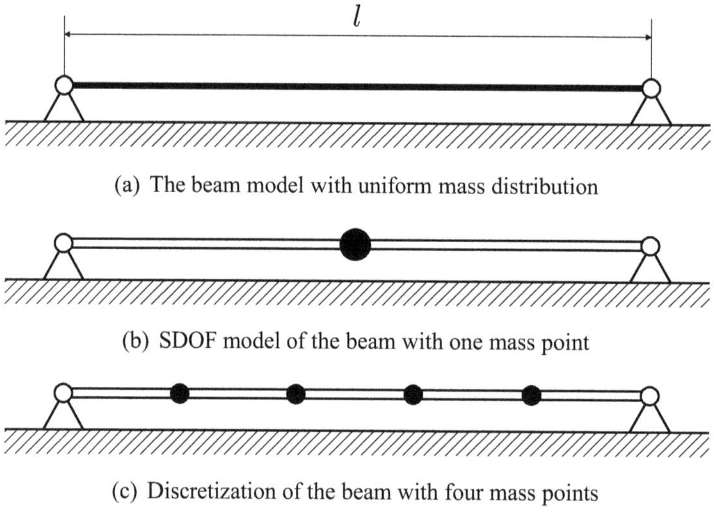

Figure 3.14. Discretization of a simply supported beam of length l subjected to lateral motions.

3.3.4 Example – Simple Estimates of Lowest Eigenfrequency of Complex Structures

In practice, the dynamic analysis can be subdivided in an eigenvibration and a forced vibration analysis. The first analysis estimates the eigenfrequencies, ω_i (and modes of vibration, if the multi-degree-of-freedom system is recognized). If $\beta_i = \bar{\omega}/\omega_i$ is close to 1 for the actual range of the excitation frequency $\bar{\omega}$ and eigenfrequencies ω_i, a forced vibration analysis considering mass, damping, stiffness, and excitation forces must be accomplished. In the following section, simplified methods for calculating the fundamental eigenfrequency of platforms is outlined, based on the expression $\omega = \sqrt{\tilde{k}/\tilde{m}}$, where $\tilde{k}$ and $\tilde{m}$ are the equivalent (generalized) stiffness and mass of the system.

In Chapter 2, the eigenfrequencies of various rigid-body, buoyant structures were estimated. Now, simplified methods to estimate the eigenfrequencies of flexible systems are exemplified.

The system in Fig. 3.15a, which is a sketch of a monotower platform, is idealized as shown in Fig. 3.15b. The tower is assumed to have a constant bending stiffness, EI. The axial and shear stiffness is for simplicity put equal to infinity. The soil stiffness is represented by a linear elastic rotational spring k_ψ. The kinematics of the lower mode of vibration of this system are specified by the horizontal displacement, u_0, at the top of the platform. The corresponding equivalent SDOF system has an equivalent stiffness, $\tilde{k}$, which can be found by establishing the relation between a fictitious horizontal (static) force, Q, and the corresponding displacement. The following relation is established when second-order effects due to the deck weight (Mg) are approximately included, but second-order effects due to the distributed weight (mg) and possible mooring stiffness, are neglected:

$$u_0 \cong \frac{Q\ell^3}{3EI} + \frac{(Mgu_o)\ \ell^2}{3EI} + \left(\frac{Q\ell}{k_\psi}\right)\ell + \frac{Mgu_0}{k_\psi}\ell. \qquad \text{(a)}$$

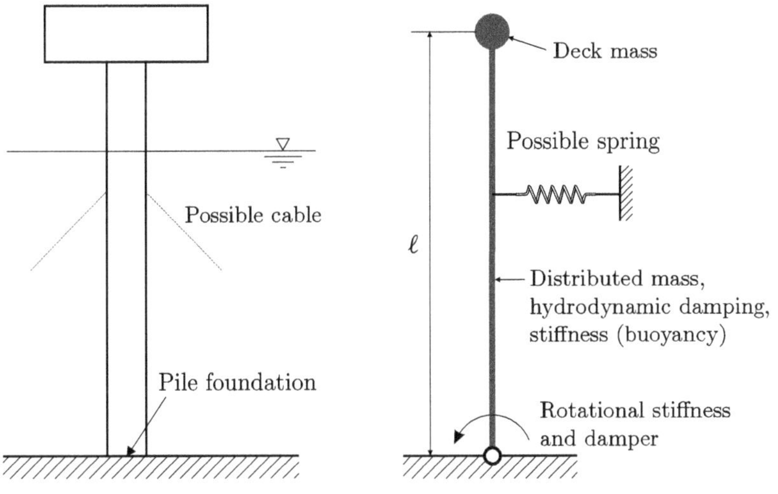

(a) System sketch (b) Model

Figure 3.15. Monotower platform.

This expression is achieved by using the formula for deflection, δ_1, at the tip of a cantilever beam of length λ subjected to a concentrated load, P: $\delta_1 = P\ell^3/(3EI)$, and the formula for deflection δ_2 at the tip of a beam at length ℓ with a rigid-body rotation $\psi = M_\psi / k_\psi$: $\delta_2 = \psi \cdot \ell = \ell \cdot M_\psi / k_\psi$. The derivation of Eq. (a) is based on an account of second-order effects relating to the deck mass M.

Hence, the $\tilde{k}$ in the relation

$$Q = \tilde{k}\, u_0 \tag{b}$$

is

$$\tilde{k} \cong \frac{3EI}{\ell^3} \frac{\left[1 - Mg\ell^2/(3EI) - Mg\, \ell/k_\psi\right]}{[1 + 3EI/(\ell\, k_\psi)]}. \tag{c}$$

If the second-order effects are neglected, the nominator will be equal to ℓ^3.

It is noted that the stiffness $\tilde{k}$ is obtained by $1/\tilde{k} = 1/k_e + 1/k_{P-\Delta}$, where k_e and $k_{P-\Delta}$ are the stiffnesses associated with elastic deformation and the second-order $(P - \Delta)$ effect, respectively.

The equivalent mass of the actual system can be taken to be

$$\tilde{m} = M + \alpha m \ell, \tag{d}$$

where α is a coefficient between 0 and 1. If the displacement of the deck mass is u_0, the average displacement and acceleration of the tower is less than $u_0(\ddot{u}_0)$. Hence, α is roughly 0.4 to 0.5.

Finally,

$$\omega^2 \cong \tilde{k}/\tilde{m} = \frac{3EI}{M\ell^3} \frac{\left[1 - Mg\ell^2/(3EI) - Mg\ell/k_\psi\right]}{[1 + 3EI/\,(\ell k_\psi)]\ [1 + \alpha\, m\ell/M]}. \tag{e}$$

Rather than using the physical, intuitive method for calculating ω, more formal techniques are used by establishing the dynamic equilibrium equation.

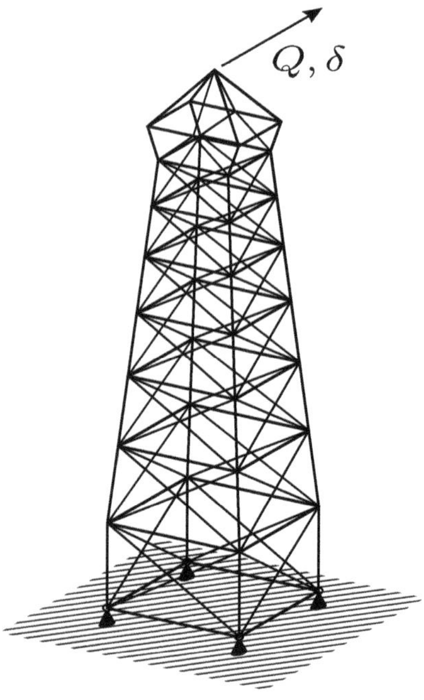

Figure 3.16. Framed platform (jacket), with a height l.

A first estimate of the fundamental eigenfrequency for the complex jacket platform in Fig. 3.16 may be obtained by determining the stiffness, $\tilde{k}_e$, by a static FE analysis; i.e., by determining the displacement δ caused by a force Q applied to the deck, where most of the mass is located. This analysis could be easily accommodated in the static FE analysis that needs to be accomplished in any case. The stiffness is then $\tilde{k}_e = Q/\delta$. In practice, this may be accomplished by introducing the load Q as distributed load in the FE load vector R, and taking δ as the average of the corresponding displacements in the direction of the loads. In this analysis, the flexibility of the structure and pile soil can be easily accommodated. It should be noted that this stiffness represents the elastic stiffness. The $P - \Delta$ effect, as well as the effect of the distributed mass and buoyancy, need to be considered separately. However, the numerator of Eq. (e) gives an indication of the $P - \Delta$-effect.

The equivalent mass of the system, as referred to the deck displacement, is taken to be

$$\tilde{m} = M_{deck} + \eta_a m_a d + \eta m l, \tag{f}$$

where m_a, η_a, and d are added mass per unit length, a factor in the range 0.4 to 0.5, and the water depth, respectively. With $\tilde{k}_e$ estimated, the "total" actual stiffness, $\tilde{k}$, can be obtained from $1/\tilde{k} = 1/\tilde{k}_e + 1/k_{P-\Delta}$. Finally, $\omega = \sqrt{\tilde{k}/\tilde{m}}$.

For a more accurate determination of eigenfrequencies of fixed platforms, it is in general necessary to apply a FE method to represent stiffness and mass of the structure, sea, and a possible pile soil.

The lowest eigenfrequencies of a ship in bending may be estimated by beam theory, by accounting for a variable stiffness EI, GA_S, and mass m.

Approximations by Using Generalized Displacements

An alternative way to discretize the behavior is to approximate the deformation of the beam with a *generalized displacement* in combination with the *principle of virtual displacements*, cf. Example 2.3.3.

The virtual work equations may be extended by the virtual work of inertia and damping forces. By using the dynamic equilibrium equation for the slender beam with axial load, Eq. (3.85), the virtual work becomes

$$\int_0^l \tilde{w}\,[q_{ext}(x,t) - m(x)\ddot{w} - c(x)\dot{w} - k(x)w - (Pw,_x),_x - (EIw,_{xx}),_{xx}]dx. \quad (3.98)$$

Now, by assuming

$$w(x) = \phi(x)Y(t)$$
$$\tilde{w}(x) = \phi(x)\tilde{Y}(t)$$

and using the same procedure by integrating by parts that was applied in connection with Eqs. (3.50)-(3.55), the following equation results, cf. Example 2.3.3:

$$\bar{m}\ddot{Y}(t) + \bar{c}\dot{Y}(t) + \bar{k}Y(t) = \bar{Q}(t) \quad (3.99)$$

where the generalized mass $\bar{m}$, damping $\bar{c}$, stiffness $\bar{k}$, and load $\bar{Q}$ are given as follows:

$$\bar{m} = \int_0^L m(x)\phi(x)^2 dx + \sum_i M_i\phi(x_i)^2, \quad (3.100)$$

$$\bar{c} = \int_0^L c(x)\phi(x)^2 dx + \sum_i C_i\phi(x_i)^2, \quad (3.101)$$

$$\bar{k} = \int_0^L k(x)\phi(x)^2 dx + \sum_i K_i\phi(x_i)^2 + \int_0^L EI(x)\left(\frac{\partial^2\phi}{\partial x^2}\right)^2 dx + \int_0^L P(x)\left(\frac{\partial\phi}{\partial x}\right)^2 dx, \quad (3.102)$$

$$\bar{Q}(t) = \int_0^L q_{ext}(x,t)\phi(x)dx + \sum_i Q_i(t)\phi(x_i). \quad (3.103)$$

It is noted that the virtual function $\tilde{Y}(t)$ is a common factor in the expression for virtual work that can be removed. In the preceding equation, contributions from discrete masses, dampers, and spring stiffnesses are added. In case of discrete properties, the integral is replaced by a sum.

The last two stiffness terms are due to the flexibility of the beam. The other terms may be present for a rigid beam.

The model expressed by Eqs. (3.99)–(3.103) is an SDOF model of a system including a flexible beam. The solution to this equation for different types of loads is carried out as explained for the SDOF in Chapter 2. The displacement at any point x along the beam is then found from $w(x,t) = \phi(x)Y(t)$.

When $w(x,t)$ is known, internal bending moments can be determined by $M = -EI\phi,_{xx}Y(t)$. The accuracy of the bending moments determined from this

Table 3.2. *Boundary conditions (BC) for beams*

Symbol	Description	Essential BC	Additional BC
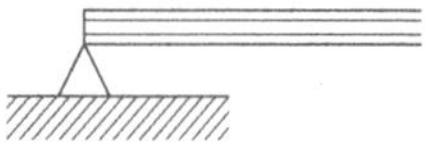	Simply supported	$\phi = 0$	$M = 0$ gives $\frac{d^2\phi}{dx^2} = 0$
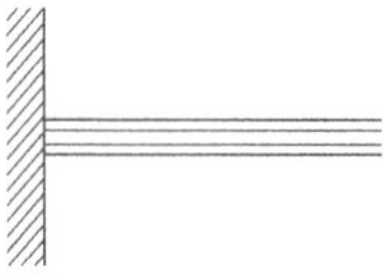	Fixed end	$\phi = 0$ and $\frac{d\phi}{dx} = 0$	None
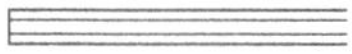	Free end	$\phi \neq 0$ and $\frac{d\phi}{dx} \neq 0$	$M = 0$ and $Q = 0$ give $\frac{d^2\phi}{dx^2} = 0$ and $\frac{d^3\phi}{dx^3} = 0$
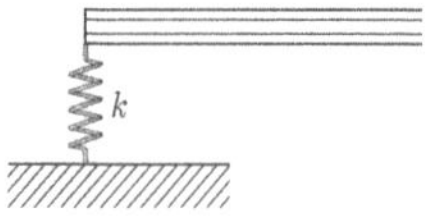	Spring	$\phi \neq 0$ and $\frac{d\phi}{dx} \neq 0$	$M = 0$ and $Q = k\phi$ give $\frac{d^2\phi}{dx^2} = 0$ and $\frac{d^3\phi}{dx^3} \approx k\phi$
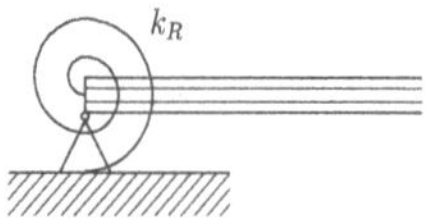	Rotational spring	$\phi = 0$	$M = k_R \frac{d\phi}{dx}$ gives $\frac{d^2\phi}{dx^2} \approx k_R \frac{d\phi}{dx}$

expression critically depends on the choice of $\phi(x)$. For beams with uniform dynamic properties and loads, we often find good generalized coordinates provided by good shape functions. See, for instance, the exact solutions given in Table 3.1. The exact functions satisfy the differential equation and all boundary conditions. When Eq. (3.99) is applied, only the so-called essential boundary conditions need to be satisfied. They are given in Table 3.2. The shape function $\phi(x)$ may, for instance, be identified by carrying out a static calculation of the beam when subjected to a spatial load pattern that corresponds to that of the dynamic loading.

For a statically determinate beam, the bending moments at different sections can be determined by equilibrium considerations, using external forces.

3.3.5 Example – Cantilever Beam

As an example, consider the cantilever beam in Fig. 3.17 with uniform mass and bending stiffness subjected to a harmonic load with uniform distribution. The exact first mode of vibration for this beam is given in Table 3.1. Here, a much simpler expression is assumed, namely,

$$w(x,t) = \phi(x)Y(t) = \frac{x^2}{L^2}Y(t). \tag{a}$$

This gives the following derivatives with respect to x:

$$\phi_{,x}(x) = \frac{2x}{L^2}; \quad \phi_{,xx}(x) = \frac{2}{L^2}. \tag{b}$$

It is observed that the curvature $\phi_{,xx}(x)Y(t)$, and hence the bending moment, is constant along the beam. This is a crude approximation, but the

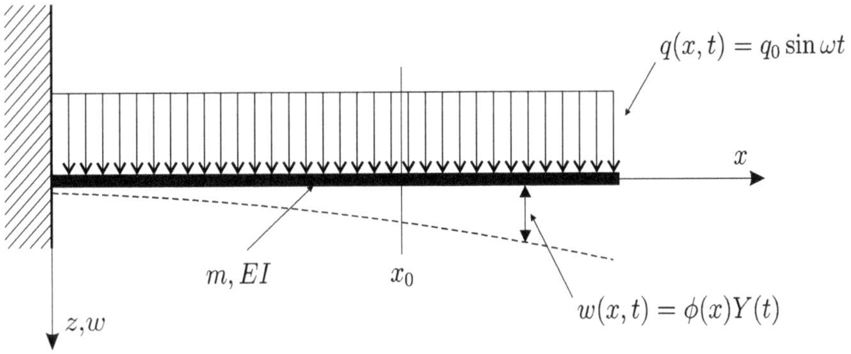

Figure 3.17. Cantilever beam with a length of L.

essential boundary conditions at the left end are satisfied. The generalized dynamic properties then become

$$\begin{aligned}
\bar{m} &= \int_0^L m(x)\phi(x)^2 dx = \frac{m}{L^4}\int_0^L x^4 dx = \frac{mL}{5} \\
\bar{c} &= \int_0^L c(x)\phi(x)^2 dx = 0.015c_{cr} = 0.03\sqrt{\bar{m}\bar{k}} \text{ (assumed)} \\
\bar{k} &= \int_0^L EI(x)\left(\frac{\partial^2\phi}{\partial x^2}\right)^2 dx = EI\left(\frac{2}{L^2}\right)^2 \int_0^L dx = \frac{4EI}{L^3} \\
\bar{Q}(t) &= \int_0^L q_{ext}(x,t)\phi(x)dx = \frac{q_0 \sin\omega t}{L^2}\int_0^L x^2 dx = \frac{q_0 L \sin\omega t}{3}.
\end{aligned} \tag{c}$$

In Eq. (c), the damping is assumed to be 1.5% of the critical damping value. If the coefficient $c(x)$ is assumed to be constant, it will be $c(x) = 5\bar{c}/L$. The expressions for generalized mass and loading show that they are one-fifth and one-third of the total mass and shear loading, respectively.

The governing dynamic equation of motion is then

$$\frac{mL}{5}\ddot{Y}(t) + 0.03\sqrt{\bar{m}\bar{k}}\dot{Y}(t) + \frac{4EI}{L^3}Y(t) = \frac{q_0 L \sin\omega t}{3}. \tag{d}$$

The free, undamped vibration is described by the mode(s) and natural frequencies of vibration. In this example, the mode is assumed. The natural frequency is estimated by

$$\omega_1^2 = \frac{\bar{k}}{\bar{m}} = \frac{4EI/(L^3)}{mL/5} = \frac{20EI}{mL^4}, \text{ or, } \omega_1 = \frac{4.47}{L^2}\sqrt{\frac{EI}{m}}. \tag{e}$$

Comparing the approximate with the accurate value obtained by solving the differential equation in Table 3.1 shows that the approximate natural frequency is about 27% higher than the exact one.

The solution of the forced vibration problem may be written as

$$
\begin{aligned}
w(x,t) &= (q_0 L/3)\phi(x)\left[(1-\beta^2)^2 + (2\xi\beta)^2\right]^{-0.5}\sin(\omega t + \theta)/\bar{k} \\
M(x,t) &= -EIw_{,xx} = -(q_0 L^2/6)\left[(1-\beta^2)^2 + (2\xi\beta)^2\right]^{-0.5}\sin(\omega t + \theta) \\
c\dot{w}(x,t) &= (0.05\beta q_0)\phi(x)\left[(1-\beta^2)^2 + (2\xi\beta)^2\right]^{-0.5}\cos(\omega t + \theta) \\
m\ddot{w}(x,t) &= (5\beta^2 q_0/3)\phi(x)\left[(1-\beta^2)^2 + (2\xi\beta)^2\right]^{-0.5}\sin(\omega t + \theta),
\end{aligned}
\tag{f}
$$

where β is the ratio of excitation and natural frequency, ξ is the damping ratio in terms of generalized coordinates, and θ is the phase angle. It is noted that the maximum value of the damping force is $0.03/\beta$ times the inertia force. The approximate bending moment obtained based on the assumed displacement is seen from Eq. (f) to be constant along the beam. Moreover, for a frequency ratio β equal to 0.5, the value is $0.22 q_0 L^2$, while the static moment due to the load q_0 varies between zero and $0.50 q_0 L^2$ along the beam.

As mentioned previously another option for a statically determinate problem is to determine the dynamic response (moment, shear force) by applying the equilibrium conditions. The main loads would then be the external loads and the inertia forces. Assuming a frequency ratio $\beta = 0.5$, and assuming the phase angle to be 0, the maximum response (bending moment) at the clamped end occurs when the excitation and inertia force are in phase:

$$
\begin{aligned}
M_{static,max} &= 0.5 q_0 L^2 \quad \text{(exact value)} \\
M_{total,max} &= 0.5 q_0 L^2 + \int_0^L m\ddot{w}(x,t) x\,dx = 0.64 q_0 L^2.
\end{aligned}
\tag{g}
$$

Hence, the dynamic amplification for this case is approximately $M_{\text{total,max}}/M_{\text{static,max}} = 1.28$.

An improved solution can be obtained by using a more accurate displacement pattern than that used in Eq. (a).

3.3.6 Example – Wind Turbines

Wind turbines for producing electric power offshore consist of a supporting structure, tower, rotor, and nacelle as shown in Fig. 3.18. The largest systems today have a rotor diameter of about 120 meters. Structural dynamics may affect the global tower response and the local response of the blades. Here, the focus is on the blades, which form cantilever beams subjected to gravity forces, variable gravity, centrifugal and centripetal loads, and a static and turbulent wind load.

The propeller blade is subjected to the following forces, as illustrated in Fig. 3.18:

- Gravitational loads – with a component mg – that varies depending on the position of the blade during the rotation
- Inertial loads, F, that occur when the rotor is accelerated or decelerated by a moment, $T = I(d\omega/dt)$, applied to the rotor shaft, where I is the moment of inertia of the rotor
- Centrifugal inertia force, F_c
- Aerodynamic loading due to a steady and turbulent wind component that produces a lift and drag force on the blade

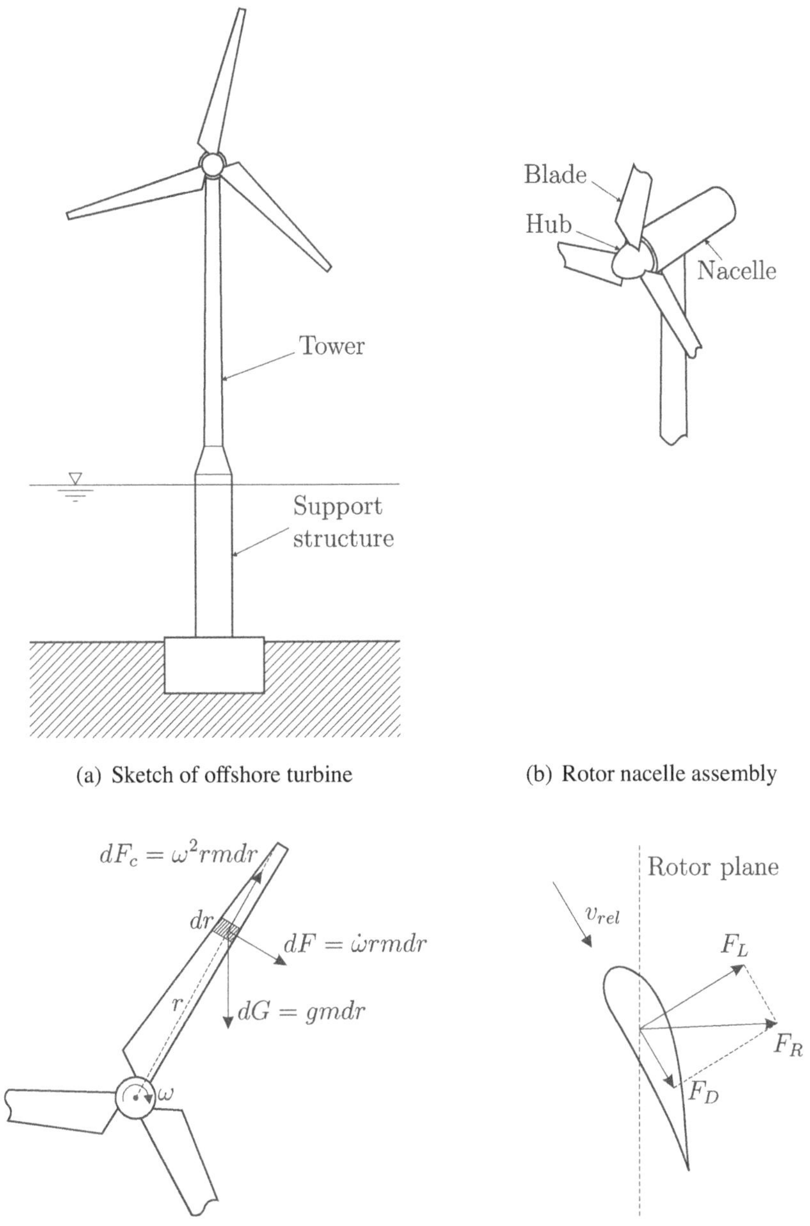

(a) Sketch of offshore turbine

(b) Rotor nacelle assembly

(c) Gravity and inertial loads on a section of the blade

(d) Cross section of rotor blade and aerodynamic forces

Figure 3.18. Bottom-fixed offshore wind turbine system.

If the rotor has a constant speed, e.g., 10 revolutions per minute, the gravity causes a sinusoidal component with about 10^7 cycles in 20 years, and contributes important fatigue loads. The aerodynamic force is caused by the flow past the structure – the blades and the tower. Besides the turbulence in the air flow, there is a variation in the steady wind speed over the height which the blade experiences during its rotation.

The forces acting on the cantilever propeller blade give rise to loads along the two principal axes of the unsymmetric cross section. Due to the need to minimize weight of the blades, they are made of lightweight materials such as fiber reinforced composites. The blades are flexible and will get large deflections that need to be accounted for. This is a nonlinear geometric effect that is not described herein. Further details on dynamic modeling of wind turbines may be found in Burton et al. (2011)

3.3.7 The Rayleigh-Ritz Method for Determining Mode Shapes and Natural Frequencies

In Section 3.3.3, it is demonstrated how an SDOF model for a beam could be established by use of generalized coordinates. The key point in this connection is the assumption of the beam displacement $w(x,t) = \phi(x)Y(t)$ as a product of an assumed spatial shape function and the time-dependent function, which is solved from the governing generalized equation of motion. This approach may be refined by assuming the displacement to be given by

$$w(x,t) = \left(\sum_i \phi_i(x)\chi_i\right) Y(t) = \left(\sum_k \psi_k(x)\right) Y(t), \tag{3.104}$$

where ϕ_i are shape functions and χ_i are weights that specify the relative contribution from different shape functions. In the following section, it is shown how approximate (undamped) mode shapes ψ_j and natural frequencies may be determined based on assumed generalized coordinates. That is, each ψ_k is a linear combination of the ϕ_i functions and a specific set of the χ_i. In this way, more accurate natural frequencies and modes may be achieved. $Y(t)$ is assumed to be a harmonic function: $Y(t) = Y_0 \sin \omega t$.

The desired shape function may be written as

$$\psi(x) = \sum_i \phi_i(x)\chi_i = \boldsymbol{\phi}^T \boldsymbol{\chi}, \tag{3.105}$$

where $\boldsymbol{\phi}^T$ is a line vector of the shape functions ϕ_i and $\boldsymbol{\chi}$ is a column vector with components χ_i. The shape functions should be linearly independent and satisfy at least the essential boundary conditions (Table 3.2).

Because the focus here is on undamped mode shapes, it is only necessary to consider the inertia and stiffness terms. The Rayleigh-Ritz method is based on equating the maximum potential and kinetic energies to express the natural frequency and minimizing the frequencies with respect to the unknown parameters χ_i. The maximum potential energy occurs when the velocity or the kinetic energy is zero, i.e., $\omega t = \pi/2$. These maxima can be expressed as follows:

$$E_p = \frac{1}{2}\bar{k}Y_0^2; \; E_k = \frac{1}{2}\omega^2 \bar{m} Y_0^2. \tag{3.106}$$

By equating the maximum potential and kinetic energy, the expression for the natural frequency becomes

$$\omega^2 = \frac{\bar{k}}{\bar{m}} = \frac{\boldsymbol{\chi}^T \mathbf{K} \boldsymbol{\chi}}{\boldsymbol{\chi}^T \mathbf{M} \boldsymbol{\chi}}, \tag{3.107}$$

where for a slender beam with axial force $P(x)$ and distributed elastic support stiffness $k(x)$ as well as discrete springs with stiffness K_r and discrete masses M_r,

$$K_{ij} = \int_0^l k(x)\phi_i(x)\phi_j(x)dx + \sum_r K_r\phi_i(x_r)\phi_j(x_r)$$

$$+ \int_0^l P(x)(\frac{\partial\phi_i}{\partial x}\frac{\partial\phi_j}{\partial x})dx + \int_0^l EI(x)(\frac{\partial^2\phi_i}{\partial x^2}\frac{\partial^2\phi_j}{\partial x^2})dx, \qquad (3.108)$$

$$M_{ij} = \int_0^l m(x)\phi_i(x)\phi_j(x)dx + \sum_r M_r\phi_i(x_r)\phi_j(x_r), \qquad (3.109)$$

or

$$\mathbf{K} = \int_0^l k(x)\boldsymbol{\phi}(x)\boldsymbol{\phi}^T(x)\,dx + \sum_r K_r\boldsymbol{\phi}(x_r)\boldsymbol{\phi}^T(x_r)$$

$$+ \int_0^l P(x)\boldsymbol{\phi}_{,x}(x)\boldsymbol{\phi}_{,x}(x)^T\,dx + \int_0^l EI(x)\boldsymbol{\phi}_{,xx}(x)\boldsymbol{\phi}_{,xx}(x)^T dx, \qquad (3.110)$$

$$\mathbf{M} = \int_0^l m(x)\boldsymbol{\phi}(x)\boldsymbol{\phi}^T(x)dx + \sum_r M_r\boldsymbol{\phi}(x_r)\boldsymbol{\phi}^T(x_r). \qquad (3.111)$$

For each mode and natural frequency, there will be a unique vector $\boldsymbol{\chi}$. It can be demonstrated that the Rayleigh quotient (Eq. (3.107)) overestimates the frequency. The solution is therefore determined by the condition

$$\frac{\partial\omega^2}{\partial\chi_i} = \frac{\partial}{\partial\chi_i}(\frac{\bar{k}}{\bar{m}}) = \frac{1}{\bar{m}}\frac{\partial\bar{k}}{\partial\chi_i} - \frac{1}{\bar{m}^2}\frac{\partial\bar{m}}{\partial\chi_i}\bar{k} = 0. \qquad (3.112)$$

This leads to the equation

$$\frac{\partial\bar{k}}{\partial\chi_i} - \omega^2\frac{\partial\bar{m}}{\partial\chi_i} = 0. \qquad (3.113)$$

Now, with $\bar{k} = \boldsymbol{\chi}^T\mathbf{K}\boldsymbol{\chi}$ and $\bar{m} = \boldsymbol{\chi}^T\mathbf{M}\boldsymbol{\chi}$, the derivatives of $\bar{k}$ and $\bar{m}$ are

$$\frac{\partial\bar{k}}{\partial\chi_i} = \mathbf{K}_{Li}\boldsymbol{\chi} + \boldsymbol{\chi}^T\mathbf{K}_{Ci} = 2\mathbf{K}_{Li}\boldsymbol{\chi}; \qquad \frac{\partial\bar{m}}{\partial\chi_i} = 2\mathbf{M}_{Li}\boldsymbol{\chi}, \qquad (3.114)$$

where the vectors $\mathbf{K}_{Li}$ and $\mathbf{M}_{Li}$ refer to line i of the $\mathbf{K}$ and $\mathbf{M}$ matrices, respectively. Similarly, $\mathbf{K}_{Ci}$ refers to column i of the $\mathbf{K}$. We get a similar result for the derivatives with respect to all variables χ_i. By collecting all equations, the resulting equation becomes

$$(\mathbf{K} - \omega^2\mathbf{M})\boldsymbol{\chi} = 0. \qquad (3.115)$$

This is the equation for an eigenvalue problem with n degrees of freedom. The absolute values of the eigenvectors are indeterminate, only the relative values between

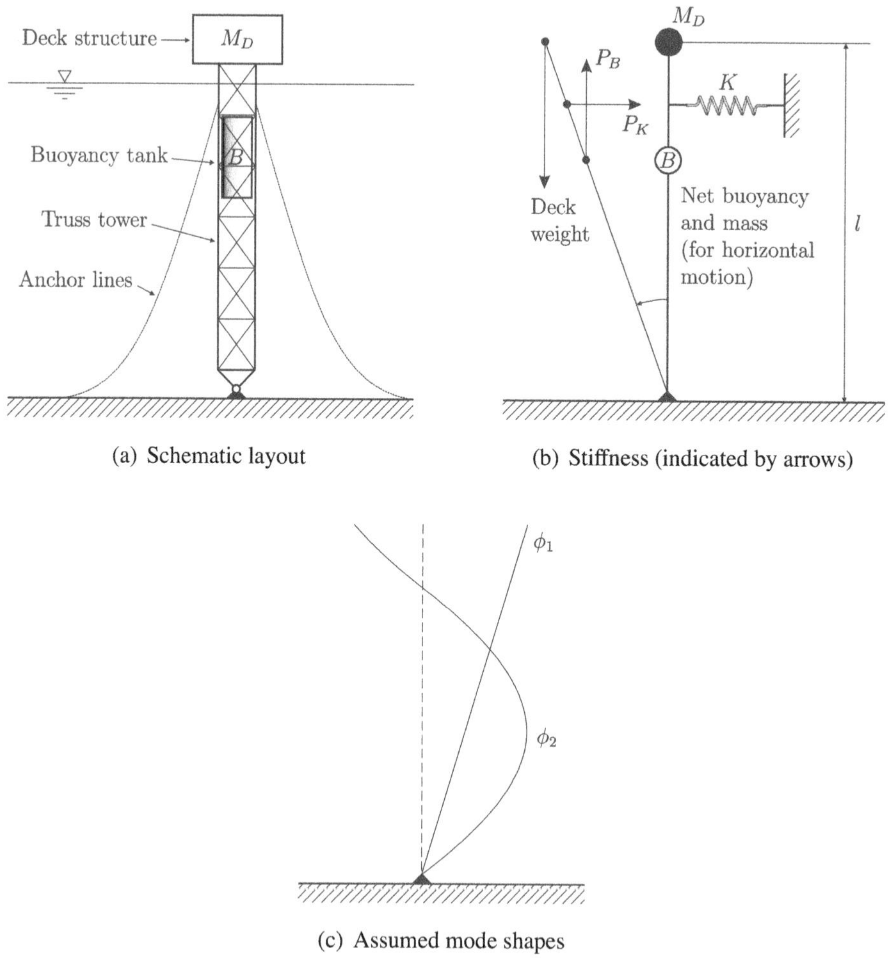

(a) Schematic layout

(b) Stiffness (indicated by arrows)

(c) Assumed mode shapes

Figure 3.19. Guyed trusswork tower.

the components in the vector $\boldsymbol{\chi}$. For each eigenvalue (natural frequency), there is an eigenvector $\boldsymbol{\chi}_k$. The corresponding mode shape is found by $\psi_k(x) = \boldsymbol{\phi}(x)^T\boldsymbol{\chi}_k = \sum \phi_i(x)\chi_{i,k}$. Solution methods are further discussed in Section 3.3.10.

3.3.8 Example – Guyed Tower

To illustrate the Rayleigh-Ritz method, consider a guyed tower, as shown in Fig. 3.19. The behavior of this tower, modeled as a rigid body, is investigated by using a shape function $\phi(x) = x/l$, which represents tower rotation about the seafloor support. However, the tower may have flexibility which needs to be accounted for. The following approximate shape functions may be used to model the behavior of the tower, taking into account both rigid-body motions and flexible deformations:

$$\phi_1(x) = x/l; \;\; \phi_2(x) = \sin\left(\pi x/(0.81l)\right). \qquad \text{(a)}$$

The first shape function corresponds to the rigid-body mode, while the second shape function represents a flexible mode with a node at the spring support. The only essential boundary condition to satisfy in this problem is to have zero motion or deflection at the support point on the seafloor. It is seen that the assumed shape functions satisfy this requirement and also zero bending moment in the support point.

The first mode ψ_1 will be dominated by the first shape function, while the relative contribution to the second mode from each shape function will especially depend on the stiffness of the mooring system and buoyancy. The larger the value of these stiffnesses, the larger the contribution from the second shape function.

For more complex problems, the method needs to be refined in a systematic manner by including more generalized coordinates, or alternatively, using the FE method (Mo and Moan, 1984). The natural modes of vibration are then an obvious choice.

3.3.9 Example – Ship Vibration

The reference case for ship vibrations is to model the ship hull as a free-free beam supported on an elastic foundation provided by the buoyancy. The springs corresponding to the buoyancy are soft and have a small effect on the eigenfrequencies of modes that involve deformation of the hull. However, the variable mass and stiffness distribution need to be accounted for in dealing with the ship hull vibrations. Although the differential equation may be difficult to solve for such cases, discretization techniques are useful to get an estimate. By the FE method, the desired accuracy can be obtained by using a sufficiently fine mesh. Finally, the effects of shear deformation and rotational inertia need to be considered.

The following (conservative) approximate estimate of the effect of shear deformation was obtained by Storhaug (2007):

$$\omega_n = \sqrt{\frac{EI}{m}a} \approx \omega_{n(b)}\sqrt{\frac{1}{1+ba_1^2}};\ a = \frac{a_1^4}{1+ba_1^2};\ a_1 = \frac{\pi}{l}(2n+1);\ b = \frac{EI}{GA_s}, \quad \text{(a)}$$

where $\omega_{n(b)}$ are the eigenfrequencies for pure bending.

For a bulk carrier with the following properties: $l = 300$ m, $I = 700$ m^4, $A_s = 2$ m^2, the frequency of the first mode is found to be 0.90 times the frequency for pure bending. For the second and third modes, the reduction factors are 0.78 and 0.67. More accurate calculations predict reduction factors that are 0.94 for the first mode and 0.83 for the 3-node mode.

3.3.10 Modal Superposition

Modal superposition is a method to determine the forced response by using particular properties of the eigenvibration problem; i.e., the solution of the dynamic equations of motion with zero external load. In this section, it is applied to a continuous system but it is also applicable to discrete dynamic models, e.g., based on FE modelling. Rather than using "general" shape functions ϕ_i, special shape functions in terms of the eigenmodes ψ_k are applied. The displacements are, hence, written as

$$w(x,t) = \sum_{k=0}^{n} \psi_k(x) v_k(t) = \boldsymbol{\psi}(x)^T \mathbf{v}(t), \quad (3.116)$$

where each $v_k(t)$ is a time-dependent weight "function" that needs to be determined by solving the resulting dynamic equations. This represents a generalization of the formulation with generalized coordinates presented previously, in the sense that the relative values of the v_ks are allowed to change with time, while they were independent of time in the formulation in Section 3.3.7.

The dynamic equations of motion for modal superposition follow the same principle as for the generalized coordinates. In the following, the principle of virtual displacement for the dynamic problem is applied to establish the governing equations considering first a system with inertia and stiffness forces. Some considerations concerning damping are made subsequently. The (scalar) virtual displacements are then expressed as follows,

$$\tilde{w}(x,t) = \sum_{r=0}^{n} \psi_r(x)\tilde{v}_r(t) = \boldsymbol{\psi}(x)^T\tilde{\mathbf{v}}(t) \equiv \tilde{\mathbf{v}}(t)^T\boldsymbol{\psi}(x). \tag{3.117}$$

The generalized dynamic equation of motion (i.e., the inertia term multiplied by the virtual displacement) will contain the term,

$$\int_0^l \tilde{w}(x,t)m(x)\ddot{w}(x,t)\,dx = \tilde{\mathbf{v}}(t)^T\left[\int_0^l m(x)\boldsymbol{\psi}(x)\boldsymbol{\psi}(x)^T\,dx\right]\ddot{\mathbf{v}}(t)$$
$$= \tilde{\mathbf{v}}(t)^T\mathbf{M}_{CON}\ddot{\mathbf{v}}(t), \tag{3.118}$$

where $\mathbf{M}_{CON} = \left(M_{ij,CON}\right)$ relates to a continuously varying mass and

$$M_{ij,CON} = \int_0^l m(x)\psi_i(x)\psi_j(x)\,dx. \tag{3.119}$$

Discrete masses contribute terms of the type

$$M_{ij,MP} = \sum_r M_r\psi_i(x_r)\psi_j(x_r). \tag{3.120}$$

The total mass matrix $\mathbf{M}$ is

$$\mathbf{M} = \mathbf{M}_{CON} + \mathbf{M}_{MP}. \tag{3.121}$$

The product of the column vector $\boldsymbol{\psi}(x)$ and line vector $\boldsymbol{\psi}(x)^T$ represents, in principle, a full matrix with components $\psi_i(x)\psi_j(x)$. However, the matrix $\mathbf{M}$ is shown to be a diagonal matrix when normal modes are applied. The diagonal elements $M_{ii} = m_i$ of the matrix $\mathbf{M}$ are different from zero. These elements are related to the modes (i) and are denoted modal masses.

The dynamic equation also consists of a stiffness term of the type

$$\int_0^l \tilde{w}_{,xx}(x,t)EI(x)w_{,xx}(x,t)dx = \tilde{\mathbf{v}}(t)^T\left[\int_0^l EI(x)\boldsymbol{\psi}_{,xx}(x)\boldsymbol{\psi}_{,xx}^T(x)dx\right]\mathbf{v}(t)$$
$$= \tilde{\mathbf{v}}(t)^T\mathbf{K}_{EI}\mathbf{v}(t), \tag{3.122}$$

where the stiffness matrix $\mathbf{K}_{EI} = \left(K_{ij,EI}\right)$ contains the terms

$$K_{ij,EI} = \int_0^l EI(x)\psi_{i,xx}(x)\psi_{j,xx}(x)dx. \tag{3.123}$$

The total stiffness matrix $\mathbf{K} = \left(K_{ij}\right)$ contains terms such as

$$K_{ij} = \int_0^l k(x)\psi_i(x)\psi_j(x)\,dx + \sum_r K_r\psi_i(x_r)\psi_j(x_r) + \int_0^l EI(x)\psi_{i,xx}\,\psi_{j,xx}\,dx \tag{3.124}$$

The resulting discrete equations may be written as follows:

$$\mathbf{M}\ddot{\mathbf{v}}(t) + \mathbf{K}\mathbf{v}(t) = \mathbf{Q}(t), \tag{3.125}$$

where the actual **M** and **K** are diagonal matrices.

The resulting, uncoupled equations may be written as follows:

$$m_i\ddot{v}_i(t) + k_i v_i(t) = Q_i(t). \tag{3.126}$$

Damping was not included in the preceding equations. If the damping along the beam was known as a function $c(x)$, a damping term, $\mathbf{C}\dot{\mathbf{q}}$, could have been included in Eq. (3.125). However, we have limited information about damping, and $c(x)$ is normally not known. Moreover, if we had this knowledge and determined the terms C_{ij} by

$$C_{ij} = \int_0^l c(x)\,\psi_i(x)\,\psi_j(x)dx, \tag{3.127}$$

$\mathbf{C} = \left(C_{ij}\right)$ would not generally be a diagonal matrix. This would imply a more demanding solution of the resulting equations than we would get with uncoupled equations.

The fact that damping values are in any case uncertain justifies including damping by a modal damping coefficient $c_i = \xi_i c_{cr} = 2\xi_i\sqrt{m_i k_i}$ in the dynamic equations. ξ_i is the damping ratio. Hence, the uncoupled equations of motions read

$$m_i\ddot{v}_i(t) + c_i\dot{v}_i(t) + k_i v_i(t) = Q_i(t). \tag{3.128}$$

This approach is also convenient because damping is often frequency dependent, and this approach allows the frequency dependence to be taken into account.

The solution of the dynamic response problem for the MDOF problem can then be found by finding the solution for different types of loads, in the same way as for an SDOF system.

3.3.11 Discussion of Forced Vibration of a Slender Beam

The modal equation of motion is now written as

$$m_i\ddot{v}_i + c_i\dot{v}_i + k_i v_i = Q_i(t), \tag{3.129}$$

with the initial conditions $w(x,0) = \dot{w}(x,0) = 0$, or $v_i(0) = \dot{v}_i(0) = 0$. With elapsing time, $v_i = v_{hi} + v_{pi} \approx v_{pi}$, where the homogeneous solution v_{hi} vanishes due to the presence of damping. Assuming Rayleigh damping, i.e., $c_i = \alpha m_i + \beta k_i$ for suitable constants α and β, the particular solution v_{pi} for mode i is

$$v_{pi}(t) = \int_0^t e^{-\xi_i \omega_i (t-s)} \frac{Q_i(s)}{m_i \omega_{id}^2} \sin \omega_{id}(t-s)\, ds, \tag{3.130}$$

where

$$\xi_i = \frac{c_i}{c_{i(cr)}} = \frac{\alpha m_i + \beta k_i}{\sqrt{m_i k_i}} = \frac{1}{2}\left[\frac{\alpha}{\omega_i} + \beta \omega_i\right], \tag{3.131}$$

$\omega_i = \sqrt{k_i/m_i}$, and $\omega_{id} = \omega_i \sqrt{1 - \xi_i^2}$.

The total displacement of the beam is then given by

$$w(x,t) = \sum_{n=1}^{\infty} \psi_i(x) v_i(t) = \sum_{n=1}^{\infty} \psi_i(x) \frac{1}{\omega_i^2 m_i} DAF_i(t), \tag{3.132}$$

where the dynamic amplification factor for mode i (relative to the static response) is given by

$$DAF_i(t) = \int_0^t \frac{\omega_i^2 Q_i(s)}{\omega_{id}^2} e^{-\xi_i \omega_i (t-s)} \sin \omega_{id}(t-s)\, ds. \tag{3.133}$$

Bending stresses in the beam can be determined by differentiating the displacements and applying the formula

$$\sigma_x = -E \frac{\partial^2 w}{\partial x^2} z,$$

where z is the distance from the neutral axis of a point in the cross section.

It is seen from Eq. (3.132) that the response, in terms of the contribution from each mode, depends on the magnitude of the dynamic load factor (DAF). If $q_{ext}(x,t) = q_s(x) g(t)$, where $q_s(x)$ represents a static load pattern and $g(t)$ is a purely time varying function, then

$$Q_i(t) = \int_0^{\ell} q_{ext}(x,t)\, \psi_i(x)\, dx = \left(\int_0^{\ell} q_s(x)\, \psi_i(x)\, dx \right) g(t). \tag{3.134}$$

It is seen that a point load acting in a node of the mode of vibration, i.e., where $\psi_i(x) = 0$, does not excite this mode because then $Q_i(t) = 0$ for all t. Moreover, a load pattern $q_s(x)$ that is distributed over the beam proportionally to a mode ψ_i will only excite this mode when $EI(x)$ and $m(x)$ are constant. This statement follows from the orthogonality properties of the modes ψ_i. It is also seen that symmetric modes are excited by a $q_s(x)$ which is symmetric (when $EI(x)$ and $m(x)$ are symmetric).

The dynamic load factor DAF_i expresses the dynamic response relative to the static one. Each mode can then be considered separately (uncoupled), and the results obtained in Chapter 2 for an SDOF system apply for each mode.

We then need to separate between two types of loading, namely,

- Impulse type of loading.
- Steady-state loading, essentially expressed in terms of Fourier series.

For impulse type of loading, the variation of the loading over the load duration t_d relative to the natural period T_i is the determining factor. If the duration is small, i.e., $t_d/T_i < 0{,}25$, it is the impulse itself that matters. For longer duration, it is the rise time, duration, etc., that would be governing parameters. For harmonic loading, $q_{ext}(x,t) = q_s(x)\sin\omega t$, the *steady-state* response in Eq. (3.132) may be expressed as

$$w(x,t) = \sum_{i=1}^{\infty} Q_{si}\psi_i(x)\frac{DAF_i}{k_i}\sin(\omega t - \varepsilon_i). \tag{3.135}$$

Here,

$$Q_{si} = \int_0^{\ell} q_s(x)\,\psi_i(x)\,dx, \tag{3.136}$$

$$\tan\varepsilon_i = \frac{2\xi_i\beta_i}{1-\beta_i^2}, \tag{3.137}$$

and

$$DAF_i = \frac{1}{\sqrt{(1-\beta_i^2)^2 + (2\xi_i\beta_i)^2}}, \tag{3.138}$$

where $\beta_i = \omega/\omega_i$ is the ratio between the excitation and natural frequency, cf. Section 2.4.

3.3.12 Formulations for Moving Loads

Another aspect of the beam model arises in connection with moving dynamic loads. Consider, for instance, a beam that is a pipe that carries an incompressible flowing fluid with an axial velocity U in the x-direction. The transverse velocity of a particle of the fluid is then

$$u_z = (\frac{\partial}{\partial t} + U\frac{\partial}{\partial x})w(x,t), \tag{3.139}$$

and the transverse acceleration is

$$a_z = (\frac{\partial}{\partial t} + U\frac{\partial}{\partial x})^2 w(x,t). \tag{3.140}$$

The transverse inertia force on a length of dx is then

$$m_z = \rho A(\frac{\partial^2}{\partial t^2} + 2U\frac{\partial^2}{\partial x\partial t} + U^2\frac{\partial^2}{\partial x^2})w(x,t). \tag{3.141}$$

The first term corresponds to the case with stationary mass distribution as treated so far. The last two terms are due to the motion. A continuous train with a uniform mass distribution will have the same effects as this fluid on the dynamic behavior.

4 Finite Element Method

4.1 Introduction

The finite element method (FEM) is a numerical approach for solving partial differential equations in an approximate manner. It provides an efficient method for discretizing a structure in space and representing the dynamic variation of loads and response in time. In this context, the focus is on using the FEM to discretize the structure in space and then establish the equations of motion by consideration of dynamic equilibrium.

Structural analysis is generally based on three principles:

- Equilibrium (local and global equilibrium in terms of stresses, stress resultants, and forces)
- Kinematic compatibility (between strains and displacements)
- Constitutive relationship (between stresses and strains/moments and curvatures)

In classical structural mechanics, these principles can be applied to establish exact relationships, e.g., between end forces (and moments) and the corresponding displacements of bars and beams. These relationships can be given in matrix form and used to discretize frames and trusses; they are commonly denoted the *matrix method*, see, e.g., Sack (1984); Felton and Nelson (1997); McGuire et al. (2000); Moan (2003). The finite element method can be used to establish such relationships by applying the principle of virtual work, minimum potential energy, or other variational principles. Both the matrix method and the FEM for bars and beams are briefly discussed. However, the FEM is more general and can be applied to analyze continuous plane stress, plate bending, shell, and solid three-dimensional structures. Only the basic principles of FEM models for linear structural behavior based on assumed displacements will be described in this context. More detailed accounts of the FEM are found in various textbooks (Cook et al., 2002; Zienkiewicz et al., 2005; Zienkiewicz and Taylor, 2006; Fish and Belytschko, 2007).

A static analysis by the FEM consists of the following steps:

1. *Discretization*: Model structure as an assemblage of finite elements.
2. *Element analysis*: Establish the element stiffness relationship.

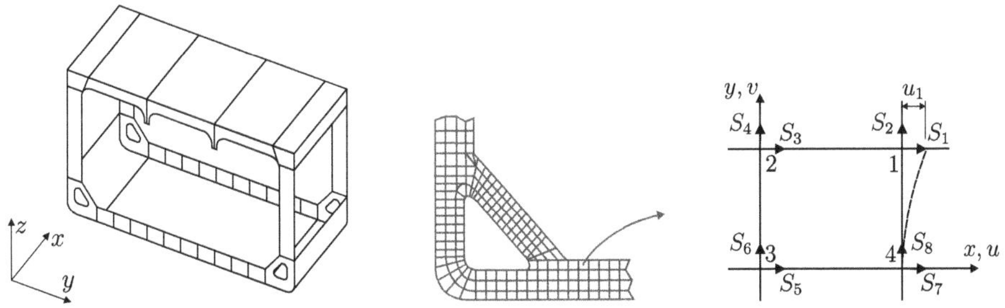

(a) Transverse frames in a ship as an example of plate structure of arbitrary shape

(b) Possible finite element model of a part of the structure (frame corner)

(c) Plane rectangular element showing nodal displacements u_i and v_i. Dashed line shows the deformation mode associated with the displacement in the local x-direction of node 1

Figure 4.1. Discretization of a structure into finite elements.

3. *System analysis*: Assemble element properties to obtain structural properties by enforcing (1) nodal point compatibility, and (2) nodal point equilibrium $\Rightarrow$ system stiffness relationship.
4. *Boundary conditions*: Impose prescribed boundary conditions by suppressing restrained nodal displacements.
5. *Solution*: Solve the modified system stiffness relationship with respect to the structure nodal displacement components.
6. *Retracking*: Compute stresses/stress resultants from nodal point displacements.

4.2 Discretization

The structure is subdivided into a number of elements. These elements are connected to each other at a number of *nodes* that are located at the element boundaries for the simplest elements with linear interpolation of the displacement (Fig. 4.1). For higher-order elements, there are often internal nodes. For frame structures, the subdivision and type of element is basically given by the structural geometry and the need to approximate the distribution of the inertia force with sufficient accuracy. For multidimensional problems such as plates, membranes, and three-dimensional bodies there is no natural subdivision. Hence, decisions are made based on consideration of several factors, e.g., simplicity of element analysis, adjustment to external boundaries, required degree of accuracy, knowledge of the type of solution, computation time and not at least the physics of the problem.

The unknowns in this model are the nodal displacements, and possibly derivatives of these. They are designated as the *degrees-of-freedom* (DOFs) of the model.

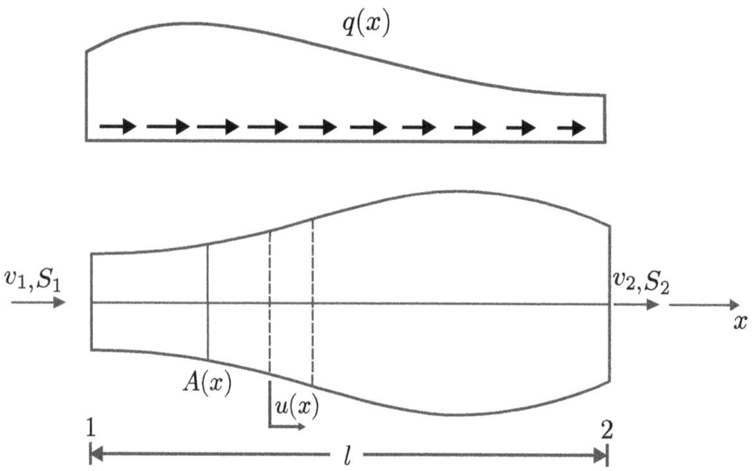

Figure 4.2. Finite element with nodal displacements v_i and forces, S_i for a bar with varying cross section area, $A(x)$.

4.3 Element Stiffness Relationship for a Bar Element

4.3.1 Matrix Method

Consider a bar element with a varying cross section $A(x)$ and distributed axial load $q(x)$, as shown in Fig. 4.2. For a bar with constant cross-sectional area $A(x) = A$, the element stiffness relation can be established based on consideration of equilibrium, compatibility, and stress-strain constitutive relationships:

$$S_1 = -S, \quad S_2 = S, \quad S = \sigma A = E\varepsilon A = EA(v_2 - v_1)/\ell. \tag{4.1}$$

That is,

$$S_1 = \frac{EA}{\ell}(v_1 - v_2),$$

$$S_2 = \frac{EA}{\ell}(-v_1 + v_2).$$

These two relationships can be written on matrix form as

$$\mathbf{S} = \begin{bmatrix} S_1 \\ S_2 \end{bmatrix} = \frac{EA}{\ell}\begin{bmatrix} 1 & -1 \\ -1 & 1 \end{bmatrix}\begin{bmatrix} v_1 \\ v_2 \end{bmatrix} = \mathbf{kv}, \tag{4.2}$$

where $\mathbf{k}$ denotes the element stiffness matrix:

$$\mathbf{k} = \frac{EA}{\ell}\begin{bmatrix} 1 & -1 \\ -1 & 1 \end{bmatrix}. \tag{4.3}$$

In this case, the element stiffness matrix could also be obtained directly by using the definition of the element stiffness relationship, $\mathbf{S} = \mathbf{kv}$, and elementary structural mechanics by the direct stiffness method. For instance, by determining the nodal forces, S_1 and S_2 for $v_1 = 1$ and $v_2 = 0$, the elements of the first column of the matrix $\mathbf{k}$ may be obtained from the nodal forces for this displacement pattern, as $k_{11} = S_1$ and $k_{21} = S_2$.

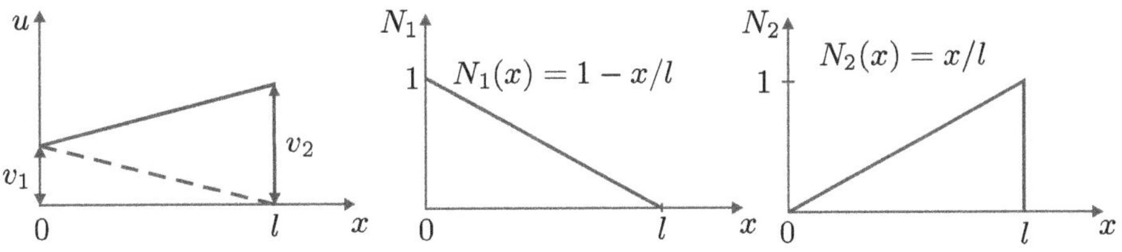

Figure 4.3. Linear interpolation over an element.

4.3.2 Finite Element Method Based on Virtual Work (Galerkin's Method) and Assumed Displacement

The element stiffness relationship is now established for a two-node truss element based on the application of the principle of virtual displacements (PVD) and an assumed displacement pattern within each element, in such a way that the displacement is continuous between elements. This approach can be generalized and used for all kinds of one-, two-, and three-dimensional elements in structural analysis.

The first step in the process is to assume a shape function for the displacements between the two known values, v_1 and v_2, at the element ends. The natural choice for interpolation between two points is a linear polynomial. The coefficients in the polynomial can be found by a fit to the two known points, but it is more convenient to use a combination of the two linear polynomials shown in Fig. 4.3. One linear function is equal to v_1 at end 1 ($x = 0$) and zero at end 2 ($x = l$), and the other linear function is zero and v_2 at the two ends. (Note that u is a displacement in the x-direction, but it is drawn as a function of x in Fig. 4.3. Hence, it is illustrated by showing the function value on a u-axis orthogonal to the x-axis.)

A displacement, $u(x)$, that varies linearly between the values v_1 and v_2 at the two ends may then be expressed as

$$u(x) = N_1(x)v_1 + N_2(x)v_2 = (1 - x/\ell)v_1 + (x/\ell)v_2. \tag{4.4}$$

This linear function may be expressed in matrix form as:

$$u(x) = \begin{bmatrix} N_1(x) & N_2(x) \end{bmatrix} \begin{bmatrix} v_1 \\ v_2 \end{bmatrix} = \mathbf{N}(x)\mathbf{v}. \tag{4.5}$$

Shape functions on this form are called Lagrange polynomials, and the procedure can be extended to higher-order, two-dimensional (plane stress), and three-dimensional elements.

According to the PVD, the assumed displacement field must be compatible with the strain field. Hence, the strain is defined as the derivative of the displacement and may thus be expressed in terms of the derivative of the shape functions:

$$\varepsilon_x = \frac{du}{dx} = (\frac{d}{dx}\mathbf{N})\mathbf{v} = \mathbf{B}\,\mathbf{v} = \begin{bmatrix} -\frac{1}{\ell} & \frac{1}{\ell} \end{bmatrix} \mathbf{v}. \tag{4.6}$$

The axial stress is obtained from Hooke's law:

$$\sigma_x = E\varepsilon_x = E\mathbf{B}\,\mathbf{v}. \tag{4.7}$$

Adopting Galerkin's method, the virtual displacement field $\tilde{u} = \tilde{u}(x)$ is conveniently interpolated with the same set of shape functions $\mathbf{N}$ as the real displacement field

$u = u(x)$. Hence, $\tilde{u} = \tilde{u}(x) = \mathbf{N}\tilde{\mathbf{v}} = \tilde{\mathbf{v}}^T\mathbf{N}^T$. Similarly, compatibility between virtual displacement and virtual strain yields $\tilde{\varepsilon}_x = \tilde{\varepsilon}_x(x) = \mathbf{B}\tilde{\mathbf{v}} = \tilde{\mathbf{v}}^T\mathbf{B}^T$.

By the PVD, the internal virtual work is expressed as the work performed by the virtual strains over stresses. The general form of this term is given by

$$\tilde{W}_i = \int_V \tilde{\boldsymbol{\varepsilon}}^T \boldsymbol{\sigma}\, dV, \tag{4.8}$$

which for a one-dimensional bar element may be simplified to

$$\tilde{W}_i = \int_\ell \tilde{\varepsilon}_x\, \sigma_x\, A(x)\, dx, \tag{4.9}$$

by considering uniaxial stress independent of the cross-section coordinate axes y and z (Remember that $\int_A dydz = A(x)$). The external virtual work done by the distributed axial load acting over the length of the bar element and the nodal point forces similarly reads

$$\tilde{W}_e^q = \int_\ell \tilde{u}\; q(x)\, dx, \tag{4.10}$$

$$\tilde{W}_e^S = \tilde{v}_1 S_1 + \tilde{v}_2 S_2 = \tilde{\mathbf{v}}^T\mathbf{S}. \tag{4.11}$$

The PVD states that the internal virtual work is equal to the external virtual work, hence,

$$\tilde{W}_i = \tilde{W}_e = \tilde{W}_e^S + \tilde{W}_e^q, \tag{4.12}$$

that is,

$$\int_\ell \tilde{\varepsilon}_x\, \sigma_x\, A(x)\, dx = \tilde{\mathbf{v}}^T\mathbf{S} + \int_\ell \tilde{u}\; q(x)dx. \tag{4.13}$$

According to the PVD, u and $\tilde{u}$, and thus $\mathbf{v}$ and $\tilde{\mathbf{v}}$, are independent fields. Remembering that $\sigma_x = E\,\mathbf{B}\mathbf{v}$, and invoking the PVD, it is obtained that

$$\tilde{\mathbf{v}}^T \int_\ell \mathbf{B}^T E\, A(x)\, \mathbf{B}\, dx\, \mathbf{v} = \tilde{\mathbf{v}}^T\mathbf{S} + \tilde{\mathbf{v}}^T \int_\ell \mathbf{N}^T q(x)\, dx. \tag{4.14}$$

Shifting all terms to the left-hand side and extracting the common factor $\tilde{\mathbf{v}}^T$, we get

$$\tilde{\mathbf{v}}^T\Big(\int_\ell \mathbf{B}^T E\, A(x)\, \mathbf{B}\, dx\, \mathbf{v} - \mathbf{S} - \int_\ell \mathbf{N}^T q(x)\, dx\Big) = 0. \tag{4.15}$$

The PVD also states that the real element displacement and the virtual displacement is independent and should be valid for an arbitrary virtual displacement. The non-trivial solution to the preceding equation is obtained by requiring that the expression within the parenthesis must be zero, that is,

$$\int_\ell \mathbf{B}^T E\, A(x)\, \mathbf{B}\, dx\, \mathbf{v} - \mathbf{S} - \int_\ell \mathbf{N}^T q(x)\, dx = 0. \tag{4.16}$$

By re-ordering the terms, we get the following element stiffness relationship:

$$\mathbf{S} = \mathbf{k}\,\mathbf{v} + \mathbf{S}^0, \tag{4.17}$$

where

$$\mathbf{k} = \int_{\ell} \mathbf{B}^T E\, A(x)\, \mathbf{B}\, dx, \tag{4.18}$$

$$\mathbf{S}^0 = -\int_{\ell} \mathbf{N}^T q(x)\, dx. \tag{4.19}$$

$\mathbf{S}^0$ is the consistent load vector because, contrary to a lumping technique, it is derived by an energy principle consistent with the element stiffness matrix $\mathbf{k}$. It may also be considered to be a fixed end force because it will define the values for the nodal point load vector, $\mathbf{S}$, when the element nodes are fixed ($\mathbf{v} = 0$). The load vector is further commented on in Section 4.3.4.

In our case we have that $\mathbf{B} = (1/\ell)\begin{bmatrix} -1 & 1 \end{bmatrix}$, and the element stiffness matrix is given as

$$\mathbf{k} = \int_0^{\ell} \mathbf{B}^T E\, A(x) \mathbf{B}\, dx = \frac{E\bar{A}}{\ell}\begin{bmatrix} 1 & -1 \\ -1 & 1 \end{bmatrix}, \tag{4.20}$$

where

$$\bar{A} = \frac{1}{\ell}\int_{\ell} A(x)\, dx, \tag{4.21}$$

which is the average area of the cross section of the truss element.

Alternatively, the cross section area could be approximated as a constant by using the area at the midpoint of the element.

The solution implies that assuming linear interpolation functions for a truss element with varying cross-sectional area is equivalent to using an element with a uniform cross section and the same total volume as the true bar over the element length.

4.3.3 Further Considerations on the Assumed Displacement for the Truss Element

Both the displacement u and the variation of the cross-sectional area may be described by a more refined polynomial.

Figure 4.4 shows the FE solution of the displacement and the stress obtained with one single element with a linear and quadratic displacement approximation over the element. Significant improvement of the accuracy is obtained by choosing a higher-order polynomial.

Note that the exact solution of the displacement for this problem is a cubic polynomial. This exact solution can also be obtained by the FEM if an element with 4 DOFs that can represent a cubic polynomial is applied.

Hence, there are two ways to improve the accuracy of a finite element method:

- Increasing the number of elements, and thus, increasing the number of DOFs.
- Increasing the order of the polynomial used to approximate the displacement within each element.

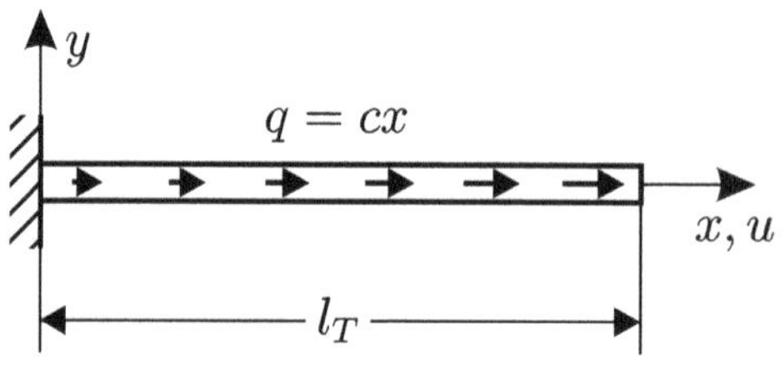

(a) Single element model

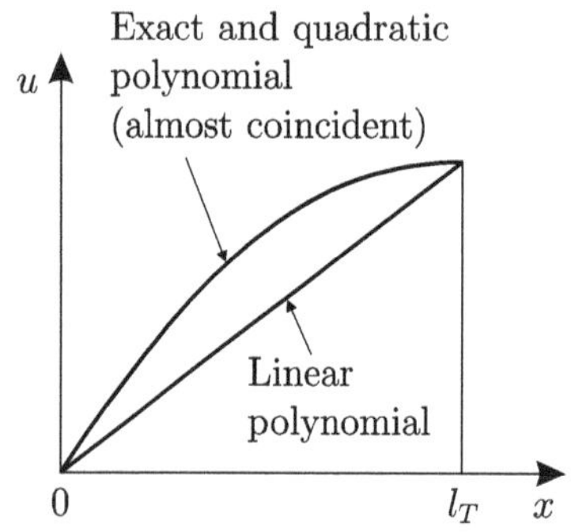

(b) Exact and approximate axial displacement u

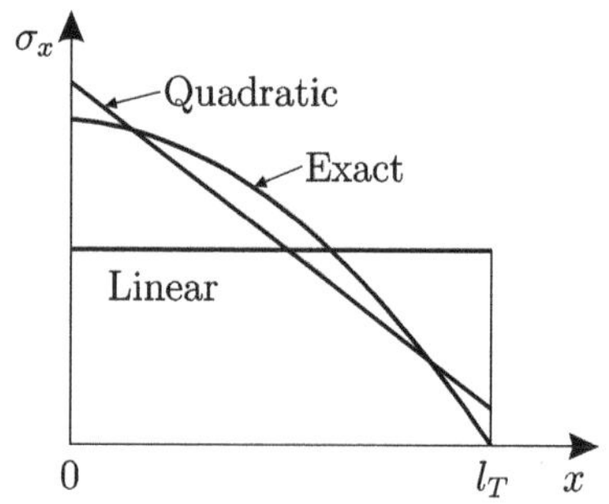

(c) Exact and approximate axial stresses based on different approximations of the axial displacement u. $\sigma_x = Eu_{,x}$

Figure 4.4. Single element model of a uniform bar under linearly varying distributed axial load of intensity $q = cx$, where c is a constant.

Usually, the accuracy is more easily (with least computer effort) increased by increasing the order of the polynomial used to represent the displacement within the element, rather than using more elements (with the same total number of DOFs).

4.3.4 Load Vector

The load vector $\mathbf{S}^0$ for an element subjected to a distributed axial loading, $q = q(x)$, is given by $\mathbf{S}^0 = -\int_\ell \mathbf{N}^T q(x)\,dx$. These loads are called *work-equivalent* loads for the following reason: the work done by nodal loads $\mathbf{S}$ in going through nodal displacements $\mathbf{v}$ is equal to work done by the distributed loading q in going through the displacement field associated with the element shape functions N_i. Such loads are called *consistent* because they are based on the same shape functions as used to calculate the element stiffness matrix. These loads are statically equivalent to the original loading because they correspond to the same resultant force with respect to an arbitrary chosen point.

An *inconsistent* load vector $\mathbf{S}^0$ is obtained as a conversion of distributed loads by maintaining static equivalence, but not work equivalence. Often this is done by *lumping* of the distributed loads.

4.3.5 Assumed Displacement by Generalized Coordinates

The displacement u in an element is so far assumed to be expressed in terms of the nodal displacements $\mathbf{v}$ by $u = \mathbf{N}\,\mathbf{v}$. The set of shape or interpolation functions, $N_i(x)$ in the $\mathbf{N}$-vector are Lagrange interpolation polynomials. This means that each $N_i(x)$ is 1 in node i and zero in all other nodes.

Alternatively, the displacement may be assumed on the form

$$u(x) = N_{q1}(x)\,q_1 + N_{q2}(x)\,q_2 + \cdots + N_{qn}(x)\,q_n = \mathbf{N}_q(x)\,\mathbf{q}. \tag{4.22}$$

The number n of generalized coordinates q_i is assumed to be equal to the number of nodal DOFs in $\mathbf{v}$. The function $N_{qi}(x)$ is usually chosen to be a polynomial (e.g., $1, x, x^2, \ldots$).

The displacement still needs to be expressed by the physical, nodal displacements, v_i. This is because the nodal displacements are used to impose the necessary compatibility between elements and the geometric boundary conditions. Hence, the vector $\mathbf{q}$ of generalized displacements needs to be related to the vector of nodal displacements $\mathbf{v}$ within a given element. This relationship can be obtained by using the definition of the nodal displacement $\mathbf{v} : v_i = u(x_i)$. Hence, the displacement in node i is equal to the value of $u(x)$ with coordinate $x = x_i$. The following relation may then be established:

$$\left\{\begin{array}{l} v_1 = u(x_1) = \mathbf{N}_q(x_1)\,\mathbf{q} \\ v_2 = u(x_2) = \mathbf{N}_q(x_2)\,\mathbf{q} \\ \vdots \\ v_n = u(x_n) = \mathbf{N}_q(x_n)\,\mathbf{q} \end{array}\right\} \Rightarrow \mathbf{v} = \mathbf{A}\mathbf{q} \Leftrightarrow \mathbf{q} = \mathbf{A}^{-1}\mathbf{v}. \tag{4.23}$$

This implies that $u(x)$ can be written as:

$$u(x) = \mathbf{N}_q\,\mathbf{q} = \mathbf{N}_q\mathbf{A}^{-1}\mathbf{v} = \mathbf{N}\,\mathbf{v}. \tag{4.24}$$

This means that the set of Lagrange polynomials $\mathbf{N}$ may be derived from $\mathbf{N}_q$ as

$$\mathbf{N} = \mathbf{N}_q\,\mathbf{A}^{-1}. \tag{4.25}$$

Obviously, for the bar elements treated so far, the Lagrange interpolation polynomials are already known. The preceding procedure, based on generalized coordinates, is advantageous for more complex elements, as shown subsequently, for beam elements.

Now, with u expressed (as before) in terms of interpolation polynomials the development of the stiffness relationship could proceed as before. However, it is also interesting to derive the element stiffness relationship in an alternative way, in terms of generalized coordinates.

First, the strain-displacement matrix $\mathbf{B}$ is established as

$$\mathbf{B} = \frac{d}{dx}\mathbf{N} = \frac{d}{dx}\mathbf{N}_q\,\mathbf{A}^{-1} = \mathbf{B}_q\,\mathbf{A}^{-1}. \tag{4.26}$$

Then,

$$\mathbf{k} = \int_{\ell} \mathbf{B}^T\,E\,A(x)\mathbf{B}\,dx = \mathbf{A}^{-T}\int_{\ell} \mathbf{B}_q^T\,E\,A(x)\mathbf{B}_q\,dx\,\mathbf{A}^{-1} = \mathbf{A}^{-T}\mathbf{k}_q\mathbf{A}^{-1}. \tag{4.27}$$

This means that the generalized stiffness matrix $\mathbf{k}_q$ could first be determined, and afterward transformed, according to Eq. (4.27). In some cases, this may be computationally more efficient than the direct approach based on interpolation polynomials. This is because $\mathbf{k}_q$ is simple to calculate, with many zero elements. To achieve an efficient formulation, $\mathbf{A}$ needs to be invertible analytically.

To illustrate the use of generalized coordinates, consider the displacement field for a two-noded straight bar element: $u(x) = q_1 + xq_2 = [1 \;\; x]\mathbf{q} = \mathbf{N}_q\mathbf{q}$. Then,

$$\left.\begin{aligned} v_1 &= 1 \cdot q_1 + 0 \cdot q_2 \\ v_2 &= 1 \cdot q_1 + \ell \cdot q_2 \end{aligned}\right\} \Rightarrow \mathbf{v} = \mathbf{A}\mathbf{q}. \tag{4.28}$$

This means that,

$$\mathbf{A} = \begin{bmatrix} 1 & 0 \\ 1 & \ell \end{bmatrix} \Rightarrow \mathbf{A}^{-1} = \begin{bmatrix} 1 & 0 \\ -\dfrac{1}{\ell} & \dfrac{1}{\ell} \end{bmatrix} \tag{4.29}$$

and

$$\mathbf{N}_q\mathbf{A}^{-1} = \begin{bmatrix} 1 & x \end{bmatrix} \begin{bmatrix} 1 & 0 \\ -\dfrac{1}{\ell} & \dfrac{1}{\ell} \end{bmatrix} = \begin{bmatrix} 1 - \dfrac{x}{\ell}, & \dfrac{x}{\ell} \end{bmatrix} = [N_1(x), N_2(x)], \tag{4.30}$$

which coincide with the linear Lagrange interpolation polynomials obtained previously by so-called direct interpolation.

Moreover,

$$\mathbf{B}_q = \begin{bmatrix} 0 & 1 \end{bmatrix}.$$

The stiffness matrix may then be obtained by Eq. (4.27). $\mathbf{k}_q$ may first be calculated and then transformed into $\mathbf{k}$ according to Eq. (4.27). In this connection, it is noted that only k_{q22} in the matrix $\mathbf{k}_q$ is different from zero.

4.4 Element Stiffness Relationship for a Beam with Uniform Lateral Load

4.4.1 Finite Element Formulation Based on Virtual Work and Assumed Displacement

Equation (3.55) may be written more concisely by using the definitions of w, $w_{,x}$, M, and Q and the nodal displacements v_i and stress resultants S_i in Fig. 3.11; i.e., in terms of Eq. (3.58). The vector of nodal forces, $\mathbf{S}$, and displacements, $\mathbf{v}$, now refer to the element DOFs no. 2, 3, 5, and 6 in Section 3.3.

By introducing a pattern of transverse (or lateral) displacements w and virtual displacements $\tilde{w}$ with corresponding nodal displacements $\mathbf{v}$ and $\tilde{\mathbf{v}}$, respectively, in the virtual work equation (Eq. (3.58)), the desired stiffness relationships are obtained.

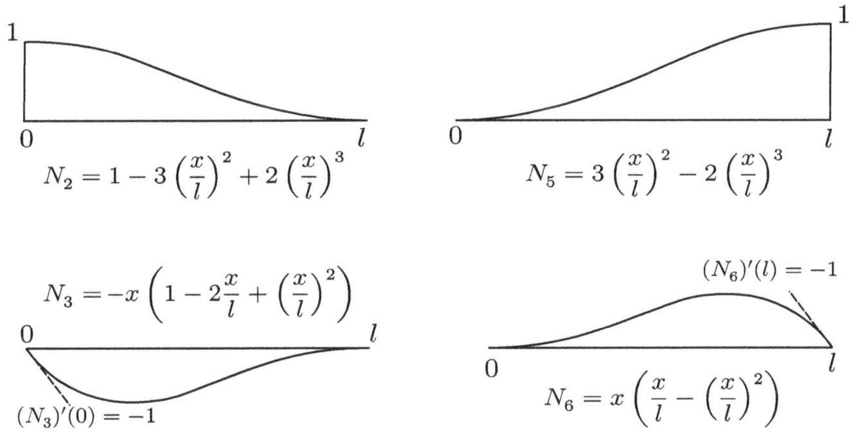

Figure 4.5. Interpolation functions for a beam based on a cubic displacement. Note that the interpolation polynomial N_i corresponds to the nodal parameters v_i $(i = 2, 3, 5, 6)$.

A reasonable assumption is to approximate the lateral displacement w by a complete cubic polynomial with four parameters (generalized coordinates), that is,

$$w = q_1 + x\,q_2 + x^2\,q_3 + x^3\,q_4$$
$$= \begin{bmatrix} 1 & x & x^2 & x^3 \end{bmatrix} \begin{bmatrix} q_1 \\ q_2 \\ q_3 \\ q_4 \end{bmatrix} = \mathbf{N}_q\,\mathbf{q}. \tag{4.31}$$

As for the bar element, $\mathbf{q}$ is a set of generalized displacements, contrary to the nodal displacements $\mathbf{v}$.

The relation between the nodal displacements $\mathbf{v}$ (defined in Fig. 3.11) and $\mathbf{q}$ can be established by

$$\mathbf{v} = \begin{bmatrix} v_2 \\ v_3 \\ v_5 \\ v_6 \end{bmatrix} = \begin{bmatrix} w(0) \\ -w,_x(0) \\ w(\ell) \\ -w,_x(\ell) \end{bmatrix} = \begin{bmatrix} 1 & 0 & 0 & 0 \\ 0 & -1 & 0 & 0 \\ 1 & \ell & \ell^2 & \ell^3 \\ 0 & -1 & -2\ell & -3\ell^2 \end{bmatrix} \begin{bmatrix} q_1 \\ q_2 \\ q_3 \\ q_4 \end{bmatrix} = \mathbf{Aq}. \tag{4.32}$$

This $\mathbf{A}$-matrix is non-singular and can be inverted as follows:

$$\mathbf{A}^{-1} = \frac{1}{\ell^3} \begin{bmatrix} \ell^3 & 0 & 0 & 0 \\ 0 & -\ell^3 & 0 & 0 \\ -3\ell & 2\ell^2 & 3\ell & \ell^2 \\ 2 & -\ell & -2 & -\ell \end{bmatrix}. \tag{4.33}$$

To check the inversion, it is verified that $\mathbf{AA}^{-1} = \mathbf{I}$.

Now, the displacement field w may be written as

$$w = \mathbf{N}_q\mathbf{q} = \mathbf{N}_q\mathbf{A}^{-1}\mathbf{v} = \mathbf{Nv}. \tag{4.34}$$

The resulting interpolation polynomials, $\mathbf{N}$, are shown in Fig. 4.5.

The approximate curvature may then be calculated by

$$\kappa = \frac{d^2w}{dx^2} = \frac{d^2}{dx^2}[\mathbf{N}_q\mathbf{q}] = \begin{bmatrix} 0 & 0 & 2 & 6x \end{bmatrix}\mathbf{q} = \mathbf{B}_q\mathbf{q} = \mathbf{B}_q\mathbf{A}^{-1}\mathbf{v}. \tag{4.35}$$

The virtual curvature is denoted $\tilde{\kappa} = \mathbf{B}_q\tilde{\mathbf{q}} = (\mathbf{B}_q\tilde{\mathbf{q}})^T = \tilde{\mathbf{q}}^T\mathbf{B}_q^T = \tilde{\mathbf{v}}^T(\mathbf{A}^{-1})^T\mathbf{B}_q^T$.

Based on the PVD,

$$\int_\ell \tilde{\mathbf{v}}^T\mathbf{A}^{-T}\mathbf{B}_q^T(EI)\mathbf{B}_q\mathbf{A}^{-1}\mathbf{v}\,dx = \tilde{\mathbf{v}}^T\mathbf{S} + \int_\ell \tilde{\mathbf{v}}^T\mathbf{N}^T q(x)\,dx, \tag{4.36}$$

where the superscript $-T$ means inversion and transposing of the matrix; $q(x)$ is the lateral load on the beam (note that $q(x)$ has nothing to do with the generalized coordinates $\mathbf{q}$).

For this relationship to apply for all possible $\tilde{\mathbf{v}}$,

$$\left(\int_\ell \mathbf{A}^{-T}\mathbf{B}_q^T\,(EI)\,\mathbf{B}_q\mathbf{A}^{-1}\,dx\right)\mathbf{v} = \mathbf{S} + \int_\ell \mathbf{N}^T q(x)\,dx \tag{4.37}$$

or

$$\mathbf{S} = \mathbf{k}\mathbf{v} + \mathbf{S}^0, \tag{4.38}$$

where

$$\mathbf{k} = \int_\ell \mathbf{A}^{-T}\mathbf{B}_q^T(EI)\mathbf{B}_q\mathbf{A}^{-1}\,dx = \mathbf{A}^{-T}\mathbf{k}_q\mathbf{A}^{-1}, \tag{4.39}$$

$\mathbf{k}_q$ denotes the generalized stiffness matrix, and $\mathbf{S}^0$ denotes the consistent load vector

$$\mathbf{S}^0 = -\int_\ell \mathbf{N}^T q(x)\,dx. \tag{4.40}$$

For the beam element with 4 DOFs, $\mathbf{k}_q$ is found to be

$$\mathbf{k}_q = EI\int_\ell \mathbf{B}_q^T\,\mathbf{B}_q\,dx = EI\begin{bmatrix} 0 & 0 & 0 & 0 \\ 0 & 0 & 0 & 0 \\ 0 & 0 & 4\ell & 6\ell^2 \\ 0 & 0 & 6\ell^2 & 12\ell^3 \end{bmatrix} \tag{4.41}$$

and

$$\mathbf{k} = \mathbf{A}^{-T}\mathbf{k}_q\mathbf{A}^{-1} = \frac{EI}{\ell^3}\begin{bmatrix} 12 & -6\ell & -12 & -6\ell \\ -6\ell & 4\ell^2 & 6\ell & 2\ell^2 \\ -12 & 6\ell & 12 & 6\ell \\ -6\ell & 2\ell^2 & 6\ell & 4\ell^2 \end{bmatrix}. \tag{4.42}$$

For instance, by using beam theory, the four entries in the first column of the element stiffness matrix $\mathbf{k}$ are obtained as the forces S_2, S_3, S_5, and S_6 corresponding to a displacement $v_2 = 1$, while the other $v_i = 0\ (i \neq 2)$. Alternatively, the flexibility relation $\mathbf{v} = \mathbf{f}\,\mathbf{S}$ may be first established and then $\mathbf{k} = \mathbf{f}^{-1}$. The first column of $\mathbf{f}$ is obtained as the displacement v_2, v_3, v_5, and v_6 corresponding to $S_2 = 1$ and the other $S_i = 0\ (i \neq 2)$.

The latter approach is convenient when shear deformation is to be accounted for. Then exactly the same stiffness matrix as obtained for slender beams based on the classical beam theory in *matrix methods* results.

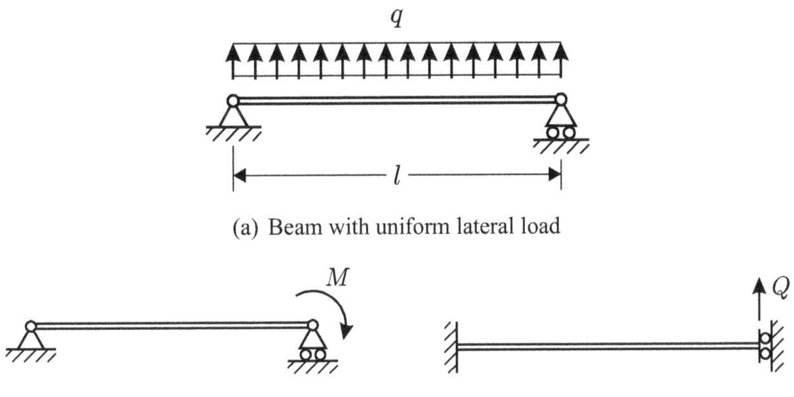

(a) Beam with uniform lateral load

(b) Beam with end moment (c) Beam with shear force

Figure 4.6. Elementary beam problems modeled by a single element.

The consistent load vector for a uniform load $q(x) = q$ in the positive z-direction is

$$\mathbf{S}^0 = -\int_\ell \mathbf{N}^T q(x)\,dx = \frac{q\ell}{12}\begin{bmatrix} -6 \\ +\ell \\ -6 \\ -\ell \end{bmatrix} \tag{4.43}$$

This consistent load vector is also known as the exact solution of the fixed-end support condition.

4.4.2 Application of Beam Element with Cubic Displacement Function

Consider first a single element model of a simply supported beam with uniform lateral load q, see Fig. 4.6. The exact solution for this problem is readily obtained from the differential equation for the beam defined in Eq. (3.49) and the corresponding boundary conditions. The exact displacement is a fourth-degree polynomial, and the exact bending moment is a quadratic polynomial $M_x = (q\ell/2)\,x(1 - x/\ell)$ on the interval $(0, \ell)$.

By using the element stiffness relationship $\mathbf{S} = \mathbf{kv}$ as the system relation with the stiffness matrix given by Eq. (4.42) (i.e., applying a single element model), and with the consistent load vector from Eq. (4.43) and then by introducing the boundary conditions $v_2 = v_5 = 0$, it is found that the rotations are $v_3 = -v_6 = -q\ell^3/(24EI)$, where ℓ is the length of the beam.

The bending moment in an arbitrary point x in the beam is then

$$\begin{aligned} M_x &= -EIw_{,xx}(x) = -EI\frac{d^2}{dx^2}\mathbf{Nv} \\ &= -EI\Big(\frac{4}{\ell} - \frac{6x}{\ell^2}\Big)v_3 - EI\Big(\frac{2}{\ell} - \frac{6x}{\ell^2}\Big)v_6 = \frac{q\ell^2}{12}. \end{aligned} \tag{4.44}$$

Hence, this finite element approximation yields a constant moment in the beam. The exact moment is zero at the two ends and $q\ell^2/8$ at the midpoint. The error in the maximum moment is hence 33%. This is the normal way to calculate internal forces in the structure in an FE analysis.

It is important to note that the analysis of a frame/truss by the matrix method is based on the following procedure:

1. Establish a particular solution ($\mathbf{r} = \mathbf{0}$): Analysis of moments and shear forces for fixed ends (no displacement of nodes)
2. Compute a complementary solution ($\mathbf{r}$): Analysis of moments and shear forces when the fixed end forces ($-\mathbf{S}^0$) are released
3. Establish the total solution by superimposing the results of steps 1 and 2

The approach commonly used in the FEM corresponds only to step 2.

If the matrix method had been used and the moments corresponding to step 1 had been determined (exactly) as done in the matrix method, and then superimposed on step 2, the exact moment distribution would have been achieved. This is because step 2 is actually solved exactly. To see that, consider the beam (element) with an end moment M in Fig. 4.6b. The exact solution of the differential equation, Eq. (3.49) with $q = 0$, corresponds to a third-degree polynomial that can be exactly represented by the assumed displacement pattern. Also, the case in Fig. 4.6c is solved exactly by a cubic polynomial. For this reason, the stiffness matrix, Eq. (4.42), is exactly calculated by the finite element with a cubic displacement. In the previous example the load vector was consistently calculated by Eq. (4.43).

If the load vector $\mathbf{S}^0$ is determined by "lumping" the transverse load q to the nodes, the load vector components become $S_2^0 = S_5^0 = -ql/2$, $S_3^0 = S_6^0 = 0$. After introducing the boundary conditions, the load vector will be zero and no displacement or moment result. This example illustrates the importance of using consistent load vectors, especially for an FE model with a single element or just a few elements.

If the finite element model is refined by employing two beam elements for the beam in Fig. 4.6a, the maximum (i.e., midspan) bending moment is found to be $M = 7q\ell^2/48$, implying an error of 16.7%. The moment at the ends is also reduced from $q\ell^2/12$ (for the FE model with one single element) to $q\ell^2/48$ (for the two element model), i.e., much closer to the exact value of zero.

Improved approximations can also be achieved by using higher-order interpolation polynomials for the element. For instance, by using an element with three nodes with parameters $(w, w_{,x})$, a fifth-degree polynomial can be described over the element.

4.4.3 Final Remarks

The bar and beam elements are special in the sense that

- exact solutions may be achieved for certain cases,
- the elements are connected only through nodes.

Hence, nodal forces get a physical meaning and global compatibility and equilibrium can be "exactly" fulfilled. In general, multidimensional elements (for plane stress, plate bending, etc.) do not share these characteristics.

For such problems, compatibility and equilibrium are fulfilled in an integrated sense. Nodal forces satisfy nodal equilibrium and are of the same accuracy as displacement. However, normally, stresses are obtained from strains that are derivatives of the displacements. The error in the displacement will be magnified by differentiating the displacements. Stresses and strains are therefore subjected to larger error than displacements in FEs based on assumed displacements. This fact is illustrated in previous sections of this chapter by FE formulations for one-dimensional problems.

Based on an assumed displacement over an element and the application of the principle of virtual displacement for a bar and beam with axial and lateral loading, respectively, the stiffness relationships of the type $\mathbf{S} = \mathbf{kv} + \mathbf{S}^0$ were derived.

The displacement patterns are expressed either by generalized coordinates or directly by interpolation polynomials. In the former case, the generalized stiffness matrix, $\mathbf{k}_q$ is first calculated and then transformed into $\mathbf{k}$.

Results obtained by using a set of linear interpolation polynomials $\mathbf{N}$ and different numbers of elements over the bar were shown. Use of more refined polynomials to approximate the displacement within the element is envisaged.

Salient features of a beam element based on cubic lateral displacement were studied.

Stiffness matrices previously derived for the bar with axial force and beam elements for bending are based on assumed linear and cubic displacement functions, respectively, and the PVD. When linear axial and cubic lateral displacement are assumed, respectively, the formulation gives exactly the same stiffness relationships as the approach applied in matrix methods based on the structural engineering principles. However, the finite element approach makes it possible to establish other formulations of bars and beams based on higher-order polynomials. Such elements are useful in combination with refined two-dimensional plane stress, plate bending, and shell elements. Moreover, when the FEM is generalized to two- and three-dimensional problems, it is not possible to obtain exact solutions. However, by using small elements, the exact solutions can be well approximated.

4.5 Stiffness Relationship for Bar Element - Beam with Axial Force

4.5.1 General

So far, the analysis is based on the assumptions that

- displacements are small, and
- the material is linear and elastic.

When the displacements are small, the equilibrium equations can be established with reference to the initial configuration. Moreover, this implies that the strains are linear functions of displacement gradients (derivatives). The linear elastic stress-strain relationship corresponds to Hooke's law. When the ultimate strength of structures is to be calculated, the assumptions about small displacements and linear material need to be modified. Often, this involves pursuing buckling and collapse. Then it may be necessary to account for the change of geometry when establishing the equilibrium equations and calculating the strains from displacements. A first step in this analysis is to consider the second-order effect of a preexisting axial force on the lateral stiffness of beams.

The stiffness relationship can be obtained by separately considering axial and lateral displacements.

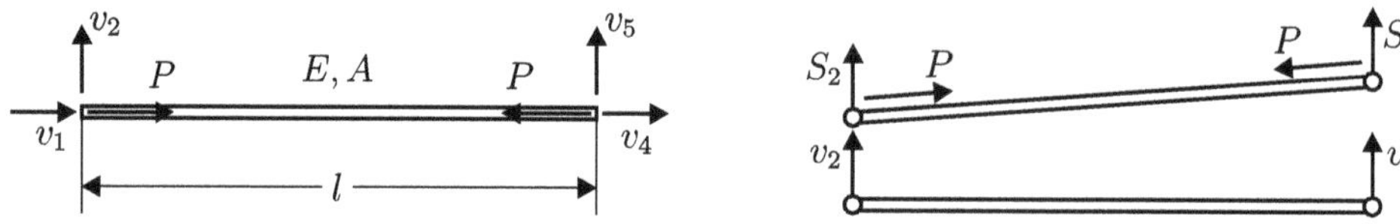

(a) Degrees of freedom (b) Bar element with lateral displacement

Figure 4.7. Bar element with 4 DOFs.

4.5.2 Bar Under Axial Loads

In Section 4.4, the stiffness relationship for a bar was derived by means of the FE procedure, based on an assumed interpolation of displacement and the PVD. Moreover, small displacements were assumed. Hence, equilibrium equations were established for the initial geometry.

This section deals with element stiffness relationships $\mathbf{S} = \mathbf{kv}$, which account for the effect of change of geometry. When such element relationships have been established, the relationship for an entire structural system can be established by the direct stiffness method:

$$\mathbf{S}^i = \mathbf{k}^i \mathbf{v}^i \Rightarrow \mathbf{R} = \mathbf{Kr}. \tag{4.45}$$

In this section, the element stiffness relationship for a bar element is established with reference to the local displacement DOFs and corresponding nodal forces, as shown in Fig. 4.7. It is assumed that the bar is initially subjected to an axial force P, which is due to other loads than $\mathbf{S}$. Rather than using a traditional FE procedure, a simple application of equilibrium, compatibility, and stress-strain relationships are used in this case.

For axial displacements, the following relations are then obtained: $S_1 = -S$, $S_4 = S$, $S = EA\varepsilon = EA(v_4 - v_1) / \ell$, or

$$S_1 = -S_4 = \frac{EA}{\ell}(v_1 - v_4). \tag{4.46}$$

Lateral displacement may be considered without reference to v_1 and v_4. Consider then the element in Fig. 4.7b with constant axial force P (positive in compression), and assume lateral displacements v_2 and v_5.

Equilibrium in the lateral direction implies $S_2 + S_5 = 0$, or $S_2 = -S_5$. Moment equilibrium about node 1 of the displaced bar with a compression force, P, implies $S_5 \ell = -P(v_5 - v_2)$ or $S_5 = \frac{P}{\ell}(v_2 - v_5)$, that is,

$$S_2 = -S_5 = \frac{P}{\ell}(-v_2 + v_5). \tag{4.47}$$

By combining Eqs. (4.46) and (4.47), the following relations are obtained:

$$\begin{bmatrix} S_1 \\ S_2 \\ S_4 \\ S_5 \end{bmatrix} = \left(\frac{EA}{\ell} \begin{bmatrix} 1 & 0 & -1 & 0 \\ 0 & 0 & 0 & 0 \\ -1 & 0 & 1 & 0 \\ 0 & 0 & 0 & 0 \end{bmatrix} - \frac{P}{\ell} \begin{bmatrix} 0 & 0 & 0 & 0 \\ 0 & 1 & 0 & -1 \\ 0 & 0 & 0 & 0 \\ 0 & -1 & 0 & 1 \end{bmatrix} \right) \begin{bmatrix} v_1 \\ v_2 \\ v_4 \\ v_5 \end{bmatrix} \tag{4.48}$$

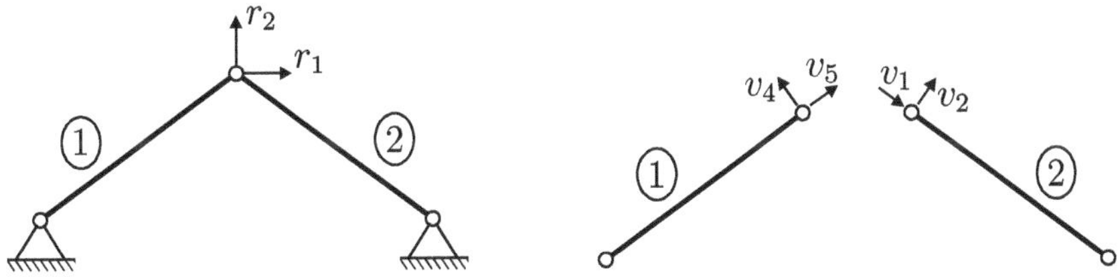

(a) Global DOFs (b) Local DOFs for each element

Figure 4.8. Two-bar system.

or

$$\mathbf{S} = (\mathbf{k} - \mathbf{k}_\sigma)\,\mathbf{v}. \tag{4.49}$$

If the axial force P, which affects $\mathbf{k}_\sigma$, is taken to be positive when it is tensile, the minus sign before $\mathbf{k}_\sigma$ changes to plus. The nodal force vector $\mathbf{S}$ and displacement vector $\mathbf{v}$ refer to the DOFs no. 1, 2, 4, and 5.

This stiffness matrix refers to a local coordinate system and needs to be transformed into a global coordinate system when it is used to assemble the global stiffness matrix, e.g., as for the two-bar system in Fig. 4.8 (see Section 4.7).

4.5.3 Beam Under Axial and Lateral Loads

In Section 4.4, the stiffness relationship was established for a beam with lateral loads by using the PVD and assumed polynomial displacements. In this section the stiffness relationship for a beam with a preexisting axial force is established based on the same principles.

The virtual work equation, when the effect of the axial load P is taken into account, is expressed as:

$$\int_\ell \tilde{\kappa}(EI)\kappa\, dx - \int_\ell P\, w_{,x}\, \tilde{w}_{,x}\, dx = \tilde{v}_2 S_2 + \tilde{v}_3 S_3 + \tilde{v}_5 S_5 + \tilde{v}_6 S_6 + \int_\ell q(x)\, \tilde{w}\, dx$$

$$= \tilde{\mathbf{v}}^T \mathbf{S} + \int_\ell q\tilde{w} dx. \tag{4.50}$$

The interpolation polynomials for a beam with 4 DOFs were established in Section 4.4 and are presented in Fig. 4.5.

The stiffness matrix may be established using generalized coordinates or interpolation polynomials. Because

$$\begin{aligned} w_{,x} &= \frac{dw}{dx} = \frac{d}{dx}(\mathbf{N}\mathbf{v}) = \mathbf{N}_{,x}\,\mathbf{v} = \mathbf{B}_\sigma \mathbf{v} \\ \kappa &= w_{,xx} = \mathbf{N}_{,xx}\,\mathbf{v} = \mathbf{B}\mathbf{v} \end{aligned} \tag{4.51}$$

and

$$\begin{aligned} \tilde{w}_{,x} &= \mathbf{B}_\sigma \tilde{\mathbf{v}} \\ \tilde{\kappa} &= \mathbf{B}\tilde{\mathbf{v}}, \end{aligned} \tag{4.52}$$

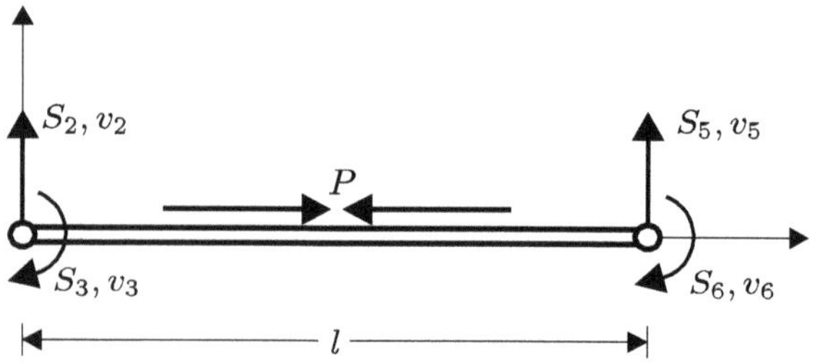

Figure 4.9. Nodal displacements and forces for a beam element. Note that the variables with index 1 and 4 correspond to axial displacements and forces, which are uncoupled from the variables describing the bending behavior.

it is found that the stiffness matrix for a 4 DOFs beam (see Figs. 3.11 and 4.9) then becomes

$$\mathbf{k}_L = EI \int_\ell \mathbf{B}^T\mathbf{B}\,dx - P \int_\ell \mathbf{B}_\sigma^T\mathbf{B}_\sigma\,dx = \mathbf{k} - \mathbf{k}_\sigma, \tag{4.53}$$

where

$$\mathbf{B} = \frac{1}{\ell}\begin{bmatrix} \frac{12x}{\ell^2} - \frac{6}{\ell} \\ -\frac{6x}{\ell} + 4 \\ -\frac{12x}{\ell^2} + \frac{6}{\ell} \\ -\frac{6x}{\ell} + 2 \end{bmatrix}, \quad \mathbf{B}_\sigma = \begin{bmatrix} \frac{6}{\ell}\left(\frac{x}{\ell}\right)^2 - \frac{6x}{\ell^2} \\ -3\left(\frac{x}{\ell}\right)^2 + 4\frac{x}{\ell} - 1 \\ -\frac{6}{\ell}\left(\frac{x}{\ell}\right)^2 + \frac{6x}{\ell^2} \\ -3\left(\frac{x}{\ell}\right)^2 + 2\frac{x}{\ell} \end{bmatrix}. \tag{4.54}$$

The first term in the expression (4.53) represents the well-known *elastic (bending) stiffness matrix* for a beam, as established in Section 4.4 and for a constant EI is given by Eq. (4.42).

The second term represents a correction to $\mathbf{k}$ and is the product of the axial load and the *geometric stiffness matrix* (initial stress matrix) $\mathbf{k}_\sigma$, which reads

$$\mathbf{k}_\sigma = P \int_\ell \mathbf{B}_\sigma^T\mathbf{B}_\sigma\,dx = \frac{P}{30\ell}\begin{bmatrix} 36 & -3\ell & -36 & -3\ell \\ -3\ell & 4\ell^2 & 3\ell & -\ell^2 \\ -36 & 3\ell & 36 & 3\ell \\ -3\ell & -\ell^2 & 3\ell & 4\ell^2 \end{bmatrix} \tag{4.55}$$

(Sometimes another notation is applied by defining $\mathbf{k}_\sigma$ by Eq. (4.55) with $P = 1$.) The stiffness matrix of Eq. (4.53), together with Eqs. (4.54) and (4.55), is approximate because the interpolation functions do not satisfy the differential equation. However, accurate results may be obtained by using several finite elements.

Equation (4.53) refers to 4 DOFs defined in Fig. 4.5. To have a complete formulation for a plane beam, the axial nodal displacements (v_1, v_4) and nodal forces (S_1, S_4) should be included. This corresponds to the relations between (v_1, v_4) and (S_1, S_4) in Eq. (4.48).

The geometric stiffness associated with v_2 and v_5 do not contribute due to the constraint provided by the rotational degrees of freedom for the plane 6 DOFs beam (Fig. 4.10). By using the stiffness relationship for the bar (or beam with axial force) (Eq. (4.48)) and that for a beam with lateral forces/bending moments (Eqs. (4.53),

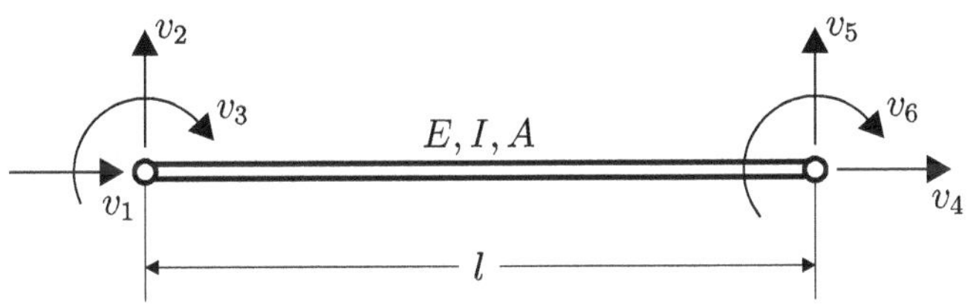

Figure 4.10. Six DOFs beam element. Positive directions of displacement variables are shown.

(4.42), (4.55)), the following stiffness matrices for a 6 DOFs beam with axial force is obtained:

$$\mathbf{k} = \begin{bmatrix} \frac{EA}{\ell} & 0 & 0 & -\frac{EA}{\ell} & 0 & 0 \\ 0 & \frac{12EI}{\ell^3} & -\frac{6EI}{\ell^2} & 0 & -\frac{12EI}{\ell^3} & -\frac{6EI}{\ell^2} \\ 0 & -\frac{6EI}{\ell^2} & \frac{4EI}{\ell} & 0 & \frac{6EI}{\ell^2} & \frac{2EI}{\ell} \\ -\frac{EA}{\ell} & 0 & 0 & \frac{EA}{\ell} & 0 & 0 \\ 0 & -\frac{12EI}{\ell^3} & \frac{6EI}{\ell^2} & 0 & \frac{12EI}{\ell^3} & \frac{6EI}{\ell^2} \\ 0 & -\frac{6EI}{\ell^2} & \frac{2EI}{\ell} & 0 & \frac{6EI}{\ell^2} & \frac{4EI}{\ell} \end{bmatrix} \tag{4.56}$$

$$\mathbf{k}_\sigma = \frac{P}{30\ell} \begin{bmatrix} 0 & 0 & 0 & 0 & 0 & 0 \\ 0 & 36 & -3\ell & 0 & -36 & -3\ell \\ 0 & -3\ell & 4\ell^2 & 0 & 3\ell & -\ell^2 \\ 0 & 0 & 0 & 0 & 0 & 0 \\ 0 & -36 & 3\ell & 0 & 36 & 3\ell \\ 0 & -3\ell & -\ell^2 & 0 & 3\ell & 4\ell^2 \end{bmatrix}. \tag{4.57}$$

The element stiffness relation for the 6 DOFs system may be written as

$$\mathbf{S} = (\mathbf{k} - \mathbf{k}_\sigma)\,\mathbf{v}, \tag{4.58}$$

with compressive P considered to be positive.

Effect of Axial Tension on Stiffness

A beam is able to carry lateral loads in bending. An axial compressive force reduces the stiffness and load-carrying capacity. When the buckling load is reached, there is no stiffness remaining and the beam cannot carry any lateral loads.

Similarly, when the beam is subjected to tensile loads, the geometric stiffness is positive and represents an increased stiffness – and load-carrying ability. This is an important effect for slender structures such as risers, tethers, and mooring lines. If the member becomes very slender, the tension becomes the dominant effect and the bending of less importance.

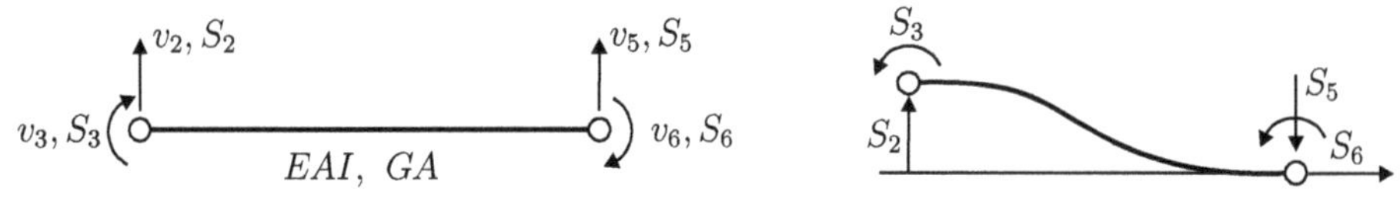

(a) Definition of positive directions of nodal displacements and forces

(b) Nodal forces in their actual direction under the displacement $v_2 = 1$

Figure 4.11. Deflection at DOF no. 2.

4.6 Stiffness Relationship for Beam with Bending and Shear Deformation

4.6.1 Matrix Method

So far, the stiffness relationship for a beam element, e.g., Eq. (4.58), has been based on slender structures, and hence neglecting the shear deformations. However, in some cases (e.g., in ship's high girders) displacements of the beam, except those due to pure axial force, creates shear deformations within the cross section, which have some effect on the element stiffness matrix.

Before looking into this effect, we recall your attention to the concept of shear deflection. The shear strain is expressed by Eqs. (3.72).

The effect of shear deformation can be derived in various ways, notably,

- using classical structural engineering beam and frame theory.
- finite element method based on the PVD and assumed displacements.

The effect of shear deformations is first illustrated by the former method. Only lateral displacements and rotations are considered because there is no coupling with axial displacements.

Then consider the beam shown in Fig. 4.11, in which its DOF no. 2 is given a displacement v_2. This deformation will have contributions from bending, v_{2b}, and shear, v_{2s}, given by

$$v_2 = v_{2b} + v_{2s}. \tag{4.59}$$

As known from the foregoing sections, the bending contribution will be given by

$$v_{2b} = \frac{\ell^3}{12EI} S_2. \tag{4.60}$$

The shear contribution is given by

$$v_{2s} = \gamma \ell = \frac{Q\ell}{GA_s} = \frac{\ell}{GA_s} S_2, \tag{4.61}$$

which results in a total lateral displacement given as

$$v_2 = \left(\frac{\ell^3}{12EI} + \frac{\ell}{GA_s} \right) S_2. \tag{4.62}$$

This implies that

$$S_2 = \frac{12EI}{(1+\alpha)\,\ell^3} v_2 \qquad \alpha = \frac{12EI}{GA_s \ell^2}. \tag{4.63}$$

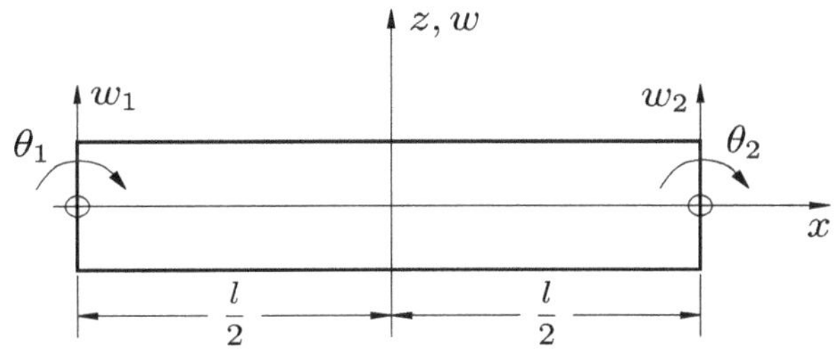

Figure 4.12. Shear deformable beam element based on linear variation of w and θ.

By repeating this procedure for the other DOFs, the element stiffness matrix, including the shear deformation, will appear as

$$\begin{bmatrix} S_2 \\ S_3 \\ S_5 \\ S_6 \end{bmatrix} = \frac{EI}{(1+\alpha)\,\ell^3} \begin{bmatrix} 12 & -6\ell & -12 & -6\ell \\ -6\ell & (4+\alpha)\,\ell^2 & 6\ell & (2-\alpha)\,\ell^2 \\ -12 & 6\ell & 12 & 6\ell \\ -6\ell & (2-\alpha)\,\ell^2 & 6\ell & (4+\alpha)\,\ell^2 \end{bmatrix} \begin{bmatrix} v_2 \\ v_3 \\ v_5 \\ v_6 \end{bmatrix}. \tag{4.64}$$

As indicated previously, an alternative derivation of the element stiffness matrix for a beam with bending and shear deformations may be based on the finite element method.

4.6.2 Finite Element Model Based on Timoshenko Beam Theory

As mentioned, the effect of shear may also be accounted for in an FE beam model, and then Timoshenko beam theory is applied. The essential assumption is that a line that is straight and normal to the beam axis of the undeformed structure is assumed to remain straight, but not necessarily normal to the beam axis after deformation. For this reason, the kinematics of a beam need to be described by the lateral deflection w of the beam axis as well as the rotation θ of the normal to the beam axis. The axial displacement u is then described by Eq. (3.67). The strains and stresses are expressed by Eqs. (3.68)–(3.70) and (3.71) and (3.72), respectively. By combining these equations with the equilibrium equation (Eq. (3.42)), the governing differential equations, Eqs. (3.75) and (3.76), were obtained.

The stiffness relationship in the finite element method is established based on assuming independent interpolation for w and θ and the PVD. Normally the same shape functions are used for both of the displacement patterns:

$$\begin{aligned} w(x) &= \mathbf{N}\,\mathbf{v}_w \\ \theta(x) &= \mathbf{N}\,\mathbf{v}_\theta, \end{aligned} \tag{4.65}$$

where $\mathbf{N}$ is a line vector while $\mathbf{v}_w$ and $\mathbf{v}_\theta$ are column vectors. The shape functions $\mathbf{N}$ are the same as those used for the truss element. Hence, for a beam element with two nodes, $\mathbf{N}$ will be the linear functions using the nodal values of w and θ in the two end nodes as parameters. Then $\mathbf{v}_w$ and $\mathbf{v}_\theta$ represents the nodal values of the lateral deflection w_i and rotation θ_i at the two nodes, see Fig. 4.12 respectively. The interpolation polynomials are

$$\mathbf{N} = \left[\frac{1}{2}(1-\xi)\,,\ \frac{1}{2}(1+\xi)\right], \tag{4.66}$$

with $\xi = 2x/\ell$.

The strains $\boldsymbol{\varepsilon}$ are then obtained by introducing Eq. (4.65) into the expressions for the strain (Eqs. (3.68)–(3.70)):

$$\begin{bmatrix} \varepsilon_x \\ \gamma_{xz} \end{bmatrix} = \begin{bmatrix} z\theta,_x \\ w,_x + \theta \end{bmatrix} = \begin{bmatrix} \mathbf{0} & z\mathbf{N},_x \\ \mathbf{N},_x & \mathbf{N} \end{bmatrix} \begin{bmatrix} \mathbf{v}_w \\ \mathbf{v}_\theta \end{bmatrix} = \mathbf{B}(x, z)\mathbf{v} \tag{4.67}$$

where

$$\mathbf{B}(x, z) = \begin{bmatrix} 0 & 0 & -z \cdot \frac{1}{\ell} & z \cdot \frac{1}{\ell} \\ -\frac{1}{\ell} & \frac{1}{\ell} & \frac{1}{2}(1-\xi) & \frac{1}{2}(1+\xi) \end{bmatrix}; \ \xi = 2x/\ell. \tag{4.68}$$

The stresses then follows from Hooke's law (Eqs. (3.71) and (3.72)):

$$\begin{aligned} \begin{bmatrix} \sigma_x \\ \tau_{xz} \end{bmatrix} &= \begin{bmatrix} E & 0 \\ 0 & G \end{bmatrix} \begin{bmatrix} \varepsilon_x \\ \gamma_{xz} \end{bmatrix} \\ &= \begin{bmatrix} E & 0 \\ 0 & G \end{bmatrix} \begin{bmatrix} \mathbf{0} & z\mathbf{N},_x \\ \mathbf{N},_x & \mathbf{N} \end{bmatrix} \begin{bmatrix} \mathbf{v}_w \\ \mathbf{v}_\theta \end{bmatrix} = \mathbf{D}\,\mathbf{B}(x, z)\,\mathbf{v}. \end{aligned} \tag{4.69}$$

The PVD is applied to an element with internal stresses $\boldsymbol{\sigma}$, nodal forces $\mathbf{S}$, and lateral load $q(x)$:

$$\int_V \tilde{\boldsymbol{\varepsilon}}^T \boldsymbol{\sigma}\, dV = \tilde{\mathbf{v}}^T \mathbf{S} + \int_\ell \tilde{w}\, q(x)\, dx. \tag{4.70}$$

If the virtual displacements are represented by the same interpolation functions as the actual displacements ($\tilde{w} = \mathbf{N}\tilde{\mathbf{v}}_w$, etc.), the stiffness matrix and load vector are obtained as

$$\mathbf{k} = \int_V \mathbf{B}^T \mathbf{D} \mathbf{B} dV - \int_{-\ell/2}^{\ell/2} \int_A \mathbf{B}^T \mathbf{D} \mathbf{B}\, dA\, dx = \begin{bmatrix} \mathbf{k}_{ww} & \mathbf{k}_{w\theta} \\ \mathbf{k}_{\theta w} & \mathbf{k}_{\theta\theta} \end{bmatrix} \tag{4.71}$$

$$\mathbf{S}^0 = -\int_\ell \mathbf{N}^T q(x)\, dx, \tag{4.72}$$

where

$$\mathbf{k}_{ww} = \frac{G\bar{A}}{\ell} \begin{bmatrix} 1 & -1 \\ -1 & 1 \end{bmatrix} \tag{4.73}$$

$$\mathbf{k}_{\theta\theta} = \mathbf{k}_{\theta\theta(1)} + \mathbf{k}_{\theta\theta(2)} = \frac{EI}{\ell} \begin{bmatrix} 1 & -1 \\ -1 & 1 \end{bmatrix} + \frac{G\bar{A}\ell}{6} \begin{bmatrix} 2 & 1 \\ 1 & 2 \end{bmatrix} \tag{4.74}$$

$$\mathbf{k}_{w\theta} = \frac{G\bar{A}}{2} \begin{bmatrix} -1 & -1 \\ 1 & 1 \end{bmatrix}, \tag{4.75}$$

where $\bar{A} = A/1.2$ and $I = Ah^2/12$ are the effective shear area and the cross-sectional moment of inertia, respectively, for a rectangular beam.

Although the shear is well represented by this model the bending performance is not acceptable unless remedies such as using selective integration, i.e., by integrating the shear strain (and work) by a one-point integration or by increasing the order of

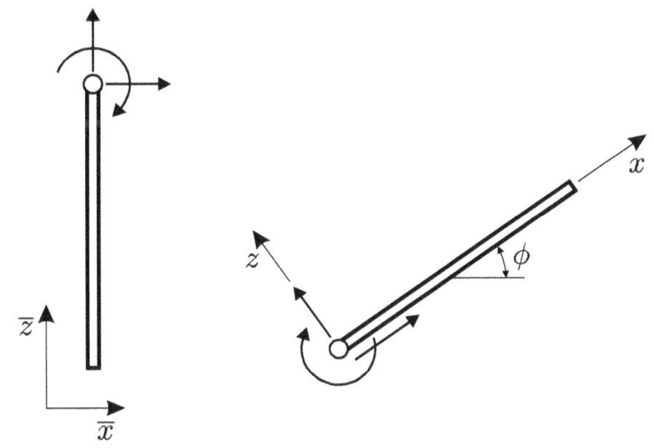

Figure 4.13. Coordinate transformation.

the polynomial for w, preferably by using a polynomial with one degree higher for w than for θ, see, e.g., (Zienkiewicz and Taylor, 2006).

4.7 Coordinate Transformations

4.7.1 General

In this section we establish the general relation between the *local* coordinate system and the *global* coordinate system. Before pursuing this aspect, it may be advantageous to explain the difference between the two. The *local* coordinate system is usually the local axes of the element, such that one of its axes passes through the center of the cross section of that element (Fig. 4.13), whereas the *global* coordinate system is a common axis system defined depending on the global structure orientation. These two coordinate systems may coincide in some cases.

The beam element is used as an example to illustrate transformation of coordinate system because it involves both translations and rotations. Only transformations in the plane are dealt with herein.

The transformation relation is of interest because the system analysis is performed in the global coordinate system. In addition, all element forces and stiffnesses are usually transformed to a common global coordinate system before they are assembled. For planar problems, the transformation is needed only for translational DOFs and not for rotations.

4.7.2 Translation

Consider Fig. 4.14a, where the displacement of a node (e.g., one end of a beam element) in the local coordinate system is illustrated. This translation can be transformed into the global coordinate system, as shown in Fig. 4.14b.

By reference to Fig. 4.14b, the following relations can be established:

$$\begin{aligned} v_1 &= \bar{v}_1 \cos\phi + \bar{v}_2 \sin\phi \\ v_2 &= -\bar{v}_1 \sin\phi + \bar{v}_2 \cos\phi. \end{aligned} \tag{4.76}$$

Denoting $\cos\phi$ by c and $\sin\phi$ by s, we can write

$$\begin{aligned} v_1 &= \bar{v}_1 c + \bar{v}_2 s \\ v_2 &= -\bar{v}_1 s + \bar{v}_2 c, \end{aligned} \tag{4.77}$$

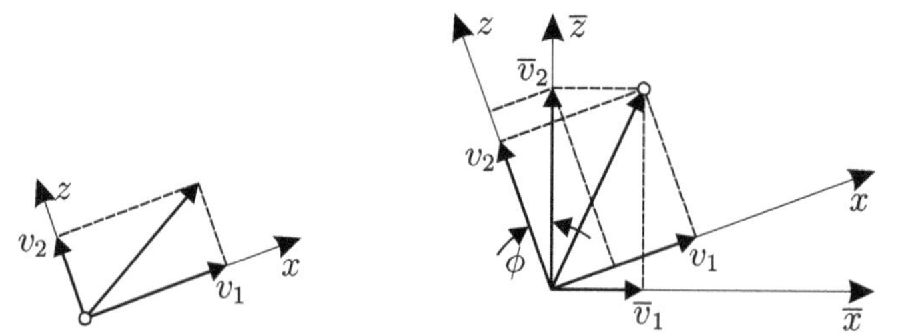

(a) Translation of a node (b) Transformation of coordinate system

Figure 4.14. Basic planar transformation.

which in matrix form becomes

$$\begin{bmatrix} v_1 \\ v_2 \end{bmatrix} = \begin{bmatrix} c & s \\ -s & c \end{bmatrix} \begin{bmatrix} \bar{v}_1 \\ \bar{v}_2 \end{bmatrix} = \mathbf{T}_o \begin{bmatrix} \bar{v}_1 \\ \bar{v}_2 \end{bmatrix}, \tag{4.78}$$

where $\mathbf{T}_o$ is the transformation matrix for a single nodal translation at one end of the element. Then, it can be easily seen that the total transformation matrix, $\mathbf{T}_g$, for a straight beam element with two nodes will appear as

$$\begin{bmatrix} v_1 \\ v_2 \\ v_3 \\ v_4 \\ v_5 \\ v_6 \end{bmatrix} \mathbf{v} = \begin{bmatrix} c & s & 0 & 0 & 0 & 0 \\ -s & c & 0 & 0 & 0 & 0 \\ 0 & 0 & 1 & 0 & 0 & 0 \\ 0 & 0 & 0 & c & s & 0 \\ 0 & 0 & 0 & -s & c & 0 \\ 0 & 0 & 0 & 0 & 0 & 1 \end{bmatrix} \begin{bmatrix} \bar{v}_1 \\ \bar{v}_2 \\ \bar{v}_3 \\ \bar{v}_4 \\ \bar{v}_5 \\ \bar{v}_6 \end{bmatrix} = \mathbf{T}_g \bar{\mathbf{v}}. \tag{4.79}$$

Nodal displacements and nodal forces are vectors and therefore transform in the same manner. Hence, $\mathbf{S} = \mathbf{T}_g \bar{\mathbf{S}}$. It can be shown that $\mathbf{T}_g^{-1} = \mathbf{T}_g^T$, which means that $\mathbf{T}_g$ is an *orthogonal* matrix. The new element stiffness matrix transformed into the global coordinate system is given by $\bar{\mathbf{S}} = \mathbf{T}_g^T \mathbf{S} = \mathbf{T}_g^T \mathbf{k}\mathbf{v} = \mathbf{T}_g^T \mathbf{k} \mathbf{T}_g \bar{\mathbf{v}}$, which implies

$$\begin{aligned} \bar{\mathbf{S}} &= \bar{\mathbf{k}} \bar{\mathbf{v}} \\ \bar{\mathbf{k}} &= \mathbf{T}_g^T \mathbf{k} \mathbf{T}_g. \end{aligned} \tag{4.80}$$

From matrix algebra, we know that the symmetry of the stiffness matrix is preserved because a matrix multiplication $\mathbf{B}^T\mathbf{A}\mathbf{B}$ preserves symmetry.

4.8 Finite Elements for Linear, Static Structural Analysis

The element stiffness relationship $\mathbf{S} = \mathbf{k}\mathbf{v} + \mathbf{S}^0$ can be developed for elements of different types of structures, using the following two components:

- assumed interpolation polynomials (shape functions).
- the principle of virtual displacement (PVD).

Based on the element stiffness relationship the system stiffness relationship, $\mathbf{R} = \mathbf{K}\mathbf{r}$ can be established, as shown in Section 4.9. The unknown global displacements $\mathbf{r}$ can then be solved for, and thus the nodal displacements $\mathbf{v}$ for each element can be calculated. The displacement $\mathbf{v}$ and stress $\boldsymbol{\sigma}$ at any point within the element can then be determined, as discussed subsequently.

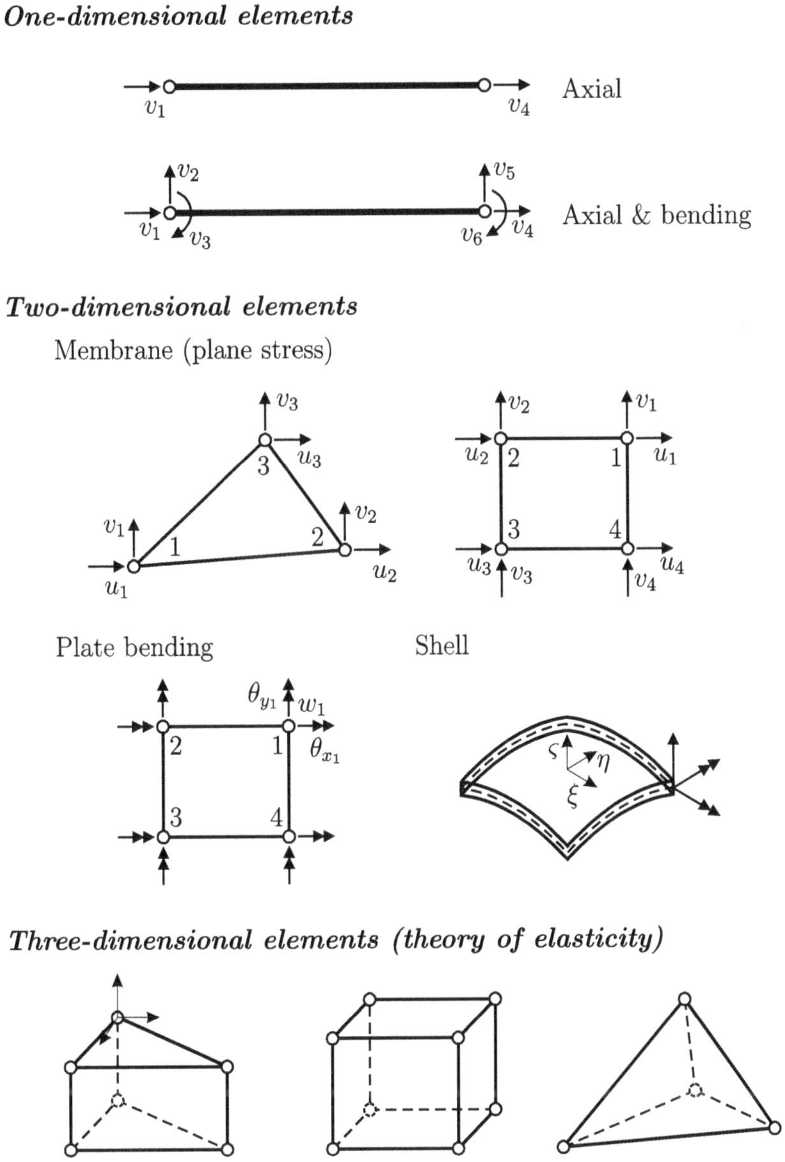

Figure 4.15. Finite elements for various types of structural mechanics problems.

Figure 4.15 gives an overview of the structural problems and the elements to be considered. There are some important aspects that need to be addressed for the two- and three-dimensional problems in this connection. They include

- How can interpolation functions for the displacements be determined?
- How can the equivalent nodal point forces be determined?
- How can the element stiffness be determined?

Although the choice of shape (interpolation) functions is straightforward for one-dimensional elements previously described, it is more challenging for two- or three-dimensional problems. This is because it may not be as easy to ensure convergence of the method by interelement compatibility, completeness of approximate displacement fields, and invariance of the element performance independent of its orientation relative to the global axis system. Other questions include

- How to evaluate the solution to get the best possible estimates of the stresses?
- How can the results be more accurate by increasing the polynomial approximation within each element or by increasing the number of elements?

These issues are treated in detail in textbooks on finite element methods (e.g., Hughes (1987); Cook et al. (2002); Zienkiewicz et al. (2005); Zienkiewicz and Taylor (2006)).

Independent of which type of problem that is addressed, it is necessary to establish the system stiffness relationship for the whole structure based on the element stiffness relationships. This is briefly explained in Section 4.9.

Next, dynamic forces acting on structures that are subjected to time-varying loads that cause inertia and damping reaction forces are accommodated in the finite element models by application of d'Alembert's principle, as described in Section 4.10.

4.9 System Stiffness Relationship for Static Problems

4.9.1 General

So far, only the element stiffness relationship is established. In *matrix methods*, it was demonstrated how the element stiffness relationship ($\mathbf{S} = \mathbf{kv} + \mathbf{S}^0$) can be used to build up the system stiffness relationship: $\mathbf{R} = \mathbf{Kr} + \mathbf{R}^0$ enforcing kinematic compatibility and equilibrium in the global nodes. This approach leads to the so-called direct stiffness method, whereby $\mathbf{K}$ and $\mathbf{R}^0$ were obtained by adding contributions from element no. j: $\mathbf{k}_j$ and $-\mathbf{S}_j^0$.

In this section, it is shown how the system stiffness relationship can be established based on assumed global interpolation polynomials and the application of the PVD.

The following formulations are general. However, the derivations are made by referring to a system of bar elements described in Section 4.3.

4.9.2 Global Interpolation Functions

The nodal displacements and forces in the global model are denoted by r_i and R_i, respectively. They are collected in the vectors $\mathbf{r}$ and $\mathbf{R}$, respectively. The global interpolation polynomial for the displacement in node no. i is defined in the same way as for individual elements, i.e., $r_i = 1$ and $r_j = 0$ for $j \neq i$.

Because the displacement in each element is approximated by a linear function in the example in Fig. 4.16, the displacement for $r_4 = 1$ only affects the adjacent two elements (nos. 3 and 4 in the example).

It is noted that the model in Fig. 4.16 is different from a global model, e.g., based on an interpolation polynomial that covers the whole bar. With 6 DOFs in the global model, a fifth-degree polynomial can be described for the whole bar. Such Lagrange polynomials are described subsequently. But the main idea with the FEM is to use local interpolation (within each element) as illustrated in Fig. 4.16.

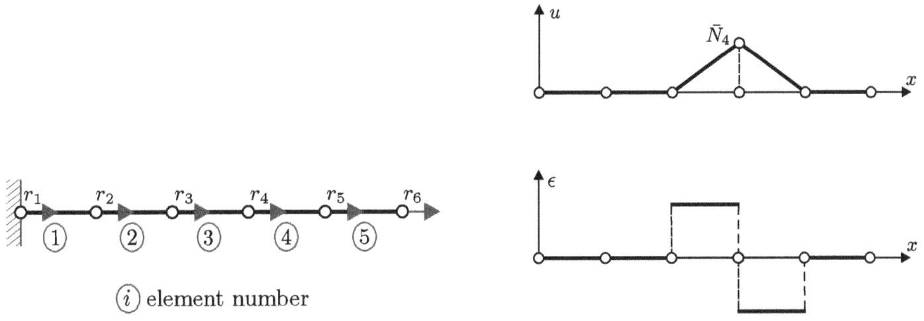

(a) Nodal displacements

(b) Upper figure: Interpolation polynomial for $r_4 = 1$, $r_i = 0$ for $i \neq 4$. Lower figure: Strain obtained by differentiating the above displacement

Figure 4.16. Example of global bar model based on elements with a linear displacement u within each element.

Let the interpolation function corresponding to $r_i = 1$, $r_j = 0$ for $j \neq i$, be denoted $\bar{N}_i$ – defined in the elements adjacent to node i, (see illustration in Fig. 4.16b).

Then the global displacement may be described by

$$u = \sum_{i=1}^{N_n} \bar{N}_i r_i = \bar{\mathbf{N}}\mathbf{r}, \tag{4.81}$$

where N_n is the number of nodes (and, in this example, number of DOFs).

Now, the element displacement DOF $\mathbf{v}$ can be related to $\mathbf{r}$ by $\mathbf{v}^k = \mathbf{a}^k\mathbf{r}$ for element no. k. Based on this relationship, the approximate u may be expressed by the element interpolation functions $\mathbf{N}^k$ for element k as

$$u = \bar{\mathbf{N}}\mathbf{r} = \sum_{k=1}^{Ne} \mathbf{N}^k\mathbf{a}^k\mathbf{r}, \tag{4.82}$$

where $\mathbf{N}^k = 0$ except for element no. k and N_e is the number of elements.

4.9.3 Principle of Virtual Displacements for System

Next, consider the PVD. By generalizing Eq. (3.41) or (4.13), the following virtual work equation is obtained:

$$\int_{V_T} \tilde{\boldsymbol{\varepsilon}}^T \boldsymbol{\sigma}\, dV = \tilde{\mathbf{r}}^T\mathbf{R} + \int_{\ell_T} \tilde{u}\, q(x)\, dx, \tag{4.83}$$

where V_T and ℓ_T refer to the whole (global) bar subjected to a distributed axial loading $q(x)$.

4.9.4 Finite Element Model for System

Based on the assumed displacement (Eq. (4.82)), the approximate strains and stresses in Eq. (4.84) can be determined. Hence, the strain is given by

$$\varepsilon = \frac{du}{dx} = \frac{d}{dx}\left[\sum_{k=1}^{N_e} \mathbf{N}^k \mathbf{a}^k \mathbf{r}\right] = \sum_{k=1}^{Ne} \mathbf{N},_x^k \mathbf{a}^k \mathbf{r} = \sum_{k=1}^{N_e} \mathbf{B}^k \mathbf{a}^k \mathbf{r} = \bar{\mathbf{B}}\mathbf{r}. \tag{4.84}$$

Although the displacement is continuous across element boundaries, the strains are usually discontinuous. This is illustrated in Fig. 4.16c for the case where the local variation of u in $\mathbf{N}^k$ is linear. The stress, σ, is obtained by

$$\sigma = E\varepsilon. \tag{4.85}$$

Now, the virtual displacements and strains are described by analogous expressions as the real ones:

$$\begin{aligned} \tilde{u} &= \bar{\mathbf{N}}\tilde{\mathbf{r}} \\ \tilde{\varepsilon} &= \bar{\mathbf{B}}\tilde{\mathbf{r}}. \end{aligned} \tag{4.86}$$

In this way, the virtual displacements satisfy the kinematic compatibility requirement.

Inserting Eqs. (4.84)-(4.86) into Eq. (4.83) yields

$$\tilde{\mathbf{r}}^T\left(\int_{V_T} \bar{\mathbf{B}}^T E\bar{\mathbf{B}} dV \mathbf{r} - \int_{\ell_T} \bar{\mathbf{N}}^T q(x)dx - \mathbf{R}\right) = 0. \tag{4.87}$$

If Eq. (4.87) is to be satisfied for an arbitrary displacement $\tilde{\mathbf{r}}$, it follows that

$$\mathbf{R} = \mathbf{K}\mathbf{r} + \mathbf{R}^0, \tag{4.88}$$

where

$$\mathbf{K} = \int_{V_T} \bar{\mathbf{B}}^T E\bar{\mathbf{B}}\, dV = \int_{\ell_T} \bar{\mathbf{B}}^T E\, A(x)\bar{\mathbf{B}}\, dx, \tag{4.89}$$

$$\mathbf{R}^0 = -\int_{\ell_T} \bar{\mathbf{N}}^T q(x)dx. \tag{4.90}$$

In the development of Eq. (4.88) based on the PVD, $\tilde{\mathbf{r}}^T\mathbf{R}$ in Eq. (4.83) represented the virtual work on the element boundary (axial force at the ends of the element). It is noted that equilibrium is achieved in an integrated sense in the nodes, despite the fact that the strain (ε) and, hence, the stress (σ) are discontinuous across the boundary between elements (in each node in the one-dimensional bar example).

The global stiffness matrix and load vector can be calculated by

$$\mathbf{K} = \sum_{k=1}^{N_e} \int_{V_k} (\mathbf{a}^k)^T (\mathbf{B}^k)^T E\, \mathbf{B}^k \mathbf{a}^k\, dV = \sum_{k=1}^{N_e} \mathbf{K}^k \tag{4.91}$$

$$\mathbf{R}^0 = -\sum_{k=1}^{N_e} \int_{\ell_k} (\mathbf{a}^k)^T (\bar{\mathbf{N}}^k)^T q(x)\, dx = \sum_{k=1}^{N_e} (\mathbf{R}^k)^0 \tag{4.92}$$

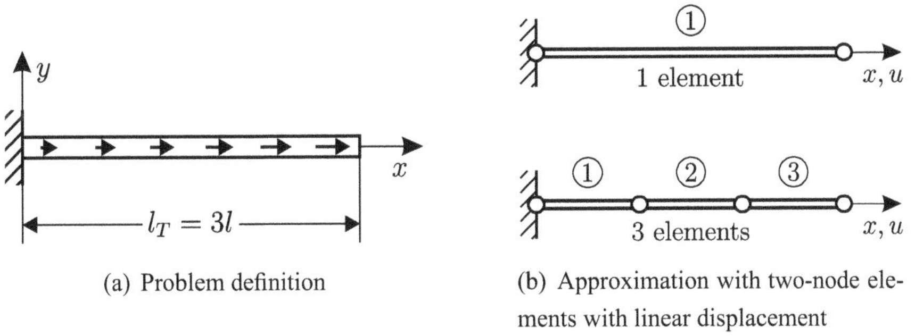

Figure 4.17. Axially loaded bar with constant cross section.

When the sequence of summation and integration is changed, the fact that $\bar{\mathbf{N}}_k \equiv 0$ outside element no. k is used.

The expressions given by Eqs. (4.91) and (4.92) correspond to those established for frame analysis by the matrix method (FEM). Hence, the direct stiffness method is again demonstrated. The matrix method allows treatment of complex structures that are impossible to analyze by analytical methods.

Equations (4.87) and (4.91) define how the global stiffness relationship can be obtained from the element stiffness relationship. The crucial issue is how the local displacement vector $\mathbf{v}^k$ or $\mathbf{v}$ relate to the global displacement $\mathbf{r}$ through the topology matrix $\mathbf{a}^k$: $\mathbf{v}^k = \mathbf{a}^k\mathbf{r}$; for each element k.

4.9.5 Example – One-Element Model

The simple example shown in Fig. 4.17 may serve to illustrate the basic features of the FE approximation (Cook et al., 2002). It is assumed that $q(x) = c\,x$, where $c > 0$ is constant. The exact solution for this case is obtained from the differential Eq. (3.31) and the boundary conditions $u(0) = 0$ and $\sigma(\ell_T) = 0$:

$$u = \frac{c}{6EA}(3\ell_T^2 x - x^3); \ \ \sigma = \frac{c}{2A}(\ell_T^2 - x^2). \tag{a}$$

The element stiffness matrix and the load vector for a single element model are then

$$\mathbf{k}^1 = \frac{EA}{3\ell}\begin{bmatrix} 1 & -1 \\ -1 & 1 \end{bmatrix} \qquad \mathbf{S}^1 = \frac{3c\ell^2}{2}\begin{bmatrix} 1 \\ 2 \end{bmatrix}. \tag{b}$$

With the boundary condition $r_1 = 0$, the solution becomes

$$\begin{bmatrix} r_1 \\ r_2 \end{bmatrix} = \begin{bmatrix} 0 \\ 9c\ell^3/(EA) \end{bmatrix} = \begin{bmatrix} 0 \\ c\ell_T^3/(3EA) \end{bmatrix}. \tag{c}$$

Hence, the approximate solution for a single element is

$$u(x) = \frac{c\ell_T^2}{3EA}x = \frac{3c\ell^2}{EA}x, \tag{d}$$

and the stress $\sigma = E\varepsilon = E\mathbf{B}\mathbf{v} = \frac{E}{\ell_T}[-1, 1]\mathbf{v}$ is constant within the element

$$\sigma = \frac{3c\ell^2}{A} = \frac{c\ell_T^2}{3A}, \tag{e}$$

while the exact solution is given by Eq. (a).

4.9.6 Example – Three-Element Model

For each element of length, ℓ, the following stiffness matrix and load vectors apply. The element stiffness matrix for all elements are:

$$\mathbf{k}^{\mathrm{k}} = \frac{EA}{\ell}\begin{bmatrix} 1 & -1 \\ -1 & 1 \end{bmatrix}. \tag{a}$$

The load vectors for each element are obtained as: $\mathbf{S}^i = -\mathbf{S}^{0i}$, that is,

$$\mathbf{S}^1 = \frac{c\ell^2}{6}\begin{bmatrix} 1 \\ 2 \end{bmatrix}; \quad \mathbf{S}^2 = \frac{c\ell^2}{6}\begin{bmatrix} 4 \\ 5 \end{bmatrix}; \quad \mathbf{S}^3 = \frac{c\ell^2}{6}\begin{bmatrix} 7 \\ 8 \end{bmatrix}. \tag{b}$$

Global equations for the three element structure can then be obtained:

$$\frac{EA}{\ell}\begin{bmatrix} 1 & -1 & 0 & 0 \\ -1 & 2 & -1 & 0 \\ 0 & -1 & 2 & -1 \\ 0 & 0 & -1 & 1 \end{bmatrix}\begin{bmatrix} r_1 \\ r_2 \\ r_3 \\ r_4 \end{bmatrix} = \frac{c\ell^2}{6}\begin{bmatrix} 1 \\ 6 \\ 12 \\ 8 \end{bmatrix}. \tag{c}$$

Introducing the boundary condition $r_1 = 0$:

$$\frac{EA}{\ell}\begin{bmatrix} 1 & 0 & 0 & 0 \\ 0 & 2 & -1 & 0 \\ 0 & -1 & 2 & -1 \\ 0 & 0 & -1 & 1 \end{bmatrix}\begin{bmatrix} r_1 \\ r_2 \\ r_3 \\ r_4 \end{bmatrix} = \frac{c\ell^2}{6}\begin{bmatrix} 0 \\ 6 \\ 12 \\ 8 \end{bmatrix}. \tag{d}$$

The solution vector is found as

$$\mathbf{r} - \frac{c\ell^3}{3EA}\begin{bmatrix} 0 \\ 13 \\ 23 \\ 27 \end{bmatrix}. \tag{e}$$

Although the nodal displacements in this one-dimensional case are exact, which is not generally true, the approximate solution deviate from the exact solution between the nodes. For instance, the approximate stress at the midpoint $(x = 3\ell/2)$ is $u(3\ell/2) = 6c\ell^3/(EA)$ whereas the exact value is 6.1875 $c\ell^3/(EA)$. However, as previously mentioned, displacements are more accurate than stresses obtained by the FEM.

The approximate stress is constant within each element and exhibits a stepwise variation over the length of the bar.

It is noted that the load vector used to obtain the displacements $\mathbf{r}$ corresponds to the fixed-end force vector with reversed sign! This is the complementary solution. Strictly speaking, the particular solution – the fixed-end case $\mathbf{r} = 0$ – should be superimposed on the complementary solution. See the approach used in the matrix method, as briefly mentioned in Sections 4.4.2 (for bars) and 4.3.2 (for beams).

If the internal stresses in, e.g., element no. 2 could be estimated and superimposed on the constant stress given in Fig. 4.18, which is obtained by the conventional FEM, then the exact total solution for element no. 2 could be estimated. It can be shown that the exact (particular) solution of the differential equation for element no. 2, given that the displacement

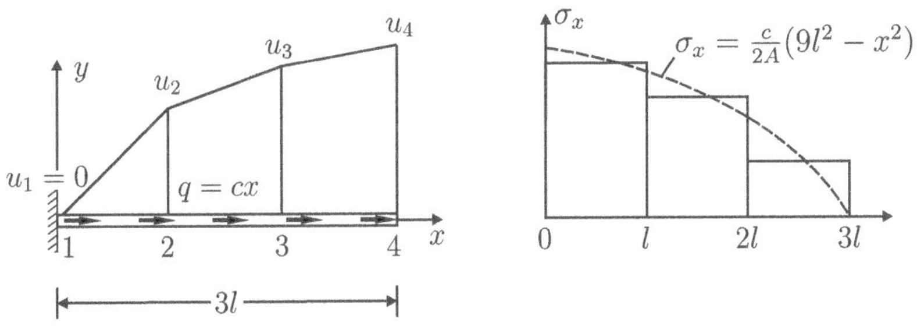

(a) Axial displacements (b) Axial stresses

Figure 4.18. Axial displacements and axial stresses in a bar under linearly varying axial load $q = cx$, modeled by three elements of equal length. The dashed line represents the exact stress.

at each node is zero, is given as

$$u = -\frac{c}{EA}\Big(\frac{1}{2}\ell\tilde{x}^2 + \frac{1}{6}\tilde{x}^3 - \frac{2}{3}\ell^2\tilde{x}\Big),$$

$$\sigma = \frac{c}{A}\Big(\frac{2}{3}\ell^2 - \ell\tilde{x} - \frac{1}{2}\tilde{x}^2\Big), \tag{f}$$

where $\tilde{x}$ denotes the local x-coordinate. This yields the total stress variation, as shown in Fig. 4.18.

If the stress due to fixed-end forces is superimposed on the approximate stress (in Fig. 4.18), the results are much closer to the exact value for this case. However, in this case, the exact particular solution is applied. If the approximate (linear) displacement is used to approximate the particular solution, it becomes equal to zero. Hence, in finite element analysis, the actual approximate solution is normally taken to be the complementary solution only.

The simple examples in Sections 4.9.5 and 4.9.6 illustrate several important general features about finite element analyses.

The largest discrepancy from the exact solution is reduced when the number of elements is increased.

The stresses at one (or a few) particular point(s) within the element are closest to the exact solution. Generally, the stresses in the interior of the element are most accurate, while the stresses at the ends (boundaries) of a given element exhibit the largest discrepancy from the exact solution.

The average of the stresses at nodal points is closer to the exact solution than the stresses at the nodes from each element individually. It is obviously a problem to find reasonable averaged stresses close to the edges of the problem because the stress will be biased due to the averaging of stresses in fewer elements that have error to one side of the exact solution.

4.9.7 Buckling Analysis of a Structural System

Buckling loads for a structure subjected to axial forces may be determined by using the element stiffness relationship Eq. (4.58). The (global) stiffness relationship for the system is assembled from the element stiffness matrices by using, e.g., the *direct stiffness method*. However, because $\mathbf{k}_\sigma$ depends on the axial force in the element,

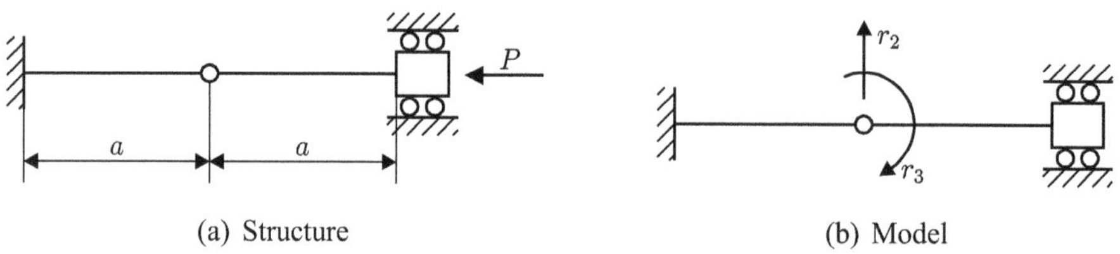

Figure 4.19. Clamped beam with axial force.

it first needs to be determined by a linear analysis. Denote the axial force in each member i by P_{0i}. The buckling load is defined by increasing all axial forces by the same factor λ_p.

The stiffness matrix for a system of beams with axial forces can be written as

$$\mathbf{R} = (\mathbf{K} - \lambda_p \mathbf{K}_\sigma)\,\mathbf{r}, \tag{4.93}$$

where λ_p is a common load factor that multiplies the axial forces P_{0i} in all elements i, i.e., $P_i = \lambda_p P_{0i}$. It is noted that the forces P_{0i} are implicit in the element stiffness matrices $\mathbf{k}_\sigma$ and hence in $\mathbf{K}_\sigma$. Moreover, Eq. (4.93) is valid when the displacements $\mathbf{r}_i$ are small. When displacements are large, a more accurate theory is required to deal with the geometrically nonlinear problem as described, e.g., by McGuire et al. (2000); Cook et al. (2002); Zienkiewicz and Taylor (2006).

If the external loads $\mathbf{R}$ are zero, the following relationship results:

$$(\mathbf{K} - \lambda_p \mathbf{K}_\sigma)\,\mathbf{r} = \mathbf{0}. \tag{4.94}$$

This represents an eigenvalue problem. The problem is analogous to the problem of calculating natural frequencies ω_i by $(\mathbf{K} - \omega^2\mathbf{M})\mathbf{r} = \mathbf{0}$. If λ_{p1} is the smallest eigenvalue, the buckling load of the system corresponds to axial forces:

$$P_{Cr(i)} = \lambda_{p1}\, P_{0i}. \tag{4.95}$$

To perform out this calculation, a computer would normally be required. However, to illustrate the approach, a simple example is described. Consider a simple beam shown in Fig. 4.19, which is clamped at both ends.

If this structure is modeled by a single element, it is not possible to obtain a solution. This is because all DOFs r_i equal zero due to the boundary conditions. Thus, the smallest FE model yielding a nontrivial solution consists of two elements.

In this case, the axial force in each element is determined to be P in a straightforward manner. The resulting stiffness relationship is obtained by the direct stiffness method. When neglecting axial deformations and introducing the boundary conditions, the stiffness relationship becomes:

$$\left\{\frac{2EI}{a^3}\begin{bmatrix} 12 & 0 \\ 0 & 4a^2 \end{bmatrix} - \frac{P}{30a}\begin{bmatrix} 72 & 0 \\ 0 & 8a^2 \end{bmatrix}\right\}\begin{bmatrix} r_2 \\ r_3 \end{bmatrix} = \begin{bmatrix} 0 \\ 0 \end{bmatrix}. \tag{4.96}$$

Introducing $\lambda_p = Pa^2/(EI)$, we may write

$$\left\{\begin{bmatrix} 720 & 0 \\ 0 & 240a^2 \end{bmatrix} - \lambda_p \begin{bmatrix} 72 & 0 \\ 0 & 8a^2 \end{bmatrix}\right\}\begin{bmatrix} r_2 \\ r_3 \end{bmatrix} = \begin{bmatrix} 0 \\ 0 \end{bmatrix}. \tag{4.97}$$

The smallest eigenvalue is $\lambda_p = 10$, corresponding to a buckling load $P_{Cr} = 10EI/a^2$, which compares closely to the exact value of $P_{Cr} = 4\pi^2 EI/(2a)^2 = \pi^2 EI/a^2$.

The work done by external forces needs to be absorbed by the internal, inertia, and (dissipative) damping forces.

Bar

The virtual work equation for a bar will then read

$$\int_{\ell} \left(\tilde{\varepsilon}_x\, \sigma_x + \tilde{u}\, \rho\, \ddot{u} + \tilde{u}\, c\dot{u}\right) A(x)\, dx = \tilde{\mathbf{v}}^T \mathbf{S} + \int_{\ell} \tilde{u}\, q(x)\, dx. \tag{4.101}$$

The FE discretization assumptions can be expressed as

$$\begin{aligned} &u(x) = \mathbf{N}(x)\mathbf{v};\ \dot{u}(x) = \mathbf{N}(x)\dot{\mathbf{v}};\ \ddot{u}(x) = \mathbf{N}(x)\ddot{\mathbf{v}} \\ &\tilde{u}(x) = \mathbf{N}(x)\tilde{\mathbf{v}}, \end{aligned} \tag{4.102}$$

where the shape functions $\mathbf{N} = \mathbf{N}(x)$ are functions of x, while the nodal displacements $\mathbf{v}$, nodal velocities $\dot{\mathbf{v}}$, and nodal accelerations $\ddot{\mathbf{v}}$ are functions of time. Substituting from Eq. (4.102) into Eq. (4.101), the equation of motion for a single FE can then be written as

$$\mathbf{m}\ddot{\mathbf{v}} + \mathbf{c}\dot{\mathbf{v}} + \mathbf{k}\mathbf{v} = \mathbf{S}^0, \tag{4.103}$$

where $\mathbf{k}$ and $\mathbf{S}^0$ are the same as before and

$$\begin{aligned} \mathbf{m} &= \int_V \rho \mathbf{N}^{\mathbf{T}} \mathbf{N}\, dV, \\ \mathbf{c} &= \int_V c \mathbf{N}^{\mathbf{T}} \mathbf{N}\, dV. \end{aligned} \tag{4.104}$$

$\mathbf{m}$ and $\mathbf{c}$ are the consistent mass and damping matrices, respectively.

By assuming linear interpolation over the element, the consistent mass matrix for a straight bar element with two nodes of length ℓ and mass density ρ (see Fig. 4.3) reads

$$\mathbf{m} = \frac{\rho A \ell}{6} \begin{bmatrix} 2 & 1 \\ 1 & 2 \end{bmatrix}, \tag{4.105}$$

while simple "lumping" of the total mass to the two ends yields,

$$\mathbf{m} = \frac{\rho A \ell}{6} \begin{bmatrix} 3 & 0 \\ 0 & 3 \end{bmatrix}. \tag{4.106}$$

By using Eq. (4.103) for elements an equation analogous to Eq. (4.100) can be established for a system of bars, i.e., a truss.

Beam under bending

In an analogous manner, as for the bar element, the virtual work equation may be applied to determine consistent matrices for the beam element. The consistent mass matrix terms for the lateral and rotational degrees of freedom of the beam undergoing *bending deformations* read

$$\mathbf{m} = \frac{\rho A \ell}{420} \begin{bmatrix} 156 & & symm. & \\ -22\ell & 4\ell^2 & & \\ 54 & -13\ell & 156 & \\ 13\ell & -3\ell^2 & 22\ell & 4\ell^2 \end{bmatrix}. \tag{4.107}$$

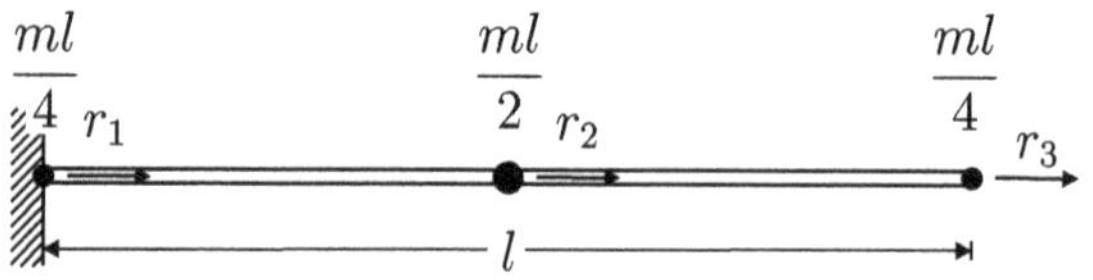

Figure 4.22. Finite element model.

while the "lumped" mass matrix assumes a diagonal form as follows:

$$\mathbf{m} = \frac{\rho A\ell}{420}\begin{bmatrix} 210 & & & symm. \\ 0 & 0 & & \\ 0 & 0 & 210 & \\ 0 & 0 & 0 & 0 \end{bmatrix}. \qquad (4.108)$$

The zero terms on the diagonal of the lumped mass matrix may cause numerical problems if this matrix needs to be factored. Then, a mass moment of inertia of the order $\rho A\ell^3/24$ may be added to ensure numerical stability without affecting the behavior noticeably (Cook et al., 2002, Sec. 11.4).

Comments

Even though "lumped" mass formulations are often less accurate than the consistent approach, i.e., requires a finer mesh, there are computational advantages, with respect to storage and number of operations in eigenvalue analysis and direct time integration with the former, that suggest use of this approach.

The mass matrix for the 6 DOFs element with axial and bending behavior can be established based on the results for the bar and lateral displacements because they are uncoupled.

If the distributed damping $c(x)$ over the element is known, the damping matrix can be determined from Eq. (4.104). However, in general, structural damping is subjected to large uncertainties and is normally accounted for in a more simple manner, e.g., by modal damping. The spatial variation of the hydrodynamic damping on marine structures may sometimes be known such that Eq. (4.104) for the damping may be relevant to apply.

It is noted that the global damping and mass matrices can be assembled from the corresponding element matrices in the same way as $\mathbf{K}$ is assembled from the individual $\mathbf{k}^i$ for each element i. This formulation needs to be improved to account for distributed inertia and damping loads. Also, it needs to be reformulated to properly account for frequency-dependent mass and damping, as described in Chapter 14.

4.10.3 Example – Bar with Free Harmonic Axial Vibrations

The case $\mathbf{R}^{ext} = \mathbf{C} = 0$ in Eq. (4.100) represents free undamped vibration. The corresponding equation may be solved by assuming harmonic vibration, i.e., all points oscillate in phase with the same frequency, that is, $\mathbf{r} = \boldsymbol{\psi} \sin \omega t$.

The bar in Fig. 4.22 is considered to have a constant axial stiffness EA and a uniform mass per unit length of $m = \rho A$. The stiffness and mass matrices for a two element model with reference to the displacements $\mathbf{r} = \left[r_1,\ r_2,\ r_3\right]^T$ becomes

$$\mathbf{K} = \frac{2EA}{\ell}\begin{bmatrix} 1 & -1 & 0 \\ -1 & 2 & -1 \\ 0 & -1 & 1 \end{bmatrix}. \tag{a}$$

Lumping the masses to the nodes, this gives

$$\mathbf{M} = \frac{m\ell}{4}\begin{bmatrix} 1 & 0 & 0 \\ 0 & 2 & 0 \\ 0 & 0 & 1 \end{bmatrix}. \tag{b}$$

By introducing the boundary condition $u = 0$, i.e., $r_1 = 0$, at the fixed end, the resulting eigenvalue problem becomes

$$\left\{\begin{bmatrix} 2 & -1 \\ -1 & 1 \end{bmatrix} - \bar{\omega}^2\begin{bmatrix} 2 & 0 \\ 0 & 1 \end{bmatrix}\right\}\begin{bmatrix} \psi_2 \\ \psi_3 \end{bmatrix} = 0, \tag{c}$$

where $\mathbf{r} = \left[0,\ \psi_2,\ \psi_3\right]^T \sin\omega t$ and

$$\bar{\omega}^2 = \omega^2\frac{m\ell^2}{8EA} = \omega^2\frac{\rho\,\ell^2}{8E}. \tag{d}$$

The determinant of the coefficient matrix in Eq. (c) is equal to zero for $\bar{\omega}_1 = 0.541$ and $\bar{\omega}_2 = 1.307$. However, only $\bar{\omega}_1 = 0.541$ gives an in-phase vibration, that is, a solution with ψ_2 and ψ_3 of the same sign. The frequency of the in-phase vibration is therefore

$$\omega = \frac{\bar{\omega}_1}{\ell}\sqrt{\frac{8E}{\rho}} = \frac{1.53}{\ell}\sqrt{\frac{E}{\rho}}, \tag{e}$$

which is 2.6% smaller than the exact value (see also Section 3.2.3).

4.10.4 Example – Slender Beam with Free Harmonic Bending Vibrations

Figure 4.23 shows a relatively slender beam for which it is only necessary to take into consideration deformations due to bending moments. Shear deformations and inertia forces due to rotation are neglected. Figure 4.23 shows the beam, the discretized mass distribution, and the 6 DOFs that must be introduced if the beam is subdivided into two elements. The stiffness matrix will then have the following form:

$$\mathbf{K} = \begin{bmatrix}
\frac{EA}{\ell} & 0 & 0 & -\frac{EA}{\ell} & 0 & 0 & 0 & 0 & 0 \\
0 & \frac{12EI}{\ell^3} & -\frac{6EI}{\ell^2} & 0 & -\frac{12EI}{\ell^3} & -\frac{6EI}{\ell^2} & 0 & 0 & 0 \\
0 & -\frac{6EI}{\ell^2} & \frac{4EI}{\ell} & 0 & \frac{6EI}{\ell^2} & \frac{2EI}{\ell} & 0 & 0 & 0 \\
-\frac{EA}{\ell} & 0 & 0 & \frac{2EA}{\ell} & 0 & 0 & -\frac{EA}{\ell} & 0 & 0 \\
0 & -\frac{12EI}{\ell^3} & \frac{6EI}{\ell^2} & 0 & \frac{24EI}{\ell^3} & 0 & 0 & -\frac{12EI}{\ell^3} & -\frac{6EI}{\ell^2} \\
0 & -\frac{6EI}{\ell^2} & \frac{2EI}{\ell} & 0 & 0 & \frac{8EI}{\ell} & 0 & \frac{6EI}{\ell^2} & \frac{2EI}{\ell} \\
0 & 0 & 0 & -\frac{EA}{\ell} & 0 & 0 & \frac{EA}{\ell} & 0 & 0 \\
0 & 0 & 0 & 0 & -\frac{12EI}{\ell^3} & \frac{6EI}{\ell^2} & 0 & \frac{12EI}{\ell^3} & \frac{6EI}{\ell^2} \\
0 & 0 & 0 & 0 & -\frac{6EI}{\ell^2} & \frac{2EI}{\ell} & 0 & \frac{6EI}{\ell^2} & \frac{4EI}{\ell}
\end{bmatrix}, \tag{a}$$

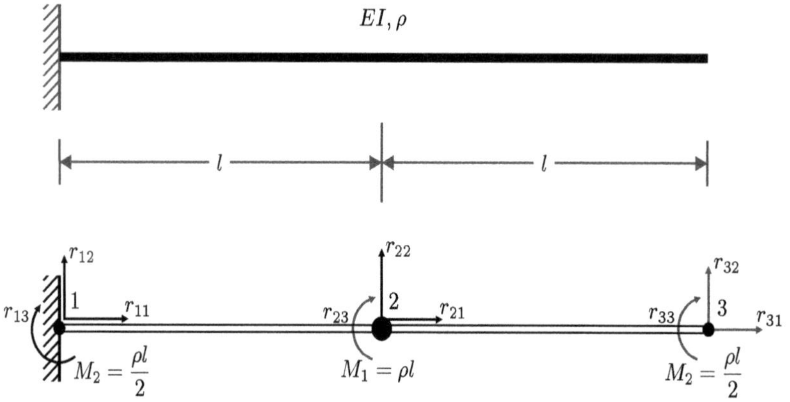

Figure 4.23. Discretization of a slender beam by two beam elements.

while the displacement vector becomes $\mathbf{r} = \left[r_{11}, r_{12}, r_{13}, r_{21}, r_{22}, r_{23}, r_{31}, r_{32}, r_{33}\right]^T$.

The boundary conditions for the clamped end at node 1 are introduced by setting $r_{11} = r_{12} = r_{13} = 0$.

Furthermore, only vertical vibration is assumed, implying that there are no axial deformations. Accordingly, r_{21} and r_{31} can also be set equal to zero. This implies that the 1st, 2nd, 3rd, 4th and 7th lines and columns of $\mathbf{K}$ can be removed. We then get

$$\mathbf{K}' = \frac{2EI}{\ell^3}\begin{bmatrix} 12 & 0 & -6 & -3\ell \\ 0 & 4\ell^2 & 3\ell & \ell^2 \\ -6 & 3\ell & 6 & 3\ell \\ -3\ell & \ell^2 & 3\ell & 2\ell^2 \end{bmatrix}, \quad \mathbf{r} = \begin{bmatrix} r_{22} \\ r_{23} \\ r_{32} \\ r_{33} \end{bmatrix}. \tag{b}$$

By introducing these restrictions, the number of DOFs is reduced to 4.

Because the rotational inertia forces are neglected, the corresponding mass matrix takes the form

$$\mathbf{M}' = \begin{bmatrix} M_1 & & & \\ & 0 & & \\ & & M_2 & \\ & & & 0 \end{bmatrix}. \tag{c}$$

For free harmonic vibration $\mathbf{r} = \mathbf{\Psi} \sin \omega t$, the corresponding eigenvalue equation reads

$$\frac{2EI}{\ell^3}\begin{bmatrix} 12 & 0 & -6 & -3\ell \\ 0 & 4\ell^2 & 3\ell & \ell^2 \\ -6 & 3\ell & 6 & 3\ell \\ -3\ell & \ell^2 & 3\ell & 2\ell^2 \end{bmatrix}\mathbf{\Psi} - \omega^2 \begin{bmatrix} \rho\ell & 0 & 0 & 0 \\ 0 & 0 & 0 & 0 \\ 0 & 0 & \frac{\rho\ell}{2} & 0 \\ 0 & 0 & 0 & 0 \end{bmatrix}\mathbf{\Psi} = \mathbf{0}. \tag{d}$$

Changing the order of the nodal DOFs r_{23} and r_{32}, Eq. (d) can be recast into the following form:

$$\frac{2EI}{\ell^3}\begin{bmatrix} 12 & -6 & 0 & -3\ell \\ -6 & 6 & 3\ell & 3\ell \\ 0 & 3\ell & 4\ell^2 & \ell^2 \\ -3\ell & 3\ell & \ell^2 & 2\ell^2 \end{bmatrix}\bar{\boldsymbol{\Psi}} - \omega^2\begin{bmatrix} \rho\ell & 0 & 0 & 0 \\ 0 & \frac{\rho\ell}{2} & 0 & 0 \\ 0 & 0 & 0 & 0 \\ 0 & 0 & 0 & 0 \end{bmatrix}\bar{\boldsymbol{\Psi}} = \mathbf{0}, \tag{e}$$

where

$$\bar{\boldsymbol{\Psi}} = \begin{bmatrix} \psi_{22} \\ \psi_{32} \\ -- \\ \psi_{23} \\ \psi_{33} \end{bmatrix} = \begin{bmatrix} \boldsymbol{\Psi}_1 \\ -- \\ \boldsymbol{\Psi}_2 \end{bmatrix}. \tag{f}$$

The first two equations can be written as

$$\frac{2EI}{\ell^3}\begin{bmatrix} 12 & -6 \\ -6 & 6 \end{bmatrix}\boldsymbol{\Psi}_1 + \frac{2EI}{\ell^3}\begin{bmatrix} 0 & -3\ell \\ 3\ell & 3\ell \end{bmatrix}\boldsymbol{\Psi}_2 - \omega^2\begin{bmatrix} \rho\ell & 0 \\ 0 & \frac{\rho\ell}{2} \end{bmatrix}\boldsymbol{\Psi}_1 = \mathbf{0}. \tag{g}$$

From the last two equations, we get

$$3\ell\begin{bmatrix} 0 & 1 \\ -1 & 1 \end{bmatrix}\boldsymbol{\Psi}_1 + \ell^2\begin{bmatrix} 4 & 1 \\ 1 & 2 \end{bmatrix}\boldsymbol{\Psi}_2 = \mathbf{0}. \tag{h}$$

Solving Eq. (h) with respect to $\boldsymbol{\Psi}_2$,

$$\boldsymbol{\Psi}_2 = -\frac{3}{\ell}\begin{bmatrix} 4 & 1 \\ 1 & 2 \end{bmatrix}^{-1}\begin{bmatrix} 0 & 1 \\ -1 & 1 \end{bmatrix}\boldsymbol{\Psi}_1 = -\frac{3}{7\ell}\begin{bmatrix} 1 & 1 \\ -4 & 3 \end{bmatrix}\boldsymbol{\Psi}_1. \tag{i}$$

If the expression in Eq. (i) for $\boldsymbol{\Psi}_2$ is inserted in Eq. (g), the eigenvalue problem can be written as

$$\left(\frac{6EI}{7\ell^3}\begin{bmatrix} 16 & -5 \\ -5 & 2 \end{bmatrix} - \omega^2\frac{\rho\ell}{2}\begin{bmatrix} 2 & 0 \\ 0 & 1 \end{bmatrix}\right)\boldsymbol{\Psi}_1 = 0. \tag{j}$$

This means that the reduction of the size of the problem by neglecting the rotational inertia is performed in the same way as for elimination of internal DOFs by application of substructuring techniques (static condensation). By introducing

$$\alpha = \omega^2\frac{7\rho\ell^4}{12EI}, \tag{k}$$

the eigenvalue problem becomes

$$\left(\begin{bmatrix} 16 & -5 \\ -5 & 2 \end{bmatrix} - \alpha\begin{bmatrix} 2 & 0 \\ 0 & 1 \end{bmatrix}\right)\boldsymbol{\Psi}_1 = 0. \tag{l}$$

By setting the determinant of the coefficient matrix in Eq. (l) equal to zero, the eigenvalues are

$$\alpha_1 = 0.36, \quad \alpha_2 = 9.64. \tag{m}$$

The frequency for the first mode (the fundamental frequency) becomes

$$\omega_1 = 0.789\sqrt{EI/\rho\ell^4} \tag{n}$$

or

$$f_1 = \frac{\omega_1}{2\pi} = 0.126\sqrt{EI/\rho\ell^4}, \quad \left(f_1^{exact} = 0.140\sqrt{EI/\rho\ell^4}\right). \tag{o}$$

The frequency for the second mode becomes

$$\omega_2 = 4.06\sqrt{EI/\rho\ell^4} \tag{p}$$

or

$$f_2 = \frac{\omega_2}{2\pi} = 0.647\sqrt{EI/\rho\ell^4}, \quad \left(f_2^{exact} = 0.877\sqrt{EI/\rho\ell^4}\right). \tag{q}$$

With two DOFs, only a rough approximation of the first two mode shapes and natural frequencies can be found. The error is least for the lowest frequency (here 10% for the first and roughly 25% for the second natural frequency). An improvement of the results would be achieved by subdividing the beam into more elements, which would reduce the uncertainty related to the assumed mass discretization. Alternatively, the distribution of mass could be modified based on experience, but this is not a generally applicable method. A better distribution of the mass is achieved by application of a consistent mass, as discussed in Section 4.10.2.

If calculations indicate that the rotational moment of inertia I has small influence compared to M, fairly good results can be achieved by setting $I = 0$. This implies that the number of unknowns in the eigenvalue problem is reduced by one-third. If increased accuracy is wanted, it is practical to increase the number of nodes in the reduced problem rather than including the rotational inertia.

The mass matrix is discussed in more detail in Section 4.10.2.

4.11 Modal Analysis

4.11.1 General

When the free vibration problem is solved, the forced dynamic response, e.g., Eq. (4.100), may be obtained by modal superposition:

$$\mathbf{r}(t) = \boldsymbol{\Psi}\mathbf{Y}(t)\ , \quad \dot{\mathbf{r}}(t) = \boldsymbol{\Psi}\dot{\mathbf{Y}}(t)\ , \quad \ddot{\mathbf{r}}(t) = \boldsymbol{\Psi}\ddot{\mathbf{Y}}(t), \tag{4.109}$$

where $\boldsymbol{\Psi}$ is a matrix representing the set of eigenmodes and $\mathbf{Y}(t)$ is a time-varying vector function. Due to the orthogonality properties of the vectors $\boldsymbol{\Psi}_i$ forming the matrix $\boldsymbol{\Psi}$, the equations to determine $\mathbf{Y}(t)$ are uncoupled for each component $Y_i(t)$ and are given by

$$\ddot{\mathbf{Y}}(t) + \mathbf{C}_g\dot{\mathbf{Y}}(t) + \boldsymbol{\Omega}^2\mathbf{Y}(t) = \boldsymbol{\Psi}^T\mathbf{R}^{ext} = \boldsymbol{\Psi}^T\mathbf{R}^s\mathbf{h}(t) = \mathbf{R}_\psi = \alpha\mathbf{h}(t). \tag{4.110}$$

Here, $\boldsymbol{\Omega} = \text{diag}(\omega_i)$ is a diagonal matrix of natural frequencies ω_i. $\boldsymbol{\Psi}^T\mathbf{K}\boldsymbol{\Psi} = \boldsymbol{\Omega}^2$, $\boldsymbol{\Psi}^T\mathbf{M}\boldsymbol{\Psi} = \mathbf{I}$, $\boldsymbol{\Psi}^T\mathbf{C}\boldsymbol{\Psi} = \mathbf{C}_g$, and $\boldsymbol{\Psi}^T\mathbf{R}^s = \alpha$. $\mathbf{R}^s$ accounts for the spatial load variation. $\mathbf{C}_g$ could be assumed to be a diagonal matrix with elements $2\xi_i\omega_i$, where ξ_i denotes the modal damping ratio for mode no. i.

From the equation $\boldsymbol{\Psi}^T\mathbf{M}\boldsymbol{\Psi} = \mathbf{I}$, it follows that $\boldsymbol{\Psi}^{-1} = \boldsymbol{\Psi}^T\mathbf{M}$. Therefore

$$\mathbf{Y} = \boldsymbol{\Psi}^{-1}\mathbf{r} = \boldsymbol{\Psi}^T\mathbf{M}\mathbf{r}. \tag{4.111}$$

Moreover,

$$\mathbf{R}_\psi = \mathbf{\Psi}^T\mathbf{R}^{ext} = \mathbf{I}\mathbf{R}_\psi = \mathbf{\Psi}^T\mathbf{M}\mathbf{\Psi}\mathbf{R}_\psi, \tag{4.112}$$

which implies that

$$\mathbf{R}^{ext} = \mathbf{M}\mathbf{\Psi}\mathbf{R}_\psi. \tag{4.113}$$

The most significant computational effort in modal analysis of large systems is associated with the eigenvalue analysis. Integration of the uncoupled modal equations is less time consuming. As compared to direct integration of the dynamic equations of motion the modal method is particularly attractive if the effect of many time-dependent load histories is to be calculated. This is because the same eigenfrequencies and modes are applied for all load histories.

If the focus is on a steady-state problem, a main question is how many modes are necessary to achieve the desired accuracy. There is no clear-cut answer to this question. However, some guidelines have been proposed, e.g., by Chopra (2001); Bathe (1996); Cook et al. (2002). One issue is that modes with a natural frequency in the range of frequencies corresponding to significant loading, when the loading is expressed by a Fourier series, need to be included. If the loading involves frequencies up to $\bar{\omega}_u$, including sum-frequency effects in the hydrodynamic loading, eigenfrequencies from the lowest and up to $\alpha\bar{\omega}_u$ should be included. As judged from the SDOF system, dynamic effects are important for α up to, say, 2 while the behavior is static for a larger α. However, this answer is simplistic because the number of modes required depends not only on the frequency content of the loading, but also on its spatial complexity, whether other responses than displacements are required, and with what accuracy. Modes retained must have frequencies that span the temporal variation of loading. Mode shapes of free vibration are unrelated to the complexity of loading, so enough modes must be included to approximate displacements associated with the spatial variation of loading and such displacement derivatives as may be needed to obtain status (stresses). Therefore, more modes are needed when the loading becomes more abrupt in time and more irregular in space, when bending moments and shear forces must be calculated, and when velocities and accelerations must be computed (Bathe, 1996; Chopra, 2001; Cook et al., 2002).

One alternative analysis strategy may be to first perform a modal analysis with the relatively few modes that represent a dynamic response and then accommodate the response of modes that behave in a static fashion. Clearly, the spatial finite element model needs to be chosen so as to represent the relevant modes. A measure of the accuracy achieved is the number of finite elements per half wavelength of the mode of oscillation. The required number depends on the assumed shape functions in the finite element, as well as of whether only displacements or also stresses are to be calculated.

4.11.2 Reduction of Modes

The solution $\mathbf{r}_r$ of Eq. (4.100) should be limited to as few modes as possible for computational efficiency. Often, dynamic effects due to inertia and damping have an effect only for some modes. This fact apparently suggests that a few modes could be applied. However, depending on the complexity of the spatial loading pattern, other modes may be crucial for accurate assessment of the response. For this reason it may be relevant to do a modal analysis and obtain the solution $\mathbf{r}_r$ before improving this solution. The improvement should be based on the error in the dynamic equilibrium equation, expressed as

$$\begin{aligned} \mathbf{R}_e &= \mathbf{R}^s\mathbf{h}(t) - \mathbf{M}\ddot{\mathbf{r}}_r - \mathbf{C}\dot{\mathbf{r}}_r - \mathbf{K}\mathbf{r}_r \\ &= \mathbf{R}^s\mathbf{h}(t) - \mathbf{M}\boldsymbol{\Psi}_r\ddot{\mathbf{Y}}_r(t) - \mathbf{C}\boldsymbol{\Psi}_r\dot{\mathbf{Y}}_r(t) - \mathbf{K}\boldsymbol{\Psi}_r\mathbf{Y}_r(t). \end{aligned} \qquad (4.114)$$

If the modes retained span an adequate frequency range, the error vector $\mathbf{R}_e$ would produce little velocity or acceleration if even more modes were retained. Accordingly, response due to the difference vector can be calculated by static analysis and added as a correction to dynamic response calculated by the foregoing modal method. By including the correction, the spatial distribution of load is given better representation, and accuracy is increased for a given number of modes. Or, equivalent accuracy is obtained using fewer modes.

The correction can be accomplished by any of several methods. One group of methods is called *mode acceleration methods* (Craig, 1981; Chopra, 2001).

Results comparable or even identical to those provided by the mode acceleration method can be obtained by *static correction*. Such a method is useful in treating a loading whose spatial complexity would require many modes, but whose temporal complexity can be represented by a smaller number of modes (Chopra, 2001). The method appears in more than one form (Maddox, 1974; Hansteen and Bell, 1988; Bathe, 1996; Dickens et al., 1997; Chopra, 2001).

The static correction $\Delta\mathbf{r}$, to be added to the $\mathbf{r}$ calculated by mode superposition, is given by the following equation:

$$\mathbf{K}\Delta\mathbf{r} = \mathbf{R}^{ext} - \mathbf{R}^{ext}_{approx}, \qquad (4.115)$$

where $\mathbf{R}^{ext}_{approx}$ can be obtained by retaining r modes in Eqs. (4.112) and (4.113); that is,

$$\mathbf{R}^{ext}_{approx} = \mathbf{M}\boldsymbol{\Psi}_r\boldsymbol{\Psi}_r^T\mathbf{R}^{ext}. \qquad (4.116)$$

Here, $\boldsymbol{\Psi}_r$ is an $n \times r$ matrix, where n and r are the number of degrees of freedom and modes used in the modal analysis, respectively. The structure must have an adequate support so that $\mathbf{K}$ in Eq. (4.115) is nonsingular. The matrix $\mathbf{K}$ needs to be reduced for equation solving only once, although $\Delta\mathbf{R}$ must be computed at each instant of time for which the correction is desired. If $\mathbf{R}^{ext}$ represents a loading of fixed spatial distribution whose intensity changes with time, $\Delta\mathbf{r}$ need only be scaled from one time instant to another. The static correction approach presented here is a simple one. The method has been applied in other forms in the previously cited literature.

In the *mode acceleration (MA) method*, an improved approximate solution is established by

$$\mathbf{r} \approx \mathbf{r}_r + \Delta\mathbf{r}_{\mathrm{MA}}, \qquad (4.117)$$

where $\Delta\mathbf{r}_{\mathrm{MA}}$ is obtained by solving the static problem as described previously.

In the mode truncation (MT) method,

$$\mathbf{r} \approx \mathbf{r}_r + \Delta\mathbf{r}_{\mathrm{MT}}, \qquad (4.118)$$

where $\Delta\mathbf{r}_{\text{MT}}$ is determined by (Dickens et al., 1997),

$$\Delta\mathbf{r}_{\text{MT}}(t) = \mathbf{P}\mathbf{Y}_p(t) \tag{4.119}$$

and

$$\mathbf{P}^T\mathbf{M}\mathbf{P}\ddot{\mathbf{Y}}_p(t) + \mathbf{P}^T\mathbf{C}\mathbf{P}\dot{\mathbf{Y}}_p(t) + \mathbf{P}^T\mathbf{K}\mathbf{P}\mathbf{Y}_p(t) = \mathbf{P}^T\left(\mathbf{R}^0 - \mathbf{R}_r\right)\mathbf{h}(t). \tag{4.120}$$

Here, $\mathbf{P} = \mathbf{X}\mathbf{Q}$ and $\mathbf{X}$ and $\mathbf{Q}$ are given by Eqs. (4.121) (4.123), respectively:

$$\mathbf{K}\mathbf{X} = \mathbf{R}^s - \mathbf{R}_r, \tag{4.121}$$

and normalizing $\mathbf{K}$ and $\mathbf{M}$ as follows:

$$\bar{\mathbf{K}} = \mathbf{X}^T\mathbf{K}\mathbf{X}, \quad \bar{\mathbf{M}} = \mathbf{X}^T\mathbf{M}\mathbf{X}. \tag{4.122}$$

Then, solve

$$(\bar{\mathbf{K}} - \bar{\omega}_p^2\bar{\mathbf{M}})\mathbf{Q} = 0. \tag{4.123}$$

$\mathbf{Y}_p(t)$ can then be obtained by Eq. (4.120) in the same manner as $\mathbf{Y}(t)$ was obtained from Eq. (4.110). Equation (4.119) then gives $\Delta\mathbf{r}_{\text{MT}}$.

The difference between the MA and MT methods is the way in which $\Delta\mathbf{r}_{\text{MA}}$ and $\Delta\mathbf{r}_{\text{MT}}$ are obtained. The MA method uses a static solution, while the MT solution is based on an approximate dynamic solution, which specializes into the MA solution if the acceleration and velocities in the nonretained modes are set equal to zero. Moreover, the computer implementation of the MT method is easier than for the MA method because it uses the same algorithms as the basic modal approach. The crucial point for both methods is that the loading should have frequencies below the highest retained modes. Then the methods are found to yield nearly the same results, except for accelerations, which were predicted more accurately by the MT method (Dickens et al., 1997).

4.12 Other Approaches for Reducing the Number of DOFs in Elastic Structures

Most of the examples included in this book are selected so as to be amenable to "hand calculation" and thus provide insight in basic principles. Dynamic models for offshore structures, ships, and other structures may include thousands of DOFs. Moreover, dynamic analysis involves more computational effort than static analyses. To reduce the computational efforts, it is sometimes helpful to reduce the number of DOFs. The DOFs in the reduced set do not need to be nodal DOFs of the FE model. They may be generalized DOFs, like those used in the classical Rayleigh-Ritz method, see, e.g., (Noor, 1994).

Various kinds of reduction techniques are envisaged. Each can be regarded as either a way of imposing an elastic constraint or as a way of providing a *reduced basis*. A *basis* is a set of linearly independent functions or vectors that can be combined in various proportions to represent or approximate other vectors. If an FE discretization is applied, the complete set of vectors could be the eigenvectors. A basis is called *reduced* if it includes fewer vectors than the complete set.

Reduction can be accomplished by methods that range from intuitive to semirigorous. We have seen in Example 4.10.4 how the eigenvalue problem for concentrated

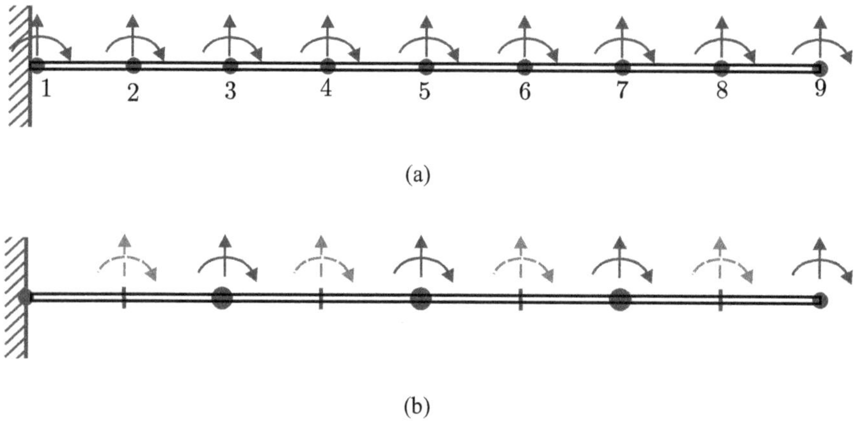

Figure 4.24. Discretization of mass.

mass can be reduced by neglecting axial deformations and by setting the rotational mass equal to zero. The latter approach can be extended by concentrating the mass associated with translational displacements to some nodes. In Fig. 4.24 two models of a cantilevered beam are shown. The axial displacements are already discarded. In Fig. 4.24(a) the mass is assigned to all 9 nodes. In the model in Fig. 4.24(b) the mass is assigned to every second node, and the displacements in the four massless nodes can be expressed by displacements in the nodes with mass. Mathematically, this operation can be described by grouping the DOFs into $\mathbf{r}_1$ and $\mathbf{r}_2$:

$$\mathbf{r} = \begin{bmatrix} \mathbf{r}_1 \\ \mathbf{r}_2 \end{bmatrix}. \tag{4.124}$$

Here, $\mathbf{r}_1$ contains the p DOFs with mass and $\mathbf{r}_2$ the remaining $(n - p)$ DOFs.

The equation of motion for free vibration,

$$\mathbf{M}\ddot{\mathbf{r}} + \mathbf{K}\mathbf{r} = 0, \tag{4.125}$$

can be regrouped accordingly as follows:

$$\begin{bmatrix} \mathbf{M}_{11} & 0 \\ 0 & 0 \end{bmatrix} \begin{bmatrix} \ddot{\mathbf{r}}_1 \\ \ddot{\mathbf{r}}_2 \end{bmatrix} + \begin{bmatrix} \mathbf{K}_{11} & \mathbf{K}_{12} \\ \mathbf{K}_{21} & \mathbf{K}_{22} \end{bmatrix} \begin{bmatrix} \mathbf{r}_1 \\ \mathbf{r}_2 \end{bmatrix} = 0. \tag{4.126}$$

The second line yields

$$\mathbf{r}_2 = -\mathbf{K}_{22}^{-1}\mathbf{K}_{21}\mathbf{r}_1, \tag{4.127}$$

which inserted in the first line gives

$$\mathbf{M}_{11}\ddot{\mathbf{r}}_1 + \left(\mathbf{K}_{11} - \mathbf{K}_{12}\mathbf{K}_{22}^{-1}\mathbf{K}_{21}\right)\mathbf{r}_1 = 0 \tag{4.128}$$

or

$$\mathbf{M}_{11}\ddot{\mathbf{r}}_1 + \bar{\mathbf{K}}_{11}\mathbf{r}_1 = 0, \tag{4.129}$$

where

$$\bar{\mathbf{K}}_{11} = \mathbf{K}_{11} - \mathbf{K}_{12}\mathbf{K}_{22}^{-1}\mathbf{K}_{21}. \tag{4.130}$$

Although $\mathbf{K}$ is usually a banded matrix, $\bar{\mathbf{K}}$ will now be a full matrix.

In the master-slave reduction, the total number of DOFs is subdivided into two groups, $\mathbf{r} = [\mathbf{r}_m, \mathbf{r}_s]^T$, where the set $\mathbf{r}_m$ is the p matrix DOFs and $\mathbf{r}_s$ the $(n - p)$ slave DOFs. Then, it is assumed that $\mathbf{r}_s$ can be expressed through a kinematic constraint that $\mathbf{r}_s = \mathbf{H}\,\mathbf{r}_m$ or

$$\mathbf{r} = \begin{bmatrix} \mathbf{r}_m \\ \mathbf{r}_s \end{bmatrix} = \begin{bmatrix} \mathbf{I} \\ \mathbf{H} \end{bmatrix} \mathbf{r}_m. \tag{4.131}$$

The initial equation of motion in terms of $\mathbf{r}$ (and its derivatives) may then be transformed into an equation in terms of $\mathbf{r}_m$. The main challenge is to determine the kinematic constraint, i.e., the $\mathbf{H}$ matrix. It is most common to let $\mathbf{H}$ be the static displacements that result when an unloaded structure is subjected to unit displacements at the master DOFs (Guyan, 1965).

Component mode synthesis (CMS) is applied under different names, such as modal synthesis, substructure synthesis, and dynamic substructuring. The method is analogous to static substructuring. The motivation for using CMS is partly to reduce the DOFs and partly the advantage of more efficient modeling. Substructuring is known to be beneficial for structures with repetitive components. Moreover, when the structural system consists of different types of components, different persons or groups can model the different substructures. This is, for instance, the case with a floating wind turbine consisting of a floating body, tower, rotor with 3 blades, etc.

Further details about various formulations are given by Craig (1981); Craig (1987); Leung (1988); Shyu et al. (1997).

In the modal method of analysis, only some of the modes with the lowest natural frequency rather than the whole set of modes are applied. It is not easy to say how many DOFs must appear in the reduced set. There should be enough DOFs to represent the lowest vibration modes because they are almost certain to be important. Few DOFs may be needed if the loading is simple in its spatial distribution and has only low-frequency content, and if only displacement response is needed. If the loading has a complicated spatial distribution or is suddenly applied, or if stress response is required, more DOFs are needed. Also, more DOFs are needed to calculate the velocity response, and still more for the acceleration response.

5 Stochastic Processes

5.1 Introduction

In Chapter 1 it is mentioned that a dynamic analysis can be carried out in two different ways, depending on how the loads are described. One alternative is a deterministic analysis, which requires that the load time history is fully known. The other alternative is a stochastic analysis, where statistical concepts are used to specify the loads. In this chapter we show why (and how) it is expedient and necessary to use statistical and probabilistic methods to describe a number of load types to which structures are subjected.

Typical examples are wind loads on a high-rise building or a suspension bridge, wave loads on an offshore structure or a floating bridge. When the loads on a structure are described in terms of statistical quantities, then the response must also be described and analyzed in terms of the same kind of quantities.

5.2 Examples of Stochastic Modeling

A typical feature of a series of physical phenomena of engineering interest is that each one of them under seemingly identical conditions exhibit quite different behaviour from one recorded experiment to the next. In addition, each recorded time history of the quantity studied is often characterized by being highly irregular.

Figure 5.1 shows three time histories of the water surface elevation measured in a laboratory wave tank. This tank is equipped with a wave maker that can be used to generate irregular waves similar to those observed on the open ocean. The three recorded wave elevations represent the same wave condition, and the three time histories do look quite similar when the overall picture is considered. However, to mimic reality the waves were generated in such a way that when the three time series are subjected to closer scrutiny, it is seen that there are significant differences between them on a local level. Another salient feature of these time histories is that they look irregular or random, as do real waves.

In principle, one can envisage that such laboratory experiments can be repeated as many times as one may want, and assume that there is an ensemble available with an infinite number of sample time histories.

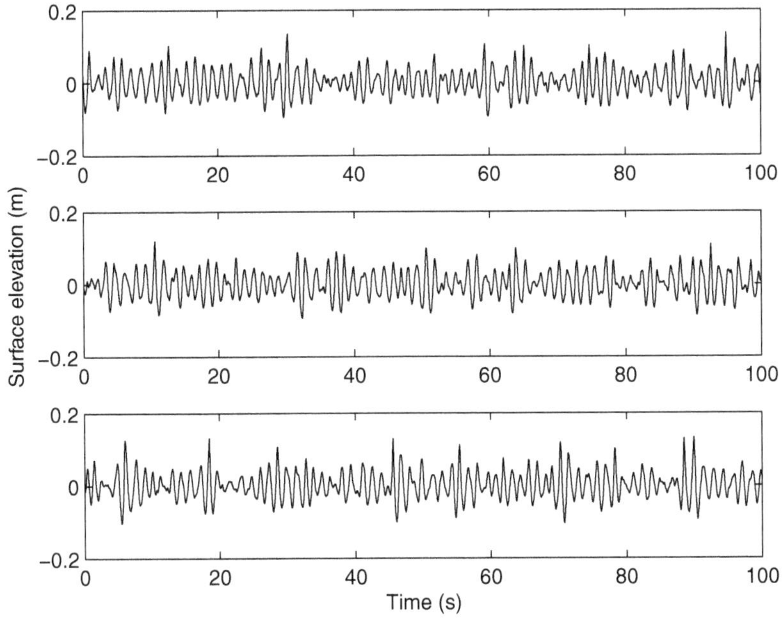

Figure 5.1. An ensemble of time histories of irregular waves measured in a laboratory wave tank.

One can imagine that for such an ensemble, the value of the surface elevation is recorded at a given point in time, say, $t = t_1$. Hence, an ensemble of elevation values is obtained. Looking at these values, one will find that they are distributed around a particular value, which is zero in Fig. 5.1, like the outcomes of a random variable. It is therefore obvious reasons to assume that the surface elevation at $t = t_1$ can be described as a random variable. Let $X(t_1)$ denote the surface elevation at $t = t_1$ modeled as a random variable. The probability distribution of $X(t_1)$ can be estimated by determining the relative number of times specific outcomes occur. Fig. 5.2 illustrates N surface elevation time histories where the value of the surface elevation has been recorded at $t = t_1$. An estimate of the cumulative distribution function (CDF) of $X(t_1)$, denoted by $F_{X(t_1)}(x) = \text{Prob}\{X(t_1) \leq x\}$ ($\text{Prob}\{E\}$ = the probability of the event E), is then

$$F_{X(t_1)}(x) \approx \frac{N[X(t_1) \leq x]}{N}, \tag{5.1}$$

where $N[X(t_1) \leq x]$ denotes the number of the observed N surface elevation time histories when the value at $t = t_1$ is less than or equal to x. If $X(t_1)$ is a genuine random variable, the following relation holds true.

$$F_{X(t_1)}(x) = \lim_{N\to\infty} \frac{N[X(t_1) \leq x]}{N}. \tag{5.2}$$

For sufficiently large N, Eq. (5.1) will therefore provide good estimates of $F_{X(t_1)}(x)$ for a suitable range of x-values.

A natural extension of this consideration is to assume that the surface elevation at any arbitrary time point t can be assumed to be a random variable $X(t)$ with a CDF $F_{X(t)}(x)$, which is dependent on t in general. The result of this is that we

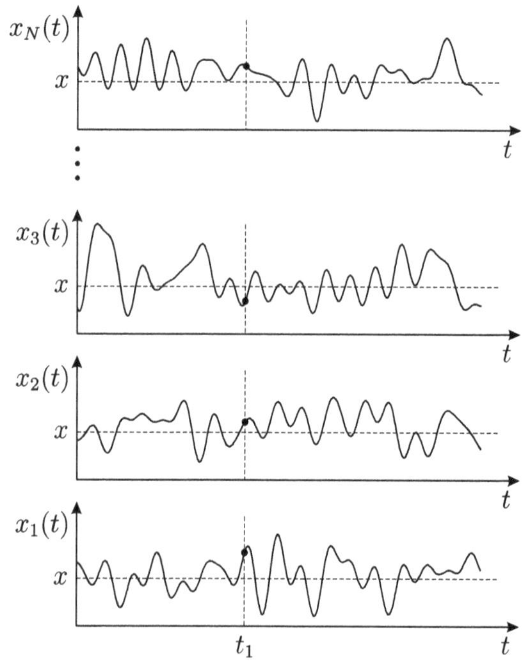

Figure 5.2. Ensemble of surface elevation time histories.

now may consider the surface elevation as a random function $X(t)$ with t as the variable argument. The expression "random function" is often replaced by the term "stochastic process," which is also the preferred notation in this book.

When a specific experiment is performed, the surface elevation at a given point in time $t = t_1$, represented by the random variable $X(t_1)$, will assume a particular value, which is the outcome of the random variable. However, the surface elevation time history considered as a stochastic process $X(t)$ will have as outcome the whole measured surface elevation time history for the specific experiment. The set of all possible surface elevation time histories that can be obtained by repeating the experiment, constitutes the sample space of the stochastic process.

A stochastic process is an abstract notion in a similar manner as a random variable. What can be observed physically are the outcomes, also usually referred to as the realizations. These will manifest themselves as specific time histories in the context considered here.

So far, we have considered physical phenomena observed as the result of a conducted experiment in a laboratory under well defined conditions. When it comes to natural phenomena that can cause loads on a structure, these have already manifested themselves for some billion years, and most likely will continue to do so for another 4-5 billion years. For every practical purpose one may therefore assume that these phenomena have been present for an infinite period of time, that is, from $t = -\infty$, and that they will continue to be present for ever, that is, to $t = \infty$.

Figure 5.3 shows part of a time history of the recorded wave elevation at a given point on the ocean surface. A prominent feature of this time series is its irregularity, which is exactly what we have tried to recreate in the surface elevation time histories of Fig. 5.1. It may therefore seem reasonable to try to also model the real wave elevation as a stochastic process. There is, however, one apparent

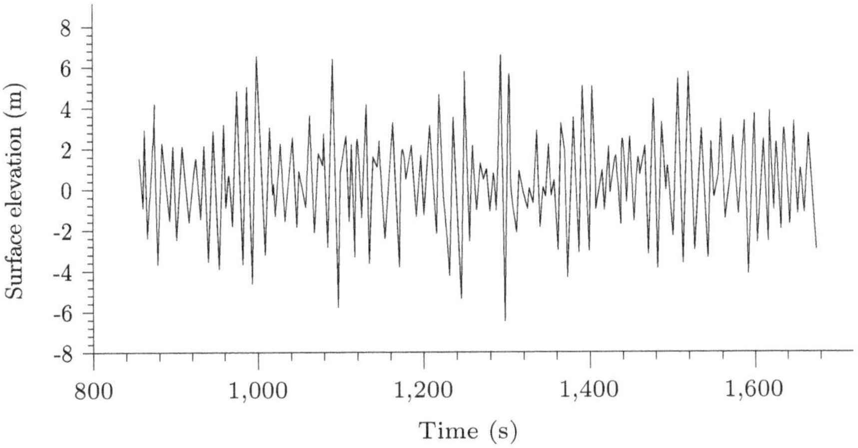

Figure 5.3. Recorded time history of the wave elevation out at sea.

complication here. It is indeed impossible for us to repeat nature's own "experiment." Nevertheless, it is possible to imagine that this can be done as a pure thought experiment. Unfortunately, this is not of much help because information about the statistical properties of the stochastic process model is obtained from an ensemble of realizations. In this particular case, only part of a single realization is available.

This difficulty can, however, be circumvented by assuming that the time history that is available is (part of) a realization of what is called an *ergodic stochastic process*. If that is a reasonable assumption, the problem is in practice solved because this assumption implies that statistical information about the process is in fact contained in each single realization of the process. As an example, the CDF $F_{X(t)}(x)$ of an ergodic process $X(t)$ is estimated by determining the relative amount of time that a realization $x(t)$ of $X(t)$ assumes values less than or equal to x, that is,

$$F_{X(t)}(x) = \lim_{T\to\infty} \frac{T[x(t) \leq x]}{T}. \tag{5.3}$$

Here T denotes the length of the time history considered, and $T[x(t) \leq x]$ denotes the total amount of time during T where $x(t) \leq x$, see Fig. 5.4

The wave elevation on the open ocean has with considerable success been modeled as an ergodic stochastic process. Partly responsible for this success is the fact that the CDF with fairly good accuracy can be assumed to be a normal distribution.

Figure 5.4. Contributions to $T[x(t) \leq x]$.

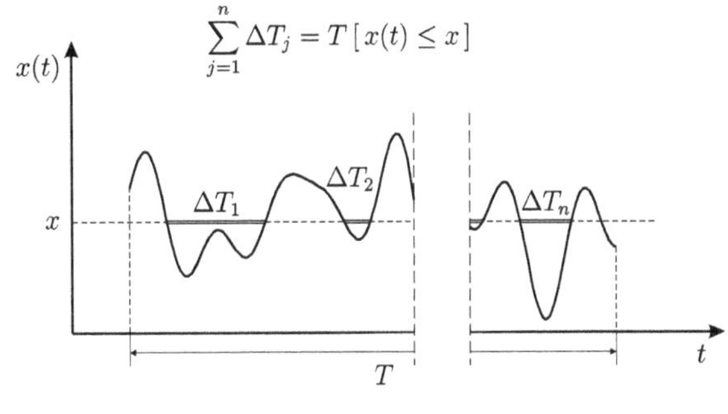

This section primarily describes a few central elements involved in modeling physical phenomena as stochastic processes. In the following, this description will be complemented with more accurate definitions and some important details that are deliberately omitted here.

5.3 Random Variable, Mean Value, and Variance

To proceed with the discussion on modeling and analysis of loads and responses by using stochastic processes, it is necessary to define this concept. Let us start by recapitulating a few key results from probability theory. Starting with the CDF $F_X(x)$ of a random variable X, which we encountered in the previous section, the corresponding probability density function (PDF), denoted by $f_X(x)$, is defined as follows

$$f_X(x) = \frac{dF_X(x)}{dx}, \tag{5.4}$$

provided, of course, that $F_X(x)$ is differentiable. This gives that

$$F_X(x) = \int_{-\infty}^{x} f_X(s)\, ds. \tag{5.5}$$

The *mean value* (expected value) of X, which is denoted by m_X, is given by

$$m_X = E[X] = \int_{-\infty}^{\infty} x\, f_X(x)\, dx. \tag{5.6}$$

If X can assume only a finite number of values $x^{(k)}$, $k = 1, \ldots, n$, the integral in Eq. (5.6) will degenerate to a finite sum:

$$m_X = \sum_{k=1}^{n} x^{(k)}\, p_k, \tag{5.7}$$

where $p_k = \text{Prob}\{X = x^{(k)}\}$.

m_X can also be expressed as an average value over all outcomes $x_1, x_2, \ldots, x_j, \ldots$, succinctly written as $\{x_j\}_{j=1}^{\infty}$, of X, which is obtained by repeating the stochastic experiment that determines X an "infinite" number of times. Then

$$m_X = \lim_{N \to \infty} \frac{1}{N} \sum_{j=1}^{N} x_j. \tag{5.8}$$

It may be instructive to illustrate the connection between the two ways of finding the mean value by having a closer look at the two Eqs. (5.7) and (5.8). Assume that N experiments have given the outcomes $x_1, \ldots, x_N$ ($N \gg n$). Because X can assume only the values $x^{(1)}, \ldots, x^{(n)}$, the outcomes are divided into n groups according to the value. The outcomes in group no. k are equal to $x^{(k)}$. Let N_k denote the number of outcomes in this group. Then, clearly,

$$\frac{1}{N} \sum_{j=1}^{N} x_j = \sum_{k=1}^{n} x^{(k)} \frac{N_k}{N}. \tag{5.9}$$

This leads to the result

$$\lim_{N\to\infty} \frac{1}{N} \sum_{j=1}^{N} x_j = \sum_{k=1}^{n} x^{(k)}\, p_k \tag{5.10}$$

because $p_k = \lim_{N\to\infty}(N_k/N)$.

In general the expected value of the random variable $g(X)$ is given as

$$m_{g(X)} = \mathrm{E}[g(X)] = \int_{-\infty}^{\infty} g(x)\, f_X(x)\, dx. \tag{5.11}$$

where $g(\cdot)$ is a suitable function such that the integral exists. As is well known, not all functions can be integrated.

By putting $g(x) = (x - m_X)^2$, the *variance* of X is obtained. It is denoted here either by $\mathrm{Var}[X]$ or σ_X^2. Hence

$$\sigma_X^2 = \mathrm{Var}(X) = \mathrm{E}[(X - m_X)^2] = \int_{-\infty}^{\infty} (x - m_X)^2\, f_X(x)\, dx. \tag{5.12}$$

$\sigma_X = \sqrt{\mathrm{Var}[X]}$ (≥ 0) is called the *standard deviation*, and it is a good measure for the spread or variability of the values/outcomes of X.

From Eq. (5.12), it is seen that the variance can also be expressed by the equation

$$\sigma_X^2 = \mathrm{E}[X^2 - 2m_X X + m_X^2] = \mathrm{E}[X^2] - m_X^2. \tag{5.13}$$

σ_X^2 calculated from a sequence of outcomes (an ensemble) $\{x_j\}_{j=1}^{\infty}$ is then

$$\sigma_X^2 = \lim_{N\to\infty} \left\{ \frac{1}{N} \sum_{j=1}^{N} x_j^2 - \left(\frac{1}{N} \sum_{j=1}^{N} x_j \right)^2 \right\}. \tag{5.14}$$

In practice, there is never an infinite set of data available so that an estimate of the mean value and the variance from Eqs. (5.8) and (5.14) will always have to be based on a finite number of outcomes; in many cases, the available number is quite limited. This may cause a substantial degree of uncertainty in the estimates, but this is something an engineer will have to live with. Fortunately, there are various techniques that can be used to quantify this uncertainty. This topic is not pursued here. Instead, the interested reader is referred to any standard book on applied statistics, e.g., Benjamin and Cornell (1978).

A dimensionless quantity, which is often used as a measure for statistical fluctuations or uncertainty, is the so-called *coefficient of variation*. For a random variable X, it is denoted here by V_X, and it is defined as

$$V_X = \frac{\sigma_X}{m_X}. \tag{5.15}$$

It is, of course, assumed that $m_X \neq 0$. A small V_X means that there are, relatively speaking, small statistical fluctuations around the mean value, and, obviously, conversely for a large V_X.

5.4 Definition of a Stochastic Process

The mathematical definition of a stochastic process, adapted to our need in this book, can be given as follows:

> *The quantity $X(t)$ is called a* stochastic process *if $X(t)$ is a random variable for each value of t in an interval (a, b).*

Note that $a = -\infty$ and/or $b = \infty$ are allowed limits for the interval.

5.4.1 Example – An Elementary Stochastic Process

Assume that X is a random variable, for example normally distributed with mean value m and standard deviation σ (> 0), that is, the PDF is

$$f_X(x) = \frac{1}{\sqrt{2\pi}\sigma} \exp\Big\{ -\frac{1}{2}\Big(\frac{x-m}{\sigma}\Big)^2 \Big\}. \tag{a}$$

If $g(t)$ is a given, real function defined for $-\infty < t < \infty$, e.g., $g(t) = \cos \omega t$, where ω is a positive constant, then

$$X(t) = X g(t) \tag{b}$$

is a stochastic process defined for $-\infty < t < \infty$. A realization $x(t)$ of this process is then given as a product of $g(t)$ with an outcome x of the random variable X, that is,

$$x(t) = x\, g(t). \tag{c}$$

Hence, if $g(t) = \cos \omega t$, the various realizations would be harmonic functions of the same period, but with different amplitudes.

The mean value $m_{X(t)}$ and standard deviation $\sigma_{X(t)}$ of $X(t)$ is given as

$$m_{X(t)} = \mathrm{E}[X g(t)] = \mathrm{E}[X]\, g(t) = m\, g(t) \tag{d}$$

and, see Eq. (5.13),

$$\begin{aligned} \sigma_{X(t)} &= \sqrt{\mathrm{E}[(X g(t))^2] - (m\, g(t))^2} = \sqrt{\mathrm{E}[X^2]\, g(t)^2 - m^2\, g(t)^2} \\ &= \sqrt{\mathrm{E}[X^2] - m^2}\, |g(t)| = \sigma\, |g(t)|. \end{aligned} \tag{e}$$

For each given value of t, $g(t)$ is a constant, that is, $X(t) = X g(t)$ is also normally distributed if $g(t) \neq 0$. The mean value and standard deviation are given by (d) and (e), that is, the PDF of $X(t)$ becomes

$$f_{X(t)}(x) = \frac{1}{\sqrt{2\pi}\sigma\, |g(t)|} \exp\Big\{ -\frac{1}{2}\Big(\frac{x - m g(t)}{\sigma\, |g(t)|}\Big)^2 \Big\}. \tag{f}$$

5.4.2 Example – A Harmonic Stochastic Process

Let X_j, $j = 1, 2$, be two normally distributed independent random variables with mean value m_j and standard deviation σ_j (> 0). Let ω be a positive constant. Then,

$$X(t) = X_1 \cos \omega t + X_2 \sin \omega t, \tag{a}$$

is a stochastic process, and for every fixed t, the random variable $X(t)$ is normally distributed. The realizations of this process are harmonic functions of amplitude and phase determined by the outcomes of the random variables X_1 and X_2. We shall see later that this type of process can be used to construct more complicated processes just as an arbitrary periodic function can be reconstructed by harmonic functions using the Fourier series technique.

5.5 Joint Probability Distributions

To analyze a stochastic process, it is not enough to know only the CDF $F_{X(t)}(x)$ for every time t. It is also necessary to know something about the connection between the values of the process at different times. This can be expressed, e.g., by the joint (multi-dimensional) CDF. Let us start with the joint CDF of $X_1 = X(t_1)$ and $X_2 = X(t_2)$ for two different times t_1 and t_2. This joint CDF is denoted by $F_{X_1 X_2}(x_1, x_2)$ and defined by the equation

$$F_{X_1 X_2}(x_1, x_2) = \text{Prob}\{X_1 \leq x_1 \text{ and } X_2 \leq x_2\}. \tag{5.16}$$

It follows that $F_{X_1 X_2}(x_1, x_2)$ expresses the probability that both events $\{X_1 \leq x_1\}$ and $\{X_2 \leq x_2\}$ occurs. If $F_{X_1 X_2}(x_1, x_2)$ is differentiable with respect to both variables x_1 and x_2, the joint PDF $f_{X_1 X_2}(x_1, x_2)$ is defined as

$$f_{X_1 X_2}(x_1, x_2) = \frac{\partial^2 F_{X_1 X_2}(x_1, x_2)}{\partial x_1 \partial x_2}. \tag{5.17}$$

Conversely,

$$F_{X_1 X_2}(x_1, x_2) = \int_{-\infty}^{x_1} \int_{-\infty}^{x_2} f_{X_1 X_2}(s_1, s_2)\, ds_1\, ds_2. \tag{5.18}$$

Because the event $\{X_2 \leq \infty\}$ is always true, then $\text{Prob}\{X_1 \leq x_1 \text{ and } X_2 \leq \infty\} = \text{Prob}\{X_1 \leq x_1\}$. This leads to the equation of marginal probability,

$$F_{X_1}(x_1) = F_{X_1 X_2}(x_1, \infty), \tag{5.19}$$

or

$$f_{X_1}(x_1) = \int_{-\infty}^{\infty} f_{X_1 X_2}(x_1, x_2)\, dx_2. \tag{5.20}$$

Two random variables X_1 and X_2 are called statistically independent, or just independent for brevity, if

$$F_{X_1 X_2}(x_1, x_2) = F_{X_1}(x_1) F_{X_2}(x_2). \tag{5.21}$$

Expressed in terms of the respective PDFs, this condition becomes

$$f_{X_1 X_2}(x_1, x_2) = f_{X_1}(x_1) f_{X_2}(x_2). \tag{5.22}$$

The expected value of $h(X_1, X_2)$ is, analogously with the case of one variable, given by the equation

$$\text{E}[h(X_1, X_2)] = \int_{-\infty}^{\infty} \int_{-\infty}^{\infty} h(x_1, x_2)\, f_{X_1 X_2}(x_1, x_2)\, dx_1\, dx_2. \tag{5.23}$$

These definitions and relations can be simply extended to any finite number of random variables $X_1, \ldots, X_n$, which may represent the values of a stochastic process

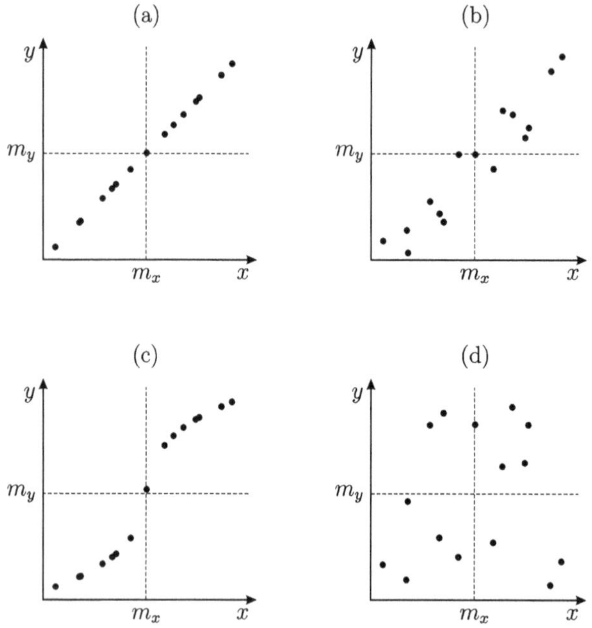

Figure 5.5. Varying degree of correlation. (a) Perfect correlation. (b) Moderate correlation. (c) Perfect nonlinear connection, but moderate correlation. (d) No correlation.

at n times $t_1, \ldots, t_n$, that is, $X_j = X(t_j)$, $j = 1, \ldots, n$. For example, the joint CDF of $X_1, \ldots, X_n$ is given as

$$F_{X_1 \ldots X_n}(x_1, \ldots, x_n) = \text{Prob}\{X_1 \leq x_1 \text{ and} \ldots \text{and } X_n \leq x_n\}. \tag{5.24}$$

In principle, one has to know the joint CDF/PDF for any choice of times $t_1 < t_2 < \ldots < t_n$ and for $n = 1, 2, \ldots$, in order to fully describe a stochastic process, and in some cases not even this suffices. For experimental data, it is in practice impossible to provide this kind of detailed information. In such situations, one has to be satisfied with far less. As we shall see, in some cases one can get quite far by knowing the joint CDF/PDF for the process at any two points in time.

5.6 Correlation

Imagine that we conduct an experiment where we want to investigate the relationship between external load on a structure and the strain at a specified location in the structure. The experiment is carried out for N different load levels, and each pair of corresponding load-strain values are plotted as illustrated in Fig. 5.5, where each plot represent a possible set of measurements. Ideally, for a linear elastic structure the plot of the measurements could be as in Fig. 5.5(a), where there is a perfect linear relationship between the load x and the strain y. The results in Fig. 5.5(d) may serve as an example of the other extreme. If the strain measurements for some reason have failed, such that only extraneous noise signals without any connection to the load have been recorded, the plotted measurements could look like Fig. 5.5(d).

In practice the measurement results would typically be somewhere between these limiting cases. In Fig. 5.5(b) there is an underlying linear trend in the x and y data, but there is a certain scattering that breaks a perfect linear relationship. This scattering may be caused by disturbances in the measurements and/or an element of randomness (statistical variation) in the relationship between the x and y values.

In Fig. 5.5(c) there is a perfect analytical relationship between x and y, but it is nonlinear.

To estimate the degree of linear connection between two random variables X and Y having mean values m_X and m_Y, respectively, a new random variable

$$\hat{Y} = kX + a\,, \tag{5.25}$$

is introduced and the following difference investigated,

$$\Delta = Y - \hat{Y}. \tag{5.26}$$

k and a are constants, which are chosen such that Δ becomes as small as possible with respect to some chosen measure of deviation. The most common choice is to use the mean square deviation measure, that is, $\mathrm{E}[\Delta^2]$, which is a function of k and a, let us call it $e = e(a, k)$. To determine the values of k and a where e is minimum, we solve the two equations $\partial e/\partial a = 0$ and $\partial e/\partial k = 0$, that is,

$$\frac{\partial e}{\partial a} = \frac{\partial}{\partial a}\mathrm{E}[\Delta^2] = -2\mathrm{E}[Y - kX] + 2a = 0\,, \tag{5.27}$$

and

$$\frac{\partial e}{\partial k} = -2\mathrm{E}[XY] + 2k\mathrm{E}[X^2] + 2a\mathrm{E}[X] = 0\,. \tag{5.28}$$

Solving these two equations with respect to a and k gives

$$a = m_Y - km_X\,, \tag{5.29}$$

and

$$k = \frac{\mathrm{E}[(X - m_X)(Y - m_Y)]}{\mathrm{E}[(X - m_X)^2]}\,. \tag{5.30}$$

Note that Eq. (5.29) gives $\mathrm{E}[\Delta] = 0$. The expression in the numerator of Eq. (5.30) is called the *covariance* of X and Y. It is denoted here by c_{XY}, that is,

$$c_{XY} = \mathrm{E}[(X - m_X)(Y - m_Y)]\,. \tag{5.31}$$

An important statistical parameter is the *correlation coefficient*, which is defined as

$$\rho_{XY} = \frac{c_{XY}}{\sigma_X \sigma_Y}\,. \tag{5.32}$$

Substituting the expressions for a and k into $\mathrm{E}[\Delta^2]$, it is obtained that

$$\mathrm{E}[\Delta^2] = \sigma_X^2 k^2 - 2c_{XY}k + \sigma_Y^2. \tag{5.33}$$

The rhs of Eq. (5.33) is a polynomial of second degree in k, which cannot have two different real roots because this would imply that $\mathrm{E}[\Delta^2]$ would become negative for the k-values between these two roots. This is impossible because $\mathrm{E}[\Delta^2] \geq 0$. Hence, $c_{XY}^2 - \sigma_X^2\sigma_Y^2 \leq 0$ because this is the expression under the square root in the formula for the roots of the polynomial. We have thereby proved the following result,

$$-1 \leq \rho_{XY} \leq 1. \tag{5.34}$$

The value of $\mathrm{E}[\Delta^2]$ at the minimum point $k = c_{XY}/\sigma_X^2$ is then

$$\mathrm{E}[\Delta^2]_{min} = \sigma_Y^2(1 - \rho_{XY}^2), \tag{5.35}$$

and it is seen that $E[\Delta^2] = 0$, which gives $Y = \hat{Y}$, if $\rho_{XY} = \pm 1$ and only then ($\sigma_Y > 0$). ρ_{XY} is therefore a direct measure of a linear relationship between X and Y.

It may be mentioned that if X and Y are independent random variables, then $\rho_{XY} = 0$. This follows from Eqs. (5.22), (5.23) and (5.31) because $E[(X - m_X)] = 0$ and $E[(Y - m_Y)] = 0$. The opposite result is, however, not true. $\rho_{XY} = 0$ does not generally imply independence. An easy example of this is obtained if X is a zero mean normal variable, and $Y = X^2$. Then $c_{XY} = E[X(X^2 - \sigma_X^2)] = E[X^3] - \sigma_X^2 E[X] = 0$, that is, $\rho_{XY} = 0$, but X and Y are clearly not independent.

The concept of correlation will play an important role in this book.

5.7 Stationary Processes

Going back to the discussion on the pressure fluctuations in Fig. 5.1, it is realized that a situation was depicted where the probability distribution did not change with time. This implies that the statistical properties of the pressure fluctuations do not change with time; they have for instance the same mean value and standard deviation for all points in time. This is a typical feature for the realizations of what is called a *stationary process*.

The definition of a stationary process can be given in several ways. We use the simplest, and the one that serves our purposes best.

> *A stochastic process $X(t)$ is called (weakly) stationary if $E[X(t)]$ and $E[X(t)X(t+\tau)]$ are both independent of t.*

This means that $E[X(t)X(t+\tau)]$ only depends on τ. This quantity is called the *autocorrelation function*, or just autocorrelation, and it is denoted here by $R_X(\tau)$. A stochastic process is (weakly) stationary if and only if the following two conditions are satisfied ($m_X = m_{X(t)}$):

$$m_X = E[X(t)] = \text{constant}, \tag{5.36}$$

and

$$R_X(\tau) = E[X(t)X(t+\tau)] = \text{function only of } \tau. \tag{5.37}$$

Assume that we have an ensemble of time histories $\{x_1(t), \ldots, x_N(t)\}$, which it seems reasonable to model as realizations of a stationary stochastic process, see Fig. 5.6. According to Eqs. (5.36) and (5.37) we then have to determine whether the estimates

$$\hat{m}_X(t) = \frac{1}{N}\sum_{j=1}^{N} x_j(t), \tag{5.38}$$

and

$$\hat{R}_X(t, t+\tau) = \frac{1}{N}\sum_{j=1}^{N} x_j(t)\,x_j(t+\tau), \tag{5.39}$$

are independent of t to within a reasonable approximation. If that is satisfied, the process can usually be modeled as a stationary process to fair approximation.

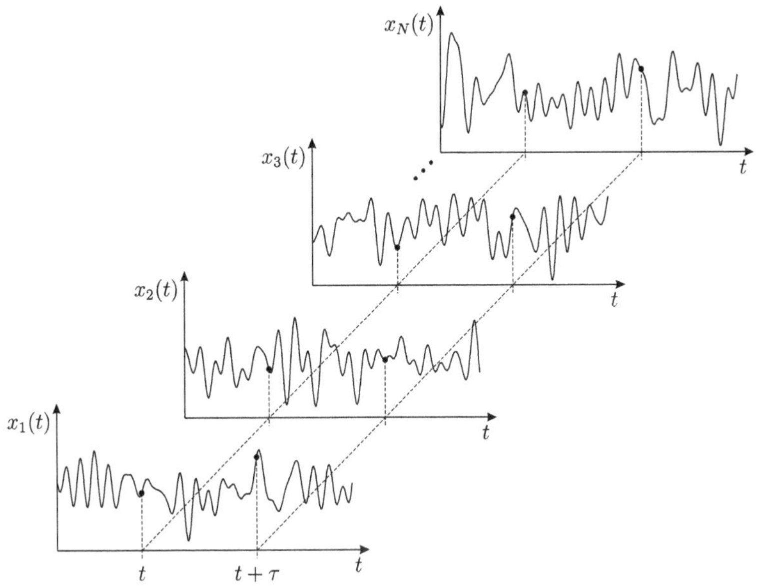

Figure 5.6. An ensemble of realizations of a stochastic process.

An equivalent definition of stationarity is obtained by using the following two conditions:

$$m_X = \mathrm{E}[X(t)] = \text{constant}, \tag{5.40}$$

and

$$C_X(\tau) = \mathrm{E}[(X(t) - m_X)(X(t+\tau) - m_X)] = \text{function only of } \tau. \tag{5.41}$$

$C_X(\tau)$ is called the *autocovariance function*, or simply the autocovariance, of $X(t)$. Note the relation $C_X(\tau) = R_X(\tau) - m_X^2$ and $C_X(\tau) = c_{X(t)X(t+\tau)}$, cf. Section 5.6.

It follows directly from Eq. (5.39) that $\sigma_X(t)^2 = C_X(0)$ = constant (independent of t. for a stationary process. This result is sufficiently important to deserve emphasis: If $X(t)$ is stationary, then it is also true that ($\sigma_X = \sigma_X(t)$)

$$\sigma_X^2 = \mathrm{E}[(X(t) - m_X)^2] = \text{constant}. \tag{5.42}$$

A stationary process is therefore characterized, among other things, by the fact that the mean value and the standard deviation do not change with time. The same holds true for the correlation between the process values at two times, which depends only on the time difference, cf. Section 5.6.

As many other mathematical models, stationarity is an idealization that will only approximately be true in practice. In particular, for natural phenomena, this may at first seem to be a poor idealization. The average wave height in the North Sea is for a fact much higher in January than in June. Fortunately, this does not mean that such processes cannot be modeled as stationary. If the statistical properties of the process change slowly compared with the typical wave periods, then the process can be idealized as "piecewise" stationary. That is, a time history of suitably chosen length is considered to be a segment of a realization of a stationary process.

A typical evolution with time of the autocovariance function $C_X(\tau)$ of a stationary process $X(t)$, which is intended to model a physical phenomenon, e.g., the

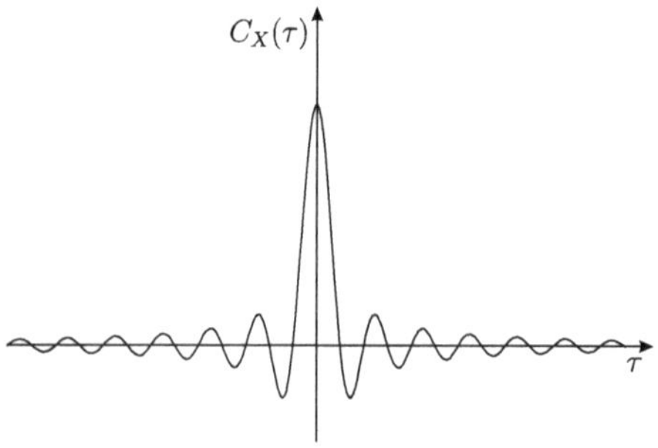

Figure 5.7. Typical autocovariance function.

fluctuations in Fig. 5.1, can be as shown in Fig. 5.7. A conspicuous feature of Fig. 5.7 is that $C_X(\tau) \to 0$ as τ increases. By scrutinizing Fig. 5.1, one may get an impression of a mean period $\bar{T}$ of the fluctuations. When $\tau \gg \bar{T}$, then very often $C_X(\tau) \approx 0$. This implies that it is difficult to predict which value a realization will assume at time $t + \tau$, even if we know the value at time t.

In Fig. 5.7, $C_X(\tau)$ is shown as an even or symmetric function, that is, $C_X(-\tau) = C_X(\tau)$. This is a general property of both the autocovariance and the autocorrelation. It is easily shown by invoking the definition of these functions.

$$\begin{aligned} R_X(-\tau) &= \mathrm{E}[(X(t)X(t-\tau)] = \mathrm{E}[(X(t-\tau)X(t)] \\ &= \mathrm{E}[(X(t-\tau)X(t-\tau+\tau)] = R_X(\tau). \end{aligned} \tag{5.43}$$

Because $C_X(\tau) = R_X(\tau) - m_X^2$, the same result follows for $C_X(\tau)$:

$$C_X(-\tau) = C_X(\tau). \tag{5.44}$$

The next two examples are important building blocks that we shall be using repeatedly.

5.7.1 Example – A Stationary Harmonic Process

A stochastic process $X(t)$ is defined by

$$X(t) = A\cos\omega t + B\sin\omega t\,, \qquad -\infty < t < \infty, \tag{a}$$

where A and B are two random variables and ω is a positive constant. An important special case of this kind of process was introduced in Example 5.4.2.

Let us see if $X(t)$ can be stationary. The mean value of the process is

$$\mathrm{E}[X(t)] = \mathrm{E}[A]\cos\omega t + \mathrm{E}[B]\sin\omega t. \tag{b}$$

The rhs of Eq. (b) can be rewritten as

$$\mathrm{E}[X(t)] = \sqrt{(\mathrm{E}[A])^2 + (\mathrm{E}[B])^2}\cos(\omega t - \theta), \tag{c}$$

where $\theta = \arctan\left(\mathrm{E}[B]/\mathrm{E}[A]\right)$ for $\mathrm{E}[A] \neq 0$, $\theta = \pi/2$ for $\mathrm{E}[A] = 0$ and $\mathrm{E}[B] > 0$, $\theta = -\pi/2$ for $\mathrm{E}[A] = 0$ and $\mathrm{E}[B] < 0$, and $\theta = 0$ for $\mathrm{E}[B] = 0$. However, this clearly implies that $\mathrm{E}[X(t)]$ is independent of t if and only if the amplitude $\sqrt{(\mathrm{E}[A])^2 + (\mathrm{E}[B])^2} = 0$, that is, $\mathrm{E}[A] = \mathrm{E}[B] = 0$. The first condition for stationarity is then satisfied if and only if A and B have mean value zero. To check the second condition, we must look at

$$\mathrm{E}[X(t)X(t+\tau)] = \mathrm{E}[A^2 \cos\omega t \cos\omega(t+\tau) + B^2 \sin\omega t \sin\omega(t+\tau)$$
$$+ AB\{\cos\omega t \sin\omega(t+\tau) + \sin\omega t \cos\omega(t+\tau)\}]$$
$$= \mathrm{E}[\frac{1}{2}A^2\{\cos(2\omega t+\omega\tau) + \cos\omega\tau\} - \frac{1}{2}B^2\{\cos(2\omega t+\omega\tau) - \cos\omega\tau\}$$
$$+ AB\sin(2\omega t+\omega\tau)]$$
$$= \mathrm{E}[\frac{1}{2}(A^2+B^2)\cos\omega\tau + \frac{1}{2}(A^2-B^2)\cos(2\omega t+\omega\tau) + AB\sin(2\omega t+\omega\tau)]$$
$$= \frac{1}{2}\mathrm{E}[A^2+B^2]\cos\omega\tau + a\cos(2\omega t+\omega\tau-\phi), \tag{d}$$

where $a = \{(\mathrm{E}[A^2-B^2])^2/4 + (\mathrm{E}[AB])^2\}^{1/2}$ and ϕ is a corresponding phase angle. In the derivation of Eq. (d), we used the trigonometric relations $2\cos\alpha\cos\beta = \cos(\alpha-\beta)+\cos(\alpha+\beta)$ and $2\sin\alpha\sin\beta = \cos(\alpha-\beta) - \cos(\alpha+\beta)$. It follows that $\mathrm{E}[X(t)X(t+\tau)]$ is independent of t if and only if $a = 0$, that is, $\mathrm{E}[A^2] = \mathrm{E}[B^2]$ and $\mathrm{E}[AB] = 0$.

Summing up: the process $X(t)$ is stationary if and only if A and B have mean value zero and the same variance, and are uncorrelated ($\rho_{AB} = 0$). We obtain that

$$m_X = 0 \tag{e}$$

and

$$R_X(\tau) = C_X(\tau) = \sigma_A^2 \cos\omega\tau\,. \tag{f}$$

For this process, it is clear that $C_X(\tau)$ does not tend to zero when τ increases. In fact, $C_X(2\pi n/\omega) = \sigma_A^2$, $n = 1, 2, \ldots$, which implies that $X(t)$ and $X(t+2\pi n/\omega)$ are fully correlated for any time t, that is, the correlation coefficient is equal to 1.0. In this special case every realization is a strictly periodic function, in particular, $x(t) = x(t+2\pi n/\omega)$ for $n = 1, 2, \ldots$, and future values of $x(t)$ are therefore known, irrespective of how far ahead in time we consider, if we know its values over one period.

When the random variables A and B are normally distributed, Eq. (a) is often rewritten in the following form:

$$X(t) = D\cos(\omega t + \epsilon)\,, \quad -\infty < t < \infty, \tag{g}$$

where the amplitude $D = \sqrt{A^2+B^2}$ is Rayleigh distributed, and the phase angle ϵ is uniformly distributed over $(0, 2\pi)$ (or $(-\pi, \pi)$), cf. Papoulis (1965).

5.7.2 Example – A Stationary Process

Example 5.7.1 can be extended in a useful way by defining a stochastic process $X(t)$ as follows.

$$X(t) = \sum_{j=1}^{N} \{A_j \cos\omega_j t + B_j \sin\omega_j t\}\,, \quad \infty < t < \infty, \tag{a}$$

where ω_j, $j = 1, \ldots, N$ are positive constants, $E[A_j] = E[B_j] = 0$, $E[A_j^2] = E[B_j^2] = \sigma_j^2$, and $E[A_j B_j] = 0$ for $j = 1, \ldots, N$. Also, $E[A_j A_k] = E[B_j B_k] = E[A_j B_k] = 0$ for $j \neq k$. When these conditions are satisfied, it is found, as in the previous example, that $X(t)$ is a stationary process with

$$m_X = 0 \tag{b}$$

and

$$R_X(\tau) = C_X(\tau) = \sum_{j=1}^{N} \sigma_j^2 \cos \omega_j \tau. \tag{c}$$

5.8 Ergodic Processes

As we have already discussed, natures own "experiment" cannot be repeated other than as a pure thought experiment. We find ourselves securely locked into the one realization that is available to us of the natural phenomenon we study, e.g., the wave elevation at a point on the ocean's surface. To use one single time history to estimate statistical parameters, it is necessary to assume that the underlying process is ergodic:

> *A stochastic process is called* ergodic *if every ensemble mean can be replaced by a time average over a single realization.*

Note that calculating an ensemble mean is equivalent to calculating an expectation value. The definition of an ergodic process implies that all statistical information about the process is contained in each and every realization. In more mathematically oriented texts, this definition of ergodicity would have to be modified somewhat. However, the definition given here is sufficiently accurate for our purposes in this book.

In Section 5.2, it is shown how the CDF was estimated as a time average on the basis of one single time history. Let us have a look at how the mean value and the autocorrelation are calculated by using time averages. For an ergodic process $X(t)$,

$$E[X(t)] = \lim_{T\to\infty} \frac{1}{2T} \int_{-T}^{T} x(s)\, ds, \tag{5.45}$$

and

$$E[X(t)\, X(t+\tau)] = \lim_{T\to\infty} \frac{1}{2T} \int_{-T}^{T} x(s)\, x(s+\tau)\, ds, \tag{5.46}$$

where $x(t)$ is an arbitrary but given realization of $X(t)$. In Eq. (5.45), the rhs is independent of t, that is, $E[X(t)] =$ constant. The rhs of Eq. (5.46) is also clearly independent of t. This means that an ergodic process is automatically stationary. In other words, for a process to be ergodic, it is necessary (but not sufficient) that it is stationary.

A numerical calculation of the mean value m_X and the autocorrelation $R_X(\tau)$ for an ergodic process on the basis of a time history $x(t)$ of length T is typically carried out by recording the values of $x(t)$ at the times t_0, $t_1 = t_0 + \Delta t, \ldots, t_N = t_0 + N\Delta t$, that is, at the time interval Δt ($N\Delta t = T$). Then,

$$m_X \approx \frac{1}{T}\sum_{j=0}^{N} x(t_j)\,\Delta t = \frac{1}{N}\sum_{j=0}^{N} x(t_j), \tag{5.47}$$

and

$$R_X(n\Delta t) \approx \frac{1}{T - n\Delta t}\sum_{j=0}^{N-n} x(t_j)\,x(t_j + n\Delta t)\,\Delta t = \frac{1}{N-n}\sum_{j=0}^{N-n} x(t_j)\,x(t_j + n\Delta t). \tag{5.48}$$

Because $x(T) = x(N\Delta t)$ is the last recorded data point, the sum in Eq. (5.48) must stop at $j = N - n$. The consequence of this is that the estimate of $R_X(n\Delta t)$ deteriorates with increasing n.

The power of the concept of an ergodic process lies in the observation that, for most of the phenomena of engineering interest that with reasonable accuracy can be modeled as stationary processes, ensemble means can be replaced by time averages over a single realization.

5.8.1 Example – An Ergodic Harmonic Process

A stochastic process $X(t)$ is defined by

$$X(t) = a\,\cos(\omega t + \Phi), \tag{a}$$

where a and ω are two positive constants, and Φ is a random variable, which is uniformly distributed over the interval $(0, 2\pi)$. That is, $f_\Phi(\phi) = 1/(2\pi)$, $0 \le \phi \le 2\pi$, where $f_\Phi(\phi)$ is the PDF of Φ. Our task is now to show that both the mean value and the autocorrelation can be obtained by time averaging over one realization. For this purpose, Eq. (a) is rewritten as

$$X(t) = a\,\cos\Phi\,\cos\omega t - a\sin\Phi\,\sin\omega t. \tag{b}$$

$X(t)$ is of the same type as the process in Example 5.7.1 with $A = a\cos\Phi$ and $B = -a\sin\Phi$. It is obtained that

$$\begin{aligned}
\mathrm{E}[A] &= a\int_0^{2\pi} \cos\phi\,\frac{d\phi}{2\pi} = 0\\
\mathrm{E}[B] &= -a\int_0^{2\pi} \sin\phi\,\frac{d\phi}{2\pi} = 0\\
\mathrm{E}[A^2] &= a^2\int_0^{2\pi} \cos^2\phi\,\frac{d\phi}{2\pi} = \frac{a^2}{2}\\
\mathrm{E}[B^2] &= a^2\int_0^{2\pi} \sin^2\phi\,\frac{d\phi}{2\pi} = \frac{a^2}{2}\\
\mathrm{E}[AB] &= -a^2\int_0^{2\pi} \cos\phi\sin\phi\,\frac{d\phi}{2\pi} = 0.
\end{aligned} \tag{c}$$

According to Example 3.3, $X(t)$ is a stationary process with $m_X = 0$ and $R_X(\tau) = \frac{a^2}{2}\cos\omega\tau$.

A realization of $X(t)$ is obtained when Φ assumes the value of an outcome, ϕ say, which is then an angle between 0 and 2π. The corresponding realization $x(t)$ of $X(t)$ is then

$$x(t) = a\cos(\omega t + \phi). \tag{d}$$

The time average of $x(t)$ is given as

$$\lim_{T\to\infty}\frac{1}{2T}\int_{-T}^{T} a\cos(\omega t+\phi)\,dt = \lim_{T\to\infty}\frac{a}{2T}\Big[\frac{1}{\omega}\sin(\omega t+\phi)\Big]_{-T}^{T} = 0 \tag{e}$$

because $|\sin(\omega T+\phi) - \sin(-\omega T+\phi)| \le 2$.

Similarly, the time average of $x(t)x(t+\tau)$ becomes

$$\begin{aligned}
&\lim_{T\to\infty}\frac{1}{2T}\int_{-T}^{T} a^2\cos(\omega t+\phi)\cos(\omega(t+\tau)+\phi)\,dt\\
&= \lim_{T\to\infty}\frac{a^2}{2T}\int_{-T}^{T}\frac{1}{2}\{\cos\omega\tau + \cos(2\omega t+\omega\tau+2\phi)\}\,dt\\
&= \frac{a^2}{2}\cos\omega\tau\lim_{T\to\infty}\frac{1}{2T}\int_{-T}^{T}dt + \frac{a^2}{2}\lim_{T\to\infty}\frac{1}{2T}\int_{-T}^{T}\cos(2\omega t+\omega\tau+2\phi)\,dt\\
&= \frac{a^2}{2}\cos\omega\tau.
\end{aligned} \tag{f}$$

Ensemble averaging and time averaging gives the same result for m_X and $R_X(\tau)$. Hence, we have demonstrated that $X(t)$ is ergodic, at least with respect to the mean value and the autocorrelation. In practice, one would often be satisfied with this and assume that the process is ergodic.

Note that the amplitude of the process in this example is constant, which is, in fact, a prerequisite for the process to be ergodic. This is in contrast to the process of Example 5.7.1, which has an amplitude that will vary from realization to realization. From the previous derivations, one can now easily convince oneself that such a process cannot be ergodic with respect to the autocorrelation.

5.8.2 Example – An Ergodic Process

Example 5.8.1 can be extended by defining a stochastic process $X(t)$ as follows.

$$X(t) = \sum_{j=1}^{N} a_j\cos(\omega_j t + \Phi_j), \tag{a}$$

where a_j and ω_j, $j = 1,\ldots,N$ are positive constants, and $\{\Phi_j\}_{j=1}^{N}$ is a set of independent random variables that are uniformly distributed over $(0, 2\pi)$. In the same way as in the previous example, it may be shown that $X(t)$ is ergodic with respect to the mean value and the autocorrelation.

5.9 Realizations of Stochastic Processes

In Chapter 16 Monte Carlo methods are discussed at some length. This method can be used for any input-output system subjected to a stochastic process such that for every realization of the input stochastic process a corresponding realization of the output process can be calculated. The basic idea underlying the Monte Carlo method is that of producing a sample of output/response time histories from a sample of input/loading time histories. This makes it possible to estimate various statistics of the output/response process based on the available sample of realizations. Hence, a key element in the Monte Carlo method is therefore the realizations of a stochastic process. Because Examples 5.7.2 and 5.8.2 represent widely used ways of representing stochastic processes, we discuss how realizations of such processes are generated, see also (Shinozuka and Jan, 1972; Shinozuka and Deodatis, 1991).

Let us start with the stationary process represented by Eq. (a) of Example 5.7.2. The first step of the procedure is to specify what kind of random variables A_j and B_j are assumed to be. A common choice in many cases is to assume that these variables are independent and normally distributed with zero mean value and known standard deviations σ_j. A realization of $X(t)$ is then obtained when we have generated a set of outcomes of the random variables A_j and B_j. In this connection, it is useful to make the following observation: if $\tilde{A}_j$ is normally distributed with zero mean and standard deviation equal to 1.0, then $A_j = \sigma_j \tilde{A}_j$ is normally distributed with mean zero and standard deviation equal to σ_j. Similarly for B_j. We may therefore write Eq. (a) of Example 4.7.2 in the form

$$X(t) = \sum_{j=1}^{N} \sigma_j \left\{ \tilde{A}_j \cos(\omega_j t) + \tilde{B}_j \sin(\omega_j t) \right\}, \tag{5.49}$$

where $\tilde{A}_j$ and $\tilde{B}_j$, $j = 1, \dots, N$, is now a set of independent, standard normally distributed variables. There are computer programs that may be used to generate independent outcomes of a standard, normally distributed variable. It is seen that we need $2N$ outcomes for our example. Specifically, assume that $\tilde{a}_j$ and $\tilde{b}_j$, $j = 1, \dots, N$ are the obtained outcomes from such a program. The corresponding realization $x(t)$ is then

$$x(t) = \sum_{j=1}^{N} \sigma_j \left\{ \tilde{a}_j \cos(\omega_j t) + \tilde{b}_j \sin(\omega_j t) \right\}. \tag{5.50}$$

This procedure can then be repeated as many times as needed to produce the requested sample size of realizations. It is, of course, understood here that each realization is generated independently of all others. We illustrate this procedure with an application to ocean waves in Example 6.3.1.

The representation given by Eq. (a) of Example 5.8.2 is also frequently used, especially when ergodic properties are desirable. Generating realizations of this process is now quite straight-forward. It is seen that, in fact, only outcomes of the uniformly distributed phase angles Φ_j, $j = 1, \dots, N$ are needed. If R_j, $j = 1, \dots, N$, denote independent random variables uniformly distributed on $(0, 1)$, usually referred to as random numbers, then clearly we may put $\Phi_j = 2\pi R_j$.

Hence, by invoking a random number generator, usually available on any computer, outcomes of Φ_j, $j = 1, \ldots, N$ can be easily generated. Let r_j, $j = 1, \ldots, N$ denote a set of independent random number outcomes. A realization $x(t)$ of $X(t)$ is then

$$x(t) = \sum_{j=1}^{N} a_j \cos(\omega_j t + 2\pi r_j). \tag{5.51}$$

6 Variance Spectrum

6.1 Introduction

In Chapter 2, Fourier analysis was used to decompose the time histories of load and response as sums or integrals of cos(·) and sin(·) terms over the frequency domain. It was shown how the transfer function gave a direct connection between the amplitudes at each frequency for the load and response, provided the system was linear and time invariant. A time history that is periodic can be decomposed as a sum over a finite or countably infinite number of frequencies. If it is not periodic, the decomposition must be expressed as an integral, see Eq. (2.92). It can be shown that this is possible only if the time history dies out with time. Regarding the realizations of stationary stochastic processes, they will not generally be periodic. Nor will they decrease with time because the variance is constant, and this is a measure of the fluctuations around the mean value. A direct frequency decomposition of the realizations of a stationary process is therefore not directly feasible. This difficulty is circumvented by using the autocovariance function of a stationary process. As mentioned previously in Section 5.7, this function will generally approach zero when its time argument increases.

6.2 Variance Spectrum

Let $X(t)$ be a stationary process with autocovariance function $C_X(\tau)$. Assume that $C_X(\tau) \to 0$ when $\tau \to \infty$ sufficiently rapidly so that $\int_{-\infty}^{\infty} |C_X(\tau)|\, d\tau$ has a finite value.

The *variance spectrum* $S_X(\omega)$ of $X(t)$ is defined as the Fourier transform of $C_X(\tau)$ as follows:

$$S_X(\omega) = \frac{1}{2\pi} \int_{-\infty}^{\infty} C_X(\tau)\, e^{-i\omega\tau}\, d\tau. \tag{6.1}$$

It was shown already in the 1930s that $C_X(\tau)$ and $S_X(\omega)$ constituted a Fourier transform pair; that is, $C_X(\tau)$ is given by the inverse Fourier transform as

$$C_X(\tau) = \int_{-\infty}^{\infty} S_X(\omega)\, e^{i\omega\tau}\, d\omega. \tag{6.2}$$

Equations (6.1) and (6.2) are often called the Wiener-Khintchine relations after the originators.

The quantity $S_X(\omega)$ is known by many names. A few names that seem appropriate to mention here are *energy spectrum*, *power spectral density*, *spectral density* or just *spectrum*. The name *variance spectrum*, which we have chosen to use, ties in directly with the interpretation of $S_X(\omega)$. By putting $\tau = 0$ in Eq. (6.2), it is obtained that

$$\sigma_X^2 = C_X(0) = \int_{-\infty}^{\infty} S_X(\omega)\, d\omega. \tag{6.3}$$

This equation shows that $S_X(\omega)$ can be interpreted as a distribution of variance along the frequency axis, provided that $S_X(\omega) \geq 0$, and in Section 6.5, we show that this is always the case. Negative frequencies are a mathematical convenience and have no real physical content. If $S_X(\omega)$ is to be interpreted as distribution of variance along the frequency axis, one would therefore expect that $S_X(-\omega) = S_X(\omega)$. This symmetry property of $S_X(\omega)$ can be shown directly from Eq. (6.1). It follows also from a rewriting of the Wiener-Khintchine relations to real form, which is of interest in itself. Because $C_X(\tau)$ is symmetric, it follows from Eq. (6.1) that,

$$\begin{aligned} S_X(\omega) &= \frac{1}{2\pi}\int_0^{\infty} C_X(\tau)\, e^{-i\omega\tau}\, d\tau + \frac{1}{2\pi}\int_0^{\infty} C_X(-\tau)\, e^{i\omega\tau}\, d\tau \\ &= \frac{1}{\pi}\int_0^{\infty} C_X(\tau)\, \cos\omega\tau\, d\tau\,, \end{aligned} \tag{6.4}$$

where Euler's relation $e^{ix} = \cos x + i \sin x$ is used. The last integral in the upper line is obtained by the change of variable $\tau \to -\tau$, while the lower line follows from the symmetry property $C_X(-\tau) = C_X(\tau)$.

From Eq. (6.4), it follows that

$$S_X(-\omega) = S_X(\omega), \tag{6.5}$$

because $\cos(-\omega t) = \cos\omega t$. Eq. (6.2) can then be rewritten as

$$C_X(\tau) = 2\int_0^{\infty} S_X(\omega)\, \cos\omega\tau\, d\omega. \tag{6.6}$$

If $S_X(\omega)$ is a reasonably nice function, the integral on the rhs of Eq. (6.6) can be approximated by a finite sum, viz.

$$C_X(\tau) = 2\int_0^{\infty} S_X(\omega)\, \cos\omega\tau\, d\omega \approx \sum_{j=1}^{N} 2S_X(\omega_j)\, \Delta\omega\, \cos\omega_j\tau, \tag{6.7}$$

for a suitable choice of $\omega_1 < \ldots < \omega_N$, and sufficiently small $\Delta\omega = (\omega_N - \omega_1)/(N - 1)$. Let us now recollect the results from Example 5.7.2 and define a stationary process

$$\tilde{X}(t) = \sum_{j=1}^{N} \left\{ A_j\, \cos\omega_j t + B_j\, \sin\omega_j t \right\}, \tag{6.8}$$

where the random variables A_j and B_j satisfy the necessary conditions for stationarity. In addition, we put $\sigma_j^2 = 2S_X(\omega_j)\Delta\omega$. Then, according to Eq. (c) of Example 5.7.2, the autocovariance of $\tilde{X}(t)$ is given as

$$C_{\tilde{X}}(\tau) = \sum_{j=1}^{N} 2S_X(\omega_j)\, \Delta\omega\, \cos\omega_j\tau. \tag{6.9}$$

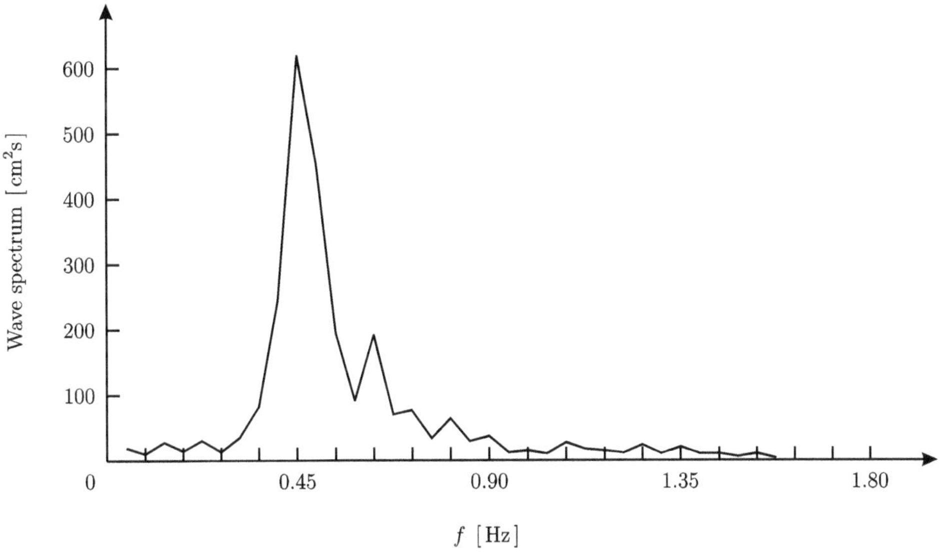

Figure 6.1. Typical "wave spectrum" from the North Sea.

We thus construct a stationary process $\tilde{X}(t)$ with approximately the same variance distribution, that is, variance spectrum, as $X(t)$. In a certain sense, $\tilde{X}(t)$ can be said to represent $X(t)$. What we have just described is a variant from a class of methods that is extensively used in practice to generate realizations of a given stationary process. To get a concrete realization of a process represented by Eq. (6.8), one has to generate outcomes of the random variables that enter the sum.

On the basis of Eq. (6.5) and the fact that negative frequencies do not really have any physical meaning, it is common practice in engineering to use the *one-sided* variance spectrum, which is denoted by $S_X^+(\omega)$ and defined as

$$S_X^+(\omega) = \begin{cases} 2S_X(\omega), & \omega \geq 0, \\ 0, & \omega < 0. \end{cases} \tag{6.10}$$

The distribution of variance is thereby concentrated to positive (physically realizable) frequencies. The variance expressed in terms of the one-sided variance spectrum is clearly

$$\sigma_X^2 = \int_0^\infty S_X^+(\omega)\, d\omega. \tag{6.11}$$

A typical variance spectrum produced from measured data for the wave elevation at a point on the sea surface at a location in the North Sea is shown in Fig. 6.1. The somewhat jagged look is mostly due to low numerical resolution.

6.3 Units of Variance Spectra

In Eq. (6.3), the units of $S_X(\omega)$ are the square of the units of $X(t)$ divided by radians per second. If $X(t)$ models the wave elevation at a location on the ocean surface and is measured in meters, then the units of $S_X(\omega)$ are m^2s/rad.

So far, we have only considered the variance spectrum as a function of circular frequency ω with units rad/s. It is also quite common to give the variance spectrum

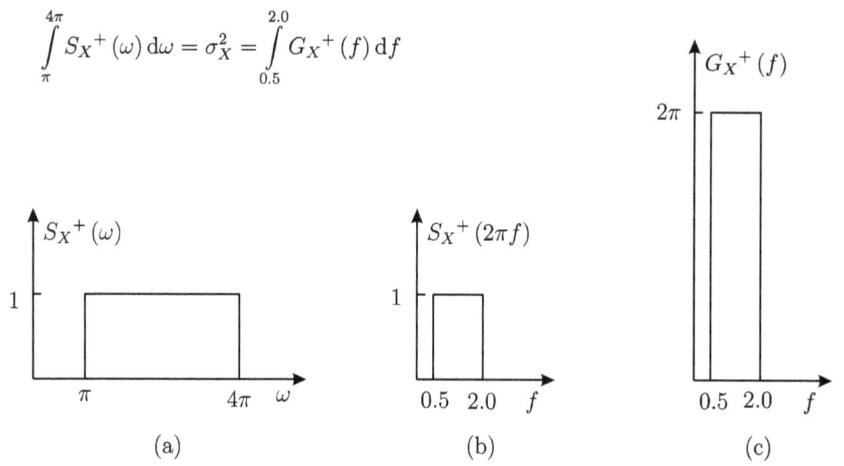

Figure 6.2. Scaling of variance spectra: a special case.

as a function of frequency f with units in Hertz (oscillations per second). The relation $\omega = 2\pi f$ implies a rescaling of the frequency axis when ω is replaced by f. It is therefore not correct to believe that $S_X(2\pi f)$ would represent the variance spectrum as a function of f; this would easily lead to the wrong variance. The situation is illustrated in Fig. 6.2 for a variance spectrum that is constant between two frequencies and zero elsewhere. Figure 6.2(a) shows $S_X^+(\omega)$ as a function of ω, and it is seen that $\sigma^2 = 3\pi$. Figure 6.2(b) shows $S_X^+(2\pi f)$ as a function of f, and now clearly $\sigma_X^2 = 1.5$. To get the correct variance, the spectral values must be multiplied by 2π, as shown in Fig. 6.2(c). The general situation is illustrated in Fig. 6.3 by approximating $S_X^+(\omega)$ with a step function, and it is seen that the same condition applies here. If $G_X^+(f)$ denotes the one-sided variance spectrum as a function of f in Hz, then

$$G_X^+(f) = 2\pi S_X^+(\omega). \tag{6.12}$$

Exactly the same relation applies to two-sided spectra. A factor of $2\pi = 6.28$ occurring erroneously in the variance can have a significant impact on the results in some cases. Thus, care should be exercised when adopting values of spectral moments by checking which kind of frequency is used in the variance spectrum.

6.3.1 Example – A Realization of a Wave Process

In this example, we return to the procedure for producing a realization of a stationary stochastic process described in Section 5.9. In particular, we illustrate how a realization of a stochastic process with a given variance spectrum can be generated. To be even more specific, we shall assume that the task is to produce an arbitrary realization of the wave elevation $X(t)$ at a given location in the North Sea with a recorded wave spectrum, as depicted in Fig. 6.1. Thus, it is assumed that this wave elevation can be represented as a stationary stochastic process $X(t)$ with a one-sided variance spectrum $G_X^+(f)$, as shown in Fig. 6.1. As mentioned in Section 6.2, there are several alternative methods that can be used to generate realizations of a stochastic process. Our choice here is to use the

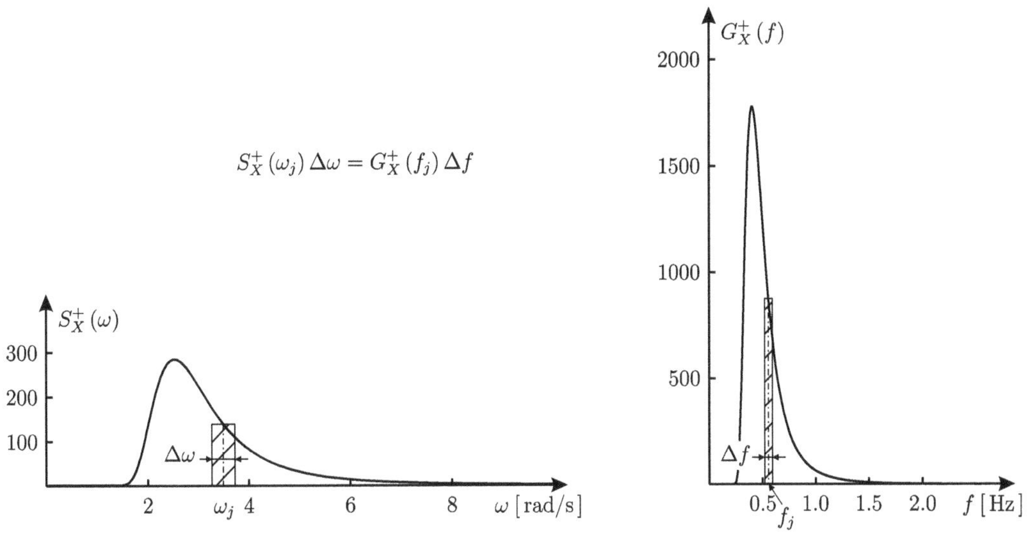

Figure 6.3. Scaling of variance spectra.

method described in Section 6.2, which amounts to approximating the process $X(t)$ with $\tilde{X}(t)$, and then generate realizations of $\tilde{X}(t)$ instead.

To proceed, it is necessary to specify what kind of random variables A_j and B_j to use in Eq. (6.8). The common choice in the case of wave processes on the open ocean is to assume that these variables are independent and normally distributed with zero mean value. To determine the standard deviation, we have to decide on a suitable discretization of the frequency axis. For many practical purposes, it would be desirable to have a discretization that would give of the order of 10^2 frequencies in the frequency range where the waves have significant energy, e.g., from 0.27 to 0.99 Hz in this case. Because the objective here is to illustrate the procedure, we shall content ourselves with a frequency increment $\Delta f = 0.09$ Hz. That gives 8 frequencies $f_1, \ldots, f_8$ ($f_1 = 0.315$ Hz, $f_{j+1} = f_j + 0.09$ Hz) in the specified frequency range. The process $\tilde{X}(t)$ can then be written as

$$\tilde{X}(t) = \sum_{j=1}^{8} \left\{ A_j \cos(2\pi f_j t) + B_j \sin(2\pi f_j t) \right\}. \tag{a}$$

Figure 6.4 shows the relevant part of the wave spectrum in Fig. 6.1 magnified along the frequency axis to clarify how the standard deviation of A_j and B_j is determined.

What remains to get an approximate realization of $X(t)$ is to generate outcomes of the random variables that enter $\tilde{X}(t)$. For this, we invoke an observation made in Section 5.9: if $\tilde{A}_j$ is normally distributed with zero mean and standard deviation equal to 1.0, then $A_j = \sigma_j \tilde{A}_j$ is normally distributed with zero mean and standard deviation equal to σ_j. Similarly for B_j. Because $\sigma_j = \sqrt{G_X^+(f_j)\,\Delta f}$, we may therefore write Eq. (a) in the form

$$\tilde{X}(t) = \sum_{j=1}^{8} \sqrt{G_X^+(f_j)\,\Delta f} \left\{ \tilde{A}_j \cos(2\pi f_j t) + \tilde{B}_j \sin(2\pi f_j t) \right\}. \tag{b}$$

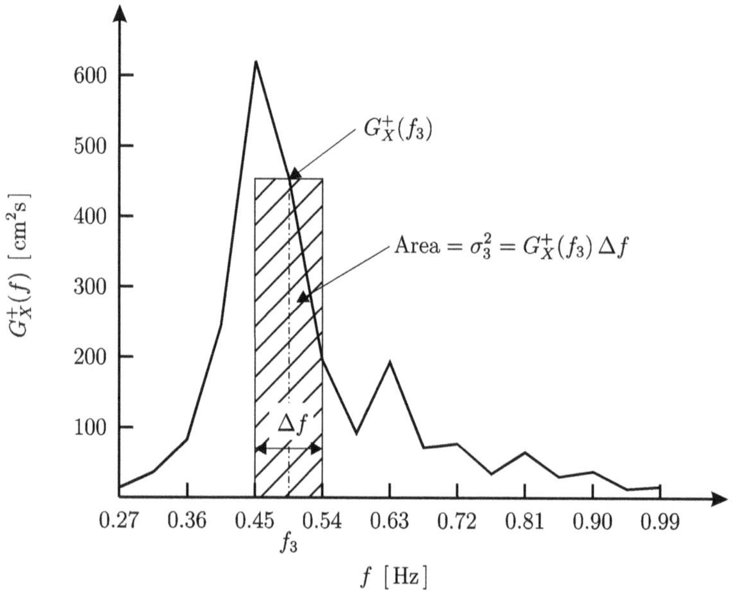

Figure 6.4. Wave spectrum from the North Sea.

where $\tilde{A}_j$ and $\tilde{B}_j$, $j = 1, \ldots, 8$, is now a set of independent, standard normally distributed variables. Computer programs are easily available for generating independent outcomes of a standard, normally distributed variable. It is seen that we need 16 for our example.

An alternative procedure is to use a table or computer program for generating (pseudo-)random numbers, which are uniformly distributed between 0 and 1. This can also be used by invoking the following result. If $\Phi(\cdot)$ denoted the CDF of a standard, normally distributed variable, and R denotes a random variable that is uniformly distributed between 0 and 1, then the random variable $Z = \Phi^{-1}(R)$ is standard and normally distributed. By generating 16 independent outcomes of R: $r_1, \ldots, r_{16}$, then $z_1 = \Phi^{-1}(r_1), \ldots, z_{16} = \Phi^{-1}(r_{16})$ will be 16 independent outcomes of a standard, normally distributed variable. This procedure is used here, and the results are shown in Table 6.1, where $\tilde{a}_j = \Phi^{-1}(r_{2j-1})$ and $\tilde{b}_j = \Phi^{-1}(r_{2j})$.

A piece of the corresponding realization is shown in Fig. 6.5, and one may get a similar impression as when observing irregular seas out on the oceans. In practice, there is often a need to generate many realizations to perform statistical

Table 6.1.

j	f_j[Hz]	$\sqrt{G_X^+(f_j)\,\Delta f}$ [cm]	r_{2j-1}	r_{2j}	$\tilde{a}_j$	$\tilde{b}_j$
1	0.315	1.90	0.10097	0.32533	−1.2760	−0.4529
2	0.405	5.45	0.76520	0.13586	0.7232	−1.0991
3	0.495	6.57	0.34673	0.54876	−0.3942	0.1225
4	0.585	3.15	0.80959	0.09117	0.8764	−1.3335
5	0.675	2.23	0.39292	0.74945	−0.2717	0.6728
6	0.765	1.56	0.37542	0.04805	−0.3175	−1.6641
7	0.855	1.44	0.64894	0.74296	0.3825	0.6525
8	0.945	1.41	0.24805	0.24037	−0.6807	−0.7051

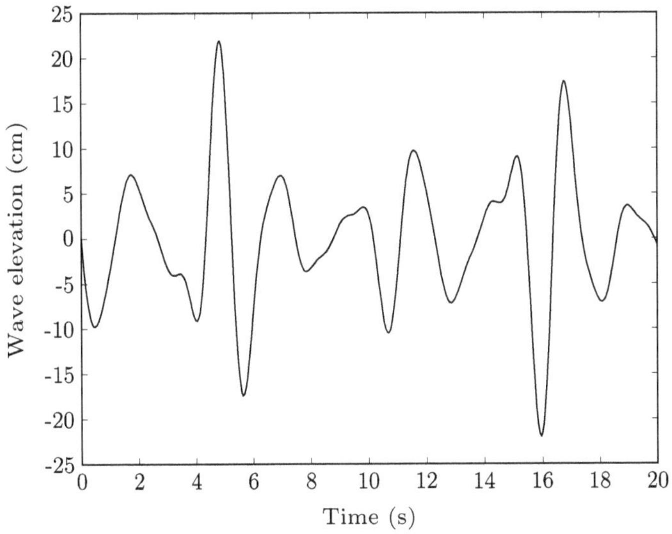

Figure 6.5. A realization of the wave elevation.

analyses. It is then necessary to repeat the procedure we just described the requisite number of times, and every time a new set of outcomes independent of the previous ones are chosen.

If we had shown a piece of the realization we have generated, which was twice as long, it would have become clear that the wave pattern is repeating itself. This is due to the way we have constructed it, which indeed makes it periodic. The period is determined by the greatest common divisor of the frequency increment and the initial frequency of the discretization of the frequency range that is chosen. In this case, the period becomes $1/0.045 = 22.2$ s. The practical consequence of this is that one must choose a discretization that is in correspondence with the required length of a realization. It may be worth mentioning that for most practically relevant discretizations, one may say that a corresponding realization will have a period that can be assumed to be given as $1/\Delta f$. One does not really avoid the problem related to periodicity by choosing an (almost) irrational ratio between the starting frequency and the frequency increment, or other "smart" tricks such as choosing the frequencies randomly within each subinterval of the discretization.

Another point worth noting is that the process $\tilde{X}(t)$ is not ergodic. If it is desirable to ensure this property, one may use the method described in Section 5.9, which was based on Examples 5.8.1 and 5.8.2. However, the difference in the practical results obtained by using this method versus the one we describe here is usually rather small if the discretization is properly done.

6.4 Examples of Variance Spectra and Autocovariances

6.4.1 Constant Autocovariance

A simple example of a stochastic process $X(t)$ is obtained by defining $X(t) = X$, where X is a random variable with mean value m_X and variance $\sigma_X^2 = 1.0$. Then

$$m_X(t) = m_X, \tag{6.13}$$

and

$$E[(X(t) - m_X)(X(t+\tau) - m_X)] = 1.0, \tag{6.14}$$

that is, $X(t)$ is a stationary process. Because $C_X(\tau) = 1.0$, the integral $\int_{-\infty}^{\infty} C_X(\tau)\, e^{-i\omega\tau}\, d\tau$ does not exist; that is, the variance spectrum cannot be defined by Eq. (6.1) without further considerations. The variance spectrum expresses a frequency decomposition of $C_X(\tau)$, but $C_X(\tau)$ is constant, which means that it contains only one frequency, namely, 0. This means that if it is at all possible to define a meaningful variance spectrum in this case, then by necessity $S_X(\omega) = 0$ for all $\omega \neq 0$. If $S_X(\omega)$ were to be interpreted as an ordinary function, it would inexorably lead to the result $1.0 = \sigma_X^2 = \int_{-\infty}^{\infty} S_X(\omega)\, d\omega = 0$. Strictly speaking, $X(t)$ can therefore not have a variance spectrum in the usual manner. However, if we allow ourselves the privilege of using an entirely formalistic approach, we are still able to install perfect meaning into such anomalies. In fact, the result turns out to be a useful concept.

Let us start by defining a "function" $\delta(\omega)$ as the Fourier transform of $C_X(\tau) = 1.0$ in a purely formal way as

$$\delta(\omega) = \frac{1}{2\pi} \int_{-\infty}^{\infty} e^{-i\omega\tau}\, d\tau. \tag{6.15}$$

Now, let $F(\omega)$ be a continuous function, which has an inverse Fourier transform $f(\tau)$, so that $f(\tau)$ and $F(\omega)$ is a Fourier transform pair. Then, formally, the following applies:

$$\begin{aligned}\int_{-\infty}^{\infty} \delta(\omega)\, F(\omega)\, d\omega &= \frac{1}{2\pi} \int_{-\infty}^{\infty} \int_{-\infty}^{\infty} F(\omega)\, e^{-i\omega\tau}\, d\omega\, d\tau \\ &= \frac{1}{2\pi} \int_{-\infty}^{\infty} f(-\tau)\, d\tau = \frac{1}{2\pi} \int_{-\infty}^{\infty} f(u)\, e^{-i0u}\, du = F(0).\end{aligned} \tag{6.16}$$

Because Eq. (6.16) must be satisfied for any such $F(\omega)$, it implies that $\delta(\omega) = 0$ for all $\omega \neq 0$. Hence, we may extend the validity of Eq. (6.16) to apply to all functions $F(\omega)$ that are continuous at $\omega = 0$.

Equation (6.15) is nothing but the definition of the variance spectrum of the chosen $X(t)$. We should therefore demand that $C_X(\tau) = 1.0$ is produced as the result of applying the inverse Fourier transform to $\delta(\omega)$, that is

$$1.0 = \int_{-\infty}^{\infty} \delta(\omega)\, e^{i\omega\tau}\, d\omega. \tag{6.17}$$

But Eq. (6.17) is just a special case of Eq. (6.16) with $F(\omega) = e^{i\omega\tau}$. $\delta(\omega)$ has therefore acquired the properties that we want the variance spectrum to have, viz.: $\delta(\omega) = 0$ for all $\omega \neq 0$, and $C_X(\tau) = 1.0 = \int_{-\infty}^{\infty} \delta(\omega)\, e^{i\omega\tau}\, d\omega$, in particular, $\sigma_X^2 = 1.0 = \int_{-\infty}^{\infty} \delta(\omega)\, d\omega$. It is not accidental that we have used the symbol $\delta(\omega)$ to denote the rhs of Eq. (6.15). The reason is that $\delta(\omega)$, as we define it here, is an example of what is often called a δ function, even if it cannot be considered as an ordinary function.

We conclude by noting the following property. Because $\delta(\omega) = 0$ for all $\omega \neq 0$, then $\delta(-\omega) = \delta(\omega)$. This implies that we may also write

$$\delta(\omega) = \frac{1}{2\pi} \int_{-\infty}^{\infty} e^{i\omega\tau}\, d\tau. \tag{6.18}$$

6.4.2 Harmonic Process

We now return to the stochastic process $X(t)$ defined in Example 5.7.1 ($\omega_0 > 0$)

$$X(t) = A\cos\omega_0 t + B\sin\omega_0 t\,, \quad \infty < t < \infty, \tag{6.19}$$

$\mathrm{E}[A] = \mathrm{E}[B] = 0$, $\mathrm{E}[A^2] = \mathrm{E}[B^2] = \sigma^2$, and $\mathrm{E}[AB] = 0$. Then $X(t)$ is stationary and

$$C_X(\tau) = \sigma^2 \cos\omega_0\tau. \tag{6.20}$$

Because clearly $C_X(\tau)$ does not decrease to zero as $\tau \to \infty$, the variance spectrum also in this case cannot be defined by simply referring to the usual Fourier transform formula of Eq. (6.1); the rhs of this equation does not exist as a standard integral. We now show that the δ function can help us out of this difficulty. Again, by a formal analysis as in the previous case, we obtain the following result by substituting from Eq. (6.20) into Eq. (6.1):

$$\begin{aligned} S_X(\omega) &= \frac{1}{2\pi}\int_{-\infty}^{\infty} \sigma^2 \cos\omega_0\tau\, e^{-i\omega\tau}\, d\tau \\ &= \frac{\sigma^2}{2}\Big\{\frac{1}{2\pi}\int_{-\infty}^{\infty} e^{-i(\omega+\omega_0)\tau}\, d\tau + \frac{1}{2\pi}\int_{-\infty}^{\infty} e^{-i(\omega-\omega_0)\tau}\, d\tau\Big\} \\ &= \frac{\sigma^2}{2}\big\{\delta(\omega+\omega_0) + \delta(\omega-\omega_0)\big\}. \end{aligned} \tag{6.21}$$

Because $\delta(\omega) = 0$ for all $\omega \neq 0$, it is seen that $\delta(\omega+\omega_0) = 0$ for all $\omega \neq -\omega_0$, and $\delta(\omega-\omega_0) = 0$ for all $\omega \neq \omega_0$. This means that $S_X(\omega)$ is concentrated at the two frequencies $\pm\omega_0$, and that is exactly what we expect because $X(t)$ contains just the (physical) frequency ω_0. It is this possibility to place the variance at just one frequency that makes the δ-function useful to us.

Exercise: Check that $\sigma^2 = \int_{-\infty}^{\infty} S_X(\omega)\, d\omega$.

Let us show that Eq. (6.2) is also satisfied, that is

$$C_X(\tau) = \int_{-\infty}^{\infty} S_X(\omega)\, e^{i\omega\tau}\, d\omega. \tag{6.22}$$

By substituting from Eq. (6.21) into Eq. (6.22) and using Eq. (6.16) for each of the two integrals that occur, it is found that

$$\int_{-\infty}^{\infty} \delta(\omega+\omega_0)\, e^{i\omega\tau}\, d\omega = \int_{-\infty}^{\infty} \delta(\omega')\, e^{i(\omega'-\omega_0)\tau}\, d\omega' = e^{-i\omega_0\tau} \tag{6.23}$$

and, similarly,

$$\int_{-\infty}^{\infty} \delta(\omega-\omega_0)\, e^{i\omega\tau}\, d\omega = e^{i\omega_0\tau}. \tag{6.24}$$

Hence,

$$\int_{-\infty}^{\infty} S_X(\omega)\, e^{i\omega\tau}\, d\omega = \sigma^2 \cos\omega_0\tau \tag{6.25}$$

which via Eq. (6.20) proves Eq. (6.22).

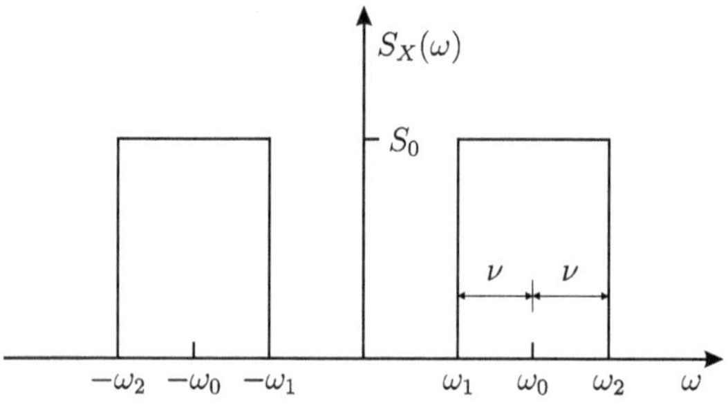

Figure 6.6. Rectangular variance spectrum.

6.4.3 Periodic Process

Let $X(t)$ be defined as in Example 5.7.2. The variance spectrum of $X(t)$ is worked out in the same way as in the previous case, and we find that

$$S_X(\omega) = \sum_{j=1}^{N} \frac{\sigma_j^2}{2} \{\delta(\omega + \omega_j) + \delta(\omega - \omega_j)\}. \tag{6.26}$$

The variance is distributed only on the (physical) frequencies $\omega_1, \ldots, \omega_N$, exactly as we would expect.

6.4.4 Rectangular Spectrum

It is now assumed that $X(t)$ is a stationary process with a rectangular variance spectrum. A rectangular spectrum can be illustrated as in Fig. 6.6. In such a case, the variance is uniformly distributed over a frequency interval, for example (ω_1, ω_2). We study the autocovariance function $C_X(\tau)$ associated with this spectrum. According to Eq. (6.6), we have that

$$C_X(\tau) = 2S_0 \int_{\omega_1}^{\omega_2} \cos \omega\tau \, d\omega = \frac{2S_0}{\tau}\left[\sin \omega_2\tau - \sin \omega_1\tau\right]. \tag{6.27}$$

There are two cases in particular that we want to investigate in some detail.

Narrow Spectrum (Narrow Banded Process)

That the spectrum is narrow simply means that $\omega_2 - \omega_1 \ll \omega_2$. It is then expedient to introduce new variables ω_0 and ν defined as, see Fig. 6.6,

$$\omega_2 = \omega_0 + \nu \text{ and } \omega_1 = \omega_0 - \nu. \tag{6.28}$$

The rhs of Eq. (6.27) can then be rewritten using the trigonometric relation $\sin(u + v) - \sin(u - v) = 2 \cos u \sin v$. That gives the result

$$C_X(\tau) = \frac{4S_0}{\tau} \cos \omega_0\tau \sin \nu\tau = 4S_0\nu \frac{\sin \nu\tau}{\nu\tau} \cos \omega_0\tau. \tag{6.29}$$

Exercise: Check that $C_X(0)$ equals the area under the spectrum, that is, the variance. (Hint: $\sin x/x \to 1$ when $x \to 0$.)

A typical plot of $C_X(\tau)$ as a function of τ is shown in Fig. 6.7. Because $\nu \ll \omega_0$, $\cos \omega_0\tau$ will vary much more rapidly with τ than $\sin \nu\tau$. The factor $4S_0\nu \frac{\sin \nu\tau}{\nu\tau}$ will serve as an envelope for $C_X(\tau)$, see Fig. 6.7. Because the spectrum is assumed to be different from zero only close to $(\pm)\omega_0$, one must expect that a realization of

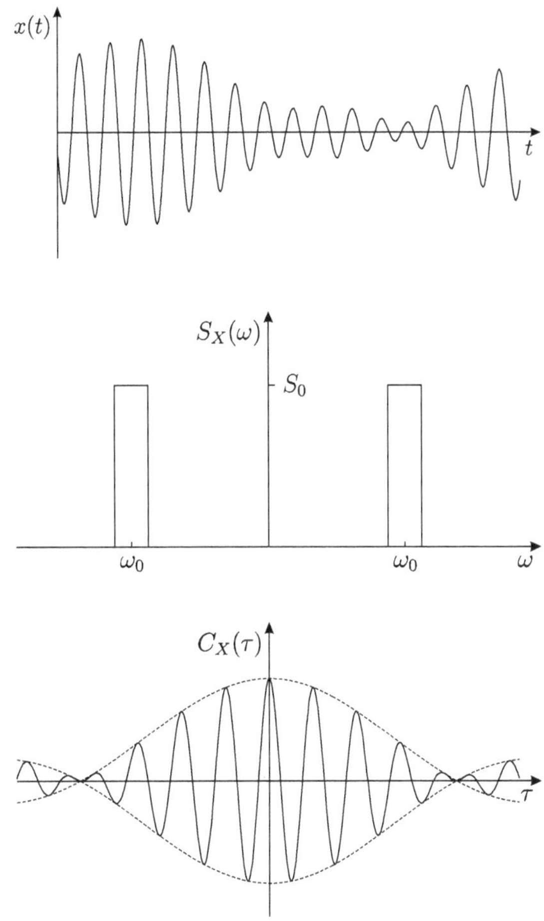

Figure 6.7. Narrow banded process.

the process $X(t)$ will behave like a sum of many harmonic components with almost the same frequency. If these components are approximately in phase at one point in time, several periods must elapse before they are out of phase. This means that high peaks will have a tendency to occur in groups, as illustrated in Fig. 6.7. Such a process is commonly referred to as a *narrow banded* process. This name is derived directly from the fact that the spectrum of such processes are concentrated in a narrow frequency band.

Broad Spectrum (Broad Banded Process)

We shall have a look at the case $\omega_1 = 0$ and $\omega_2 \gg 1$. $\omega_1 = 0$ in (6.27) leads to the result

$$C_X(\tau) = \frac{2S_0}{\tau} \sin \omega_2\tau = 2\, S_0\, \omega_2 \, \frac{\sin \omega_2\tau}{\omega_2\tau}. \tag{6.30}$$

$C_X(\tau)$ in this case typically behaves as shown in Fig. 6.8. It is seen that $C_X(\tau)$ for a broad spectrum decreases to zero with increasing τ much more rapidly than for a narrow spectrum. This reflects the fact that the statistical dependence between the values of the process at two different time instants is much weaker for a broad banded than for a narrow banded process. We can get an impression of this by

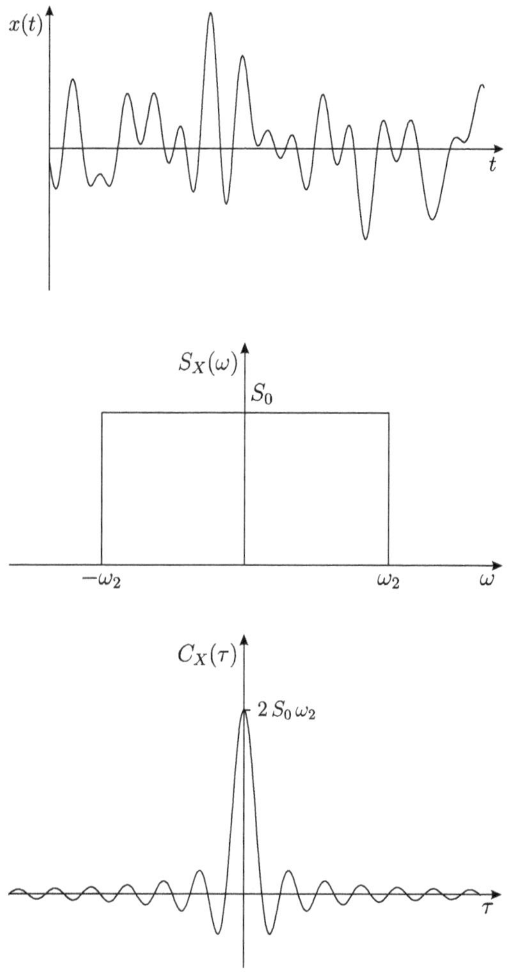

Figure 6.8. Broad banded process.

comparing a typical realization of a broad banded process, as shown in Fig. 6.8 with one of a narrow banded process, as shown in Fig. 6.7.

An important special case of a broad banded process is obtained by letting $\omega_2 \to \infty$. This means that

$$S_X(\omega) = S_0\,, \qquad \text{for all } \omega. \tag{6.31}$$

A process with such a spectrum would then contain all frequencies. A spectrum of this type is therefore often called a *white noise* spectrum, in analogy with white light, which contains "all" frequencies. To determine the corresponding autocovariance, we face the same problem as in the example in Subsection 6.4.1. The only difference is that ω and τ have switched roles. Hence, $C_X(\tau)$ can be expressed by the following equation

$$C_X(\tau) = S_0 \int_{-\infty}^{\infty} e^{i\omega\tau}\, d\omega = 2\pi\, S_0\, \delta(\tau). \tag{6.32}$$

This means that $C_X(\tau) = 0$ for all $\tau \neq 0$; that is, the correlation between values of the process at two different times is zero, irrespective of how close the times are.

A *white noise process* will also have infinite variance because the area under the spectral density becomes infinitely large. In reality, however, such a process cannot be fully realized. In practice, one would call the process white noise if ω_2 is much greater than all other relevant frequencies. This point is revisited and discussed in Chapter 8.

6.5 The Variance Spectrum Directly from the Realizations

When we defined the variance spectrum, it was mentioned that a realization $x(t)$ of a stationary process $X(t)$ does not have a Fourier transform because it does not decrease toward zero for large t. Intuitively, one would nevertheless expect that most of the information regarding the frequency content of $x(t)$ should be contained in a finite section of $x(t)$ if this section is large enough. A section of $x(t)$ can be defined as follows:

$$x_T(t) = \begin{cases} x(t), & 0 \le t \le T, \\ 0, & \text{elsewhere.} \end{cases} \tag{6.33}$$

Because $x_T(t)$ is zero outside a finite interval, it has a Fourier transform, viz.

$$X_T(\omega) = \frac{1}{2\pi} \int_{-\infty}^{\infty} x_T(t)\, e^{-i\omega t}\, dt = \frac{1}{2\pi} \int_0^T x(t)\, e^{-i\omega t}\, dt. \tag{6.34}$$

It would seem natural to expect that there is a connection between $X_T(\omega)$ and the variance spectrum $S_X(\omega)$. And it turns out that the connection is, in fact, quite simple. We write it in the following way: assume that $x_j(t)$, $j = 1, 2, \ldots$ are realizations of a stationary process. Then the following equation applies:

$$S_X(\omega) = \lim_{T\to\infty} \lim_{N\to\infty} \frac{2\pi}{TN} \sum_{j=1}^{N} |X_{j,T}(\omega)|^2, \tag{6.35}$$

where $X_{j,T}(\omega)$ denotes the Fourier transform of $x_{j,T}(t)$, which equals $x_j(t)$ for $0 \le t \le T$ and zero elsewhere. From Eq. (6.35), it is also immediately seen that $S_X(\omega) \ge 0$.

Equation (6.35) is based on the availability of an ensemble of realizations. As previously discussed, there are many situations where only one realization is available. By assuming that the process is ergodic, it can be shown that the variance spectrum can be determined in the following way. Let $x(t)$ denote a realization of the assumed ergodic process $X(t)$. Define a set of truncated Fourier transforms $X_{(j),T}(\omega)$ as

$$X_{(j),T}(\omega) = \frac{1}{2\pi} \int_{(j-1)T}^{jT} x(t)\, e^{-i\omega t}\, dt. \tag{6.36}$$

It is then obtained that

$$S_X(\omega) = \lim_{T\to\infty} \lim_{N\to\infty} \frac{2\pi}{TN} \sum_{j=1}^{N} |X_{(j),T}(\omega)|^2, \tag{6.37}$$

Equation (6.35) or (6.37) can be regarded as the basis for the *fast Fourier transform* (FFT) algorithm for calculating spectra. This method, developed around 1965, has assumed a dominating position among numerical methods for calculating Fourier transforms. This is primarily due to the fact that the method is much faster than traditional methods. An extensive discussion of the FFT method for calculating Fourier transforms is given by Newland (1991).

7 Environmental Loads

7.1 Introduction

A brief review of the various types of "dynamic" loads acting on marine structures is given. In general, the intensity of loads varies in both space and time. The description needs to address this variability.

There are multiple ways to classify loads. Usually, account is made of what kind of physical phenomenon causes them, the uncertainty of the loads (this will affect the safety (load) factor that is to be applied), etc. Design codes contain a load classification. The following list covers major types of loads for marine structures, which could have dynamic structural effects:

- Functional loads due to the operation of machinery to extract energy (wave and wind) and the operation of cranes
- Environmental loads due to wind, wave, current, ice, and earthquakes

Environmental phenomena vary over time. If the variation of the corresponding environmental load over time synchronizes with a natural period of the structure, dynamic amplification of the response may occur. Typical natural periods of oscillation are a few seconds for bottom-supported structures and up to 300 seconds for moored floating structures. Load fluctuations with time may, hence, be separated into (see, e.g., (Larrabee and Cornell, 1979)):

- Macro-scale variations, which do not affect the structural response
- Micro-scale variations, which may directly influence the structural response

Macro-scale description may, for instance, be

- average wind velocity over a 10-minute period
- tidal current
- significant wave height and peak period of a wave spectrum
- peak ground acceleration of an earthquake motion

The macro-scale variation of the loading environment can be studied without regard for the structure involved.

The micro-scale variation has an influence on the interaction between the loading and structural response. The time variation then may also be so rapid that the load intensity at various locations of the structure has a phaselag. It is the micro-scale fluctuations that give rise to structural dynamic effects.

7.2 Hydrodynamic Loads, Added Mass, and Damping

At a given moment, the real wind generated sea surface may be represented by a combination of regular waves by considering waves with different amplitudes and periods (wave lengths) from different directions. The kinematics (pressure, particle velocities, and accelerations) of regular waves are expressed by the sea surface elevation by various wave theories. The Airy theory is commonly used because it assumes linearity between the kinematic quantities and wave height. This is the approach used in Chapter 8. More refined nonlinear theories (Stokes 5th order etc.) have also been developed (see, e.g., (Sarpkaya and Isaacson, 1981)).

The sea state in a short-term period, typically 3 hours, is commonly assumed to be described by a zero-mean, ergodic Gaussian process completely specified by a wave spectrum. As discussed in Chapter 8, various analytical formulations for the wave spectrum are applied. In North Sea waters, the JONSWAP spectrum is recommended. For open seas, the Pierson-Moskowitz spectrum is relevant. Other formulations with up to six parameters have been proposed (Ochi, 1998).

In a long-term period, the variation of the seastate is slower than the short-term fluctuations. For instance, the dominant period of high and low pressure movements in the North Atlantic and the North Sea is of the order of four days, while typical wave periods are less than, say, 10 to 15 seconds.

The long-term variation of the sea is, therefore, often approximated by a compound locally stationary Gaussian process. By neglecting the transient effects, the process can be treated as a series of stationary, zero-mean Gaussian processes given by the wave spectral parameters; the significant wave height, H_s; and wave peak period, T_p. The long-term seastate is then specified by the relative frequency of occurrence of H_s, T_p and other parameters.

Data for the long-term model of the waves can be generated in three alternative ways, namely: (1) by instrumental observations, (2) by visual observations, (3) by hindcasting based on wind data and a model for wind wave generation.

In general, the effects of waves and currents on marine structures are obtained as vector superposition of all forces on the individual structural elements. If relevant, the subsequent response, e.g., the motion of the structure, also needs to be considered. To calculate hydrodynamic forces, it is necessary to integrate the pressure field over the wetted surface of the structure. The main force components are (Faltinsen, 1990; Clauss et al., 1991):

- *Froude-Krylov force*: Pressure effects due to undisturbed incident waves
- *Diffraction force*: Pressure effects due to presence of the structure
- *Hydrodynamic* "added" *mass and potential damping force*: Pressure effects due to motion of structural components in an ideal fluid
- *Viscous drag force*: Pressure effects due to relative velocity between water particles and structural components

For fixed slender structures, the Froude-Krylov force and the diffraction force are approximated by a single inertia term, and the viscous force makes up the drag term. In this case, it is assumed that the water particle velocity and acceleration in the region of the structure do not differ significantly from the values at the cylinder axis. This assumption is only acceptable for small structural diameters D as compared to the wave length; i.e., for $D/\lambda < 0.2$.

When the wave acts on a structure, the latter will be set in motion, which will set up waves radiating away from it. Reaction forces are then set up in the fluid that are proportional to acceleration and velocity of the structure, respectively. These are added mass (inertia) and potential damping forces due to wave generation. In addition, viscous (drag) forces are set up.

If the structure moves, the total inertia force acting on a slender member of the structure may then be established as the same Froude-Krylov force as that acting on a fixed structure, together with the added mass force associated with the relative acceleration between fluid and structure. The drag force may be established by replacing the particle velocity with the relative velocity. Hence, the total force normal to the axis of the member may be written as

$$\begin{aligned} q_n &= \rho\, d\forall \cdot a_n + (C_m - 1)\rho\, d\forall \cdot (a_n - \ddot{x}_n) \\ &\quad + \frac{1}{2} C_d \rho\, dA \cdot (v_n - \dot{x}_n)\, |v_n - \dot{x}_n| \\ &= C_m \rho\, d\forall \cdot a_n - (C_m - 1)\, \rho\, d\forall \cdot \ddot{x}_n \\ &\quad + \frac{1}{2} C_d \rho\, dA \cdot (v_n - \dot{x}_n)\, |v_n - \dot{x}_n| \end{aligned} \tag{7.1}$$

where ρ is the density of the fluid; C_m and C_d are the inertia and drag force coefficients, respectively; and v_n and a_n are, respectively, the wave particle velocity and acceleration perpendicular to the member. dA and $d\forall$ are the exposed area and displaced water per unit length. Equation (7.1) is an example of Morison's equation (Morison et al., 1950; Sarpkaya and Isaacson, 1981).

It should be observed that the applicability of the relative motion formulation of the drag force requires sufficiently large motion amplitudes of the structure. Moreover, it is important to note that the relative velocity formulation of the drag produces both an excitation and a damping force. Depending on which effect is dominant, the drag coefficient should be appropriately determined in design analyses.

A crucial issue in applying Morison's equation is the determination of C_d and C_m. Extensive data from laboratory experiments indicate a general range of 0.6 to 1.2 for C_d and 1.2 to 2.0 for C_m, depending on flow conditions (as measured by the KC and Re numbers) and surface roughness (Sarpkaya and Isaacson, 1981). When applying hydrodynamic coefficients to calculate loading on platforms consisting of many members, additional uncertainties are encountered and should actually be reflected in the coefficients. Moreover, the coefficients C_d and C_m depend on parameters that would vary in time and space in irregular waves. Under such circumstances, the coefficients are chosen to adequately represent the loading in view of the wave kinematics formulation used. The ISO 19902 (2007) specification for calculating loads on jacket platforms may serve to illustrate this point.

For (dynamic) spectral or time-domain analysis of surface piercing framed structures in random Gaussian waves and use of modified Airy (Wheeler) kinematics with no account of kinematics factor, the hydrodynamic coefficients should, in absence of more detailed documentation, be taken to be (NORSOK N-003, 2007): $C_d = 1.0$ and $C_m = 2.0$.

With larger structural diameters, the incident wave is significantly disturbed by the structure. Assuming linear wave theory, the steady-state wave field then results from the interference of the incident wave and the body, and may be derived from the superposition of the potentials of the undisturbed incident wave and an induced wave field of the same frequency, generated by and radiating from the body. Here, viscous forces are of less significance, because the ratio of wave height to structural diameter remains sufficiently small. Although analytical solutions are obtained for a simple cylinder, numerical methods need to be applied in general, see, e.g., (Clauss et al., 1991; Faltinsen, 1990).

7.2.1 Nonlinear Features of Morison Type Loads

The drag force in Morison's equation, Eq. (7.1), is nonlinear in the particle velocity. The particle velocity is proportional to wave height according to the linear theory, cf. Chapter 8. Moreover, the fact that the drag force is nonlinear will introduce higher order harmonics in the force associated with a regular, periodic wave. Neglecting the motion of the structure, the drag, and inertia force on a completely submerged horizontal member caused by a regular wave with particle velocity $v_x = \sin(\omega t)$ and acceleration $a_x = \cos(\omega t)$ (Mo, 1983; Mo and Moan, 1984) are

$$q_d \propto v_x|v_x| = \sin(\omega t)|\sin(\omega t)| \tag{7.2}$$

$$\approx 0.85\sin(\omega t) - 0.17\sin(3\omega t) - 0.02\sin(5\omega t)\ \ldots, \tag{7.3}$$

and

$$q_i \propto a_x = \cos(\omega t). \tag{7.4}$$

When a harmonic wave of finite height passes a structure, forces on a horizontal or a segment of a vertical member in the "splash zone" may exhibit an on-off type of variation in time. This occurs because parts of the structure in the "splash zone" may occasionally be out of the water. Clearly, by expanding these forces in Fourier series, it is observed that there will be higher order harmonic components in the overall load on the structure. This effect will be more pronounced when drag forces are predominant. Also, drag forces are more important in an extreme seaway than in a moderate one.

To illustrate this point more explicitly, consider a cylinder piercing the wave surface. When the velocity is assumed to be constant above the mean water level (MWL) and equal to the velocity at the MWL, the drag and inertia forces (Mo, 1983) are

$$q_d \propto v_x\,|v_x| \cdot H[\zeta(t)] = \sin(\omega t)|\sin(\omega t)| \cdot H[\sin(\omega t)] \tag{7.5}$$

$$\approx 0.25 + 0.424\sin(\omega t) - 0.25\cos(2\omega t) - 0.085\sin(3\omega t)\ldots, \tag{7.6}$$

and

$$q_i \propto a_x \cdot H[\zeta(t)] = \cos(\omega t) \cdot H[\sin(\omega t)] \tag{7.7}$$

$$= 0.5\cos(\omega t) + 0.424\sin(\omega t) + 0.170\sin(\omega t)\ldots, \tag{7.8}$$

where $H[\cdot]$ is the Heaviside unit function defined by $H[x] = 0$ for $x < 0; = 1/2$ for $x = 0; = 1$ for $x > 0$. This fact is also reflected in the probabilistic description of the complete Morison equation, which is discussed in Section 11.2.2.

Moreover, if a current velocity, v_c, is added vectorially to the wave particle velocity in Eq. (7.1), the nature of the (drag) force will be affected. Consider, for simplicity, two wave components of an irregular seaway with velocity amplitudes of v_{x1} and v_{x2}, respectively, together with a current v_c. The drag force during that part of the wave cycle for which the drag force is positive, may be obtained as

$$\begin{aligned} q_d &\propto (v_{x1}\cos\beta_1 + v_{x2}\cos\beta_2 + v_c)^2 \\ &= v_c^2 + \tfrac{1}{2}(v_{x1}^2 + v_{x2}^2) + \tfrac{1}{2}v_{x1}^2\cos(2\beta_1) + \tfrac{1}{2}v_{x2}^2\cos(2\beta_2) \\ &\quad + 2v_c\,(v_{x1}\cos\beta_1 + v_{x2}\cos\beta_2) \\ &\quad + v_{x1}v_{x2}\,(\cos(\beta_1+\beta_2) + \cos(\beta_1-\beta_2)) \end{aligned} \tag{7.9}$$

in which $\beta_j = \omega_j t + \varepsilon_j$, and ε_j is a (random) phase angle.

Clearly, this expansion can be extended to comprise all frequencies, ω_j, in a random sea.

7.2.2 Nonlinear Loads on Large Volume Structures

Higher order terms in the potential theory to account for finite wave elevation also cause time-variant sum and difference frequency forces on large volume structures in (irregular) waves. For instance, the second order term of the surface elevation for the deterministic wave and the quadratic velocity terms in Bernoulli's equation based on the first order potential will contribute second order force components. The term in Bernoulli's equation is somewhat analogous to the velocity squared term in the Morison equation. Although the second order sum and difference frequency wave forces are much smaller than the first order forces in magnitude, they may still excite resonant motions of a floating structure leading to large responses, cf. Chapter 11.

Ringing is a particular phenomenon associated with nonlinear transient inertia wave loads and load effects that occur in steep, high waves encountering structural components extending above the still water level (Krokstad et al., 1996). Structural responses to these loads may be dynamically amplified and cause increased extreme response (ringing) for structures consisting of large diameter shafts and having natural periods in the range from 2 to 8 seconds. These load effects have been of concern for monotowers, gravity base and TLP platforms. Ringing loads depend on the wave shape and particle kinematics close to the water surface and are highly nonlinear, and it is generally difficult to distinguish impact/slamming phenomena from higher order effects.

The ringing phenomenon is illustrated in Figs. 7.1 and 7.2, which show results obtained during model tests of the Heidrun TLP, which is a large concrete tension leg platform designed for use in the North Sea (Munkeby, 1996). Figure 7.1 shows a measured irregular wave profile for a severe sea state. It can be seen that the wave profile contains a steep wave with a peak at around 880 seconds. This steep wave causes ringing loads on the platform and the resulting ringing response event is clearly reflected in a corresponding tether tension event shown in Fig. 7.2 (Naess et al., 2012).

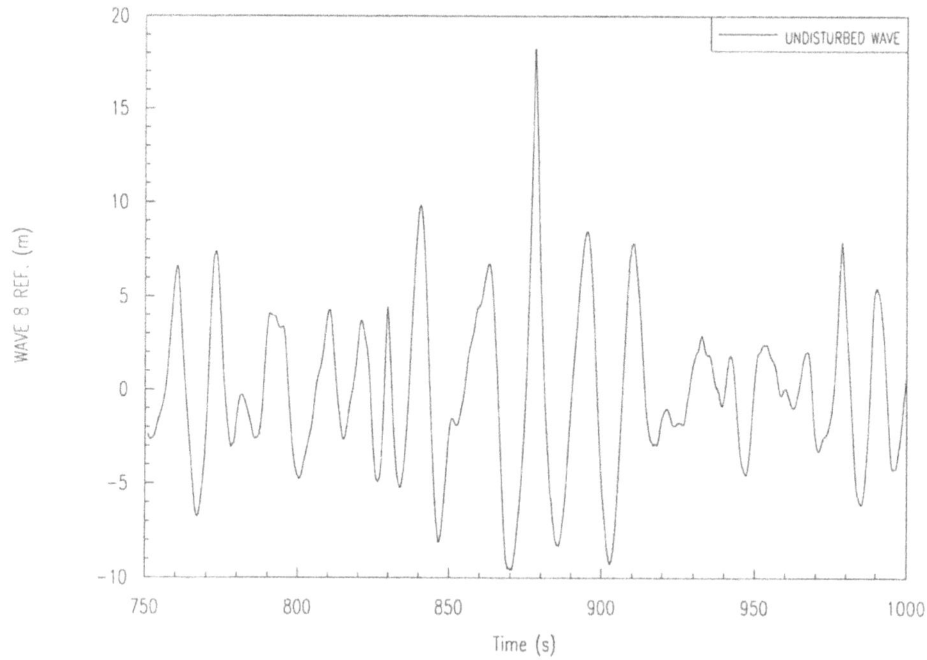

Figure 7.1. Short time series sample of wave elevation containing a steep wave.

In connection with this type of loading, wave impact and buoyancy effects can occur. Because these forces are nearly vertical, they contribute little to the base shear and overturning moment of a platform. However, slamming forces may need to be considered in local member design.

Slamming forces can also occur on platform members overhanging the end of the barge while the platform is being towed, or on members that strike the water first during side launching of platforms.

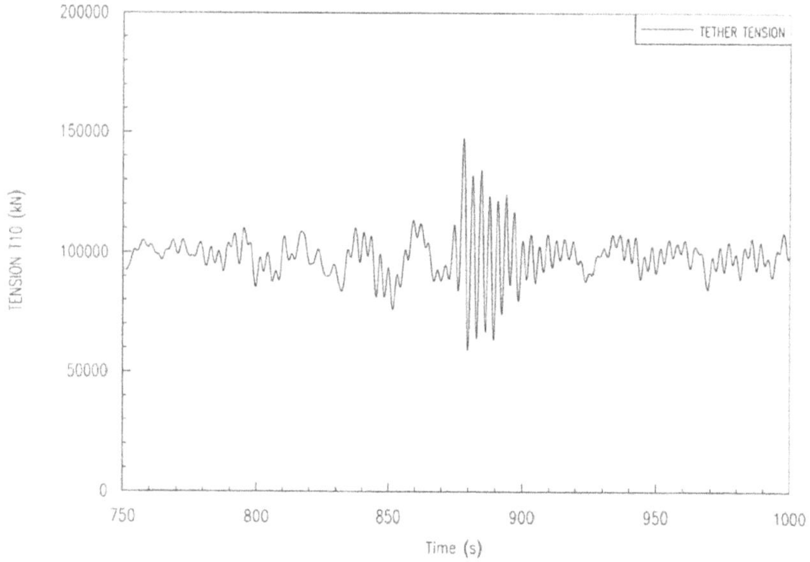

Figure 7.2. Tether tension with a ringing event caused by the steep wave in Fig. 7.1.

In theory, slamming forces are impulsive. If this is the case, the member may be dynamically excited. In the real world, the slamming forces may not be impulsive because of the three-dimensional shape of the sea surface, the compressibility of air trapped between the member and the sea surface, and the separated nature of water near the free surface. For instance, there will be a phase difference in the slamming forces on long members caused by a realistic seaway, because the time for building up the maximum slamming pressure is of the order of milliseconds.

The slamming force F_s per unit length can be calculated from the equation

$$F_s = \frac{1}{2} C_s \, \rho \, D \, v^2 , \tag{7.10}$$

where v is the component of water particle velocity normal to the member axis at impact. The coefficient C_s may lie between 0.5 and 1.7 times its theoretical value of π, depending on the rise time and natural frequency of the elastically mounted cylinder in tests (Sarpkaya and Isaacson, 1981). Sarpkaya and Isaacson recommend that if a dynamic response analysis is performed, the theoretical value of $C_s = \pi$ can be used; otherwise, a value of $C_s = 5.5$ should be used. Recommendations for design are given in DNV (2010b).

7.2.3 Effect of Phase Angle on Wave Forces

An important aspect of wave force calculations on complex structures is the phase difference of the forces on different members, in particular, those piercing the water surface. For a regular wave of length, λ, the phase angle, θ, is $\theta = 2\pi \Delta x / \lambda$, in which Δx and λ are the distance between the locations and the wave length, respectively. The difference in phase may lead to global "amplification" or cancellation of forces. In particular, it is beneficial to establish the cancellation effect at wave frequencies close to fundamental eigenfrequencies of the structure.

7.2.4 Mass, Damping, and Stiffness

Mass is contributed by structural and contained mass as well as by the added hydrodynamic mass. For a slender cylinder, the latter mass is usually taken to be that of the displaced water. The added mass for large volume structures (e.g., caissons of floating gravity structures, floating bridges) is frequency dependent and has to be determined by potential theory for the relevant modes of behaviour. Particular attention should be paid to structural components that are close to the surface, relative to their size. Ogilvie (1963) and Vugts (1970) give data for an infinite cylinder moving horizontally at a certain distance below the water surface. Yeung (1989) determined added mass for a vertical cylinder and for an infinite ship-shaped cross-section moving in the water surface.

Damping may be contributed by the structure, water, and possible soil, and is subjected to significant uncertainties. The structural damping (Adams and Barltrop, 1991) in a welded steel structure may be of the order 0.2% to 0.5% of the critical damping, and for concrete structures that are stressed so that micro-cracks occur, it may be of the order 0.5% to 1.5%. The structural damping of platforms or submerged bridges may be about 1% – as referred to pure structural modes of vibration.

Obviously, the energy take-off in wave energy converters or wind turbines during operation obviously implies a significant damping of the system.

The hydrodynamic damping stems from generation of waves (radiation damping) and from viscous damping. The first source is determined from potential theory and is given for the special cases mentioned by Ogilvie (1963), Vugts (1970), and Yeung (1989); it exhibits strong dependence on frequency and submergence. For significant drag damping to occur, vortex shedding must take place. The drag damping will be small if the Keulegan-Carpenter number, KC, is below, say, 2. Hence, drag damping will be small for large-diameter vertical columns in platforms and submerged bridges. The corresponding damping ratio may be less than 0.1%. Similarly, the potential (radiation) damping is relatively small compared to drag damping for platform structures consisting of slender members. For floating bridges, wave difference frequency (slow drift) excitation may be of importance. Both drag damping and second order (slow drift) potential damping are quite small at the excitation frequencies. For ships in roll motion, artificial mechanisms may be used to increase the damping.

For moored structures, mooring lines imply internal (structural) and drag damping in catenary lines due to their transverse velocity when they are dynamically loaded.

For structures fixed to the seabed, if the soil or rock is activated during vibration response, it will contribute radiation and hysteric (material) damping. Soil damping for (embedded) plate and pile foundations is discussed by Moan et al. (1976), Moan et al. (1977), Adams and Barltrop (1991), and Wolf (1994). Soil damping, especially in rocking motion, is frequency dependent. If a nonlinear soil model is used, the hysteric damping will be implicitly included in the analysis.

If the damping of the structure or the soil is given with reference to a pure structural or foundation mode of vibration, the damping should be appropriately modified when it is included in an interaction mode. It is, for instance, shown by Moan et al. (1976) that the contribution from the structural damping ratio (ξ_s) to the damping ratio ξ for the first mode of a simple flexible tower rocking on soil

$$\xi = (\omega/\omega_s)^3\,\xi_s\,, \tag{7.11}$$

where ω and ω_s are the natural frequencies of a tower on flexible and rigid soil, respectively.

Similarly, the damping ratio (ξ_{wet}) for the structure referred to a wet system (including the effect of added mass) can be expressed by the damping ratio of the "dry" structure (ξ_{dry}) as follows:

$$\xi_{wet} = \xi_{dry}\left(m^*_{dry}\,\omega^*_{dry}/m^*_{wet}\,\omega^*_{wet}\right) \tag{7.12}$$

where m^* and ω^* are generalized mass and frequency, respectively.

In particular, the assessment of damping and soil stiffness is susceptible to significant uncertainties. The only way to assess damping is through full scale measurements. Hence, in-service measurements are useful in justifying the assumptions made in design analyses. Hoen et al. (1993), for instance, show that the (total) modal damping ratios are about 2% for the first three modes of gravity platforms. Karunakaran et al. (1997) found total damping ratios between 0.6% to 1.5% for a jacket with natural periods around 1 second. These references also provide information about assumed versus observed soil stiffness.

Stiffness is also contributed by the structure, water, and soil (rock). Linear elastic structural models are usually applied, except for possible catenary mooring lines. The water provides buoyancy that will influence the stiffness of a bridge supported by pontoons, but would be negligible for bottom-supported platforms. For a moored floating structure, there is no stiffness from buoyancy in the surge, sway, and yaw motions. The stiffness would then be provided solely by the mooring system. The soil is of importance for bottom-supported platforms and may be modeled by equivalent linear properties or by a more refined nonlinear material model. Even if soil stiffness properties are frequency dependent the low frequency of water loading implies that the dynamic stiffness is close to the static values.

The mass, damping, and stiffness properties presented previously refer to ultimate and fatigue limit state criteria, based primarily on linear elastic global models. However, if prediction of the ultimate global capacity is required in connection to survival check in accidental limit states, models that account more properly for nonlinear effects need to be applied.

Under such circumstances framed structures and possible piles are modeled with beam elements including strain hardening nonlinear material and geometric effects. Plasticity may be efficiently incorporated with plastic hinges. Pile-soil interaction may be modeled by nonlinear spring elements along the piles with cyclic (hysteric) behaviour. Structural damping for elastic behaviour and radiation damping in the soil should be explicitly incorporated, while hysteric loss in the structure and soil are implicitly included by this model. Further details about this nonlinear modeling may be found in Stewart (1992); Søreide and Amdahl (1994); Hellan (1995); Nadim and Dahlberg (1996); and Azadi (1998). In particular, the assessment of damping and soil stiffness is susceptible to significant uncertainties.

7.3 Wind Loads

A brief summary of wind loads is given herein. More details are found in Simiu and Scanlan (1996) and Dyrbye and Hansen (1997). An introduction to wind forces on and energy capture by wind turbines is given by Manwell et al. (2002).

Similarly to wave loads on floating bodies, wind loads on structures are caused by complex fluid dynamic phenomena that are generally difficult to calculate with high accuracy. The most widely used engineering approach to wind loading on structures is therefore based on the following observation: when a stream of air flows with constant velocity W, it will generate a force on a flat plate of area A, placed orthogonal to the flow direction, which is proportional to AW^2. The proportionality constant is, with reasonable accuracy, independent of the area A. The wind force on a plate orthogonal to the wind flow can therefore be characterized by a net wind pressure p_w, which is commonly written as

$$p_w = \frac{1}{2}\rho_a C_w W^2 , \tag{7.13}$$

where ρ_a is the mass density of air and C_w denotes a wind pressure coefficient. The value of ρ_a is approximately 1.25 kg/m^3. However, the possible increase of ρ_a due to water spray up to a height of 20 to 30 meters above MWL should be noted. The total wind pressure induced force on the plate is then $p_w\, A$. If the plate has an angle θ to the wind direction that deviates from the orthogonal, that is, 90 degrees, the wind force becomes $p_w\, A \sin\theta$, where $A \sin\theta$ is the projected area onto the plane orthogonal to the wind direction. A common approximation of the wind force on a more general structure is to express it as the product of the wind pressure with the projected area onto the plane orthogonal to the wind direction, or an appropriate variation of this.

C_w is normally determined under controlled stationary wind flow conditions in a wind tunnel. C_w usually depends on the Reynolds number Re (Sarpkaya and Isaacson, 1981). A drag coefficient of 0.7 and 1.2 is typical for cylinders. The major problem in determining wind forces on complex real structures is the significant interaction between the wind loads on various structural parts.

Natural wind can be considered to consist of a mean wind component and a fluctuating, or gusty, wind component, which is generated by the turbulence in the flow field. The turbulence causes gusty wind speed components in all three spatial directions. However, the mean wind speed is usually much larger than the typical gust wind fluctuations, especially at offshore locations. As a consequence, structures exposed to natural wind will experience a fluctuating wind load, but with the dominating force component in the direction of the mean wind. Otherwise, the magnitude and character of the wind load will depend on the size and shape of the structure, and its orientation relative to the mean wind direction. The discussion in this book is limited to cases where the wind load on the structure in the mean wind direction can be calculated largely in terms of an effective wind pressure $p_w(t)$, which can be expressed as

$$p_w(t) = \frac{1}{2}\rho_a C_w W(t)^2 , \qquad (7.14)$$

where $W(t)$ denotes the total wind speed in the mean wind direction; that is, $W(t) = \overline{W} + U(t)$; $\overline{W}$ is the mean wind speed; and $U(t)$ is the fluctuating, gusty along-wind component. The average value of $U(t)$ is zero. By adopting the assumptions discussed in Section 8.3, the spatial dependence of $\overline{W}$ is only through the vertical coordinate, while $U(t)$ can be considered homogeneous in space and time for each short-term condition. For the sake of completeness, it should be mentioned that when the wind load is given in terms of Eq. (7.14), which derives from flow separation effects, it has been tacitly assumed that the inertia forces due to the relative fluid-body accelerations can be neglected. Because these forces are about two orders of magnitude smaller than the drag forces, they can be neglected in practical applications.

Discussions of wind load models of the kind described previously are found in Davenport (1977) and Dyrbye and Hansen (1997). For discussions of other wind load phenomena, such as vortex shedding, galloping, flutter, etc., the reader may consult Simiu and Scanlan (1996) or Dyrbye and Hansen (1997).

The modeling of $W(t)$ as a stochastic process is discussed in Section 8.3, and various wind spectra for engineering calculations are described. Hence, the wind loads on structures are also modeled as stochastic processes. Let us start by writing down the expression for the wind load process on a structure of small dimensions compared to the significant "wavelengths" $2\pi\overline{W}/\omega$ in the wind gusts. Assuming that the projected area of the structure orthogonal to the mean wind direction is A, this requires that $\omega\sqrt{A}/(2\pi\overline{W}) \ll 1$ for frequencies ω where the wind spectrum has significant values. If this is the case, the wind load on the structure can be written as

$$F_w(t) = \frac{1}{2}\rho_a C_w A W(t)^2 = \frac{1}{2}\rho_a C_w A\big(\overline{W}^2 + 2\overline{W}U(t) + U(t)^2\big). \tag{7.15}$$

Generally, $\big(U(t)/\overline{W}\big)^2 \ll 1$. We may therefore adopt the approximation,

$$F_w(t) \approx \frac{1}{2}\rho_a C_w A\overline{W}^2 + \rho_a C_w A\overline{W}U(t). \tag{7.16}$$

The wind load is seen to consist of a mean drag force

$$\overline{F}_w = \mathrm{E}[F_w(t)] = \frac{1}{2}\rho_a C_w A\overline{W}^2, \tag{7.17}$$

and a fluctuating, zero-mean force

$$F_g(t) = \rho_a C_w A\overline{W}U(t). \tag{7.18}$$

Within this approximation, the (one-sided) power spectral density of the wind load process $F_w(t)$ is then related to the wind spectrum as follows:

$$S_F^+(\omega) = (\rho_a C_w A\overline{W})^2\, S_U^+(\omega) \tag{7.19}$$

or

$$S_F^+(\omega) = 4\frac{\overline{F}_w^2}{\overline{W}^2}\, S_U^+(\omega). \tag{7.20}$$

For cases when the size of the structure is not small in the previous sense, it is necessary to make some adjustment for the reduced spatial correlation of the wind forces over the structure. For massive or bluff-body structures, this is conveniently achieved through the introduction of an aerodynamic admittance function $[\chi(\omega\sqrt{A}/(2\pi\overline{W}))]^2$ into the rhs of Eq. (7.19) or (7.20) as follows:

$$S_F^+(\omega) = 4\frac{\overline{F}_w^2}{\overline{W}^2}\left[\chi\Big(\frac{\omega\sqrt{A}}{2\pi\overline{W}}\Big)\right]^2 S_U^+(\omega). \tag{7.21}$$

When $\omega\sqrt{A}/(2\pi\overline{W}) \to 0$, then $\chi\big(\omega\sqrt{A}/(2\pi\overline{W})\big)^2 \to 1$, and when $\omega\sqrt{A}/(2\pi\overline{W}) \to \infty$, it is anticipated that $\chi\big(\omega\sqrt{A}/(2\pi\overline{W})\big)^2 \to 0$. For flat plates and prismatic bodies, the following empirical formula was proposed (Davenport, 1977)

$$\chi(x) = \frac{1}{1 + (2x)^{4/3}}. \tag{7.22}$$

For a thorough discussion of aerodynamic admittance functions, cf. Dyrbye and Hansen (1997).

However, this approach would not generally be appropriate for long slender structures exposed to the wind, like, e.g., long-span bridges. Spatial correlation effects of the wind forces over the structure are then commonly incorporated into the analysis through the cross-spectral density of the wind speed at different locations. This is necessary to obtain a more correct picture of the wind load process in space and time, which is needed in order to calculate the dynamic response of the structure, for example, by a modal analysis (Davenport, 1977; Simiu and Scanlan, 1996; Dyrbye and Hansen, 1997).

Finally, it is noted that the calculation of wind loads totally depends on having accurate C_w coefficients. When new structural concepts are encountered, it is normally required that wind tunnel tests are performed to determine the wind load. The calculation of wind loads on the rotor blades of wind turbines obviously requires a more detailed approach (Manwell et al., 2002).

The statistics of wind loads and response are discussed in Section 11.4.

7.4 Ice Loads

Ice loads represent the dominant load for ice-going ships and structures in the arctic environment. Prediction of ice loads are associated with a significant uncertainty. Arctic structures encounter various ice conditions in their lifetime, such as level ice, broken ice, ice ridges, and ice bergs, and there are also many kinds of structural forms and corresponding interactions between ice and structure. For a ship, the variability of ice loads increases due to its possible operation in a large geographic area and the variation of its forward speed. Moreover, different kinds of failure mode (e.g., creep, cracking, buckling, spalling crushing) may occur separately or in combination, and hence affect the ice loading (Sanderson, 1988; Muhonen, 1996; Jordaan, 2001; Masterson et al., 2007).

The ice loads imposed on ships and structures exhibit a random variation in both time and space. Generally, ice load acting on a ship or other structures can be classified into two types: total or global load, and local load (pressure). Global loads affect the overall motion and stability of structures. Local loads affect areas from 1 m^2 up to as much as 100 m^2.

When considering the local load, the pattern of decreasing ice pressure with increasing contact area has emerged as evidenced in Fig. 7.3.

In level ice condition, the ice loads due to continuous ice breaking have a clear stochastic character and thus can be considered a random process. The frequency of interaction between the structure and ice, i.e., the number of interactions per unit time, plays an important role.

There are three different analytical types of local ice load models: *continuum ice load models*, *stochastic ice load models* and *discrete ice load models*.

Figure 7.4 shows the idealized time histories of the ice forces due to intermittent crushing and bending and the definition of ice resistance.

The random periodic ice loading, see Fig. 7.4, may cause dynamic amplification in slender structures, and stochastic models for ice loads have been addressed, e.g., by Kärnä et al. (2004).

Most of the research on ice load in the past decades focused on the calculation of extreme static ice load in engineering. Only the formula of the extreme static ice load is included in the current codes. It has been demonstrated by many researchers that the ice force on a conical structure is smaller than the force on a cylindrical structure of similar size, e.g., Sanderson (1988) and Määttänen (1996). The main reason for the reduction in the ice force is that a well-designed cone can change the ice failure mode from crushing to bending. Brown and Määttänen (2009) found that both peak values of forces and ice-induced vibrations are reduced by use of

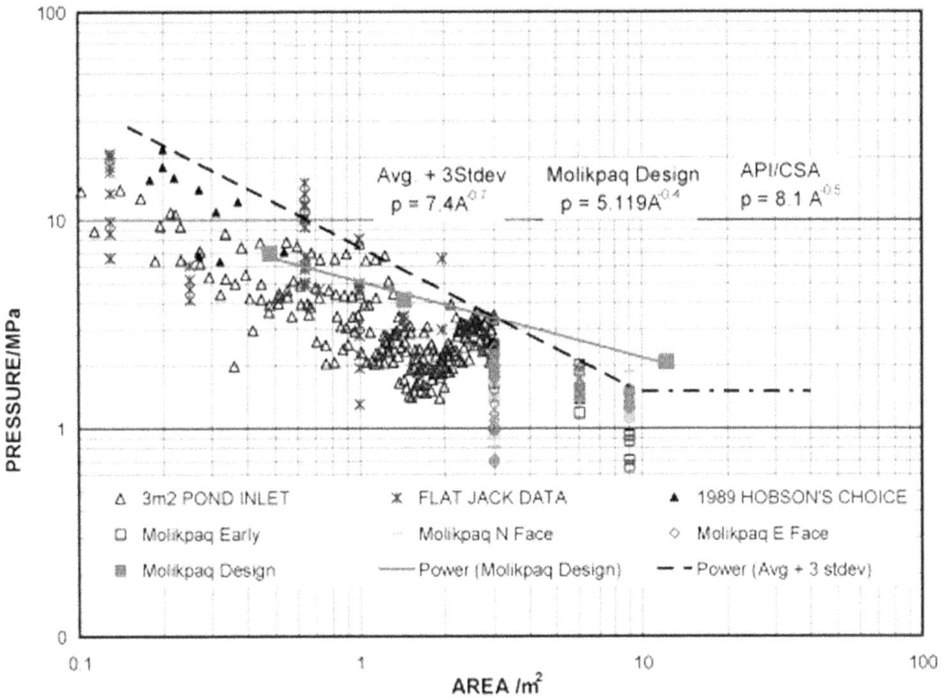

Figure 7.3. Observed pressure-area relationship (Masterson et al., 2007).

a conical structure. Yue et al. (2007) observed vibrations even for conical structures. In general the ice-induced response can be determined by time or frequency domain methods. In a frequency domain analysis, the excitation is then modeled as a sinusoidal pseudo-excitation, and the response characterized by a transfer function.

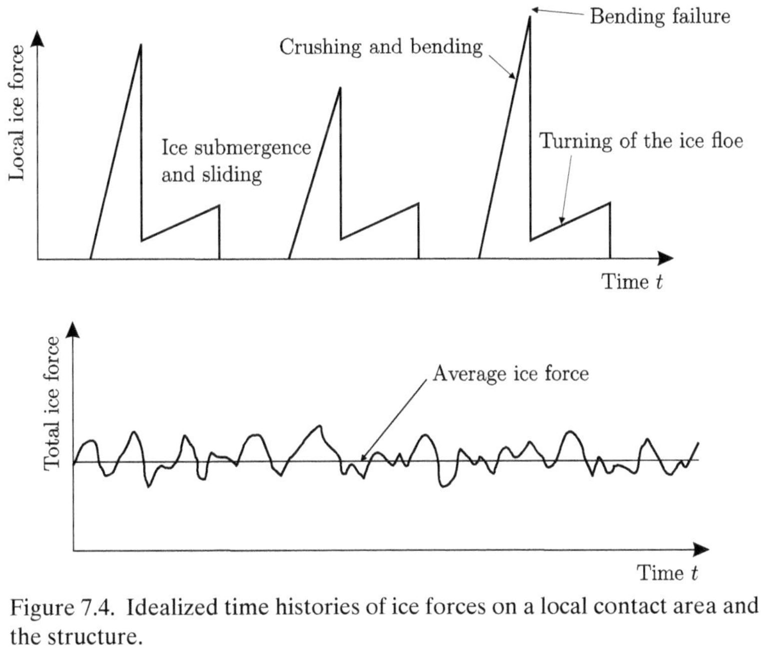

Figure 7.4. Idealized time histories of ice forces on a local contact area and the total force on the structure.

Sundararajan and Reddy (1973) initiated the application of a frequency domain analysis of dynamic ice actions. Kärnä et al. (2004) derived a formula for spectral density functions of the crushing ice force on vertical structures. Based on full-scale field tests and observations, Gravesen et al. (2005), Yue et al. (2007), and Liu et al. (2009) proposed the ice force spectrum on the narrow conical structure in the form of the power spectral density (PSD) shown in Eq. (7.23):

$$S^{+}(f) = \frac{A\bar{F}_0^2 \bar{T}^{-\delta}}{f^{\gamma}} \exp\left(-\frac{B}{\bar{T}^{\alpha} f^{\beta}}\right) \tag{7.23}$$

Here, A and B are constants, $\bar{F}_0$ is the force amplitude on the structure, $\bar{T} = L_b/V$ is the ice force period of the conical structure and f is the frequency, L_b is the ice breaking length, $L_b = kh$, and V is the velocity. k is the ratio of ice thickness to ice breaking length. Observations in the Bohai Gulf show that $A = 10$, $B = 5.47$, $\alpha = 0.64$, $\beta = 0.64$, $\gamma = 3.5$, and $\delta = 2.5$, and k typically assumed values between 4 and 10. F_b in Eq. (7.23) is given by Eq. (7.24) (Liu et al., 2009)

$$\bar{F}_0 = C\sigma_f h^2 \left(\frac{D}{L_c}\right)^{0.34}, \tag{7.24}$$

where C is a constant, σ_f is the ice bending strength (0.7 MPa), h is the ice thickness, D is the diameter of the ice cone, and L_c is the characteristic length:

$$L_c = \left(\frac{Eh^3}{12g\rho_w}\right)^{0.25}, \tag{7.25}$$

where E is the ice elastic modulus (0.5 GPa), g is the gravitational acceleration and ρ_w is the density of the water.

The ice velocity and ice thickness are two important parameters that need to be described as the random variables based on the field observations. The probability density functions of velocity and ice thickness are commonly assumed to follow a Rayleigh distribution and lognormal distribution, respectively (Liu et al., 2009).

In practical engineering, a stationary or an ergodic hypothesis is generally employed for random vibration, and in most vibration problems, the system acting as a filter has at least one resonant frequency at which large amplitudes can be generated by small input. A reasonable approximation is then to assume that the response, e.g., deck acceleration, is a narrow band Gaussian process. The response spectrum may then be determined by means of the force-response transfer function and force spectrum.

In broken ice condition, the ice loads are smaller than in level ice. When the floes are large, their inertia is great, and thus there is a limit size of floe after which the floe can be considered as an ice field.

The irregular features of ice ridge are difficult to define. A number of simplifications and idealizations have to be made before one can even start to think of mathematical methods in determining the ridge resistance.

The first phase of resistance occurs between the bow of the ship and the ridge, and the second phase occurs between the parallel middle body of the ship and the ridge. The parallel middle body experiences a resistance component mainly of a static nature not pushing the ice aside anymore but introducing frictional losses due to ice leaning against the sides and bottom of the ship (Keinonen, 1977).

Impacts with icebergs could also imply a dynamic response. Because there are no full-scale experiences with forces exerted by icebergs, a simplified approach is based on several assumptions (Sanderson, 1988).

Forces due to moving floes of limited size represent scenarios that combine features of continuous ice floes and icebergs.

7.5 Seismic Loads

In some offshore areas, earthquake loading governs the design structures with stiff connections to the seabed. It is noted that the vertical force in joints in articulated towers and tethers in TLPs may be affected. This section briefly describes the earthquake phenomena and the load on structures they cause. More detailed information may be found in, e.g., Dowrick (1977) and Housner and Jennings (1982).

Earthquakes especially occur in zones between the large tectonic plates in the earth's crust (interplate earthquakes). These zones are easy to observe in the Pacific Ocean area, Himalaya to Turkey, the Balkans, Greece and Italy. In Scandinavia and the North Sea area, there is no such zone. In the latter areas, so-called intraplate earthquakes occur. Although interplate earthquakes are most often associated with the horizontal movements (strike-ship faults), the intraplate quake is often characterized by vertical movements (thrust faults).

An earthquake gives rise to horizontal and vertical motions with a duration of typically 15 to 30 seconds. The motion (acceleration) history exhibits a random character due to the nature of the mechanism causing the earthquake, and the wave propagation, reflection and deflection. The ground movements caused by earthquakes can have damaging effects such as direct movement of structures, ground surface faults, tidal waves, flooding, and so on. Although these effects are of concern, only the first effect, the direct motion of structures, is considered in this chapter.

The earthquake motions in the base rock are transmitted through soil layers to the structure and create

- inertia forces due to the accelerations
- damping forces due to the motions in water

The actual motions forced on the structure will be magnified or reduced compared to the bedrock motions, depending on the stiffness, etc., of the soil layers.

An accurate evaluation of seismic response requires a dynamic analysis in which the stochastic dynamic character of the loading and the structure is taken into account. Often, design codes prescribe simplified methods of analysis that have been calibrated against more accurate analyses.

In case of seismic loads, building codes, for instance, specify a total horizontal (or vertical) inertia load due to earthquakes, which depends on the seismic activity in the area and the dynamic properties of the structure (notably, the fundamental period of vibration, T), cf., e.g., the Uniform Building Code (UBC, 1988).

Total horizontal force depends, among other things, on the seismic activity at the site, structure mass, and the natural period of the structure. The total horizontal force is distributed over the height of the structure. Although the simplified approach described previously gives some insight into the parameters of influence on earthquake loads on tower-type platforms, it should be noted that the method is specific for buildings on land. The approach should therefore be applied with caution to other structures. The loading design analysis is described by horizontal/vertical acceleration time histories.

8 Random Environmental Processes

8.1 Introduction

The focus of this chapter is on ocean waves and wind, modeled as random processes (or random fields). The ocean climate is continuously changing, implying that the wave and wind processes are nonstationary in the long run, even on the time scale of hours. However, for modeling purposes, it is common practice to assume that the long term evolution of the wave and wind conditions can be considered as a sequence of 3 hour stationary pieces. This is obviously a simplification, but it is useful and sufficiently accurate for most engineering purposes. The benefit of this kind of modeling is that the tools developed for stationary processes can be used also for long term assessment of the loads and responses of a marine structure. Of particular importance is the possibility to apply the concept of a spectral density (or variance spectrum) for the characterization and representation of the wave and wind processes.

8.2 Ocean Waves

8.2.1 Wave Process

The design of structures to withstand the impact of environmental forces such as wind and ocean waves, typical of ships and ocean structures, has to invariably take into account the irregular or random nature of these kinds of loads. The standard approach to the statistical modeling of ocean waves is to assume that the ocean surface constitutes a stochastic wave field that can be assumed stationary in time and homogeneous in space. However, in engineering practice, stationarity is assumed only for limited periods of time, like three hours, which is then referred to as the short-term description of the wave field. This says that we are assuming steady-state wave conditions for specified intervals of time with no transition periods between different sea states. This is clearly an approximation, but it seems to work well for most offshore design applications.

Usually, the wave process is assumed to be Gaussian, which gives a reasonably good approximation to reality. This assumption implies that the wave elevation at a given point in time and at a specified location on the sea surface can be represented by a Gaussian random variable. The Gaussian probability distribution therefore plays a central role in the assessment of loads and responses of offshore structures to ocean waves. It should be emphasized, however, that although the assumption about a Gaussian wave field is acceptable in many cases, especially for calculating the statistics of the dynamic response of compliant offshore structures, this is not always the case. It is a recognized fact that the crests of ocean waves are generally larger than the corresponding troughs. For problems relating to wave slamming into the deck structure of offshore platforms, this effect is important.

A Gaussian sea surface implies that the stationary wave field is fully described by a directional wave spectrum and a dispersion relation specifying the relationship between the (scalar) wave number and the wave frequency. For deep water waves, which are assumed here, this relationship is given by the formula

$$\omega^2 = g\,k \tag{8.1}$$

where ω denotes (circular) wave frequency, k denotes the wave number, and g equals the gravitational acceleration.

The wave elevation $X(t;x,y)$, relative to an inertial frame of reference (x, y, z), with $z = 0$ as the mean free surface, which is specified by a directional wave spectrum $S_X^+(\omega,\theta)$, can then be approximately represented as follows:

$$\begin{aligned} X(t;x,y) \\ &= \sum_{j=1}^{n}\sum_{k=1}^{m}\sqrt{S_X^+(\omega_j,\theta_k)\,\Delta\omega_j\,\Delta\theta_k}\Big[A_{jk}\cos\Big(\omega_j t - \frac{\omega_j^2}{g}(x\cos\theta_k + y\sin\theta_k)\Big) \\ &\quad + B_{jk}\sin\Big(\omega_j t - \frac{\omega_j^2}{g}(x\cos\theta_k + y\sin\theta_k)\Big)\Big], \end{aligned} \tag{8.2}$$

where $0 < \omega_1 < \ldots < \omega_n$ and $-\pi \le \theta_1 < \ldots < \theta_m \le \pi$ are discretizations of the frequency axis and the directional range, respectively. $\{A_{jk}\}$ and $\{B_{jk}\}$ are independent sets of independent zero-mean Gaussian variables of unit variance.

The directional wave spectrum employed in ocean and offshore engineering is usually written as $S_X^+(\omega,\theta) = S_X^+(\omega)\,D(\omega,\theta)$, where $S_X^+(\omega)$ is chosen from a list of standard wave spectra and $D(\omega,\theta)$ is a directional function. This is discussed in the next section.

In complex number notation, Eq. (8.2) can be rewritten in a more convenient and compact form as

$$\begin{aligned} X(t;x,y) = \Re\Big\{\sum_{j=1}^{n}\sum_{k=1}^{m}\sqrt{2\,S_X^+(\omega_j,\theta_k)\,\Delta\omega_j\,\Delta\theta_k} \\ \times C_{jk}\exp\Big(\mathrm{i}\omega_j t - \mathrm{i}\frac{\omega_j^2}{g}(x\cos\theta_k + y\sin\theta_k)\Big)\Big\} \end{aligned} \tag{8.3}$$

where $\Re\{z\}$ denotes the real part of the complex number z, $\mathrm{i} = \sqrt{-1}$, and $C_{jk} = (A_{jk} + \mathrm{i}B_{jk})/\sqrt{2}$. Note that this makes each C_{jk} a complex, zero-mean Gaussian variable of unit variance.

The ocean surface wave elevation is usually specified relative to some chosen reference point, e.g., the origin of the inertial coordinate system. Let $X(t) = X(t;0,0)$

denote the wave elevation at the reference point. Hence,

$$X(t) = \sum_{j=1}^{n}\sum_{k=1}^{m}\sqrt{S_X^+(\omega_j, \theta_k)\,\Delta\omega_j\,\Delta\theta_k}\left[A_{jk}\cos(\omega_j t) + B_{jk}\sin(\omega_j t)\right]$$
$$= \Re\Big\{\sum_{j=1}^{n}\sum_{k=1}^{m}\sqrt{2\,S_X^+(\omega_j, \theta_k)\,\Delta\omega_j\,\Delta\theta_k}\,C_{jk}\,e^{i\omega_j t}\Big\}. \tag{8.4}$$

Note that an equivalent version of Eq. (8.4) is the following expression, which is also widely quoted:

$$X(t) = \sum_{j=1}^{n}\sum_{k=1}^{m}\sqrt{S_X^+(\omega_j, \theta_k)\,\Delta\omega_j\,\Delta\theta_k}\,R_{jk}\cos(\omega_j t - \varepsilon_{jk}) \tag{8.5}$$

where $R_{jk}^2 = A_{jk}^2 + B_{jk}^2$; that is, $\{R_{jk}\}$ is a set of independent Rayleigh distributed variables, which is independent of the set of independent random variables $\{\varepsilon_{jk}\}$. Each ε_{jk} is uniformly distributed over the interval $(0, 2\,\pi)$, or equivalently, $(-\,\pi, \pi)$, cf. Example 5.7.1.

A simplified, approximate version of this formula is used extensively in offshore engineering:

$$\tilde{X}(t) = \sum_{j=1}^{n}\sum_{k=1}^{m}\sqrt{2\,S_X^+(\omega_j, \theta_k)\,\Delta\omega_j\,\Delta\theta_k}\,\cos(\omega_j t - \varepsilon_{jk}). \tag{8.6}$$

Note that $\tilde{X}(t)$ is never exactly a Gaussian random variable because the Rayleigh variables R_{jk} have been replaced by the constant $\sqrt{2}$; however, due to the central limit theorem, it can be expected that $\tilde{X}(t)$ is approximately Gaussian when nm is large enough.

Each term in the sums of the preceding expressions is a periodic function. Hence, there is a limit to how long the simulation time should be when such an expression is used. When a constant $\Delta\omega$ is used, the simulated wave elevation will exactly repeat itself after a period of $2\pi/\Delta\omega$. If we let $\Delta\omega_{\min}$ denote the smallest and $\Delta\omega_{\max}$ the largest of the $\Delta\omega_j$, $j = 1, \ldots, n$, then the time interval of validity of the preceding expressions will be approximately between $(0, 2\pi/\Delta\omega_{\max})$ and $(0, 2\pi/\Delta\omega_{\min})$, with $(0, 2\pi/\Delta\omega_{\max})$ as a first, simple estimate.

The expression for the wave elevation or any other quantity that is linearly related to it can be obtained at any other relevant point from the expression given in Eq. (8.4) by using the linear transfer function that specifies the relationship. Let the linear transfer function relating the wave elevation at the reference point from a regular wave train propagating in the θ direction to the desired quantity related to the point (x, y, z) $(z \le 0)$ be denoted by $\hat{H}(\omega; \theta)$. Using this transfer function, the corresponding stochastic process, $Y(t; x, y, z)$, at the point (x, y, z) corresponding to the wave elevation process $X(t)$, is then calculated as follows:

$$Y(t; x, y, z) = \Re\Big\{\sum_{j=1}^{n}\sum_{k=1}^{m}\hat{H}(\omega_j; \theta_k)\sqrt{2\,S_X^+(\omega_j, \theta_k)\,\Delta\omega_j\,\Delta\theta_k}\,C_{jk}\,e^{i\omega_j t}\Big\}. \tag{8.7}$$

The spectral density $S_Y^+(\omega)$ of the stationary process $Y(t; x, y, z)$ is given approximately by the following equation:

$$S_Y^+(\omega) = \sum_{k=1}^{m} |\hat{H}(\omega; \theta_k)|^2 \, S_X^+(\omega, \theta_k) \, \Delta\theta_k. \tag{8.8}$$

Let us look at two examples. For simplicity, we consider the case of no current. In the case of a regular wave train propagating in the direction θ, the linear transfer function relating the wave elevation at the reference point to the water particle velocity in the x-direction at the point (x, y, z) $(z \leq 0)$ is

$$\hat{H}_{XV_x}(\omega; \theta) = \omega \cos\theta \, e^{\frac{\omega^2}{g} z} \, e^{-\mathrm{i}\frac{\omega^2}{g}(x\cos\theta + y\sin\theta)}, \tag{8.9}$$

when the x-direction corresponds to $\theta = 0$. According to Eq. (8.7), the water particle velocity $V_x(t; x, y, z)$ in the x-direction at the point (x, y, z) corresponding to the wave elevation process $X(t)$ is then calculated simply as follows:

$$V_x(t; x, y, z) = \Re\Big\{ \sum_{j=1}^{n}\sum_{k=1}^{m} \omega_j \cos\theta_k \sqrt{2\, S_X^+(\omega_j, \theta_k)\, \Delta\omega_j \, \Delta\theta_k} \times C_{jk}\, e^{\frac{\omega_j^2}{g} z} \, e^{\mathrm{i}\left[\omega_j t - \frac{\omega_j^2}{g}(x\cos\theta_k + y\sin\theta_k)\right]} \Big\}. \tag{8.10}$$

Similarly, the water particle acceleration $A_x(t; x, y, z)$ in the x-direction at the point (x, y, z) is given by the following expression:

$$A_x(t; x, y, z) = \Re\Big\{ \sum_{j=1}^{n}\sum_{k=1}^{m} i\, \omega_j^2 \cos\theta_k \sqrt{2\, S_X^+(\omega_j, \theta_k)\, \Delta\omega_j \, \Delta\theta_k} \times C_{jk}\, e^{\frac{\omega_j^2}{g} z} \, e^{\mathrm{i}\left[\omega_j t - \frac{\omega_j^2}{g}(x\cos\theta_k + y\sin\theta_k)\right]} \Big\}. \tag{8.11}$$

According to Eq. (8.8), the corresponding spectral densities are as follows:

$$S_{V_x}^+(\omega) = \sum_{k=1}^{m} \omega^2 \cos^2\theta_k \, e^{\frac{2\omega^2}{g} z} \, S_X^+(\omega, \theta_k)\, \Delta\theta_k \tag{8.12}$$

and

$$S_{A_x}^+(\omega) = \sum_{k=1}^{m} \omega^4 \cos^2\theta_k \, e^{\frac{2\omega^2}{g} z} \, S_X^+(\omega, \theta_k)\, \Delta\theta_k. \tag{8.13}$$

8.2.2 Wave Spectra

As pointed out previously, the directional wave spectrum $S^+(\omega, \theta)$ is usually written as a product of two factors: $S^+(\omega, \theta) = S^+(\omega) \cdot D(\omega, \theta)$, where $\int D(\omega, \theta)\, d\theta = 1.0$. $S^+(\omega)$ is generally referred to as the wave spectrum, while $D(\omega, \theta)$ specifies the directional distribution of the wave energy. We do not list all ocean wave spectra proposed over the years, but rather focus on a class of wave spectra that seems to have acquired a certain prominence. The basic form, often referred to as a Pierson-Moskowitz type of spectrum, is given as

$$S^+(\omega) = \frac{A}{\omega^5} \exp\big(- \frac{B}{\omega^4}\big), \qquad (8.14)$$

where A and B are constants given in terms of other parameters. The spectra belonging to this class are supposed to characterize fully developed sea states. The following three spectral shapes are commonly used:

- *Pierson-Moskowitz spectrum* – This type is specified by $A = \alpha g^2$, where the Phillips constant $\alpha = 0.0081$, and $B = 1.25\omega_p^4$, where ω_p is the peak frequency at which $S(\omega)$ is a maximum. $\omega_p = \omega_p(U)$ is a function of a reference mean wind speed measured at 19.5 meters above mean water level, and it is the only parameter needed to specify the wave spectrum, see, e.g., Chakrabarti (2005), chapter 3.
- *Modified Pierson-Moskowitz spectrum* – This case differs from the previous one by taking $A = \frac{5}{16} H_s^2 \omega_p^4$, where the significant wave height H_s now enters as an additional parameter.
- *ISSC spectrum* – The wave spectrum proposed by the International Ship Structures Congress (ISSC) is given by $A = 0.1107 H_s \overline{\omega}^4$ and $B = 0.4427\overline{\omega}^4$, where $\overline{\omega}$ denotes the average wave frequency, that is, $\overline{\omega} = m_1/m_0$, cf. Eq. (8.31).

A spectral model that has become popular in the offshore industry is the *JONSWAP spectrum* (Hasselmann et al., 1973). It is a modification of the Pierson-Moskowitz spectrum in the following way,

$$S^+(\omega) = \frac{\tilde{\alpha} g^2}{\omega^5} \exp\big(- 1.25\frac{\omega_p^4}{\omega^4}\big) \cdot \gamma^{a(\omega)}, \qquad (8.15)$$

where the peakedness parameter γ appears with an exponent

$$a(\omega) = \exp\big(- \frac{(\omega - \omega_p)^2}{2\tilde{\sigma}^2 \omega_p^2}\big), \qquad (8.16)$$

and

$$\tilde{\sigma} = \begin{cases} \tilde{\sigma}_a = 0.07\,, & \text{for} \quad \omega \leq \omega_p\,; \\ \tilde{\sigma}_b = 0.09\,, & \text{for} \quad \omega > \omega_p. \end{cases} \qquad (8.17)$$

In the original formulation, the parameter $\tilde{\alpha}$ is related to a fetch parameter and to the mean wind speed. However, for offshore applications, a modified version is often adopted. In particular, for North Sea projects, the following form is often adopted:

$$\tilde{\alpha} = 3.25 \cdot 10^{-3} H_s^2 \omega_p^4 (1 - 0.287 \ln\gamma) = 5.058 \frac{H_s^2}{T_p^4} (1 - 0.287 \ln\gamma). \qquad (8.18)$$

The range of values for the peakedness parameter γ was typically between 1 and 7 with 3.3 as an average value (Hasselmann et al., 1973). Most of the storm conditions in the North Sea have γ values around 3 or less. Note that for $\gamma = 1$, the JONSWAP spectrum reduces to the modified Pierson-Moskowitz spectrum. Figure 8.1 plots JONSWAP spectra for $\gamma = 1$ and 3, illustrating the markedly more peaked form of the wave spectrum for higher values of γ.

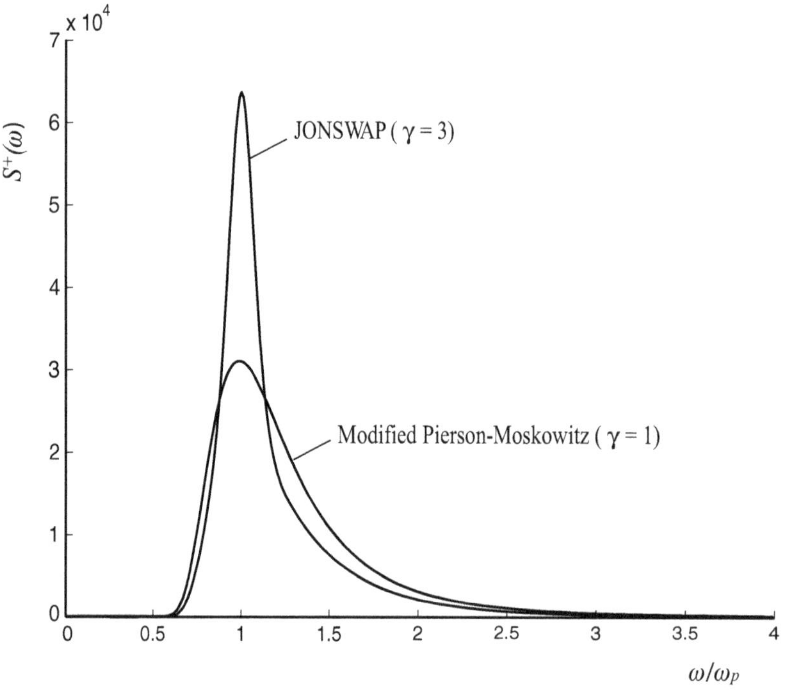

Figure 8.1. Plots of JONSWAP spectra for $\gamma = 1$ and 3.

If the value of γ is not specified, the following rule has been suggested,

$$\gamma = \begin{cases} 5, & \text{for } \frac{T_p}{\sqrt{H_s}} \leq 3.6; \\ \exp\left(5.75 - 1.15\frac{T_p}{\sqrt{H_s}}\right), & \text{for } \frac{T_p}{\sqrt{H_s}} > 3.6. \end{cases} \tag{8.19}$$

For a discussion of other alternative formulas relating the peakedness parameter to sea state parameters, the reader may consult (Torsethaugen and Haver, 2004).

It should be noted that in the tail of large frequencies, every one of the preceding ocean wave spectra decays as ω^{-5}. This derives from an analysis presented in (Phillips, 1958). However, the same author modified his analysis a quarter of a century later to reach the conclusion that ω^{-4} would represent a more correct decay behaviour (Phillips, 1984). We do not enter that discussion here, other than to recommend that the reader take note of it.

All wave spectra presented previously are unimodal, or single-peak, spectra. For some offshore applications, such as fatigue assessment, it is important for some locations to also account for the presence of swell, originating from distant weather systems, that may be present simultaneously with more locally generated waves. This situation was described by a class of two-peaked spectra (Torsethaugen and Haver, 2004), which have adopted the ω^{-4} tail decay.

It has been observed that the directional distribution $D(\omega, \theta)$ is generally a function of both frequency and direction. However, for practical purposes, one usually adopts the approximation $D(\omega, \theta) = D(\theta)$; that is, the frequency dependence of the directional function is neglected. It seems that the most frequently adopted directional dependence is obtained by a power of a cosine function. We list two versions here. The first one provides a full 360-degree distribution; it is given as

$$D(\theta) = C(s)\cos^{2s}\left(\frac{\theta - \theta_0}{2}\right), \quad |\theta - \theta_0| \leq \pi \tag{8.20}$$

where $C(s)$ is a normalizing constant ensuring that $\int D(\omega, \theta)\, d\theta = 1.0$, and θ_0 is the main wave direction. It is found that

$$C(s) = \frac{1}{2\sqrt{\pi}} \frac{\Gamma(s+1)}{\Gamma(s+0.5)}, \tag{8.21}$$

where Γ denotes the Gamma function.

The second version, which provides a distribution over 180 degrees, takes the form

$$D(\theta) = C'(s)\cos^{2s}\left(\theta - \theta_0\right), \quad |\theta - \theta_0| \leq \pi/2, \tag{8.22}$$

while it is zero otherwise. It can be shown that $C'(s) = 2C(s)$.

8.2.3 The Distribution of the Wave Surface Elevation

The $\{A_{jk}\}$ and $\{B_{jk}\}$ in Eq. (8.4) are independent sets of independent Gaussian variables, and the wave elevation $X(t)$ is a linear sum in these variables. The sum of independent Gaussian variables is again a Gaussian variable. Therefore, $X(t)$ is also a Gaussian variable for each point in time t. In fact, $X(t)$ is a stationary Gaussian process. From the fact that $E[A_{jk}] = E[B_{jk}] = 0$, it follows that $E[X(t)] = 0$. The variance is obtained from the spectral density by the formula

$$\sigma_X^2 = Var[X(t)] = \int_{-\pi}^{\pi}\int_0^{\infty} S_X^+(\omega, \theta)\, d\omega\, d\theta \approx \sum_{j=1}^{n}\sum_{k=1}^{m} S_X^+(\omega_j, \theta_k)\, \Delta\omega_j\, \Delta\theta_k. \tag{8.23}$$

The statistical distribution of the wave elevation $X = X(t)$ is therefore determined by the PDF , denoted by $f_X(x)$, of a normal or Gaussian random variable of mean value zero and standard deviation σ_X, which is given by the equation (Papoulis, 1965)

$$f_X(x) = \frac{1}{\sqrt{2\pi}\sigma_X} \exp\left(-\frac{x^2}{2\sigma_X^2}\right). \tag{8.24}$$

The corresponding CDF , denoted by $F_X(x)$, is given as follows:

$$F_X(x) = \text{Prob}[X(t) \leq x] = \int_{-\infty}^{x} \frac{1}{\sqrt{2\pi}\sigma_X} \exp\left(-\frac{s^2}{2\sigma_X^2}\right) ds \tag{8.25}$$

where $\text{Prob}[E]$ denotes the probability of the event E.

A Gaussian random variable of mean value zero and standard deviation 1.0 is referred to as a standard Gaussian variable. Its PDF is often denoted by $\phi(x)$ and its CDF by $\Phi(x)$.

Figure 8.2 plots the PDF and the CDF of the standard Gaussian variable as given by Eq. (8.24) with $\sigma_X = 1.0$.

As with X, $V_x = V_x(t; x, y, z)$ and $A_x = A_x(t; x, y, z)$ are both zero-mean Gaussian random variables for every fixed time t. To fully specify their PDFs, we only

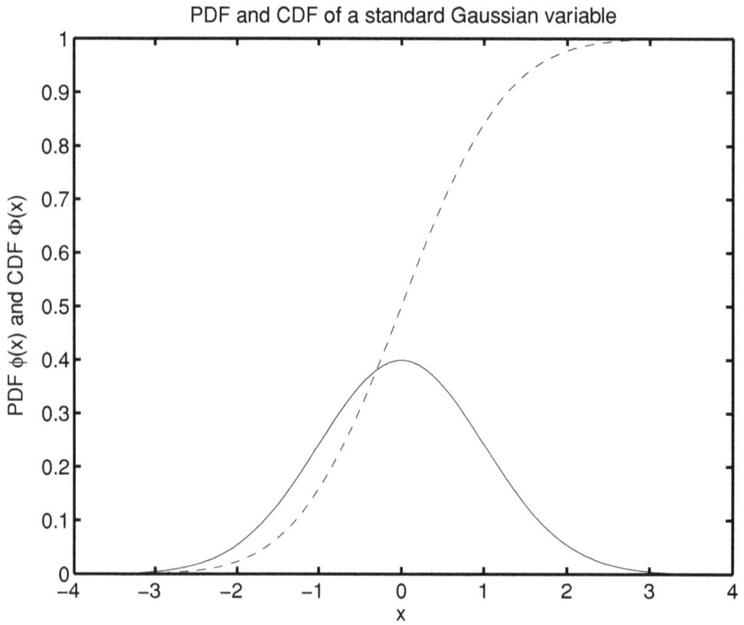

Figure 8.2. Plots of the PDF (solid line) and CDF (dashed line) of a normalized Gaussian variable.

need to calculate their variance, which is given by the expressions

$$\sigma_{V_x}^2 = \mathrm{Var}[V_x] = \int_0^\infty S_{V_x}^+(\omega)\,d\omega$$
$$\approx \sum_{j=1}^{n}\sum_{k=1}^{m} \omega_j^2 \cos^2\theta_k \, e^{\frac{2\omega_j^2}{g}z}\, S_X^+(\omega_j,\theta_k)\,\Delta\omega_j\,\Delta\theta_k \tag{8.26}$$

and

$$\sigma_{A_x}^2 = \mathrm{Var}[A_x] = \int_0^\infty S_{A_x}^+(\omega)\,d\omega$$
$$\approx \sum_{j=1}^{n}\sum_{k=1}^{m} \omega_j^4 \cos^2\theta_k \, e^{\frac{2\omega_j^2}{g}z}\, S_X^+(\omega_j,\theta_k)\,\Delta\omega_j\,\Delta\theta_k \tag{8.27}$$

8.2.4 The Distribution of Wave Crests

As seen, the stochastic model adopted for the water waves leads to a normally distributed sea surface elevation. In many practical applications, it is not the instantaneous surface elevation that is of most interest, but rather the individual waves. So, there are two immediate questions to answer: what is the statistical distribution of the wave heights? What is the distribution of wave crests?

Our answers are based on the notion of an average or mean-level upcrossing rate of a stationary stochastic process, cf. Section 10.2. The average number of times per unit of time that the zero-mean, stationary Gaussian process $X(t)$ crosses the level a with a positive slope, which we denote by $\nu_X^+(a)$, can be given by the formula, cf. Section 10.4,

$$\nu_X^+(a) = \frac{\sigma_{\dot{X}}}{2\,\pi\sigma_X} \exp\Big(-\frac{1}{2}\frac{a^2}{\sigma_X^2}\Big), \tag{8.28}$$

where $\sigma_X^2 = \mathrm{Var}[X]$ and $\sigma_{\dot{X}}^2 = \mathrm{Var}[\dot{X}]$. If $S_X^+(\omega)$ denotes the one-sided spectral density of $X(t)$, then

$$\sigma_X^2 = \int_0^\infty S_X^+(\omega)\, d\omega, \tag{8.29}$$

and

$$\sigma_{\dot{X}}^2 = \int_0^\infty \omega^2\, S_X^+(\omega)\, d\omega. \tag{8.30}$$

The following expression is a widely adopted notation for the spectral moments

$$m_j = \int_0^\infty \omega^j\, S_X^+(\omega)\, d\omega, \qquad j = 0, 1, 2, \ldots \tag{8.31}$$

It is quite common to express the spectral density in terms of frequency f in Hz. If $G_X^+(f)$ denotes the appropriate spectral density in Hz, the following relation must be satisfied to preserve variance: $S_X^+(\omega)\, d\omega = G_X^+(f)\, df$. Thus, the spectral moments

$$\tilde{m}_j = \int_0^\infty f^j\, G_X^+(f)\, df, \qquad j = 0, 1, 2, \ldots \tag{8.32}$$

satisfy the relation $m_j = (2\pi)^j\, \tilde{m}_j$.

It is seen that $\nu_X^+(a)$ decreases quickly (with σ_X as measuring unit) on either side of the mean value zero, where it assumes its largest value

$$\nu_X^+(0) = \frac{\sigma_{\dot{X}}}{2\,\pi\sigma_X} = \frac{1}{2\,\pi}\sqrt{\frac{m_2}{m_0}} = \sqrt{\frac{\tilde{m}_2}{\tilde{m}_0}} \tag{8.33}$$

$\nu_X^+(0)$ is called the *mean zero-upcrossing rate.* A corresponding parameter one often meets is its inverse, commonly referred to as the *mean zero-crossing period* T_z, which is then defined as

$$T_z = \frac{2\,\pi\sigma_X}{\sigma_{\dot{X}}} = 2\,\pi\sqrt{\frac{m_0}{m_2}} = \sqrt{\frac{\tilde{m}_0}{\tilde{m}_2}}. \tag{8.34}$$

Hence, it is important to know which spectral moments have been calculated.

Let X_p denote the size/height of an arbitrary peak of $X(t)$ relative to the zero level, that is, the wave crest height. X_p becomes a random variable. For a narrow band process, where there is typically only one peak between an upcrossing and the subsequent downcrossing of zero, a reasonable definition of the probability distribution of X_p is obtained by the relation, see also Section 10.3,

$$\mathrm{Prob}(X_p > a) = \frac{\nu_X^+(a)}{\nu_X^+(0)}, \quad (a \geq 0). \tag{8.35}$$

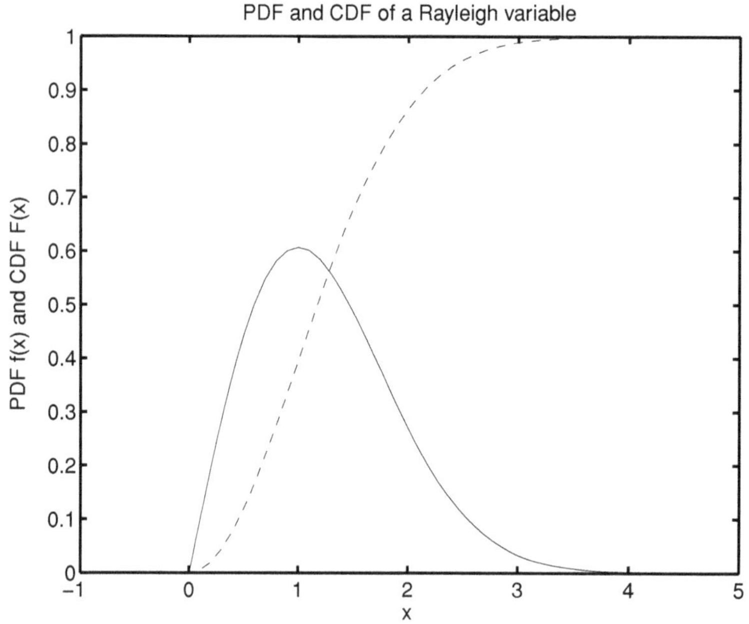

Figure 8.3. Plots of the PDF (solid line) and CDF (dashed line) of the Rayleigh variable X_p with $\sigma_X = 1.0$.

The corresponding CDF then becomes,

$$F_{X_p}(a) = 1 - \frac{\nu_X^+(a)}{\nu_X^+(0)}, \quad (a \geq 0). \tag{8.36}$$

By substituting from Eq. (8.28), it is found that

$$F_{X_p}(a) = 1 - \exp\big(-\frac{1}{2}\frac{a^2}{\sigma_X^2}\big), \quad (a \geq 0). \tag{8.37}$$

The PDF is obtained as

$$f_{X_p}(a) = \frac{a}{\sigma_X^2}\exp\big(-\frac{1}{2}\frac{a^2}{\sigma_X^2}\big), \quad (a \geq 0). \tag{8.38}$$

Hence, the wave crest height X_p in a Gaussian sea way $X(t)$ is a *Rayleigh* distributed random variable (Papoulis, 1965) if $X(t)$ is a narrow band process. If the individual waves are defined as zero crossing waves, Eqs. (8.37) and (8.38) can be adopted as valid expressions for the statistical distribution of wave crest heights for non-narrow band wave processes.

The mean value and variance of X_p are $E[X_p] = \sqrt{\pi/2}\,\sigma_X$ and $\text{Var}[X_p] = 2(1 - \pi/4)\,\sigma_X^2$, respectively. Figure 8.3 plots the PDF of the Rayleigh variable as given by Eq. (8.38) with $\sigma_X = 1.0$.

To avoid heavy wave loads on the deck structure of offshore platforms, the wave crest height becomes a parameter of particular importance. Already, for some time, the knowledge that wave crests of real ocean waves tend to be higher above the mean water level than the subsequent trough is below has been taken into account in practical design work. What this means in practical terms is that our Rayleigh approximation to the distribution of wave crest heights is not accurate enough for some applications. To account for the observation that the ocean surface process is typically positively skewed with higher crests and shallower troughs than expected under the Gaussian assumption, an empirical correction to the Rayleigh distribution of the wave crests was already presented more than 30 years ago by Jahns and Wheeler (1972). Their proposed distribution can be written as,

$$F_{X_p}(a) = 1 - \exp\left\{ -8\frac{a^2}{h_s^2}[1 - \beta_1\frac{a}{d}(\beta_2 - \frac{a}{d})]\right\}, \quad (a \geq 0). \tag{8.39}$$

Here, h_s denotes the significant wave height, d denotes the water depth, and β_1 and β_2 are empirical coefficients. $\beta_1 = 4.37$ and $\beta_2 = 0.57$ are recommended by Haring and Heideman (1978). Compare Eq. (8.37) with Eq. (8.39) for the narrow band limit case where $h_s = 4\sigma_X$.

A more recent proposal is due to Forristall (2000), which is based on simulation data for a full second order model of the wave elevation. From the analysis of these data, he cites a two-parameter Weibull distribution as the short-term model for the wave crest heights:

$$F_{X_p}(a) = 1 - \exp\left\{ -(\frac{a}{\alpha_F h_s})^{\beta_F}\right\}, \quad (a \geq 0), \tag{8.40}$$

where h_s is the significant wave height, and the two parameters α_F and β_F are given as follows:

$$\alpha_F = 0.3536 + 0.2892\, s_1 + 0.1060\, Ur, \tag{8.41}$$

$$\beta_F = 2 - 2.1597\, s_1 + 0.0968\, Ur^2, \tag{8.42}$$

for the case of long-crested waves. Similar expressions are given for short-crested waves (Forristall, 2000). s_1 is a measure of steepness and is given by,

$$s_1 = \frac{2\pi h_s}{g t_1^2}, \tag{8.43}$$

where t_1 is the mean wave period determined by the first two moments of the wave spectrum. The Ursell number Ur, which is a measure of the influence of the water depth on the nonlinearity of waves, is given by the formula,

$$Ur = \frac{h_s}{k_1^2 d^3}, \tag{8.44}$$

where k_1 is the wave number corresponding to wave period t_1, and d is the water depth.

8.2.5 The Distribution of Wave Heights

The statistical distribution of the wave height, denoted by H, is somewhat more involved than for the wave crest height because the usual definition of the wave height involves a wave crest and the subsequent wave trough, which occurs half a wave period later. This time delay introduces some difficulties that are greatly simplified in the case of a narrow banded wave process. This simplification is achieved by adopting the approximation that the wave crest and the subsequent wave trough have the same size. Hence, for a narrow banded wave process, the approximation

$H = 2\,X_p$ is often adopted. In fact, for many years the most commonly adopted CDF for the crest-to-trough wave heights in a random seaway has been the following Rayleigh distribution

$$F_H(h) = 1 - \exp\big(- \frac{h^2}{8\,\sigma_X^2}\big), \quad (h \geq 0). \tag{8.45}$$

Its simplicity and the fact that it agrees reasonably well with observations are probably the reasons for its widespread use. However, high-quality ocean wave data have shown significant differences between Eq. (8.45) and the empirical distributions (Haring et al., 1976; Forristall, 1978). To improve the fit, Forristall proposed to use the following two-parameter Weibull distribution ,

$$F_H(h) = 1 - \exp\big\{ - \frac{1}{\alpha_f}\big(\frac{h}{\sigma_X}\big)^{\beta_f}\big\}, \quad (h \geq 0), \tag{8.46}$$

with parameters $\alpha_f = 8.42$ and $\beta_f = 2.126$.

The observation that the preceding Rayleigh distribution does not fit real wave data is not surprising. In fact, in his now classical paper of 1952, Longuet-Higgins shows that the CDF of the crest-to-trough wave heights in a Gaussian seaway of narrow but finite bandwidth is given by

$$F_H(h) = 1 - \exp\big(- \frac{h^2}{\bar{h}^2}\big), \quad (h \geq 0), \tag{8.47}$$

where $\bar{h}$ denotes the root-mean-square (rms) crest-to-trough wave height (Longuet-Higgins, 1952).

For an infinitely narrow band process , $\bar{h}^2 = 8\,\sigma_X^2$. In the case of finite bandwidth (Longuet-Higgins, 1980; Larsen, 1981; Boccotti, 1982), $\bar{h}^2 < 8\,\sigma_X^2$. Equation (8.45) therefore overpredicts the true values, which is exactly what has been observed. Longuet-Higgins (1980) showed that good agreement can be achieved between Eq. (8.47) and the empirical distribution Eq. (8.46) if $\bar{h}$ is chosen properly. To account for the effect of finite bandwidth, Naess (1985b) proposed the following CDF for the crest-to-trough wave height:

$$F_H(h) = 1 - \exp\big(- \frac{h^2}{4\sigma_X^2(1 - r')}\big), \quad (h \geq 0), \tag{8.48}$$

where $r' = R_X(T_d/2)/\sigma_X^2$, and the autocorrelation function $R_X(\tau)$ is given as

$$R_X(\tau) = \int_0^\infty S_X(\omega)\, \cos \omega\tau \; d\omega. \tag{8.49}$$

T_d denotes the dominant wave period chosen so that $R_X'(T_d) = 0$; that is, T_d denotes the time when the autocorrelation function $R_X(\tau)$ assumes its first maximum after $\tau = 0$.

For most sea states of some severity and without significant swell, it was observed that r' assumes values in the interval $-0.75 \leq r' \leq -0.65$ (Naess, 1985a). Good agreement is obtained between the distribution given by Eq. (8.48) and the empirical distribution Eq. (8.46) based on storm data from the Gulf of Mexico if the parameter $r' = -0.71$, which is a reasonable value to expect. Experience with storm data from the North Sea indicates a typical peakedness parameter in the JONSWAP spectrum slightly less than 3, which corresponds to $r' \approx -0.71$ (Naess, 1985a). A discussion

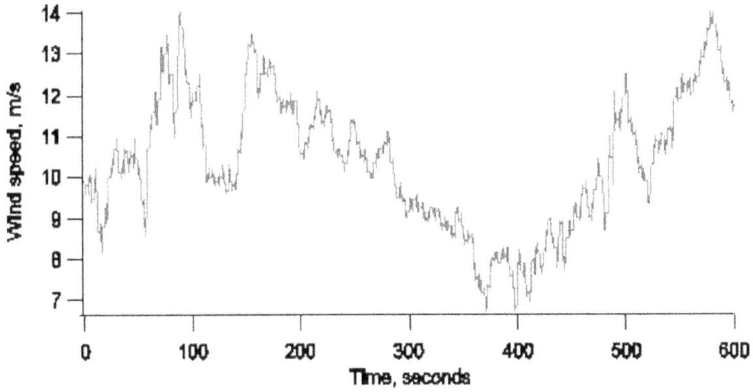

Figure 8.4. Sample wind data for a short-term period (Manwell et al., 2002).

of the accuracy of various proposed wave height distributions in mixed sea states, including those presented here, is presented by Rodriguez et al. (2002).

8.3 Wind

8.3.1 Wind Speed

The main variations in wind speed and direction are related to variations of temperature. Periodic trends can thus be expected to correspond to the yearly cycle of seasons, the meteorological cycle (of about 4 days' duration), and the daily cycle. There is also a fourth periodic trend which is related to the turbulence generated in the air flow by friction with the earth's surface. This fact can be illustrated by the spectral representation of wind speed.

The main environmental parameter that determines wind loads on structures is the wind speed. This section focuses on how the wind speed varies in space and time.

The wind velocity may be idealized as a superposition of a stochastic high frequency gust velocity and a slowly varying mean wind, which carries the bulk of the energy. The short term variation of the wind speed, i.e., over 10 minutes or less, may appear as shown in Fig. 8.4. This may also be readily seen from the power spectral density of wind velocity shown in Fig. 8.5, which reveals that the wind contains fluctuations on two vastly different time scales, i.e., macro-meteorological and micro-meteorological fluctuations separated by a spectral gap. The macrometeorological fluctuations, which primarily govern the mean wind speed, refer to motions of weather systems as a whole. The micro-meteorological fluctuations, in contrast, refer to atmospheric turbulence (gusts) induced by mechanical and/or thermal effects.

The mean wind (macrometeorological) is assumed to flow parallel to the sea surface (but the velocity varies over the height) while the micro-meteorological fluctuations are three-dimensional.

In a macroscale, it is necessary to distinguish between different wind phenomena, especially between "continuous current" winds and hurricanes and tornadoes. Platform accidents in the Gulf of Mexico have been caused by hurricanes, notably, hurricane Andrew in 1992 and Katrina in 2005.

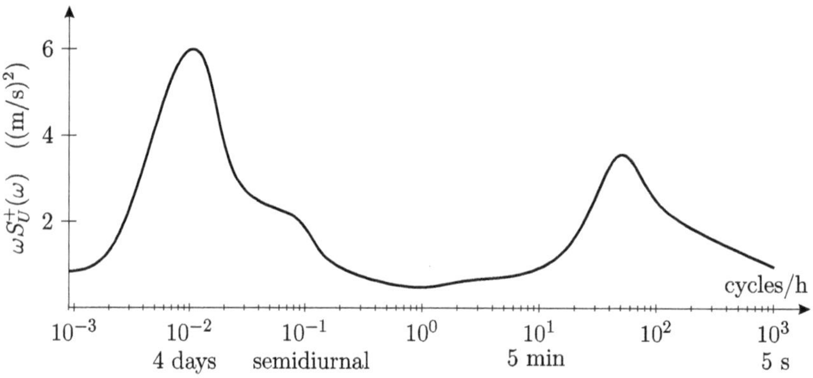

Figure 8.5. Van der Hoven's power spectral density of horizontal wind velocity near the ground for an extensive frequency range (Van der Hoven, 1957).

A sustained wind velocity is the temporal average over a period of some minutes, e.g., 1-3 minutes. A gust velocity is typically the average over a few seconds, e.g., 3 seconds. The wind speed varies with the height, and different wind speed profiles have been suggested.

8.3.2 Wind Shear and Turbulence

Turbulence in the wind is caused by dissipation of the wind's kinetic energy into thermal energy via the creation and destruction of progressively smaller eddies (or gusts). Turbulent wind may have a relatively constant mean over time periods of an hour or more, but over shorter times (minutes or less) it may be quite variable. The wind's variability superficially appears to be quite random, but it actually has distinct features. These features are characterized by a number of statistical properties (Manwell et al., 2002):

- Turbulence intensity
- Wind speed probability density functions
- Autocorrelation
- Integral time scale/length scale
- Power spectral density function
- Cross-correlation function

Turbulent wind consists of longitudinal, lateral, and vertical components. The longitudinal component, in the prevailing wind direction, is designated $\tilde{U}$. It generally depends on both temporal and spacial coordinates; however, for convenience, this is suppressed. $\tilde{U}$ is commonly considered as consisting of a short-term mean wind, for example, $\bar{U}$, with a superimposed fluctuating wind of zero mean, U, added to it. Thus,

$$\tilde{U} = \overline{U} + U. \tag{8.50}$$

The lateral wind can be decomposed into a mean and fluctuating component in a similar manner.

The mean wind speed $\overline{U}$ depends on the height z and also on the length τ of the time averaging. Based on dimensional analysis, it can be shown that the dependence of $\overline{U}$ on z is logarithmic. The dependence of $\overline{U}$ on z and τ has, for instance, been given in (NORSOK N-003, 2007).

$$\overline{U}(z,\tau) = \overline{U}(z,\tau_r)\,[1 - 0.41\,I_U(z)\ln(\tau/\tau_r)]\,. \tag{8.51}$$

Here, $\tau_r = 3600$ s, and

$$\overline{U}(z,\tau_r) = \overline{U}(z_r,\tau_r)\,[1 + C\ln(z/z_r)]\,, \tag{8.52}$$

where $z_r = 10$ m and $C = 5.73 \cdot 10^{-2}(1 + 0.15\overline{U}(z_r,\tau_r))$. The turbulence intensity factor $I_U(z)$ in Eq. (8.51) is discussed at some length below. In (NORSOK N-003, 2007) it is given by the formula:

$$I_U(z) = 0.06\left[1 + 0.043\,\overline{U}(z_r,\tau_r)\right](z/z_r)^{-0.22}. \tag{8.53}$$

Another representation used in engineering calculations, which neglects the dependence on averaging time, is a power law dependence as follows,

$$\overline{U}(z) = \overline{U}_{10}\left(\frac{z}{10}\right)^{\alpha}, \tag{8.54}$$

where $\overline{U}_{10}$ is a reference mean value at height 10 meters, and α is a parameter whose value depends on the type of terrain. For offshore applications, a typical value chosen is $\alpha = 0.14$. The wind velocity average over 10 to 60 minutes at 10 meters above the ground level is, in general, the basis for the long-term macrometeorological wind specification.

The most basic measure of turbulence is the turbulence intensity I_U. It is defined by the ratio of the standard deviation of the wind speed to the mean, that is, $I_U = \sigma_U/\bar{U}$. In this calculation, both the mean and standard deviation σ_U are calculated over a time period longer than that of the turbulent fluctuations, but shorter than periods associated with other types of wind speed variations (such as diurnal effects), typically 10 minutes or 1 hour. Turbulence intensity is frequently in the range of 0.1 to 0.4. In general, the highest value of turbulence intensities occur at the lowest wind speeds, but the lower limiting value at a given location will depend on the specific terrain features and surface conditions at the site. Because σ_U varies slightly with height and $\bar{U}$ increases with height in the wind boundary layer, $I_U = I_U(z)$ decreases with the height z. For flat terrain, approximately,

$$I_U(z) = 1/\ln(z/z_0), \tag{8.55}$$

where $z_0 =$ the roughness length, which can be interpreted as the characteristic size of a vortex being formed due to the friction between the flowing air and the ground surface. For open sea without waves $z_0 = 10^{-4}$, see (Dyrbye and Hansen, 1997). Because $I_U(z_0) = \infty$, Eq. (8.55) can only provide a reasonable value of turbulence intensity for heights significantly larger than z_0.

The probability density function that best describes the turbulent wind speed is the Gaussian distribution, even though the tail can deviate from Gaussian.

The autocorrelation for the wind speed at two different times t_1 and t_2 provides some information about the size of the wind speed at time $t_2 - t_1$ into the future given its current value. Figure 8.6 shows the autocorrelation for the data in Fig. 8.4. It is seen that the normalized autocorrelation function decay from a value of 1.0 at lag zero to a value of zero, and then tend to take on small positive or negative

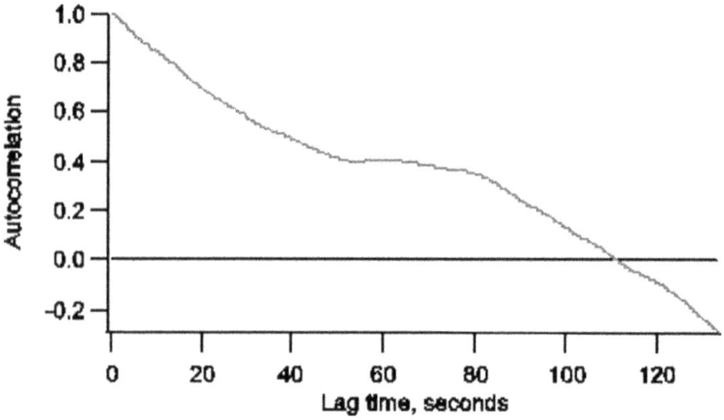

Figure 8.6. Autocorrelation function of sample wind data (Manwell et al., 2002).

values as the lag increases. The average time over which wind speed fluctuations are significantly correlated with each other is found by integrating the autocorrelation from zero lag to the first zero crossing. This is called the *integral time scale of the wind process*, and may be used to express the typical size of the turbulent eddies, or *integral length scales*. This connection is obtained via Taylor's hypothesis about convected "frozen turbulence," which links the spatial and temporal correlation functions (Batchelor, 1953; Dyrbye and Hansen, 1997).

8.3.3 Wind Spectra

A number of power spectral density functions are used as models in wind energy engineering when representative turbulence power spectral densities are unavailable for a given site (Simiu and Scanlan, 1996). They typically satisfy the asymptotic range condition of decaying such as $\omega^{-5/3}$ for increasing ω, which has both a theoretical and an experimental basis (Simiu and Scanlan, 1996). Note that, contrary to tradition, we express wind spectra in terms of circular frequencies. However, from the scaling properties of spectra, we know that $\omega S_U^+(\omega) = fG_U^+(f)$, cf. Section 6.3 for notation and properties. The significance of this equality will soon become clear when the formulas for the various spectra are given. Among the most cited along-wind spectra (for the wind speed at $z = 10$ m) are the following:

Davenport spectrum

$$\frac{\omega S_U^+(\omega)}{\delta \overline{U}_{10}^2} = \frac{4\theta^2}{(1+\theta^2)^{4/3}}, \tag{8.56}$$

Harris spectrum

$$\frac{\omega S_U^+(\omega)}{\delta \overline{U}_{10}^2} = \frac{4\theta}{(2+\theta^2)^{5/6}}. \tag{8.57}$$

Here the dimensionless variable θ is given by the equation,

$$\theta = \frac{\omega L_u}{2\pi \overline{U}_{10}} = \frac{f L_u}{\overline{U}_{10}}, \quad 0 < \theta < \infty, \tag{8.58}$$

where L_u is an integral length scale taken to be 1200 meters in the Davenport spectrum, and 1800 meters in the Harris spectrum. δ is the surface drag coefficient referred to $\overline{U}_{10}$. For offshore locations, we assume that $\delta = 0.001$. However, none of these spectra have been derived from analysis of wind speed data recorded offshore; instead they are based on data obtained at onshore locations. Their application to response analysis of offshore structures may therefore be questioned. In particular, the low frequency range, which is relevant for large floating structures, may be inaccurately modeled by these spectra.

Kaimal spectrum is also frequently used in applications. It can be written in the following way, cf. Dyrbye and Hansen (1997),

$$\frac{\omega S_U^+(\omega)}{\sigma_U^2} = \frac{6.8\theta}{(1+10.2\theta)^{5/3}}, \tag{8.59}$$

where σ_U^2 is the variance of $U(t)$ at reference height 10 meters. For a discussion of how to calculate θ for this spectrum, cf. Dyrbye and Hansen (1997). This form of the Kaimal spectrum is used in Eurocode 1.

The American Petroleum Institute (API) has issued a recommended practice document (API, 2000) that specifies the wind speed spectrum in the following way:

$$\frac{\omega S_U^+(\omega)}{\sigma_U(z)^2} = \frac{\omega/\omega_p}{(1+1.5\,\omega/\omega_p)^{5/3}}, \tag{8.60}$$

where ω_p is a "peak" frequency, and $\sigma_U(z)^2$ is the variance of $U(t)$, which is not assumed to be independent of z. This is reflected in the following condition on the values of ω_p:

$$0.01 \leq \frac{\omega_p\, z}{\overline{U}(z)} \leq 0.1, \tag{8.61}$$

where the value 0.025 is usually chosen. The standard deviation of the wind speed is given by

$$\sigma_U(z) = \begin{cases} 0.15\,\overline{U}(z)\left(\dfrac{z_s}{z}\right)^{0.125} & : \; z \leq z_s, \\ 0.15\,\overline{U}(z)\left(\dfrac{z_s}{z}\right)^{0.275} & : \; z > z_s, \end{cases} \tag{8.62}$$

where z_s is the thickness of the *surface layer*, and is taken as 20 meters.

Figure 8.7 shows plots of three of the spectra discussed here. Note the marked difference between the values of the three spectra in the low frequency range. This may have significant influence on the predicted wind induced response of moored offshore structures.

The turbulence spectrum describes the temporal variation of turbulence at a given point. Its spatial variation in the vertical and lateral directions can be modeled by the cross-correlation of the wind. The correlation decreases with increasing distance between two points $\mathbf{r}_1$ and $\mathbf{r}_2$, and also decreases for increasing frequency. They can be described by the root-coherence functions, $\mathrm{Coh}(\mathbf{r}_1, \mathbf{r}_2, \omega)$, given as

$$\mathrm{Coh}(\mathbf{r}_1, \mathbf{r}_2, \omega) = \frac{\left|S_U(\mathbf{r}_1, \mathbf{r}_2, \omega)\right|}{\sqrt{S_U(\mathbf{r}_1, \omega) S_U(\mathbf{r}_2, \omega)}}. \tag{8.63}$$

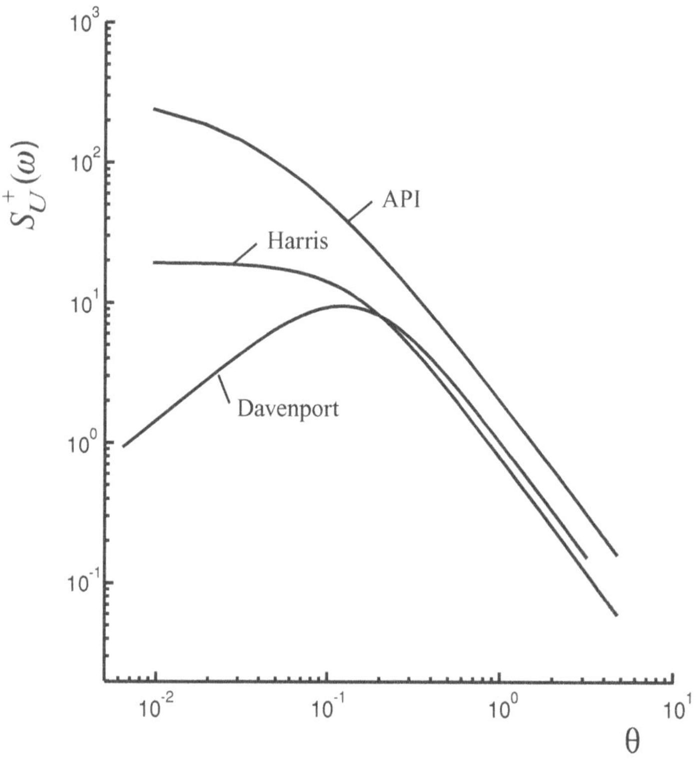

Figure 8.7. Plots of wind spectra. $\theta = \omega L_u / 2\pi \overline{U}_{10}$ for the Davenport and Harris spectra, $\theta = \omega/\omega_p$ for the API spectrum.

where $S_U(\mathbf{r}_1, \mathbf{r}_2, \omega)$ denotes the cross-spectrum, while $S_U(\mathbf{r}_1, \omega)$ and $S_U(\mathbf{r}_2, \omega)$ denote the auto-spectra of the wind speed at the two locations $\mathbf{r}_1$ and $\mathbf{r}_2$, cf. Sections 9.7 and 9.10. For the Kaimal spectrum, $\text{Coh}(\mathbf{r}_1, \mathbf{r}_2, \omega)$ is given by IEC (1999) as

$$\text{Coh}(\Delta\mathbf{r}, \omega) = \exp\left[-8.8\Delta\mathbf{r}\sqrt{\left(\frac{0.22}{L}\right)^2 + \left(\frac{\omega}{2\pi\bar{U}}\right)^2}\right], \tag{8.64}$$

where $\Delta\mathbf{r} = |\mathbf{r}_2 - \mathbf{r}_1|$ is the distance between the two points.

The characteristics of turbulence described so far refer to open landscape. Topography and structures cause wakes and affect turbulence characteristics.

9 Response Spectrum

9.1 Introduction

In many cases the loads that a structure is exposed to can be modeled as a (piecewise) stationary process. This applies, for example, to the wind loads on a building or the wave loads on an offshore structure. In this chapter, we assume that the transfer between the load and the response can be modeled as a linear, time-invariant system; that is, the system can be characterized by a transfer function. It is shown that, in this case, the response will become a stationary process if the load is modeled as a stationary process. One of the most important results in this chapter is the relation between the variance spectrum of the response (the response spectrum) and the variance spectrum of the load (the load spectrum). It is demonstrated that this relationship is completely determined by the transfer function.

9.2 Representation of the Response Process

Let $F(t)$ denote a stochastic load process. Assume that $F(t)$ acts on a linear, time-invariant system, which has an impulse response function $h_{FX}(t)$, cf. Section 2.8. For each realization $f(t)$ of $F(t)$, we get a corresponding realization $x(t)$ of the response process, which we denote by $X(t)$. According to Eq. (2.107), we have that

$$x(t) = \int_{-\infty}^{\infty} h_{FX}(s)\, f(t-s)\, ds = \int_{0}^{\infty} h_{FX}(s)\, f(t-s)\, ds \tag{9.1}$$

because $h_{FX}(s) = 0$ for $s < 0$. Equation (9.1) establishes the connection between the realizations of the load process and the corresponding realizations of the response process. This connection is described by the following equation:

$$X(t) = \int_{0}^{\infty} h_{FX}(s)\, F(t-s)\, ds\,. \tag{9.2}$$

The way we define Eq. (9.2), it should be interpreted as a relation between all corresponding pairs of realizations of $F(t)$ and $X(t)$. It may be noted that there are other alternative ways of defining such integrals, cf., e.g., Lin (1967); Soong and Grigoriu (1993).

A small remark on notation may be useful. The impulse response function, or the transfer function, which determines the connection between the load and the response, is completely specified by the properties of the linear system and is completely independent of any given load. When we write $h_{FX}(t)$, the index FX is to be understood only as a visual indicator for the connection between $F(t)$ and $X(t)$. If $Y(t)$ is the response to a load process $G(t)$ acting on the same linear system, then $h_{GY}(t) \equiv h_{FX}(t)$.

9.3 Mean Value of the Response Process

Assume that $f_1(t), \ldots, f_N(t)$ is a sequence of realizations of $F(t)$, and let $x_1(t), \ldots, x_N(t)$ denote the corresponding response realizations. Then

$$\frac{1}{N}\sum_{j=1}^{N} x_j(t) = \frac{1}{N}\sum_{j=1}^{N}\int_0^{\infty} h_{FX}(s)\, f_j(t-s)\, ds$$

$$= \int_0^{\infty} h_{FX}(s)\,\{\frac{1}{N}\sum_{j=1}^{N} f_j(t-s)\}\, ds. \tag{9.3}$$

This leads to the relation

$$\mathrm{E}[X(t)] = \lim_{N\to\infty}\frac{1}{N}\sum_{j=1}^{N} x_j(t) = \int_0^{\infty} h_{FX}(s)\,\{\lim_{N\to\infty}\frac{1}{N}\sum_{j=1}^{N} f_j(t-s)\}\, ds$$

$$= \int_0^{\infty} h_{FX}(s)\,\mathrm{E}[F(t-s)]\, ds \tag{9.4}$$

where we assume that the operations of integration and taking limits of sums can be interchanged.

In the following, we suppose that $F(t)$ is stationary. Then, in particular, $m_F = \mathrm{E}[F(t)] = \text{constant}$. This implies that

$$\mathrm{E}[X(t)] = m_F \int_0^{\infty} h_{FX}(s)\, ds. \tag{9.5}$$

The rhs of Eq. (9.5) is clearly independent of t, that is, $m_X = \mathrm{E}[X(t)] = \text{constant}$. Let $H_{FX}(\omega)$ be the transfer function that corresponds to the impulse response function $h_{FX}(t)$. Then, according to Eq. (2.112) ($h_{FX}(t) = 0, t < 0$)

$$H_{FX}(0) = \int_0^{\infty} h_{FX}(s)\, ds. \tag{9.6}$$

Equation (9.5) can then be written as

$$m_X = H_{FX}(0)\, m_F. \tag{9.7}$$

If we let $H_{FX}(\omega)$ be the transfer function for the linear vibration system described by Eq. (2.39), $H_{FX}(\omega)$ is given by Eq. (2.71). That is, $H_{FX}(0) = 1/k$, and $m_X = m_F/k$. Expressed in words, Eq. (9.7) can be stated as follows:

The mean value of the response equals the mean value of the load multiplied by the system response to a static load of unit size.

From Eq. (9.7) it is seen that if the excitation has mean value zero, the response also has mean value zero.

9.4 Autocovariance of the Response Process

Here, and for the rest of this chapter, we assume that the load process $F(t)$ is stationary. As we have just seen, $m_F = 0$ implies that $m_X = 0$. This makes it convenient to extract the mean value by defining a load process $F'(t) = F(t) - m_F$, which clearly has mean value zero. Let us now introduce the response $X'(t)$ to the load process $F'(t)$. By combining Eqs. (9.2) and (9.4), it is obtained that

$$\begin{aligned} X'(t) &= \int_0^\infty h_{FX}(s)\, F'(t-s)\, ds \\ &= \int_0^\infty h_{FX}(s)\, F(t-s)\, ds - \int_0^\infty h_{FX}(s)\, m_F\, ds = X(t) - m_X, \end{aligned} \tag{9.8}$$

and it is also seen that $X'(t)$ has mean value zero. $F(t)$ and $F'(t)$ obviously have the same autocovariance. The same applies to $X(t)$ and $X'(t)$. We may therefore suppose that $F(t)$ and $X(t)$ already at the initial stage have mean value zero, which is done in the following if nothing is specifically said to the contrary.

Let $f_1(t), \ldots, f_N(t)$ and $x_1(t), \ldots, x_N(t)$ be as in the previous section. Then

$$\begin{aligned} x_j(t)\, x_j(t+\tau) &= \int_0^\infty h_{FX}(s_1)\, f_j(t-s_1)\, ds_1 \cdot \int_0^\infty h_{FX}(s_2)\, f_j(t+\tau-s_2)\, ds_2 \\ &= \int_0^\infty \int_0^\infty h_{FX}(s_1) h_{FX}(s_2)\, f_j(t-s_1)\, f_j(t+\tau-s_2)\, ds_1\, ds_2. \end{aligned} \tag{9.9}$$

Similarly, as for the passage from Eq. (9.3) to Eq. (9.4), it is obtained that

$$\begin{aligned} \mathrm{E}[X(t)\, X(t+\tau)] &= \lim_{N\to\infty} \frac{1}{N} \sum_{j=1}^{N} x_j(t)\, x_j(t+\tau) \\ &= \int_0^\infty \int_0^\infty h_{FX}(s_1) h_{FX}(s_2) \Big\{ \lim_{N\to\infty} \frac{1}{N} \sum_{j=1}^{N} f_j(t-s_1)\, f_j(t+\tau-s_2) \Big\}\, ds_1\, ds_2 \\ &= \int_0^\infty \int_0^\infty h_{FX}(s_1) h_{FX}(s_2) \mathrm{E}[F(t-s_1)\, F(t+\tau-s_2)]\, ds_1\, ds_2 \\ &= \int_0^\infty \int_0^\infty h_{FX}(s_1) h_{FX}(s_2)\, C_F(\tau+s_1-s_2)\, ds_1\, ds_2. \end{aligned} \tag{9.10}$$

The expression to the rhs of the last equality sign in Eq. (9.10) shows that $\mathrm{E}[X(t)\, X(t+\tau)]$ is independent of t because $F(t)$ is assumed to be stationary. This, together with the result from the previous paragraph, implies that $X(t)$ becomes a stationary process. The autocovariance $C_X(\tau)$, which in this case equals the autocorrelation $R_X(\tau)$ because we have assumed zero mean value, is then given as

$$C_X(\tau) = \int_0^\infty \int_0^\infty h_{FX}(s_1) h_{FX}(s_2)\, C_F(\tau+s_1-s_2)\, ds_1\, ds_2. \tag{9.11}$$

The fact that $X(t)$ becomes a stationary process can be generalized to apply to any linear, time-invariant system, that is, a system that can be characterized by a transfer function. That this represents a more general situation than the one just considered is discussed in Section 9.9. Briefly stated, we now have the following result:

> *The response of a linear, time-invariant system to a stationary load process is also stationary.*

9.5 Response Spectrum

By using Eq. (6.1) together with Eq. (9.11), we may establish the connection between the variance spectrum $S_X(\omega)$ of the response process $X(t)$ and the variance spectrum $S_F(\omega)$ of the load process $F(t)$. According to Eq. (6.1), $S_X(\omega)$ is given as

$$S_X(\omega) = \frac{1}{2\pi} \int_{-\infty}^{\infty} C_X(\tau)\, e^{-i\omega\tau}\, d\tau. \tag{9.12}$$

By substituting for $C_X(\tau)$ from Eq. (9.11) and changing the integration order, we obtain that

$$S_X(\omega) = \int_0^{\infty} h_{FX}(s_1) \int_0^{\infty} h_{FX}(s_2) \cdot \frac{1}{2\pi} \int_{-\infty}^{\infty} C_F(\tau + s_1 - s_2)\, e^{-i\omega\tau}\, d\tau\, ds_2\, ds_1. \tag{9.13}$$

In the inner integral, we change the integration variable to $\theta = \tau + s_1 - s_2$, $d\theta = d\tau$. This gives the result

$$\begin{aligned} S_X(\omega) &= \int_0^{\infty} h_{FX}(s_1) \int_0^{\infty} h_{FX}(s_2) \cdot \frac{1}{2\pi} \int_{-\infty}^{\infty} C_F(\theta)\, e^{-i\omega\theta}\, d\theta\, e^{i\omega(s_1 - s_2)} ds_2\, ds_1 \\ &= \int_0^{\infty} h_{FX}(s_1)\, e^{i\omega s_1}\, ds_1 \int_0^{\infty} h_{FX}(s_2)\, e^{-i\omega s_2}\, ds_2 \cdot S_F(\omega) \\ &= H_{FX}(-\omega)\, H_{FX}(\omega)\, S_F(\omega). \end{aligned} \tag{9.14}$$

Because $(e^{-ix})^* = e^{ix}$ and $h_{FX}(t)$ is a real function, it follows that

$$\begin{aligned} H_{FX}(-\omega) &= \int_0^{\infty} h_{FX}(t)\, e^{i\omega t}\, dt = \int_0^{\infty} h_{FX}(t)\, (e^{-i\omega t})^*\, dt \\ &= \left(\int_0^{\infty} h_{FX}(t)\, e^{-i\omega t}\, dt \right)^* = H_{FX}(\omega)^*. \end{aligned} \tag{9.15}$$

We have now proven the following simple but important relationship between the response spectrum $S_X(\omega)$ and the load spectrum $S_F(\omega)$:

$$S_X(\omega) = |H_{FX}(\omega)|^2\, S_F(\omega). \tag{9.16}$$

Note carefully that Eq. (9.16) does not contain any information about phase shift between the load and the response, only amplitude amplification. This information resides in the transfer function $H_{FX}(\omega)$ itself, but not in $|H_{FX}(\omega)|$. This is seen immediately when the polar representation of a complex number is used because we may then write $H_{FX}(\omega) = |H_{FX}(\omega)|\, e^{i\psi_{FX}(\omega)}$, where $|H_{FX}(\omega)|$ determines amplitude amplification and $\psi_{FX}(\omega)$ gives the phase shift at frequency ω.

Equation (9.16) does not change if we use one-sided spectra instead because

$$S_X^+(\omega) = |H_{FX}(\omega)|^2\, S_F^+(\omega). \tag{9.17}$$

This case is illustrated in Fig. 9.1.

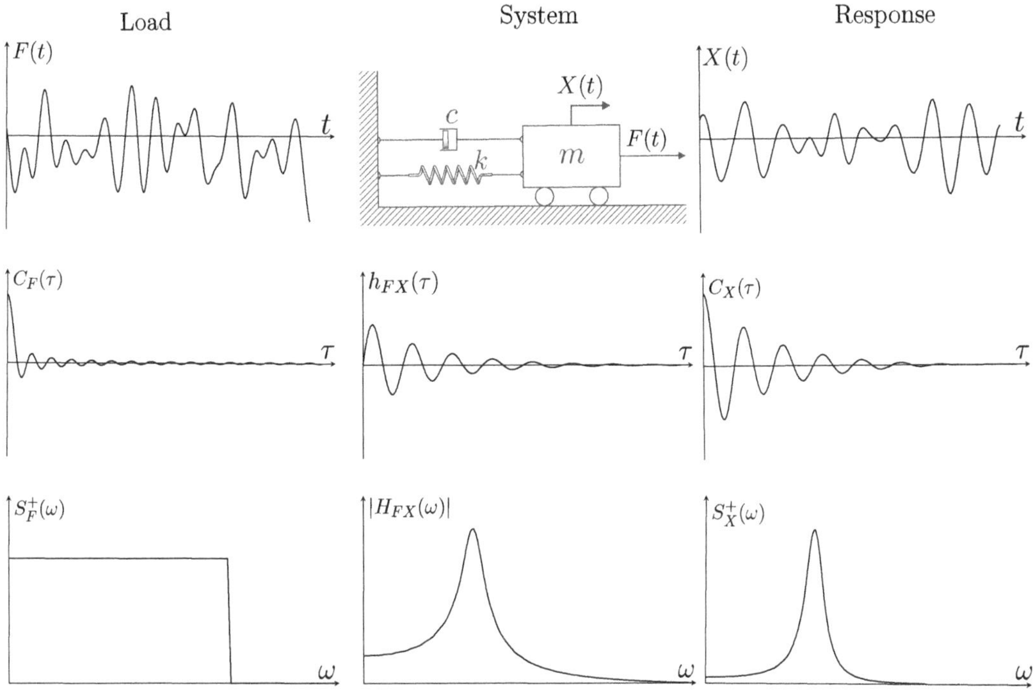

Figure 9.1. Transfer relations between load and response.

The mean value and variance are particularly important statistical parameters. The mean value of the response is given by Eq. (9.7). The corresponding variance is found by using Eq. (6.3) and Eq. (9.16) or Eq. (9.17). Expressed in terms of the two-sided spectrum, it is obtained that

$$\sigma_X^2 = \int_{-\infty}^{\infty} |H_{FX}(\omega)|^2 \, S_F(\omega) \, d\omega. \tag{9.18}$$

Alternatively,

$$\sigma_X^2 = \int_{0}^{\infty} |H_{FX}(\omega)|^2 \, S_F^+(\omega) \, d\omega. \tag{9.19}$$

9.5.1 Example – Response Spectra for a Crane Vessel

The calculation procedure of transfer functions and response spectra is illustrated by a somewhat simplified example of an offshore installation operation. Figure 9.2 shows a crane vessel that is installing subsea equipment on the ocean floor. It is assumed that the vessel is able to maintain its horizontal position by using an automatic positioning system. It is therefore assumed that the vertical motion of the equipment module is due to the heave and roll motion of the vessel if the crane is locked.

To be able to carry out the response calculations of this system, we have to know how to calculate the forces on the crane vessel. We limit ourselves to the wave forces, and assume that these hydrodynamic forces are linearly related to the waves; that is, they are described by a linear transfer function. Assume for simplicity, the situation of long-crested, beam sea waves propagating in the

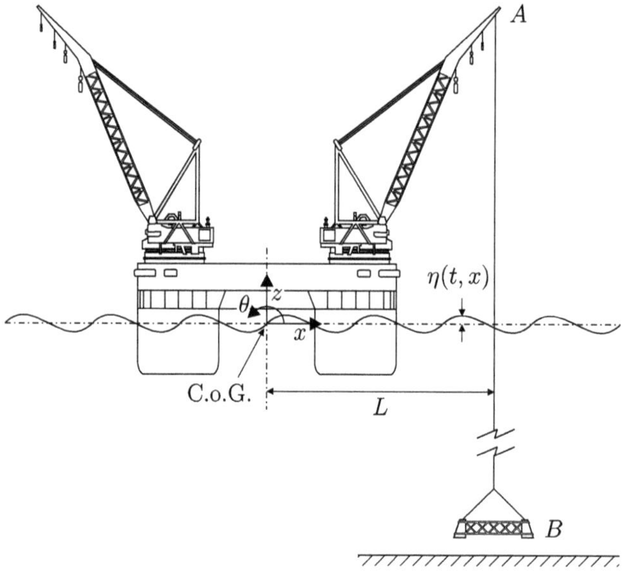

Figure 9.2. A crane vessel for installation of subsea equipment.

x-direction. This means that a regular ocean wave of unit amplitude represented as

$$\eta(t) = Re\{e^{i(\omega t - kx)}\} \quad (k = \omega^2/g) \tag{a}$$

gives a vertical heave force on the vessel, referred to the center of gravity CG ($x = 0$), which can be written as

$$f_z(t) = Re\{H_z(\omega)\, e^{i\omega t}\} \tag{b}$$

and a corresponding roll moment about CG given by

$$m_\theta(t) = Re\{H_\theta(\omega)\, e^{i\omega t}\}\,. \tag{c}$$

It is thereby seen that $H_z(\omega)$ is the transfer function between the waves (referred to $x = 0$) and the heave force, while $H_\theta(\omega)$ is the transfer function between the waves and the roll moment. It is assumed that the installation operation can only be carried out in small to moderate seas so that the rolling motions of the vessel can be assumed to be small. Hence, the following linear equations of motion for the heave and roll response $z(t)$ and $\theta(t)$, respectively, are adopted

$$M\,\ddot{z}(t) + c_z\,\dot{z}(t) + k_z\,z(t) + \int_0^\infty g_z(s)\,z(t-s)\,ds = f_z(t) \tag{d}$$

and

$$I\,\ddot{\theta}(t) + c_\theta\,\dot{\theta}(t) + k_\theta\,\theta(t) + \int_0^\infty g_\theta(s)\,\theta(t-s)\,ds = m_\theta(t)\,. \tag{e}$$

$f_z(t)$ and $m_\theta(t)$ denote the general heave force and roll moment on the vessel, both referred to CG. M, I, c_j, k_j $(j = z, \theta)$ are positive constants, and their interpretation should need no further elaboration. The reason for the Duhamel integral in each of the two equations of motion is due to the fluid structure interaction. This gives rise to an added mass term for the heave motion and an added mass moment of inertia for the roll motion. The motion of the vessel itself will also generate waves emanating from the vessel. This will induce added damping. Both effects are taken care of by the two Duhamel integrals. This connection is brought to the fore by taking the Fourier transform of the two impulse response functions $g_z(t)$ and $g_\theta(t)$. It can be shown that

$$\int_0^\infty g_j(t)\, e^{-\mathrm{i}\omega t} dt = -\omega^2 A_j(\omega) + \mathrm{i}\,\omega\, B_j(\omega) \quad (j = z, \theta) \tag{f}$$

where $A_j(\omega)$ and $B_j(\omega)$ are frequency dependent added mass and damping terms, respectively, due to the interaction effects between the fluid and the floating vessel. Both are real functions of the frequency ω, and computer programs to calculate these terms are standard commercial software.

On the basis of the equations of motion for the heave and roll motions, there is a linear, time-invariant relationship between the wave elevation process at $x = 0$ and the corresponding heave and roll response of the crane vessel. Let us determine the associated linear transfer functions, denoted by $G_z(\omega)$ and $G_\theta(\omega)$. By exploiting the standard recipe for calculating transfer functions, we know that by putting $\eta(t) = e^{i\omega t}$, it is obtained that $z(t) = G_z(\omega)\, e^{i\omega t}$ and $f_z(t) = H_z(\omega)\, e^{i\omega t}$. By substituting into Eq. (d), it is found that

$$\Big(-\omega^2 M + \mathrm{i}\,\omega\, c_z + k_z + \int_0^\infty g_z(s)\, e^{-\mathrm{i}\omega s} ds \Big)\, G_z(\omega)\, e^{\mathrm{i}\omega t} = H_z(\omega)\, e^{\mathrm{i}\omega t}. \tag{g}$$

Combining this equation with Eq. (f), it follows that

$$G_z(\omega) = \frac{H_z(\omega)}{-\omega^2 \big(M + A_z(\omega)\big) + \mathrm{i}\,\omega \big(c_z + B_z(\omega)\big) + k_z}. \tag{h}$$

Analogously, it is found that

$$G_\theta(\omega) = \frac{H_\theta(\omega)}{-\omega^2 \big(I + A_\theta(\omega)\big) + \mathrm{i}\,\omega \big(c_\theta + B_\theta(\omega)\big) + k_\theta}. \tag{i}$$

The vertical motion $v(t)$ of the crane top can be written as $v(t) = z(t) + L\,\theta(t)$, assuming small roll angles justifying the approximation $\sin\theta = \theta$. Assuming that the horizontal distance between CG and the crane top is kept constant, the relationship between the wave elevation process at $x = 0$ and the crane top motions is linear and time invariant, which is determined by the transfer function

$$H_{\eta v}(\omega) = G_z(\omega) + L G_\theta(\omega). \tag{j}$$

Let the wave elevation process at $x = 0$ be modeled as a zero-mean stationary Gaussian process $H(t)$ with a spectral density $S_H(\omega)$. Then the vertical motion of the crane top is also a stationary Gaussian process, which we denote by $V(t)$. The spectral density $S_V(\omega)$ of $V(t)$ is given as follows:

$$S_V(\omega) = |H_{\eta v}(\omega)|^2\, S_H(\omega) = |G_z(\omega) + L G_\theta(\omega)|^2\, S_H(\omega). \tag{k}$$

To finally arrive at the vertical motion response $u(t)$ of the equipment module at the end of a wire rope, an equation of motion of the module has to be established. The wire between the crane top A and the equipment module B is assumed to be modeled as a massless, linear spring with stiffness k. The total mass of the equipment module, including added mass due to the vertical motions, is denoted by m. The damping associated with the vertical motions of

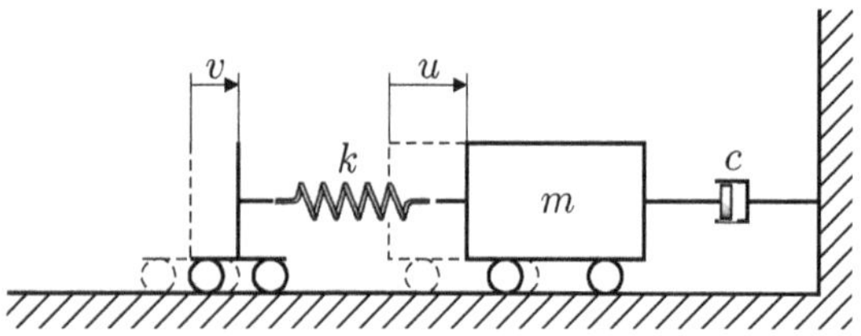

Figure 9.3. A principle sketch of the dynamic system.

the module is assumed to be modeled with reasonable accuracy as linear viscous damping with a damping coefficient c. It is assumed that external forces on the module are only exerted by the wire, and that horizontal motions of the module are restricted by guides. A principle sketch of the resulting dynamic system is shown in Fig. 9.3.

The equation of motion for the vertical displacement response $u(t)$ of the module is then

$$m\ddot{u}(t) + c\dot{u}(t) + ku(t) = kv(t), \tag{l}$$

which gives a linear, time-invariant connection between $v(t)$ and $u(t)$. The corresponding transfer function, denoted by $H_{vu}(\omega)$, is then given as

$$H_{vu}(\omega) = \frac{k}{-\omega^2 m + \mathrm{i}\omega c + k}. \tag{m}$$

Therefore, if $H(t)$ is a Gaussian process as shown previously, the vertical response of the module $U(t)$ also becomes a stationary Gaussian process with a variance spectrum $S_U(\omega)$ given as

$$S_U(\omega) = |H_{vu}(\omega)|^2\, S_V(\omega) = |H_{vu}(\omega)|^2\, |G_z(\omega) + LG_\theta(\omega)|^2\, S_H(\omega). \tag{n}$$

This example is continued in Example 9.13.1 at the end of this chapter.

On the basis of the response spectrum, one can compute several statistical quantities that are important for assessing the response. So far, we have limited ourselves to the standard deviation; however, in Chapter 10, we see that the spectrum is also crucial for estimating extreme responses.

9.6 Cross-Covariance

In the previous section, we saw that the response spectrum does not contain any information about the phase shift between load and response. We now show that such information can be obtained by introducing a quantity called *cross-covariance*.

Let $X(t)$ and $Y(t)$ be two stationary processes with zero mean values. If the covariance $c_{X(t)Y(t+\tau)} = \mathrm{E}[X(t)Y(t+\tau)]$ is independent of t, $X(t)$ and $Y(t)$ are said to be jointly stationary. That two processes $X(t)$ and $Y(t)$ are jointly stationary means that both $X(t)$ and $Y(t)$ are stationary, and that the covariance $c_{X(t)Y(t+\tau)}$ is independent of t.

We can now define a function $C_{XY}(\tau)$ as follows:

$$C_{XY}(\tau) = \mathrm{E}[X(t)Y(t+\tau)], \tag{9.20}$$

or $C_{XY}(\tau) = \mathrm{E}[(X(t) - m_X)(Y(t+\tau) - m_Y)]$ in the case of non-zero mean values. $C_{XY}(\tau)$ is called the cross-covariance function or just the cross-covariance of $X(t)$ and $Y(t)$. The prefix "cross" indicates that the covariance is between two different processes, as opposed to "auto," which indicates that only one process is involved.

If $(x_1(t), y_1(t)), \ldots, (x_N(t), y_N(t)), \ldots$ is a sequence of corresponding pairs of realizations of $X(t)$ and $Y(t)$, the cross-covariance can be calculated as

$$C_{XY}(\tau) = \lim_{N\to\infty} \frac{1}{N} \sum_{j=1}^{N} x_j(t) y_j(t+\tau). \tag{9.21}$$

Assume that $F(t)$ and $G(t)$ are two jointly stationary load processes with mean value zero. Each acts on a linear system with impulse response function $h_{FX}(t)$ and $h_{GY}(t)$, respectively. Let $X(t)$ and $Y(t)$ denote the corresponding response processes. From the results of Section 9.4 it follows that $X(t)$ and $Y(t)$ are stationary. We now show that they are also jointly stationary.

By carrying through a derivation entirely analogous to that leading to Eq. (9.10), it is obtained that

$$\mathrm{E}[X(t)\,Y(t+\tau)] = \int_0^\infty \int_0^\infty h_{FX}(s_1) h_{GY}(s_2)\, C_{FG}(\tau + s_1 - s_2)\, ds_1\, ds_2. \tag{9.22}$$

This shows that the covariance $c_{X(t)Y(t+\tau)}$ is independent of t, and hence that $X(t)$ and $Y(t)$ are jointly stationary with cross-covariance function

$$C_{XY}(\tau) = \int_0^\infty \int_0^\infty h_{FX}(s_1) h_{GY}(s_2)\, C_{FG}(\tau + s_1 - s_2)\, ds_1\, ds_2. \tag{9.23}$$

By putting $G(t) = F(t)$ and $h_{FX}(t) = h_{GY}(t)$, which gives $Y(t) = X(t)$, it follows that Eq. (9.11) is a special case of Eq. (9.23). It may, of course, also be shown that $F(t)$ and $X(t)$ are jointly stationary, and that $C_{FX}(\tau) = \int_0^\infty h_{FX}(s)\, C_F(\tau - s)\, ds$.

9.7 Cross-Spectrum

The cross-spectrum $S_{XY}(\omega)$ of two jointly stationary processes $X(t)$ and $Y(t)$ is defined as the Fourier transform of the cross-covariance, that is,

$$S_{XY}(\omega) = \frac{1}{2\pi} \int_{-\infty}^{\infty} C_{XY}(\tau)\, e^{-i\omega\tau}\, d\tau. \tag{9.24}$$

To ensure that the rhs of Eq. (9.24) is well defined, it is commonly required that $\int_{-\infty}^{\infty} |C_{XY}(\tau)|\, d\tau < \infty$. In contrast to the variance spectrum, the cross-spectrum is generally a complex function of ω. We may therefore write

$$S_{XY}(\omega) = Co_{XY}(\omega) - i\, Qu_{XY}(\omega), \tag{9.25}$$

where $Co_{XY}(\omega) = \Re S_{XY}(\omega)$ is called the *co-spectrum* and $Qu_{XY}(\omega) = -\Im S_{XY}(\omega)$ is called the *quadrature-spectrum*. The minus sign in Eq. (9.25) is just a convention. The reason is simply the following representation of $Qu_{XY}(\omega)$. By introducing $e^{-i\omega t} = \cos \omega t - i \sin \omega t$ in Eq. (9.24), it is found that

$$Co_{XY}(\omega) = \frac{1}{2\pi} \int_{-\infty}^{\infty} C_{XY}(\tau)\, \cos \omega\tau\, d\tau, \tag{9.26}$$

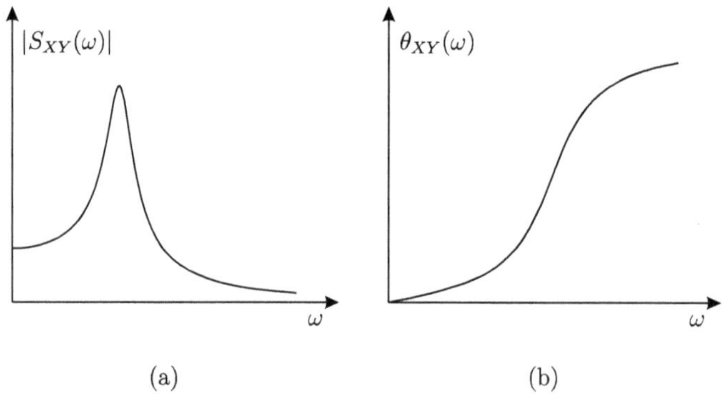

Figure 9.4. An example of a cross-spectrum. (a) Modulus spectrum. (b) Phase spectrum.

and

$$Qu_{XY}(\omega) = \frac{1}{2\pi} \int_{-\infty}^{\infty} C_{XY}(\tau) \sin \omega\tau \, d\tau. \tag{9.27}$$

From Eq. (9.26), it follows that $Co_{XY}(-\omega) = Co_{XY}(\omega)$, while $Qu_{XY}(-\omega) = -Qu_{XY}(\omega)$. This leads to the result

$$S_{XY}(-\omega) = Co_{XY}(\omega) + \mathrm{i}\, Qu_{XY}(\omega) = S_{XY}(\omega)^*. \tag{9.28}$$

It can be shown that $C_{XY}(\tau)$ and $S_{XY}(\omega)$ is a Fourier transform pair. Hence, the following result is also obtained

$$C_{XY}(\tau) = \int_{-\infty}^{\infty} S_{XY}(\omega)\, e^{\mathrm{i}\omega\tau} \, d\omega. \tag{9.29}$$

By putting $\tau = 0$ in Eq. (9.29), it is found that

$$\mathrm{E}[X(t)\, Y(t)] = C_{XY}(0) = \int_{-\infty}^{\infty} S_{XY}(\omega)\, d\omega = \int_{-\infty}^{\infty} Co_{XY}(\omega)\, d\omega \tag{9.30}$$

because $Qu_{XY}(\omega)$ is antisymmetric. The co-spectrum can therefore be interpreted as a decomposition over frequency of the covariance $c_{X(t)Y(t)}$.

In the literature, one can also find the cross-spectrum represented in polar form, for example,

$$S_{XY}(\omega) = |S_{XY}(\omega)|\, e^{-\mathrm{i}\theta_{XY}(\omega)}, \tag{9.31}$$

where the *modulus spectrum* $|S_{XY}(\omega)|$ is the length of the complex function $S_{XY}(\omega)$, that is,

$$|S_{XY}(\omega)| = \sqrt{Co_{XY}(\omega)^2 + Qu_{XY}(\omega)^2}, \tag{9.32}$$

while $\theta_{XY}(\omega)$ is the *phase spectrum* given by

$$\theta_{XY}(\omega) = \arctan\left\{Qu_{XY}(\omega)/Co_{XY}(\omega)\right\}. \tag{9.33}$$

The phase spectrum contains information about an average phase shift at each frequency ω between corresponding pairs of realizations of $X(t)$ and $Y(t)$. If $0 \leq \theta_{XY}(\omega) \leq \pi$, $X(t)$ is ahead of $Y(t)$ at the frequency ω. This is due to the minus sign in front of $\theta_{XY}(\omega)$ in Eq. (9.31). A typical cross-spectrum is shown in Fig. 9.4.

We now return to the situation from the previous section where $F(t)$ and $G(t)$ are two jointly stationary processes with zero mean value, acting on two linear systems with impulse response functions $h_{FX}(t)$ and $h_{GY}(t)$, respectively. It was shown that the corresponding response processes $X(t)$ and $Y(t)$ also became jointly stationary. Let us calculate the cross-spectrum $S_{XY}(\omega)$ of $X(t)$ and $Y(t)$. By using a derivation entirely analogous to that leading to Eq. (9.14), it is found that

$$S_{XY}(\omega) = H_{FX}(\omega)^* \, H_{GY}(\omega) \, S_{FG}(\omega), \tag{9.34}$$

where $H_{FX}(\omega)$ and $H_{GY}(\omega)$ are the transfer functions that correspond to $h_{FX}(t)$ and $h_{GY}(t)$, respectively. It is seen that Eq. (9.16) is a special case of Eq. (9.34), which is obtained by putting $G(t) = F(t)$ and $H_{GY}(\omega) = H_{FX}(\omega)$, leading to $Y(t) = X(t)$.

Of particular interest is the relation we can derive from Eq. (9.34) by putting $G(t) = F(t)$ and $H_{FX}(\omega) \equiv 1$, that is, $X(t) = F(t)$:

$$S_{FY}(\omega) = H_{FY}(\omega) \, S_F(\omega). \tag{9.35}$$

This implies that the cross-spectrum between the load and the response of a linear system can be used to provide information about both the amplitude amplification and the phase shift between these because Eq. (9.35) provides us with the opportunity to estimate the transfer function via estimates of the cross-spectrum and the variance spectrum in the following way:

$$H_{FY}(\omega) = \frac{S_{FY}(\omega)}{S_F(\omega)}, \quad (S_F(\omega) > 0). \tag{9.36}$$

9.8 Cross-Spectrum Directly from Realizations

In Section 6.5, we see how the variance spectrum of a stationary process can be calculated directly from the realizations of the process. In a similar way, the cross-spectrum can be calculated directly from corresponding pairs of realizations of two jointly stationary processes $X(t)$ and $Y(t)$. Let $(x_1(t), y_1(t)), (x_2(t), y_2(t)), \ldots$ be a sequence of corresponding realizations of $X(t)$ and $Y(t)$. It can then be shown that

$$S_{XY}(\omega) = \lim_{T\to\infty} \lim_{N\to\infty} \frac{2\pi}{TN} \sum_{j=1}^{N} X_{j,T}(\omega)^* \, Y_{j,T}(\omega), \tag{9.37}$$

where $Y_{j,T}(\omega)$ denotes the Fourier transform of $y_{j,T}(t)$, where the notation is as defined in Section 6.5. As in Section 6.5, the case of two (jointly) ergodic processes will lead to an equation that resembles Eq. (9.37), but is based on a single realization of each process.

As for the variance spectrum, Eq. (9.37) constitutes the basis for estimating cross-spectra by using an FFT algorithm.

9.9 Spectra and Cross-Spectra of Differentiated Processes

Let $X(t)$ be a stationary process with possibly non-zero mean value. We say that the process $X(t)$ is differentiable if all realizations are differentiable. In this way, we may define a new stochastic process whose realizations are the derivatives of the realizations of $X(t)$. This process is called "the derivative of $X(t)$" and is denoted by

$\dot{X}(t) = dX(t)/dt$. This definition can be easily extended to any order of differentiability.

To take the derivative of a function is clearly a linear operation ($d[\alpha\, f(t) + \beta\, g(t)]/dt = \alpha\, df(t)/dt + \beta\, dg(t)/dt$), which is also time invariant (the operation itself does not change with time). Let us determine the transfer function $H_{X\dot{X}}(\omega)$. According to our recipe, $\dot{X}(t) = H_{X\dot{X}}(\omega)\, e^{i\omega t}$ when $X(t) = e^{i\omega t}$. Hence, we obtain

$$H_{X\dot{X}}(\omega)\, e^{i\omega t} = \frac{dX(t)}{dt} = \frac{d(e^{i\omega t})}{dt} = i\,\omega\, e^{i\omega t}, \tag{9.38}$$

which gives the result

$$H_{X\dot{X}}(\omega) = i\,\omega. \tag{9.39}$$

The first thing to note is that $H_{X\dot{X}}(\omega)$ does not have an (inverse) Fourier transform ($|H_{X\dot{X}}(\omega)| \to \infty$ when $\omega \to \infty$), which means that we cannot find any impulse response function for this particular system. The transfer function therefore provides a more general description of a linear, time invariant system than the impulse response function. This may serve as an explanation to the concluding remarks of Section 9.4.

It follows from Section 9.4 that $\dot{X}(t)$ is a stationary process. However, we cannot use Eq. (9.11) to calculate the autocovariance $C_{\dot{X}}(\tau)$ of $\dot{X}(t)$ because we do not have any impulse response function. We derive $C_{\dot{X}}(\tau)$ in an indirect way, and for that we need the variance spectrum of $\dot{X}(t)$. According to Eq. (9.16) it is given as

$$S_{\dot{X}}(\omega) = |i\,\omega|^2\, S_X(\omega) = \omega^2\, S_X(\omega). \tag{9.40}$$

From Eq. (6.2), we then get

$$C_{\dot{X}}(\tau) = \int_{-\infty}^{\infty} S_{\dot{X}}(\omega)\, e^{i\omega\tau}\, d\omega = \int_{-\infty}^{\infty} \omega^2\, S_X(\omega)\, e^{i\omega\tau}\, d\omega. \tag{9.41}$$

where we assume that $\int_{-\infty}^{\infty} \omega^2\, S_X(\omega)\, d\omega < \infty$. By differentiating twice with respect to τ on both sides of Eq. (6.2), we also get that

$$\begin{aligned}\frac{d^2 C_X(\tau)}{d\tau^2} &= \frac{d^2}{d\tau^2}\int_{-\infty}^{\infty} S_X(\omega)\, e^{i\omega\tau}\, d\omega = \int_{-\infty}^{\infty} S_X(\omega)\, \frac{d^2 e^{i\omega\tau}}{d\tau^2}\, d\omega \\ &= -\int_{-\infty}^{\infty} \omega^2\, S_X(\omega)\, e^{i\omega\tau}\, d\omega.\end{aligned} \tag{9.42}$$

By comparing Eqs. (9.41) and (9.42), it is seen that

$$C_{\dot{X}}(\tau) = -\frac{d^2 C_X(\tau)}{d\tau^2}, \tag{9.43}$$

because both functions have the same Fourier transform, viz. $\omega^2\, S_X(\omega)$, and because Fourier transform pairs are unique, it follows that Eq. (9.43) is correct. The minus sign in Eq. (9.43) is explained by the fact that $C_X(\tau)$ has a maximum at $\tau = 0$. The second derivative is therefore negative at $\tau = 0$. Since $\sigma_{\dot{X}}^2 = C_{\dot{X}}(0) \geq 0$, it follows that the minus sign is necessary on the rhs of Eq. (9.43).

We know from before that the output (response) process of a linear, time-invariant system is stationary if the input (load) process is stationary. It can also be shown that the two processes are jointly stationary. In particular, so are $X(t)$ and $\dot{X}(t)$. Let us have a look at the cross-spectrum $S_{X\dot{X}}(\omega)$ between $X(t)$ and $\dot{X}(t)$. According to Eq. (9.35), it is given by the equation

$$S_{X\dot{X}}(\omega) = \mathrm{i}\,\omega\, S_X(\omega). \tag{9.44}$$

It is seen that $S_{X\dot{X}}(\omega)$ is purely imaginary, that is, $Co_{X\dot{X}}(\omega) = 0$ and $Qu_{X\dot{X}}(\omega) = -\omega\, S_X(\omega)$. This means that the phase spectrum $\theta_{X\dot{X}}(\omega) = 3\pi/2$, which implies that $\dot{X}(t)$ is 90 degrees ahead of $X(t)$ at each frequency ω. This agrees with the observation that $d\cos\omega t/dt = -\omega \sin\omega t = \cos(\omega t + \pi/2)$ is 90 degrees ahead of $\cos\omega t$.

Using Eq. (9.29) together with Eq. (9.44), it is obtained that

$$C_{X\dot{X}}(\tau) = \int_{-\infty}^{\infty} \mathrm{i}\,\omega\, S_X(\omega)\, e^{\mathrm{i}\omega\tau}\, d\omega. \tag{9.45}$$

This leads to the result

$$\mathrm{E}[X(t)\dot{X}(t)] = C_{X\dot{X}}(0) = \mathrm{i}\int_{-\infty}^{\infty} \omega\, S_X(\omega)\, d\omega = 0 \tag{9.46}$$

because $\omega\, S_X(\omega)$ is an odd (antisymmetric) function of ω. This implies that $X(t)$ and $\dot{X}(t)$ are uncorrelated random variables at each point in time t, that is, $\rho_{X(t)\dot{X}(t)} = 0$.

If $\dot{X}(t)$ is differentiable as we define it, we get the second derivative $\ddot{X}(t)$. By repeating a linear operation twice, the result is still a linear operation. That is, $X(t) \to \ddot{X}(t)$ defines a linear (and time invariant) operation. The transfer function $H_{X\ddot{X}}(\omega)$ is determined by $H_{X\ddot{X}}(\omega)\, e^{\mathrm{i}\omega t} = d^2 e^{\mathrm{i}\omega t}/dt^2 = (\mathrm{i}\omega)^2\, e^{\mathrm{i}\omega t}$. This gives

$$H_{X\ddot{X}}(\omega) = -\,\omega^2. \tag{9.47}$$

By using these results, the following expressions are obtained:

$$S_{\ddot{X}}(\omega) = \omega^4\, S_X(\omega), \tag{9.48}$$

$$S_{X\ddot{X}}(\omega) = -\,\omega^2\, S_X(\omega), \tag{9.49}$$

$$S_{\dot{X}\ddot{X}}(\omega) = \mathrm{i}\,\omega^3\, S_X(\omega). \tag{9.50}$$

The following variances and covariances are also obtained:

$$\sigma_{\dot{X}}^2 = \int_{-\infty}^{\infty} \omega^2\, S_X(\omega)\, d\omega, \tag{9.51}$$

$$\sigma_{\ddot{X}}^2 = \int_{-\infty}^{\infty} \omega^4\, S_X(\omega)\, d\omega, \tag{9.52}$$

$$\mathrm{E}[\dot{X}(t)\ddot{X}(t)] = \mathrm{i}\int_{-\infty}^{\infty} \omega^3\, S_X(\omega)\, d\omega = 0, \tag{9.53}$$

and

$$\mathrm{E}[X(t)\ddot{X}(t)] = -\int_{-\infty}^{\infty} \omega^2\, S_X(\omega)\, d\omega = -\sigma_{\dot{X}}^2. \tag{9.54}$$

9.10 Coherence Function

For two jointly stationary processes $X(t)$ and $Y(t)$ the *coherence function* $\gamma_{XY}(\omega)^2$ is defined by the equation

$$\gamma_{XY}(\omega)^2 = \frac{|S_{XY}(\omega)|^2}{S_X(\omega)\,S_Y(\omega)}, \quad (\text{for } S_X(\omega)\cdot S_Y(\omega) > 0). \tag{9.55}$$

It can be shown that $|S_{XY}(\omega)|^2 \le S_X(\omega)\,S_Y(\omega)$, and we choose to define $\gamma_{XY}(\omega) \ge 0$. Hence, $0 \le \gamma_{XY}(\omega) \le 1$.

Let $X(t)$ and $Y(t)$ be the responses of two linear, time-invariant systems with the same load $F(t)$. If $F(t)$ is stationary, $X(t)$ and $Y(t)$ are jointly stationary (put $G(t) = F(t)$ in Section 9.6). If $H_{FX}(\omega)$ and $H_{FY}(\omega)$ denote the transfer functions for the two systems, we get that $S_{XY}(\omega) = H_{FX}(\omega)^*\,H_{FY}(\omega)\,S_F(\omega)$, $S_X(\omega) = |H_{FX}(\omega)|^2\,S_F(\omega)$, and $S_Y(\omega) = |H_{FY}(\omega)|^2\,S_F(\omega)$. The coherence function for $X(t)$ and $Y(t)$ is then

$$\gamma_{XY}(\omega)^2 = \frac{|H_{FX}(\omega)^*\,H_{FY}(\omega)\,S_F(\omega)|^2}{|H_{FX}(\omega)|^2\,S_F(\omega)\,|H_{FY}(\omega)|^2\,S_F(\omega)} = 1. \tag{9.56}$$

In particular, $\gamma_{FX}(\omega) = 1$ for all ω. Hence, the coherence between the load and the response for a linear, time-invariant system is identically equal to 1.0. The coherence function can therefore be used to test for linear connections.

In practice, if $0 < \gamma_{XY}(\omega) < 1$, this will mainly be due to one or more of the following three causes:

- The system is not linear
- The response $X(t)$ is not only the result of $F(t)$, but is also caused by other simultaneous loads.
- The measurements of the response $X(t)$ are corrupted by noise.

If $C_{XY}(\tau) = 0$ for all τ, Eq. (9.24) implies that $S_{XY}(\omega) = 0$, and thereby $\gamma_{XY}(\omega) = 0$ for all ω. In particular, $\gamma_{XY}(\omega) = 0$ if $X(t)$ and $Y(t+\tau)$ are independent random variables for all t and τ.

9.11 Response to "White Noise"

Let $X(t)$ be the response of a linear system with transfer function $H_{FX}(\omega)$ to a stationary load process $F(t)$. Of particular importance for the assessment of the size of the response is the mean value m_X and the standard deviation σ_X. The variance can be calculated from the expression, cf. Eq. (9.18),

$$\sigma_X^2 = \int_{-\infty}^{\infty} |H_{FX}(\omega)|^2\,S_F(\omega)\,d\omega. \tag{9.57}$$

In general, the expression on the rhs of Eq. (9.57) must be calculated numerically. However, many problems in vibration theory are associated with structures that have little damping ($\xi \ll 1.0$). This means that $|H_{FX}(\omega)|^2$ is very narrow around the resonance frequency ω_r, which implies that the main contribution to the integral in Eq. (9.57) generally comes from a small interval around ω_r. Figure 9.5 shows $|H(\omega)|$, where $H(\omega)$ is given by Eq. (2.71). For $\xi = 0.01$, $|H(\omega_r)|^2 \approx 2500|H(0)|^2$. If $S_F(\omega)$ varies much slower than $|H_{FX}(\omega)|^2$ as a function of ω, it is often possible for good approximation to replace $S_F(\omega)$ under the integral sign in (9.57) with the

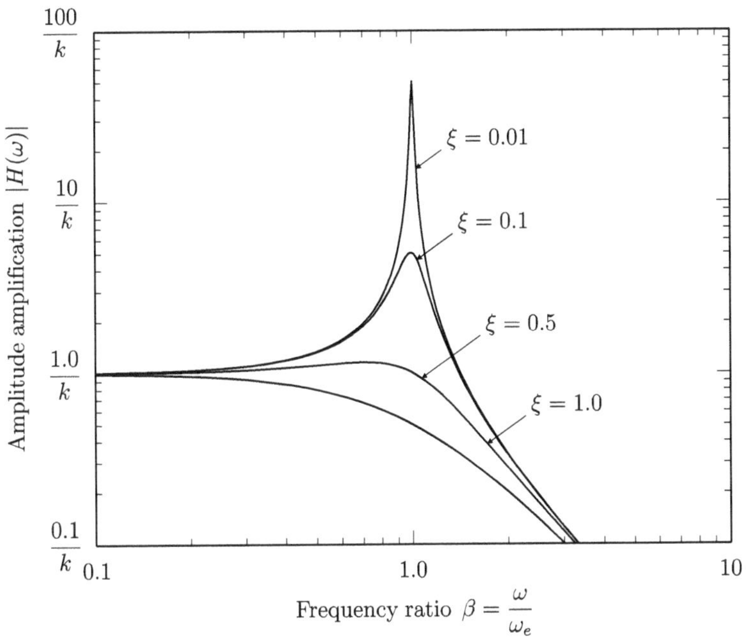

Figure 9.5. Amplitude amplification for various values of relative damping.

constant $S_0 = S_F(\omega_r)$. By this, we have effectively replaced the real load process $F(t)$ with white noise, which is precisely characterized by a constant variance spectrum. This procedure of replacing the input spectrum by an appropriate constant, is often referred to as the *white noise approximation*. An illustration of when the white noise approximation is acceptable and when it is more questionable is shown in Fig. 9.6.

One may also note that a typical feature of the response spectrum of a weakly damped system is that it is narrow banded. This follows from the fact that the response spectrum to a large extent is determined by $|H_{FX}(\omega)|^2$, and when this function is "narrow," then the same is generally true of the response spectrum.

With the white noise approximation, the variance becomes

$$\sigma_X^2 = S_0 \int_{-\infty}^{\infty} |H_{FX}(\omega)|^2 \, d\omega. \tag{9.58}$$

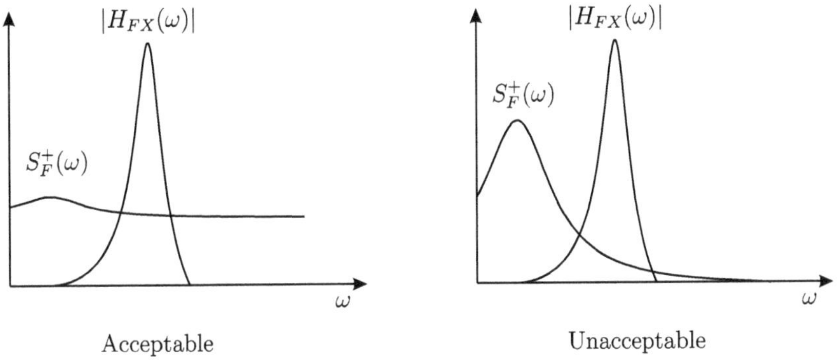

Figure 9.6. White noise approximation.

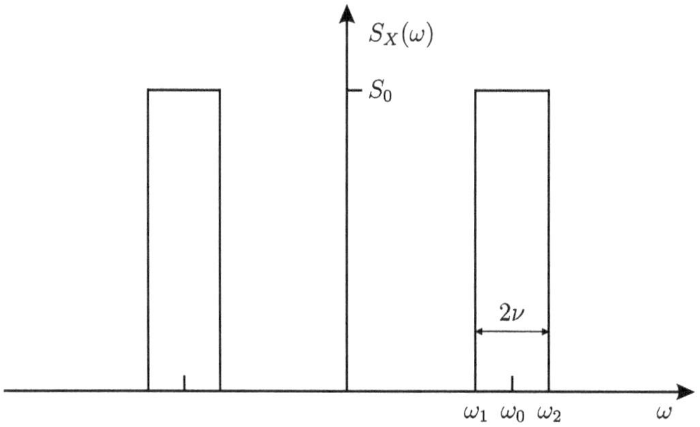

Figure 9.7. Narrow banded rectangular load spectrum.

In Appendix A a few frequently appearing forms of $H_{FX}(\omega)$ and the corresponding integrals $\int_{-\infty}^{\infty} |H_{FX}(\omega)|^2\, d\omega$ are listed. These are taken from Newland (1991).

Applied to our standard transfer function $H(\omega)$ given by Eq. (2.71), it is obtained that

$$\sigma_X^2 = S_0 \int_{-\infty}^{\infty} |H(\omega)|^2\, d\omega = \frac{\pi}{2} \cdot \frac{S_0}{m^2\, \xi\, \omega_e^3}. \tag{9.59}$$

It can be shown that the autocovariance function that corresponds to the spectrum $S_0\, |H(\omega)|^2$ is given as

$$C_X(\tau) = \frac{\pi}{2} \cdot \frac{S_0}{m^2\, \xi\, \omega_e^3}\, e^{-\xi\, \omega_e |\tau|} \Big(\cos \omega_d \tau + \xi\, \frac{\omega_e}{\omega_d}\, \sin \omega_d |\tau| \Big). \tag{9.60}$$

9.12 Response to a Narrow-Banded Load Process

We assume the same situation as in the previous section, except that the load process is now assumed to be narrow banded. By the help of a simplified calculation, we illustrate three important load cases. It is assumed for simplicity that the load spectrum is rectangular, as shown in Fig. 9.7, and that $H_{FX}(\omega) = H(\omega)$. Then

$$\sigma_X^2 = 2 S_0 \int_{\omega_0 - \nu}^{\omega_0 + \nu} |H(\omega)|^2\, d\omega. \tag{9.61}$$

The integral can be solved exactly, but for our purpose it is more expedient to find approximate solutions for these three cases.

Damping Controlled Response

Let us consider the case $\omega_1 < \omega_e < \omega_2$. This situation is illustrated in Fig. 9.8. To simplify the calculations somewhat, we assume that $\omega_e = \omega_0$. With $\omega = \omega_0(1+\varepsilon)$, ε will vary between $-\nu/\omega_0$ and ν/ω_0, and $d\omega = \omega_0 d\varepsilon$. For the subsequent calculation, it is practical to rewrite $|H(\omega)|^2$ to the form

$$|H(\omega)|^2 = \frac{1}{m^2\{(\omega_e^2 - \omega^2)^2 + (2\xi\omega_e\omega)^2\}}. \tag{9.62}$$

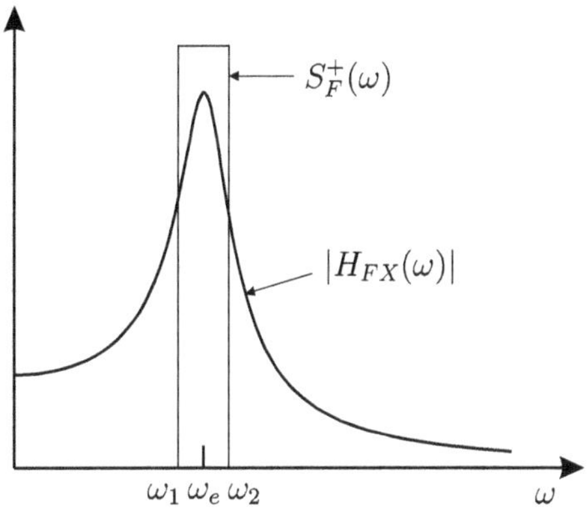

Figure 9.8. Damping controlled response.

By substituting $\omega = \omega_0(1+\varepsilon)$ and $\omega_e = \omega_0$, it is obtained that

$$(\omega_e^2 - \omega^2)^2 + (2\xi\omega_e\omega)^2 = \omega_0^4\,\varepsilon^2\,(4 + 4\varepsilon + \varepsilon^2) + 4\,\xi^2\,\omega_0^4(1 + 2\varepsilon + \varepsilon^2)$$
$$\approx 4\,\omega_0^4(\varepsilon^2 + \xi^2) \tag{9.63}$$

when it is assumed that $\varepsilon_{\max} = \nu/\omega_0 \ll 1.0$. This leads to the result

$$\sigma_X^2 \approx \frac{2S_0}{m^2}\int_{-\nu/\omega_0}^{\nu/\omega_0} \frac{\omega_0\,d\varepsilon}{4\,\omega_0^4(\varepsilon^2+\xi^2)} = \frac{S_0}{m^2\,\omega_0^3}\int_0^{\nu/\omega_0}\frac{d\varepsilon}{\varepsilon^2+\xi^2}$$
$$= \frac{S_0}{m^2\,\omega_0^3\,\xi}\arctan\left(\frac{\nu}{\omega_0\,\xi}\right). \tag{9.64}$$

Because $\arctan(x) \to \pi/2$ when $x \to \infty$ ($\arctan(10) \approx 0.94 \cdot \pi/2$), it is seen that when $\nu \gg \omega_0\,\xi$, then Eq. (9.64) will give approximately the same answer as Eq. (9.59), which is the expression we found for the white noise load process. In cases with low damping, the white noise approximation may prove to be useful even if the load spectrum is narrow banded, as long as the condition $\nu \gg \omega_0\,\xi$ is satisfied.

Stiffness Controlled (Quasi-Static) Response

In this case it is assumed that $\omega_e \gg \omega_0$, see Fig. 9.9. This means that $\omega_e \gg \omega$ for $\omega_1 \leq \omega \leq \omega_2$. Hence, we find that

$$(\omega_e^2 - \omega^2)^2 + (2\xi\omega_e\omega)^2 \approx \omega_e^4, \tag{9.65}$$

which gives

$$\sigma_X^2 \approx \frac{2S_0}{m^2}\int_{\omega_1}^{\omega_2}\frac{1}{\omega_e^4}\,d\omega = \frac{4\,S_0\,\nu}{m^2\,\omega_e^4} = \frac{4\,S_0\,\nu}{k^2} = \frac{\sigma_F^2}{k^2} = |H_{FX}(0)|^2\,\sigma_F^2. \tag{9.66}$$

This is the same result that we would have obtained in the static case ($X = F/k$). The response is therefore called *stiffness controlled* or *quasi-static*.

Mass Controlled Response

Here, we assume that $\omega_e \ll \omega_0$, see Fig. 9.10. This gives

$$(\omega_e^2 - \omega^2)^2 + (2\xi\omega_e\omega)^2 \approx \omega^4, \tag{9.67}$$

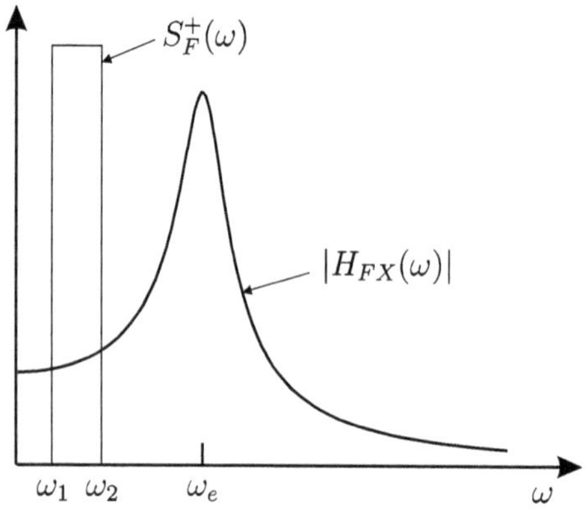

Figure 9.9. Stiffness controlled response.

for $\omega_1 \leq \omega \leq \omega_2$. From this, we find that

$$\sigma_X^2 \approx \frac{2S_0}{m^2} \int_{\omega_1}^{\omega_2} \frac{1}{\omega^4}\, d\omega = \frac{2\, S_0}{3m^2}\left(\frac{1}{\omega_1^3} - \frac{1}{\omega_2^3}\right) \approx \frac{4\, S_0\, \nu}{m^2\, \omega_0^4} = \frac{\sigma_F^2}{m^2\, \omega_0^4}. \tag{9.68}$$

The last approximation applies because

$$\frac{\omega_2^3 - \omega_1^3}{\omega_2^3\, \omega_1^3} = \frac{\omega_2 - \omega_1}{\omega_2^3\, \omega_1^3}\,(\omega_2^2 + \omega_1\, \omega_2 + \omega_1^2) \approx \frac{2\nu}{\omega_0^6}\, 3\omega_0^2 = \frac{6\nu}{\omega_0^4}. \tag{9.69}$$

In this case, it is only m of the parameters of the vibration system that enter Eq. (9.68). The response therefore becomes mass controlled. It is also seen that σ_X rapidly decreases toward zero when ω_0 increases.

Because $\nu \ll \omega_0$, usually $S_0\, \omega_0 > 4\, S_0\, \nu = \sigma_F^2$. Then $\pi\, S_0/(2m^2\omega_0^3\xi) > \sigma_F^2/(k^2\xi)$ $(\omega_0 = \omega_e)$. For mass controlled response, $\omega_0 > \omega_e$, that is, $\sigma_F^2/(m^2\omega_0^4) < \sigma_F^2/(m^2\omega_e^4) = \sigma_F^2/(k^2)$. Hence, it is seen that the damping controlled response is larger than the other two cases by an of order of magnitude at least, given by the factor $1/\sqrt{\xi}$.

Equations (9.64), (9.66), and (9.68) can be used for approximate calculations when the load spectrum is narrow banded.

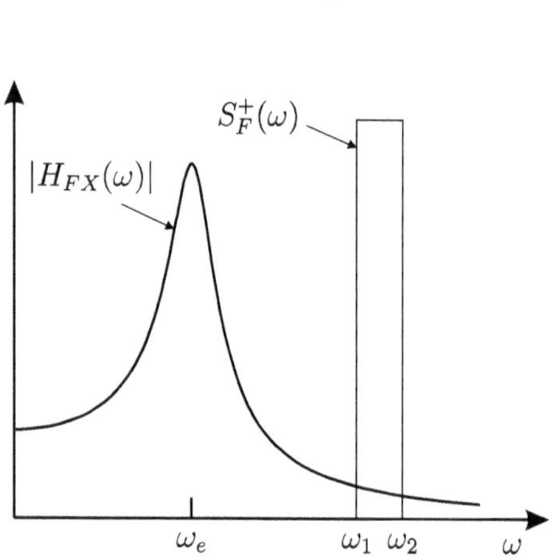

Figure 9.10. Mass controlled response.

9.13 Response to Random Wave Loads

9.13.1 Example (Continuation of Example 9.5.1)

Let us return to Example 9.5.1 and look more closely at a few topics. In practice, most calculations of statistical response quantities have to be performed using a computer. However, one should try to get a qualitative picture of the situation that makes it possible to assess the numerical results that are produced. We now illustrate a few elements of such a qualitative analysis, where the goal is to provide a coarse picture of the response spectrum $S_U(\omega)$ of the vertical motions of the equipment module.

To accomplish this goal, we need some information. It is assumed that the relative damping ξ_z in heave of the crane vessel is approximately 10%, and that the natural period in heave $T_z = 6$ s ($\omega_z = 1.05$ rad/s). Similarly, it is assumed that the relative damping $\xi_\theta \approx 3\%$ and the natural period $T_\theta = 14$ s ($\omega_\theta = 0.45$ rad/s) for roll. In addition, the following simplifications are assumed:

1. $$S_H(\omega) = \frac{\alpha\,\omega^2}{1+\omega^6}$$
   where α is a constant.
2. $$\left| \frac{H_z(\omega_z)\cdot k_\theta}{L\cdot H_\theta(\omega_\theta)\cdot k_z} \right| = 1.0$$
3. $H_z(\omega)$ and $H_\theta(\omega)$ varies sufficiently slowly with ω to be considered as constants with reasonable accuracy.
4. The parameters $A_j(\omega)$ and $B_j(\omega)$, $j = z, \theta$, can be considered as constants.

Because $\xi_z \approx 0.1$ and $\xi_\theta \approx 0.03$, there is relatively large amplification at resonance. On the basis of simplification no. 3, this means that $|G_z(\omega)|$ and $|G_\theta(\omega)|$ both typically look like the transfer function sketched in Fig. 9.8. As the two frequencies ω_z and ω_θ are well separated, the transfer function $H_{\eta\upsilon}(\omega) = G_z(\omega) + L\,G_\theta(\omega)$ in the vicinity of a given natural frequency ω_j is mainly determined by the corresponding transfer function $G_j(\omega)$ ($j = z, \theta$). In accordance with this, we assume that $|H_{\eta\upsilon}(\omega_z)| \approx |G_z(\omega_z)|$ and $|H_{\eta\upsilon}(\omega_\theta)| \approx |L\,G_z(\omega_\theta)|$.

A coarse picture of $|H_{\eta\upsilon}(\omega)|$ as a two-peaked function is now starting to develop, but we still lack information on the relative size of the two peaks. By substitution into Eqs. (h) and (i) from Example 9.5.1, it is obtained that $|G_z(\omega_z)| = |H_z(\omega_z)|/(2\,\xi_z\,k_z)$ and $|G_\theta(\omega_\theta)| = |H_\theta(\omega_\theta)|/(2\,\xi_\theta\,k_\theta)$. This, together with simplification no. 2, gives $|H_{\eta\upsilon}(\omega_z)|/|H_{\eta\upsilon}(\omega_\theta)| \approx |G_z(\omega_z)|/|L\,G_z(\omega_\theta)| \approx \xi_\theta/\xi_z \approx 0.3$. Finally, we find that $S_V(\omega_z)/S_V(\omega_\theta) \approx 0.3^2\,S_H(\omega_z)/S_H(\omega_\theta) \approx 0.4$. A qualitative sketch of $S_V(\omega) = |H_{\eta\upsilon}(\omega)|^2\,S_H(\omega)$ is given in Fig. 9.11.

We finally make the connection to the equipment module. The total mass m of the equipment module, including added mass due to the vertical motions, is assumed to be $5\cdot 10^5$ kg. The damping associated with the vertical motions of the module is estimated as $c = 6\cdot 10^4$ Ns/m. The linear stiffness of the wire between the crane top A and the equipment module B is calculated to be $k = 5\cdot 10^6$ N/m.

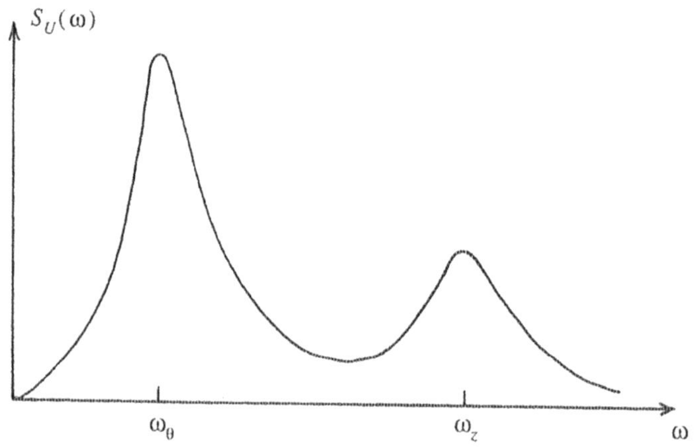

Figure 9.11. Rough sketch of the response spectrum of the crane top.

These parameter values gives a relative damping $\xi_u \approx 0.02$ and a natural frequency $\omega_u \approx 3.2$ rad/s, which is much larger than ω_z. If we draw a sketch of $|H_{vu}(\omega)|^2$ in Fig. 9.11, it would be apparent that a good approximation is obtained by assuming that there is a quasi-static relation between the vertical motions of the crane top and the equipment module, and because $|H_{vu}(0) = 1|$, this gives $S_U(\omega) \approx S_V(\omega)$. This means that, in practice, the wire can be considered as infinitely stiff with respect to the vertical response of the equipment module, that is, $U = V$.

9.14 Response Spectra of MDOF Linear Systems

9.14.1 Transfer Function Matrix of an MDOF Response Process

The goal of this section is to establish a relation similar to Eq. (9.16) for a linear MDOF dynamic system, cf. Lutes and Sarkani (2004). To do that, we need to determine the transfer function matrix. Each entry of this matrix is equal to the transfer function between an individual component of the force vector and an individual component of the response vector. The derivations are carried out in the context of a basic MDOF linear dynamic system of the form

$$\mathcal{L}\big[\mathbf{X}(t)\big] = \mathbf{M}\ddot{\mathbf{X}}(t) + \mathbf{C}\dot{\mathbf{X}}(t) + \mathbf{K}\mathbf{X}(t) = \mathbf{F}(t)\,, \tag{9.70}$$

where $\mathbf{M}$, $\mathbf{C}$, and $\mathbf{K}$ are constant $n \times n$ matrices, and $\mathbf{X}(t) = \big(X_1(t), \ldots, X_n(t)\big)^T$ and $\mathbf{F}(t) = \big(F_1(t), \ldots, F_n(t)\big)^T$ are n-dimensional response and force (column) vectors, respectively. To determine the transfer function between $F_i(t)$ and $X_j(t)$, denoted by $H_{ij}(\omega) = H_{F_i X_j}(\omega)$, for every i, j, a procedure analogous to that described in Chapter 2 can be adopted. For the MDOF system used here, it leads to the following equation,

$$\mathcal{L}\big[\mathbf{H}(\omega)\, e^{\mathrm{i}\omega t}\big] = \mathbf{I}\, e^{\mathrm{i}\omega t}\,, \tag{9.71}$$

where the transfer function matrix $\mathbf{H}(\omega) = \big(H_{ij}(\omega)\big)^T$, and $\mathbf{I}$ denotes the $n \times n$ unit matrix. Note that $\mathbf{H}(\omega)$ is defined as the transpose of the matrix $\big(H_{ij}(\omega)\big)$. This is due to our convention that when writing $H_{ij}(\omega)$, the index i indicates the input/load while the index j indicates the output/response. By substituting $\mathbf{X}(t) = \mathbf{H}(\omega)\, e^{i\omega t}$ and $\mathbf{F}(t) = \mathbf{I}\, e^{i\omega t}$ into Eq. (9.70), it is obtained that

$$\mathbf{H}(\omega) = \left[-\omega^2\mathbf{M} + i\omega\mathbf{C} + \mathbf{K} \right]^{-1}. \tag{9.72}$$

9.14.2 Covariance Spectral Matrix

A vector process $\mathbf{X}(t) = \left(X_1(t), \ldots, X_n(t)\right)^T$ is called *stationary* if all component processes are stationary and also jointly stationary. This means that, under suitable conditions, the variance and cross spectra of all component processes exist. They can conveniently be collected in the covariance spectral matrix $\mathbf{S}_{\mathbf{X}}(\omega) = \left(S_{X_i X_j}(\omega)\right)$, where the diagonal terms are just the variance spectra, that is, $S_{X_i X_i}(\omega) = S_{X_i}(\omega)$.

To derive the relation between the spectral matrix of the vector force process and the spectral matrix of the vector response, it is expedient to use a result similar to Eq. (6.35). Assume that $\mathbf{f}_j(t)$, $j = 1, 2, \ldots$ are realizations of the stationary vector process $\mathbf{F}(t)$. Then the following equation applies

$$\mathbf{S}_{\mathbf{F}}(\omega) = \lim_{T\to\infty} \lim_{N\to\infty} \frac{2\pi}{TN} \sum_{j=1}^{N} \mathbf{F}_{j,T}(\omega)\, \mathbf{F}_{j,T}(\omega)^{*T}, \tag{9.73}$$

where $\mathbf{F}_{j,T}(\omega)$ denotes the Fourier transform of $\mathbf{f}_{j,T}(t)$, which equals $\mathbf{f}_j(t)$ for $0 \leq t \leq T$ and zero elsewhere.

For each of the truncated force process realizations $\mathbf{f}_{j,T}(t)$, there will be a corresponding response realization given as follows:

$$\mathbf{x}_{j,T}(t) = \int_0^\infty \mathbf{h}(\tau)\, \mathbf{f}_{j,T}(t-\tau)\, d\tau, \tag{9.74}$$

where $\mathbf{h}(\tau)$ denotes the impulse response matrix, and $\mathbf{H}(\omega)$ and $\mathbf{h}(\tau)$ constitutes a Fourier transform pair. In particular, the following equation obtains

$$\mathbf{H}(\omega) = \int_0^\infty \mathbf{h}(\tau)\, e^{-i\omega\tau}\, d\tau. \tag{9.75}$$

It follows from Eq. (9.74) that the Fourier transform of $\mathbf{x}_{j,T}(t)$ can be calculated as follows:

$$\begin{aligned}
\mathbf{X}_{j,T}(\omega) &= \frac{1}{2\pi}\int_0^\infty \mathbf{x}_{j,T}(t)\, e^{-i\omega t}\, dt = \frac{1}{2\pi}\int_0^\infty \int_0^\infty \mathbf{h}(\tau)\mathbf{f}_{j,T}(t-\tau)\, e^{-i\omega t}\, d\tau\, dt \\
&= \int_0^\infty \mathbf{h}(\tau)\, e^{-i\omega\tau} \frac{1}{2\pi}\int_\tau^\infty \mathbf{f}_{j,T}(t-\tau)\, e^{-i\omega(t-\tau)}\, dt\, d\tau \\
&= \int_0^\infty \mathbf{h}(\tau)\, e^{-i\omega\tau} \frac{1}{2\pi}\int_0^\infty \mathbf{f}_{j,T}(s)\, e^{-i\omega s}\, ds\, d\tau = \mathbf{H}(\omega)\, \mathbf{F}_{j,T}(\omega).
\end{aligned} \tag{9.76}$$

This last equation together with Eq. (9.73) can now be used to establish the following important relation:

$$\begin{aligned}\mathbf{S_X}(\omega) &= \lim_{T\to\infty}\lim_{N\to\infty}\frac{2\pi}{TN}\sum_{j=1}^{N}\mathbf{X}_{j,T}(\omega)\,\mathbf{X}_{j,T}(\omega)^{*T}\\ &= \lim_{T\to\infty}\lim_{N\to\infty}\frac{2\pi}{TN}\sum_{j=1}^{N}\mathbf{H}(\omega)\,\mathbf{F}_{j,T}(\omega)\,\mathbf{F}_{j,T}(\omega)^{*T}\mathbf{H}(\omega)^{*T}\\ &= \mathbf{H}(\omega)\lim_{T\to\infty}\lim_{N\to\infty}\frac{2\pi}{TN}\sum_{j=1}^{N}\mathbf{F}_{j,T}(\omega)\,\mathbf{F}_{j,T}(\omega)^{*T}\mathbf{H}(\omega)^{*T}\\ &= \mathbf{H}(\omega)\,\mathbf{S_F}(\omega)\,\mathbf{H}(\omega)^{*T}. \end{aligned} \tag{9.77}$$

To exemplify the use of Eq. (9.77), the expression for the variance spectrum $S_{X_i}(\omega)$ of a response component $X_i(t)$ is found to be given as

$$S_{X_i}(\omega) = \sum_{k=1}^{n}\sum_{l=1}^{n} H_{ki}(\omega)H_{li}(\omega)^* S_{F_kF_l}(\omega). \tag{9.78}$$

Similarly, the cross-spectrum $S_{X_iX_j}(\omega)$ between the response components $X_i(t)$ and $X_j(t)$ can be expressed as

$$S_{X_iX_j}(\omega) = \sum_{k=1}^{n}\sum_{l=1}^{n} H_{ki}(\omega)H_{lj}(\omega)^* S_{F_kF_l}(\omega). \tag{9.79}$$

Let $\mathbf{m_F} = \left(m_{F_1},\dots,m_{F_n}\right)^T = \mathrm{E}[\mathbf{F}(t)]$ and $\mathbf{m_X} = \left(m_{X_1},\dots,m_{X_n}\right)^T = \mathrm{E}[\mathbf{X}(t)]$. Then, by a derivation similar to that leading to Eq. (9.7), it can be shown that

$$\mathbf{m_X} = \mathbf{H}(0)\,\mathbf{m_F}. \tag{9.80}$$

This implies that the mean value m_{X_i} of the response component $X_i(t)$ is given as

$$m_{X_i} = \sum_{j=1}^{n} H_{ji}(0)m_{F_j}. \tag{9.81}$$

9.14.3 Example – Response Spectra of a 2DOF System

To illustrate the use of some of the material developed in this section, we discuss a 2DOF linear dynamic system as illustrated in Fig. 9.12(a), cf. Section 3.2. In this example it is assumed that $F_1(t)$ and $F_2(t)$ are two independent zero-mean Gaussian processes. From the free body diagram, it follows that the equations of motion of the two bodies of mass m_1 and m_2 become

$$m_1\,\ddot{Y}_1 + c_1(\dot{Y}_1 - \dot{Y}_2) + k_1(Y_1 - Y_2) = F_1(t) \tag{a}$$

and

$$m_2\ddot{Y}_2 + c_1(\dot{Y}_2 - \dot{Y}_1) + c_2\dot{Y}_2 + k_1(Y_2 - Y_1) + k_2Y_2 = F_2(t). \tag{b}$$

The response processes $Y_1(t)$ and $Y_2(t)$ become zero-mean jointly stationary Gaussian processes.

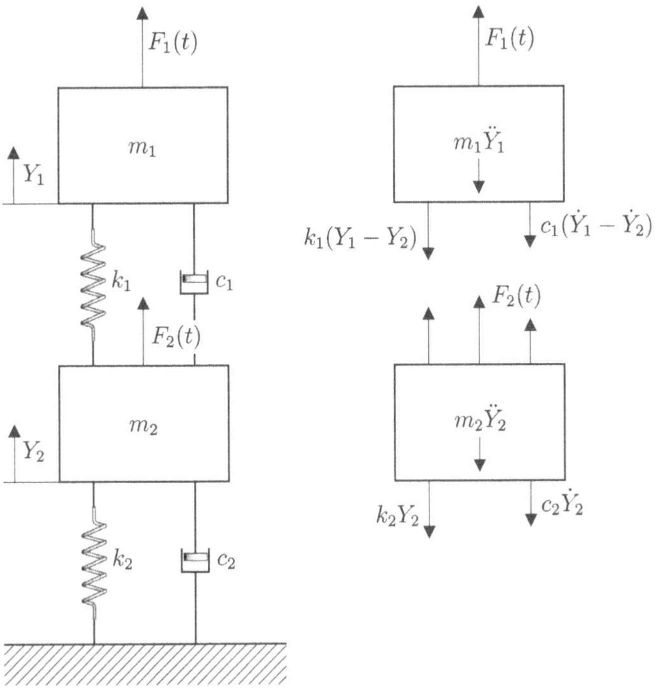

(a) 2DOF model (b) Free body diagram

Figure 9.12. Two degrees of freedom system.

It is now assumed that the issue at hand is a concern about the relative displacement response $Y_2 - Y_1$, which is investigated in terms of its variance. To focus on this response, we introduce the transformed vector process $\mathbf{X}(t) = \left(X_1(t), X_2(t)\right)^T = \left(Y_1(t), Y_2(t) - Y_1(t)\right)^T$. In terms of the transformed processes, the equations of motion are as follows:

$$m_1 \ddot{X}_1 - c_1\dot{X}_2 - k_1 X_2 = F_1(t)\,, \tag{c}$$

and

$$m_2\ddot{X}_1 + m_2\ddot{X}_2 + c_2\dot{X}_1 + (c_1 + c_2)\dot{X}_2 + k_2X_1 + (k_1 + k_2)X_2 = F_2(t)\,. \tag{d}$$

Written in matrix form,

$$\begin{bmatrix} m_1 & 0 \\ m_2 & m_2 \end{bmatrix}\ddot{\mathbf{X}} + \begin{bmatrix} 0 & -c_1 \\ c_2 & c_1 + c_2 \end{bmatrix}\dot{\mathbf{X}} + \begin{bmatrix} 0 & -k_1 \\ k_2 & k_1 + k_2 \end{bmatrix}\mathbf{X} = \mathbf{F}(t)\,. \tag{e}$$

From Eq. (9.72), it follows that the transfer function matrix is determined by the equation,

$$\mathbf{H}(\omega) = \left[-\omega^2\begin{bmatrix} m_1 & 0 \\ m_2 & m_2 \end{bmatrix} + \mathrm{i}\omega\begin{bmatrix} 0 & -c_1 \\ c_2 & c_1 + c_2 \end{bmatrix} + \begin{bmatrix} 0 & -k_1 \\ k_2 & k_1 + k_2 \end{bmatrix}\right]^{-1}\,. \tag{f}$$

By calculating the inverse matrix on the rhs of this equation, it is obtained that

$$\mathbf{H}(\omega) = \begin{bmatrix} H_{11}(\omega) & H_{21}(\omega) \\ H_{12}(\omega) & H_{22}(\omega) \end{bmatrix} = \frac{1}{\Delta(\omega)}\begin{bmatrix} \Delta_{11}(\omega) & \Delta_{12}(\omega) \\ \Delta_{21}(\omega) & \Delta_{22}(\omega) \end{bmatrix}, \tag{g}$$

where

$$\Delta(\omega) = \omega^4 m_1 m_2 - i\omega^3 \left((m_1 + m_2)c_1 + m_1 c_2\right)$$
$$- \omega^2 \left((m_1 + m_2)k_1 + m_1 k_2 + c_1 c_2\right) + i\omega\, c_2 k_1 + k_1 k_2 \,, \tag{h}$$

$$\Delta_{11}(\omega) = -\omega^2 m_2 + i\omega\,(c_1 + c_2) + k_1 + k_2 \,, \qquad \Delta_{22}(\omega) = -\omega^2 m_1 \,, \tag{i}$$

$$\Delta_{12}(\omega) = \omega^2 m_2 - i\omega\, c_2 - k_2 \,, \qquad \Delta_{21}(\omega) = i\omega\, c_1 + k_1 \,. \tag{j}$$

Using Eq. (9.78) we can determine the expression for the variance spectrum of $X_2(t)$:

$$S_{X_2}(\omega) = \sum_{k=1}^{2} H_{k2}(\omega) H_{k2}(\omega)^* S_{F_k}(\omega)$$
$$= |H_{12}(\omega)|^2 S_{F_1}(\omega) + |H_{22}(\omega)|^2 S_{F_2}(\omega) \,, \tag{k}$$

where

$$H_{12}(\omega) = \frac{\Delta_{21}(\omega)}{\Delta(\omega)} \,, \qquad H_{22}(\omega) = \frac{\Delta_{22}(\omega)}{\Delta(\omega)} \,. \tag{l}$$

Note that the formula for the response spectrum of $X_2(t)$ does not contain any cross-spectra between $F_1(t)$ and $F_2(t)$. This is due to the assumed independence.

The variance of the response $X_2(t) = Y_2(t) - Y_1(t)$ can now be calculated as follows:

$$\mathrm{Var}[X_2(t)] = \int_{-\infty}^{\infty} \left|\frac{\Delta_{21}(\omega)}{\Delta(\omega)}\right|^2 S_{F_1}(\omega)\, d\omega + \int_{-\infty}^{\infty} \left|\frac{\Delta_{22}(\omega)}{\Delta(\omega)}\right|^2 S_{F_2}(\omega)\, d\omega \,, \tag{m}$$

which, in general, would have to be calculated numerically. If the force spectra could be approximated by white noise spectra, e.g., for sufficiently low damping, then the formulas in Appendix A could be used to find an explicit formula for the variance.

10 Response Statistics

10.1 Introduction

In Chapter 9 we derive results that make it possible for us to calculate the mean value and the standard deviation of the response of a linear, time-invariant system if we know the load spectrum and the transfer function of the system. Even if such information can be important enough, it is in many cases of limited value. In connection with the design of a structure to withstand wave loads, it would be necessary to estimate the largest response and the associated stresses or strains in the structure over its specified lifetime. Such estimates cannot generally be calculated on the basis of the mean value and the standard deviation, or for that matter, on the basis of the response spectrum.

In this chapter, we provide an introduction to the calculation of extreme responses and response statistics. "Extreme" here means "the largest," interpreted in a way that follows from the context. The background for this is that collapse of a structure, or part of a structure, is often assumed to take place either by first time exceedance of a capacity limit, or by fatigue, which is caused by accumulated damage due to repeated stress cycles at relatively moderate stress levels. We see that an important key to being able to handle these problems in practice lies in the calculation of the average number of times that the response process crosses a particular level during a given time interval.

Although the methods described in this chapter have been used extensively over several decades for prediction of the response statistics of marine structures, the prerequisites for their application are often not satisfied, not even approximately. Under such circumstances, the Monte Carlo simulation method may offer a viable alternative for prediction of response statistics. The Monte Carlo approach is described in Chapter 16.

10.2 Average Rate of Level Crossings

Let us assume that the forces at a given location in a structure can be modeled as a stationary stochastic process $X(t)$. We now want to calculate how often an arbitrary realization of $X(t)$ can be expected to exceed a given force level a. An equivalent formulation of the same problem is the following: what is the average number of

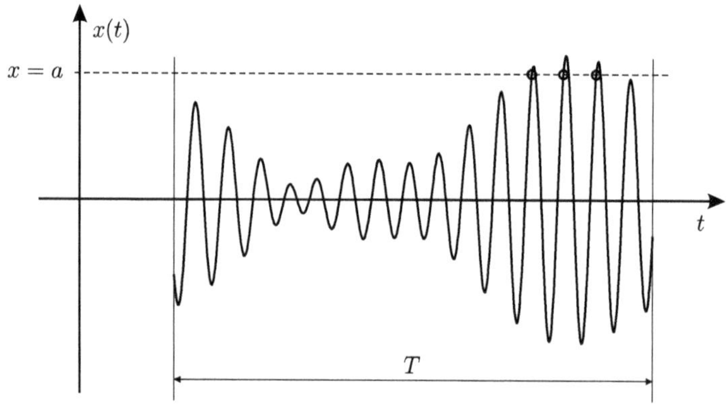

Figure 10.1. A realization of a narrow banded process with three upcrossings of the level a during time T.

a-upcrossings per unit time by $X(t)$? An a-upcrossing means that the level a is exceeded with positive slope. Figure 10.1 shows part of a realization of a narrow banded process, where there are three upcrossings of the indicated level a. Each upcrossing is marked with a small circle in Fig. 10.1.

Let $N^+(a, \Delta t)$ denote the random number of times that $X(t)$ upcrosses the level a during the time interval $(t, t + \Delta t)$. We now assume that $X(t)$ is differentiable and that the realizations of $\dot{X}(t)$ are continuous. This means that $x(t + \delta) \approx x(t) + \dot{x}(t)\,\delta$ for $0 \leq \delta \leq \Delta t$ (Δt small) for any realization $x(t)$ of $X(t)$. In other words, any realization $x(t)$ can be approximated by a straight line in the interval $(t, t + \Delta t)$. This implies that $x(t)$ crosses the level a at most once in this interval, see Fig. 10.2. The conditions for one upcrossing of the level a in the interval $(t, t + \Delta t)$ then become

$$x(t) \leq a \tag{10.1}$$

and

$$x(t + \Delta t) \approx x(t) + \dot{x}(t)\,\Delta t > a. \tag{10.2}$$

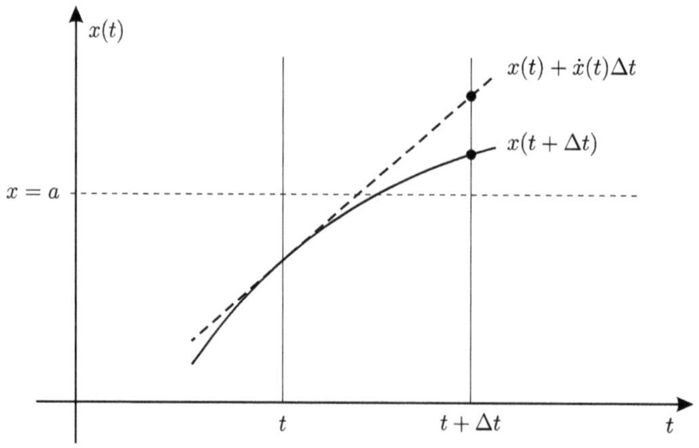

Figure 10.2. Local approximation by a straight line at an upcrossing.

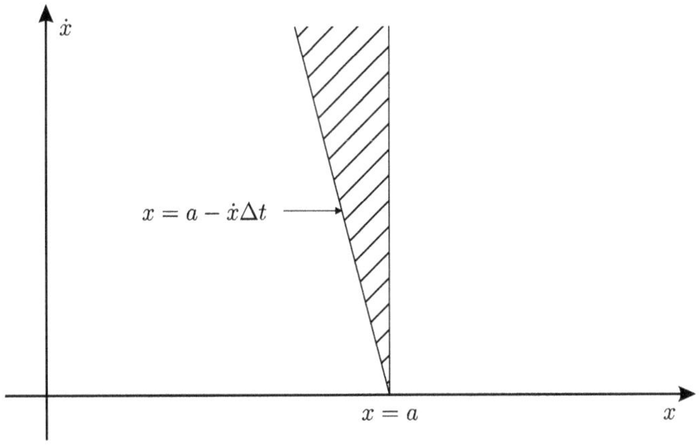

Figure 10.3. Domain of integration in the $(x, \dot{x})$-plane which gives an upcrossing of the level a.

To satisfy Eqs. (10.1) and (10.2), it is clearly necessary that $\dot{x}(t) > 0$. We can therefore write the conditions for one upcrossing in the following way:

$$a - \dot{x}(t)\,\Delta t < x(t) \leq a \tag{10.3}$$

and

$$\dot{x}(t) > 0\,. \tag{10.4}$$

This region in the $(x, \dot{x})$-plane is shown in Fig. 10.3. It is seen that to calculate the probability for an upcrossing, we need the joint PDF $f_{X(t)\dot{X}(t)}(x, \dot{x})$ of $X(t)$ and $\dot{X}(t)$. The way we define stationarity, $f_{X(t)\dot{X}(t)}(x, \dot{x})$ is not automatically independent of t. However, this is often the case, so we assume this to be true and write simply $f_{X\dot{X}}(x, \dot{x})$ to indicate this independence. From Eqs. (10.3) and (10.4), we obtain that

$$\text{Prob}\{N^+(a, \Delta t) = 1\} = \int_0^\infty \int_{a-\dot{x}\Delta t}^a f_{X\dot{X}}(x, \dot{x})\,dx\,d\dot{x}. \tag{10.5}$$

For small Δt,

$$\int_{a-\dot{x}\Delta t}^a f_{X\dot{X}}(x, \dot{x})\,dx = \dot{x}\,\Delta t\; f_{X\dot{X}}(a, \dot{x}), \tag{10.6}$$

provided that $f_{X\dot{X}}(x, \dot{x})$ is continuous. This implies that

$$\text{Prob}\{N^+(a, \Delta t) = 1\} = \Delta t \int_0^\infty \dot{x}\; f_{X\dot{X}}(a, \dot{x})\,d\dot{x}. \tag{10.7}$$

Because we assume that the realizations can be approximated locally by a straight line, then $p_n = \text{Prob}\{N^+(a, \Delta t) = n\}$ is negligible for $n = 2, 3, \ldots$ compared to p_1 for sufficiently small Δt. It follows from Eq. (5.7), which is correct also for $n = \infty$, that

$$\begin{aligned} \text{E}[N^+(a, \Delta t)] &= \sum_{n=0}^\infty n\,p_n \approx 0 \cdot p_0 + 1 \cdot p_1 + 2 \cdot 0 + 3 \cdot 0 + \ldots \\ &= p_1 = \Delta t \int_0^\infty \dot{x}\; f_{X\dot{X}}(a, \dot{x})\,d\dot{x}. \end{aligned} \tag{10.8}$$

The expected (or average) number of a-upcrossings per unit of time, which we denote by $\nu_X^+(a)$, is then given by the following expression:

$$\nu_X^+(a) = \lim_{\Delta t \to 0} \frac{1}{\Delta t} \mathrm{E}[N^+(a, \Delta t)] = \int_0^\infty \dot{x}\, f_{X\dot{X}}(a, \dot{x})\, d\dot{x}. \tag{10.9}$$

$\nu_X^+(a)$ is referred to by several names. In this book we use mostly average (or mean) (a-)upcrossing rate and average (or mean) (a-)upcrossing frequency. It is seen that $\nu_X^+(a)$ depends only on the level a. Because $\nu_X^+(a)$ is independent of t, $\mathrm{E}[N^+(a, T)] = \nu_X^+(a)\, T$ for any value of T, cf. Eq. (10.8), which was derived under the assumption that Δt is small.

Equation (10.9) is a useful formula. It is often referred to as the Rice formula after its creator Rice (1954). To get a feeling for the physical content of the Rice formula, one may note that the rhs of Eq. (10.9) expresses a sort of expectation value of the slope at upcrossing of the level a coupled with the probability of being at that level. The greater the average slope at a given level, the more often an arbitrary realization will upcross that level. In other words, large average positive slope of the time histories implies shorter cycles and thereby more frequent level crossings. At the same time one must expect that the number of upcrossings of a given level is coupled to the probability of reaching that level. The Rice formula therefore appears to have a fairly plausible form when it is subjected to closer scrutiny.

In same way that we define an a-upcrossing, we can define an a-downcrossing. An a-downcrossing implies that the level a is passed with negative slope. The expected number of a-downcrossings per unit of time of a stationary process $X(t)$ is denoted by $\nu_X^-(a)$. To derive the formula for $\nu_X^-(a)$, we make the following observation: it is obvious that an a-downcrossing for the process $X(t)$ is equivalent with a $(-a)$-upcrossing for the process $Y(t) = -X(t)$. Hence $\nu_X^-(a) = \nu_Y^+(-a)$. From the relation $Y(t) = -X(t)$ follows that $f_{Y\dot{Y}}(y, \dot{y}) = f_{X\dot{X}}(-y, -\dot{y})$. This, together with Eq. (10.9), gives

$$\begin{aligned}\nu_X^-(a) = \nu_Y^+(-a) &= \int_0^\infty \dot{y}\, f_{Y\dot{Y}}(-a, \dot{y})\, d\dot{y} = \int_0^\infty \dot{y}\, f_{X\dot{X}}(a, -\dot{y})\, d\dot{y} \\ &= \int_0^{-\infty} \dot{x}\, f_{X\dot{X}}(a, \dot{x})\, d\dot{x} = -\int_{-\infty}^0 \dot{x}\, f_{X\dot{X}}(a, \dot{x})\, d\dot{x}.\end{aligned} \tag{10.10}$$

A common way of rewriting Eq. (10.10) is

$$\nu_X^-(a) = \int_{-\infty}^0 |\dot{x}|\, f_{X\dot{X}}(a, \dot{x})\, d\dot{x}. \tag{10.11}$$

An a-crossing is either an a-upcrossing or an a-downcrossing. The expected number of a-crossings per unit of time, denoted by $\nu_X(a)$, must satisfy the equation $\nu_X(a) = \nu_X^+(a) + \nu_X^-(a)$. From Eqs. (10.9) and (10.11), it is then obtained that

$$\nu_X(a) = \int_{-\infty}^\infty |\dot{x}|\, f_{X\dot{X}}(a, \dot{x})\, d\dot{x}. \tag{10.12}$$

For a stationary process, any a-upcrossing must by necessity be followed by an a-downcrossing, and conversely. If that was not the case, the realizations would have a mean drift in the positive or negative direction; that is, the mean value would not

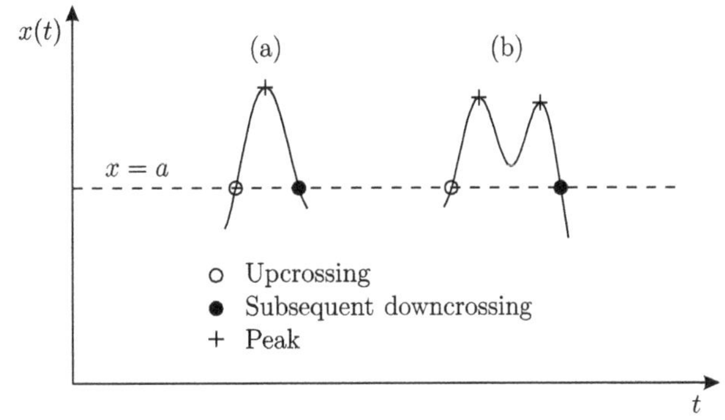

Figure 10.4. (a) Peak between an upcrossing and a subsequent downcrossing of the level a. (b) Two peaks between an upcrossing and a subsequent downcrossing.

be constant. For a stationary process, the following relations apply:

$$\nu_X^-(a) = \nu_X^+(a) = \frac{1}{2}\,\nu_X(a). \tag{10.13}$$

10.3 Statistical Distribution of Peaks of a Narrow-Banded Process

Assume that $X(t)$ is a stationary process with zero mean value, which is also narrow banded. What characterizes a realization of a narrow banded process is that the amplitude and length (period) of subsequent cycles vary slowly, as illustrated in Fig. 10.1. This implies that almost invariably, there is only one maximum or peak value between an upcrossing and a subsequent downcrossing of any level a, see Fig. 10.4. Because we assumed zero mean value, the mean number of peaks per unit of time will therefore be approximately equal to the mean rate of zero-upcrossings, that is, $\nu_X^+(0)$.

Let X_p denote the size or height of an arbitrary peak of $X(t)$. X_p becomes a random variable. The probability distribution of X_p for a narrow banded process $X(t)$ with zero mean value is now defined as

$$\text{Prob}\{X_p > a\} = \frac{\nu_X^+(a)}{\nu_X^+(0)}, \quad (a \geq 0). \tag{10.14}$$

The CDF $F_{X_p}(a)$ is then given as

$$F_{X_p}(a) = 1 - \frac{\nu_X^+(a)}{\nu_X^+(0)}, \quad a \geq 0, \tag{10.15}$$

while $F_{X_p}(a) = 0$ for $a < 0$.

If $m_X \neq 0$, we get the definition $F_{X_p}(a) = 1 - \nu_X^+(a)/\nu_X^+(m_X)$ for $a \geq m_X$ and $F_{X_p}(a) = 0$ for $a < m_X$. Because $\nu_X^+(a)$ is assumed to equal the mean number of peaks per unit of time above the level a, then clearly $\nu_X^+(a)$ will decrease with increasing a. For all processes of interest to us, we may assume that $\nu_X^+(a) \to 0$ when $a \to \infty$. This implies that $F_{X_p}(a)$ gets the properties that a CDF must have; that is, $F_{X_p}(a)$ is a nondecreasing function for increasing a, $F_{X_p}(a) \to 0$ when $a \to -\infty$ and $F_{X_p}(a) \to 1$ when $a \to \infty$. We tacitly assume that $\nu_X^+(a)$ gets its largest value when $a = m_X$, which is usually the case. As we see, it is true for a Gaussian process. Note that it must also apply to processes that are characterized by having only one maximum between an upcrossing and a subsequent downcrossing of the mean value level, that is, for infinitely narrow banded processes. The assumption made is therefore quite reasonable.

If $\nu_X^+(a)$ can be differentiated with respect to a, we get the PDF for peaks, assuming that $m_X = 0$,

$$f_{X_p}(a) = -\frac{1}{\nu_X^+(0)} \frac{d\nu_X^+(a)}{da}, \quad a \geq 0, \tag{10.16}$$

while $f_{X_p}(a) = 0$ for $a < 0$. Equations (10.15) and (10.16) are sometimes called the "peak formulas." We emphasize that they only apply to narrow banded processes.

For the sake of completeness, it should be mentioned that the peak distributions can also be defined in the general case. However, this would require that the exact mean number of peaks per unit of time were used in definitions. The expressions we would then find are of limited practical use because they are difficult, if not impossible, to calculate. The simplifications that are sometimes introduced to make the expressions amenable to calculations are often, in fact, ill defined.

10.4 Average Upcrossing Frequency and Statistical Distribution of Peaks of a Gaussian Process

A random variable X is normally distributed if the PDF of X is given as

$$f_X(x) = \frac{1}{\sqrt{2\pi}\sigma_X} \exp\left\{-\frac{1}{2}\left(\frac{x - m_X}{\sigma_X}\right)^2\right\} \tag{10.17}$$

for $\sigma_X > 0$. If $\sigma_X = 0$, then $X = m_X$. In this case we may consider X as a degenerate normal variable.

Two random variables X and Y are called *jointly normally distributed* if the joint PDF of X and Y is given by the equation

$$f_{XY}(x, y) = \frac{1}{2\pi\,\sigma_X\sigma_Y\sqrt{1-\rho_{XY}^2}} \exp\left\{-\frac{1}{2(1-\rho_{XY}^2)}\left[\left(\frac{x-m_X}{\sigma_X}\right)^2 - 2\rho_{XY}\left(\frac{x-m_X}{\sigma_X}\right)\left(\frac{y-m_Y}{\sigma_Y}\right) + \left(\frac{y-m_Y}{\sigma_Y}\right)^2\right]\right\} \tag{10.18}$$

where ρ_{XY} is the correlation coefficient for X and Y, cf. Eq. (5.32). To be precise, in Eq. (10.18) it is assumed that $|\rho_{XY}| < 1$ and that $\sigma_X > 0$, $\sigma_Y > 0$.

If X and Y are uncorrelated, $\rho_{XY} = 0$ by definition. For this case, Eq. (10.18) assumes the form

$$\begin{aligned} f_{XY}(x, y) &= \frac{1}{2\pi\,\sigma_X\sigma_Y} \exp\left\{-\frac{1}{2}\left[\left(\frac{x-m_X}{\sigma_X}\right)^2 + \left(\frac{y-m_Y}{\sigma_Y}\right)^2\right]\right\} \\ &= \frac{1}{\sqrt{2\pi}\,\sigma_X} \exp\left\{-\frac{1}{2}\left(\frac{x-m_X}{\sigma_X}\right)^2\right\} \cdot \frac{1}{\sqrt{2\pi}\,\sigma_Y} \exp\left\{-\frac{1}{2}\left(\frac{y-m_Y}{\sigma_Y}\right)^2\right\} \\ &= f_X(x) \cdot f_Y(y). \end{aligned} \tag{10.19}$$

According to the condition in Eq. (5.22), X and Y are (statistically) independent variables. We have therefore shown that two uncorrelated normally distributed (real) variables are automatically independent. However, one should make a note of the fact that this does not apply to other types of random variables.

A stochastic process $X(t)$ is called *Gaussian* or *normally distributed* if the random variable $Z = \sum_{j=1}^{n} c_j\, X(t_j)$ is normally distributed for any (arbitrary) choice of $n(=1, 2, \ldots)$, constants $c_1, \ldots, c_n$, and times $t_1, \ldots, t_n$. If $Y(t)$ is the response of a linear, time-invariant system where the input process $F(t)$ is Gaussian, $F(t)$ and $Y(t)$ are jointly normally distributed variables for any time t, and $Y(t)$ also becomes a Gaussian process.

If $X(t)$ is a stationary and differentiable Gaussian process, $X(t)$ and $\dot{X}(t)$ are jointly normally distributed for any t. In Section 9.9, it was shown that $\rho_{X(t)\,\dot{X}(t)} = 0$; hence, $X(t)$ and $\dot{X}(t)$ are independent variables, and we have that

$$\begin{aligned} f_{X(t)\dot{X}(t)}(x, \dot{x}) &= f_{X(t)}(x) \cdot f_{\dot{X}(t)}(\dot{x}) \\ &= \frac{1}{2\pi\,\sigma_X\sigma_{\dot{X}}} \exp\left\{-\frac{1}{2}\left[\left(\frac{x - m_X}{\sigma_X}\right)^2 + \left(\frac{\dot{x}}{\sigma_{\dot{X}}}\right)^2\right]\right\} \end{aligned} \tag{10.20}$$

where we have used the fact that $m_{\dot{X}} = 0$ for the derivative $\dot{X}(t)$ of any stationary and differentiable process $X(t)$. This can be seen as follows:

$$\mathrm{E}[\dot{X}(t)] = \lim_{N\to\infty} \frac{1}{N} \sum_{j=1}^{N} \dot{x}_j(t) = \frac{d}{dt} \lim_{N\to\infty} \frac{1}{N} \sum_{j=1}^{N} x_j(t) = \frac{d}{dt}\,\mathrm{E}[X(t)]. \tag{10.21}$$

Because $m_X = \mathrm{E}[X(t)] =$ constant for a stationary process, it follows immediately that $m_{\dot{X}} = 0$. Note also that $f_{X(t)\dot{X}(t)}(x, \dot{x})$ is independent of t.

Let us calculate the mean upcrossing rate $\nu_X^+(a)$. Substituting from Eq. (10.20) into Eq. (10.9) gives

$$\begin{aligned} \nu_X^+(a) &= \int_0^\infty \frac{\dot{x}}{2\pi\,\sigma_X\sigma_{\dot{X}}} \exp\left\{-\frac{1}{2}\left[\left(\frac{a - m_X}{\sigma_X}\right)^2 + \left(\frac{\dot{x}}{\sigma_{\dot{X}}}\right)^2\right]\right\} d\dot{x} \\ &= \frac{1}{2\pi}\frac{\sigma_{\dot{X}}}{\sigma_X} \exp\left\{-\frac{1}{2}\left(\frac{a - m_X}{\sigma_X}\right)^2\right\} \int_0^\infty \frac{\dot{x}}{\sigma_{\dot{X}}} \exp\left\{-\frac{1}{2}\left(\frac{\dot{x}}{\sigma_{\dot{X}}}\right)^2\right\} d\left(\frac{\dot{x}}{\sigma_{\dot{X}}}\right) \\ &= \frac{1}{2\pi}\frac{\sigma_{\dot{X}}}{\sigma_X} \exp\left\{-\frac{1}{2}\left(\frac{a - m_X}{\sigma_X}\right)^2\right\}. \end{aligned} \tag{10.22}$$

It is seen that $\nu_X^+(a)$ decreases rapidly (with σ_X as reference scale) at each side of the mean value, where it assumes its largest value

$$\nu_X^+(m_X) = \frac{1}{2\pi}\frac{\sigma_{\dot{X}}}{\sigma_X}. \tag{10.23}$$

In many situations one would prefer to define the origin so that $m_X = 0$. The expression on the rhs of Eq. (10.23) is therefore often referred to as the mean zero upcrossing rate, under the tacit assumption that $m_X = 0$. Another corresponding parameter that is often met in the literature is the mean zero-crossing period T_z, which is defined by

$$T_z = (\nu_X^+(0))^{-1} = 2\pi\,\frac{\sigma_X}{\sigma_{\dot{X}}}. \tag{10.24}$$

For a stationary Gaussian process $X(t)$ with mean value zero, $\nu_X^+(a)$ is completely determined by the two standard deviations σ_X and $\sigma_{\dot{X}}$. If the variance spectrum $S_X(\omega)$

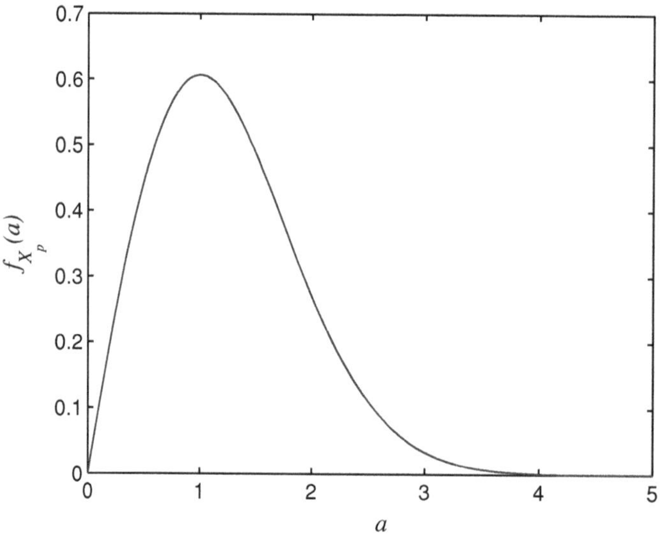

Figure 10.5. The PDF $f_{X_p}(a)$ with $\sigma_X = 1.0$ of Rayleigh distributed peaks.

of $X(t)$ is known, σ_X and $\sigma_{\dot{X}}$ can be calculated by using the formulas

$$\sigma_X^2 = \int_{-\infty}^{\infty} S_X(\omega)\, d\omega \tag{10.25}$$

and

$$\sigma_{\dot{X}}^2 = \int_{-\infty}^{\infty} \omega^2\, S_X(\omega)\, d\omega. \tag{10.26}$$

If $X(t)$ is also assumed to be narrow banded, we may calculate the PDF $f_{X_p}(a)$ of the peaks of $X(t)$. From Eqs. (10.16) and (10.22), we find that ($m_X = 0$)

$$f_{X_p}(a) = \begin{cases} \frac{a}{\sigma_X^2} \exp\left\{-\frac{a^2}{2\sigma_X^2}\right\}, & a \geq 0; \\ 0, & a < 0. \end{cases} \tag{10.27}$$

A PDF of this type is called a *Rayleigh density*, and X_p becomes a Rayleigh distributed variable. An example of $f_{X_p}(a)$ is shown in Fig. 10.5.

Gaussian processes have great practical significance. This is primarily due to the following two reasons. In many cases, important physical phenomena that give rise to loads on structures can be modeled as Gaussian processes with a reasonable degree of accuracy. Moreover, weakly damped structures usually make the response more Gaussian than the load. In addition comes the fact that a Gaussian process is particularly amenable to analytical treatment.

10.5 Extreme Values

In connection with a structure and its function, one would often distinguish between two types of structural failure: reversible and irreversible.

Reversible failure. The response is too large compared with explicitly stated operational limit values, but without causing damage to the structure. Example: The dynamic response at the top of a high-rise building gets so large during certain wind conditions that people will start to feel uncomfortable. Such limit values are usually called *serviceability limit states*.

Irreversible failure. The structure is weakened or collapses due to buckling, ruptures, or corrosion. Such failure modes can often be described as an exceedance of an *ultimate limit state*. Similarly, a failure due to fatigue damage will relate to a *fatigue limit state*.

At the beginning of this chapter we put up two hypotheses for structural failure (primarily irreversible): (1) failure due to first exceedance of a critical level (a capacity limit), and (2) fatigue failure. In the following, we are mainly concerned with the first type of failures.

Our starting point is that we have modeled a response quantity as a stationary stochastic process. We now want to calculate the probability distribution of the largest value of the response process $X(t)$ during a specified time period. We also want to determine the probability distribution of the time to the first exceedance of a given response level. These problems are difficult to solve exactly, but by simplifying somewhat, one may often find reasonably accurate approximate solutions.

Let us denote the largest value that $X(t)$ assumes during the time T by $M(T)$. That is, $M(T) = \max\{X(t); 0 \leq t \leq T\}$. Also, let $\Theta(a)$ denote the time to the first exceedance of the level a. $M(T)$ and $\Theta(a)$ are random variables. If it is convenient to emphasize that $M(T)$ and $\Theta(a)$ refer to the process $X(t)$, we write $M_X(T)$ and $\Theta_X(a)$. We have that

$$\text{Prob}\{M(T) \leq a\} = \text{Prob}\{\Theta(a) > T\} \tag{10.28}$$

because both events $\{M(T) \leq a\}$ and $\{\Theta(a) > T\}$ express the same, namely, that there are no exceedances of the level a during time T. Let us call this event $\mathcal{E}$. Then $\mathcal{E} = \{X(t) \leq a \text{ for all } t \in (0, T)\}$, but this event can also be expressed as $\mathcal{E} = \{X(0) \leq a \text{ and } N^+(a, T) = 0\}$. This is so because if $X(0) \leq a$ and there are no subsequent upcrossings of a, there can be no exceedances. Hence, $\text{Prob}\{\mathcal{E}\} = \text{Prob}\{X(0) \leq a \text{ and } N^+(a, T) = 0\} \to \text{Prob}\{N^+(a, T) = 0\}$ when $a \to \infty$ because of the law of marginal probability, cf. Eq. (5.19). When we write $a \to \infty$, it means that we let a assume values that are large compared to typical values for the process considered; it should not be strictly interpreted as meaning that a grows beyond all limits. Because the extreme values in most cases are much larger than the typical values, we introduce the approximation $\text{Prob}\{\mathcal{E}\} = \text{Prob}\{N^+(a, T) = 0\}$.

To determine $\text{Prob}\{N^+(a, T) = 0\}$, we shall introduce the following simplifying assumption: upcrossings of high levels are statistically independent events. If the process $X(t)$ is not too narrow banded, this is a reasonable approximation. This simplification implies that the random number of upcrossings in an arbitrary time interval of length T is Poisson distributed with parameter $\text{E}[N^+(a, T)] = \nu_X^+(a)\,T$. In particular, this leads to the result

$$\text{Prob}\{N^+(a, T) = 0\} = \exp\{-\nu_X^+(a)\,T\}. \tag{10.29}$$

A derivation of Eq. (10.29) is given in Appendix B.

From Eq. (10.29), we then obtain that

$$F_{M(T)}(a) = \text{Prob}\{M(T) \leq a\} = \exp\{-\nu_X^+(a)\,T\}, \quad (a \to \infty), \tag{10.30}$$

and

$$F_{\Theta(a)}(\theta) = 1 - \text{Prob}\{\Theta(a) > \theta\} = 1 - \exp\{-\nu_X^+(a)\,\theta\}, \quad (a \to \infty), \tag{10.31}$$

Often, $\nu_X^+(a)\,T \ll 1$ for a relevant level a and time interval $(0, T)$, such that the probability of exceedance of a during the time T can be approximated as

$$\text{Prob}\{\text{Exceedance}\} = F_{\Theta(a)}(T) \approx \nu_X^+(a)\,T \tag{10.32}$$

because $e^x \approx 1 + x$ for $|x| \ll 1$.

We succeeded in deriving an approximate expression for the CDF of the extreme value $M(T)$ and for the time to first passage $\Theta(a)$ for large values of a, and we saw that these distributions are determined by the mean level-crossing rate $\nu_X^+(a)$. The only significant simplification that has been adopted is the assumption that upcrossings of high levels are independent. Regarding the response of a lightly damped structure, the response maxima will have a tendency to occur in clumps. In particular, large peaks will tend to occur simultaneously as illustrated in Fig. 10.1. The assumption about independent upcrossings will then tend to be less valid. In such cases, Eqs. (10.30) and (10.31) may give significant deviations from the correct values, but always on the safe side in the sense that Eq. (10.30), for instance, leads to larger extreme value estimates than the correct ones.

In some situations, the relevant extreme values will be connected to the smallest or minimum values of a process. Such a case can, however, be easily recast to a study of maximum values by observing that $\min\{X(t); 0 \le t \le T\} = -\max\{-X(t); 0 \le t \le T\}$.

10.6 Classical Extreme Value Theory

The derivation of an approximate extreme value distribution as given in the previous section is a typical engineering approach, and it leads to a flexible and often accurate approximation. The approach used in classical extreme value theory is different. It starts by looking at a sequence of independent and identically distributed (iid) random variables $X_1, X_2, \ldots$ with common distribution function $F_X(x)$. The extreme value of a finite number $X_1, \ldots, X_n$ is then $M_n = \max\{X_1, \ldots, X_n\}$. The distribution of M_n can be easily derived as

$$\begin{aligned} F_{M_n}(x) &= \text{Prob}(M_n \le x) = \text{Prob}(X_1 \le x, \ldots, X_n \le x) \\ &= \text{Prob}(X_1 \le x) \cdot \ldots \cdot \text{Prob}(X_n \le x) = \left(F_X(x)\right)^n. \end{aligned} \tag{10.33}$$

This relation is not very helpful in practice. Small discrepancies in the estimates of $F_X(x)$ can lead to substantial discrepancies in the values of $\left(F_X(x)\right)^n$. In classical extreme value theory, one proceeds by studying the behaviour of $\left(F_X(x)\right)^n$ as $n \to \infty$, but with a twist. Obviously, for any x such that $F_X(x) < 1$, $\left(F_X(x)\right)^n \to 0$ as $n \to \infty$. This necessitates a rescaling. Specifically, instead of studying M_n, one introduces a renormalized version of M_n:

$$M_n^* = \frac{M_n - b_n}{a_n} \tag{10.34}$$

for suitable sequences of constants $a_n > 0$ and b_n that are chosen to stabilize the location and scale of M_n^* as $n \to \infty$. It is then proven that there are, in fact, only three types of limiting distributions for this renormalized M_n^*. This is the famous extremal types theorem (Leadbetter et al., 1983), which can be expressed as follows.

If there exist sequences of constants $a_n > 0$ and b_n such that

$$\text{Prob}\Big(\frac{M_n - b_n}{a_n} \le x\Big) \to G(x), \quad n \to \infty, \tag{10.35}$$

where $G(x)$ is a nondegenerate distribution function, then $G(x)$ belongs to one of the following three families,

$$\text{I} \quad G(x) = \exp\Big\{-\exp\Big[-\Big(\frac{x-b}{a}\Big)\Big]\Big\}, \quad -\infty < x < \infty; \tag{10.36}$$

$$\text{II} \quad G(x) = \begin{cases} 0 & , \quad x \le b, \\ \exp\Big\{-\Big(\frac{x-b}{a}\Big)^{-\alpha}\Big\} & , \quad x > b; \end{cases} \tag{10.37}$$

$$\text{III} \quad G(x) = \begin{cases} \exp\Big\{-\Big(\frac{b-x}{a}\Big)^{\alpha}\Big\} & , \quad x < b, \\ 1 & , \quad x \ge b; \end{cases} \tag{10.38}$$

for parameters $a > 0$, b and for families II and III, $\alpha > 0$. These three types of extreme value distributions are also commonly referred to as Gumbel, Fréchet, and Weibull, respectively. Note that the Weibull distribution given here is not the same as the commonly known Weibull distribution, which corresponds to the type III extreme value distribution for minima. Also, carefully note that even if the Weibull distribution is the only type of extreme value distribution with a finite upper limit on its values, this does not mean that extremes of limited data must follow this distribution. For such data, it may very well happen that the rescaling constants $a_n \to 0$ as n increases.

It may be verified that it is, in fact, possible to express all three types of extreme value distributions in a common form, which is known as the generalized extreme value (GEV) distribution. This is achieved as follows:

$$G(x) = \exp\Big\{-\Big[1 + \gamma\Big(\frac{x-\mu}{\sigma}\Big)\Big]_+^{-1/\gamma}\Big\}, \tag{10.39}$$

where the parameters satisfy $-\infty < \mu < \infty$, $\sigma > 0$, $-\infty < \gamma < \infty$, and $[z]_+ = \max(z, 0)$. This distribution has three parameters: a location parameter μ, a scale parameter σ, and a shape parameter γ. The type II distributions correspond to $\gamma > 0$, while type III corresponds to $\gamma < 0$. The case $\gamma = 0$ must be interpreted as a limiting case when $\gamma \to 0$, which leads to the Gumbel distribution:

$$G(x) = \exp\Big\{-\exp\Big[-\Big(\frac{x-\mu}{\sigma}\Big)\Big]\Big\}, \quad -\infty < x < \infty. \tag{10.40}$$

For statistical inference on experimental data, this unified form has the advantage that the data themselves determine which type of distribution is appropriate, thereby avoiding a prior subjective judgment about any specific tail behaviour. The uncertainty in the estimated value of γ is also a reflection of the uncertainty about the correct distribution for the data.

The results cited previously are derived under the assumption of a sequence of iid variables. For most of the applications in this book, we do not consider a sequence of random variables $X_1, X_2, \ldots,$ but rather a stochastic process in continuous time $X(t)$. To fit into the framework discussed in this section, one could envisage sampling the considered process at discrete time points and obtaining a sequence $X_j = X(t_j)$, $j = 1, 2, \ldots$. In practice, this is often done by extracting local peak values from an observed realization of the process. However, in general, the resulting random variables are not independent. Under suitable conditions, the extremal types theorem still holds true. Unfortunately, no satisfactory general theory of extremes is available for the case of continuous time stochastic processes, but some results have been proven (Leadbetter et al., 1983).

In this book, we largely use the approximation to the extreme value distribution derived in the previous section, and in Section 16.6.3, we describe a general method for estimation of extreme values of stochastic processes for which the mean upcrossing rate function can be empirically determined by Monte Carlo simulation, or some other method.

10.7 Extreme Values of Gaussian Processes

Let $X(t)$ be a stationary Gaussian process with a mean level upcrossing rate given by Eq. (10.22). For simplicity, it is assumed that $m_X = 0$. In any case, because changing the mean value is equivalent to a constant shift of all realizations, nothing is lost by this assumption.

The CDF of $M(T)$ (≥ 0) for large values of a is then, according to Eqs. (10.22) and (10.30), given by the expression

$$F_{M(T)}(a) = \exp\Big\{ - \nu_X^+(0)\, T \exp\Big(- \frac{a^2}{2\sigma_X^2}\Big)\Big\}, \quad (a \to \infty), \tag{10.41}$$

where the mean zero upcrossing rate enters, that is,

$$\nu_X^+(0) = \frac{1}{2\pi}\frac{\sigma_{\dot{X}}}{\sigma_X}. \tag{10.42}$$

The PDF of $M(T)$, $f_{M(T)}(a)$, can be calculated from Eq. (10.41) by $f_{M(T)}(a) = dF_{M(T)}(a)/da$, and it is given as follows (for large values of a):

$$f_{M(T)}(a) = \frac{a}{\sigma_X^2}\, \nu_X^+(0)\, T \exp\Big(- \frac{a^2}{2\sigma_X^2}\Big) \exp\Big\{ - \nu_X^+(0)\, T \exp\Big(- \frac{a^2}{2\sigma_X^2}\Big)\Big\}. \tag{10.43}$$

Assuming that Eq. (10.43) is valid for all values of a, Fig. 10.6 shows the PDF of $M(T)$ for various values of $\nu_X^+(0)\, T$.

A quantity of particular interest in connection with design of structures is the level $a = \xi_p = \xi_p(T)$, which with probability p is not exceeded during the time T, that is,

$$F_{M(T)}(\xi_p) = p. \tag{10.44}$$

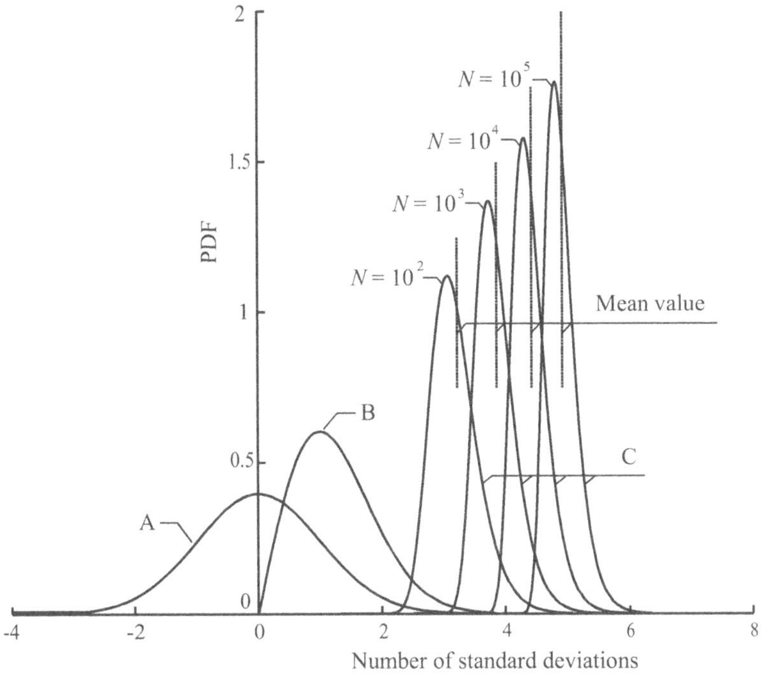

Figure 10.6. Various PDFs for a stationary Gaussian process $X(t)$. A: PDF of $X(t)$. B: PDF of the peaks of $X(t)$ (narrow banded case). C: PDF of $M(T)$ for various values of $N = \nu_X^+(0)\,T$.

Because $p = \exp(\ln p)$, it follows from Eq. (10.41) that

$$\exp\left(-\frac{\xi_p^2}{2\sigma_X^2}\right) = -\frac{\ln p}{\nu_X^+(0)\,T}. \tag{10.45}$$

This leads to the formula

$$\xi_p(T) = \sigma_X \sqrt{2\ln\left(\frac{\nu_X^+(0)\,T}{\ln(1/p)}\right)}. \tag{10.46}$$

This formula can then be used to find the response level that has a probability of 1% of being exceeded ($p = 0.99$) during time T.

The most probable extreme value, denoted by $\hat{\xi} = \hat{\xi}(T)$, is given to good approximation by the formula

$$\hat{\xi} = \sigma_X \sqrt{2\,\ln\left(\nu_X^+(0)\,T\right)} \tag{10.47}$$

$\hat{\xi}$ is the value where the PDF of the extreme value $M(T)$ attains its maximum. It follows that $\hat{\xi} \approx \xi_{0.37}$ because $\ln(1/0.37) \approx 1.0$ ($e^{-1} \approx 0.37$).

Specific values of the quantiles ξ_p/σ_X for various numbers of zero upcrossings are listed in Table 10.1. Apart from providing a set of useful reference values, Table 10.1 also clearly illustrates the slow increase in typical extreme values with increasing time for a stationary Gaussian process.

Another quantity that is often used as a measure of extreme values is the expected largest value during a given time T, that is, $\mathrm{E}[M(T)]$. Let us calculate

Table 10.1. *Table of ξ_p/σ_X-values*

$\nu_X^+(0)\,T$	10	100	1,000	10,000
$p = 0.37$	2.1460	3.0349	3.7169	4.2919
$p = 0.5$	2.3105	3.1533	3.8143	4.3765
$p = 0.9$	3.0176	3.7028	4.2797	4.7876
$p = 0.95$	3.2474	3.8924	4.4448	4.9357
$p = 0.99$	3.7156	4.2908	4.7975	5.2556

this expected value. We have that

$$\mathrm{E}[M(T)] = \int_0^\infty a\,\frac{dF_{M(T)}(a)}{da}\,da. \tag{10.48}$$

It is now convenient to introduce a new integration variable η defined by the equation $F_{M(T)}(a) = e^{-\eta}$, which implicitly defines a as a function of η. Note that $a = 0\,(\infty)$ corresponds to $\eta = \infty\,(0)$. We obtain that

$$\frac{dF_{M(T)}(a)}{da}\,da = \frac{dF_{M(T)}(a)}{d\eta}\,\frac{d\eta}{da}\,da = \frac{dF_{M(T)}(a)}{d\eta}\,d\eta = -e^{-\eta}\,d\eta. \tag{10.49}$$

Substituted into Eq. (10.48), this leads to the equation

$$\mathrm{E}[M(T)] = \int_0^\infty a(\eta)\,e^{-\eta}\,d\eta, \tag{10.50}$$

where $a = a(\eta)$ is a function of η. The way η is defined, $\eta = \nu_X^+(0)\,T\,\exp\{-a^2/(2\sigma_X^2)\}$. Our focus is on large a-values, that is, small η-values, and we obtain by solving with respect to a:

$$\begin{aligned} a &= \sigma_X\sqrt{2\ln(\nu_X^+(0)\,T) - 2\ln\eta} \\ &= \sigma_X\sqrt{2\ln(\nu_X^+(0)\,T)}\,\Big\{1 - \frac{\ln\eta}{\ln(\nu_X^+(0)\,T)}\Big\}^{1/2} \\ &= \sigma_X\sqrt{2\ln(\nu_X^+(0)\,T)}\,\Big\{1 - \frac{\ln\eta}{2\ln(\nu_X^+(0)\,T)} - \frac{(\ln\eta)^2}{8(\ln(\nu_X^+(0)\,T))^2} + \dots\Big\}. \end{aligned} \tag{10.51}$$

It can be shown that that the main contribution to the integral in Eq. (10.50) comes from small η-values, and we find that

$$\begin{aligned} \mathrm{E}[M(T)] &\approx \sigma_X\sqrt{2\ln(\nu_X^+(0)\,T)}\,\Big\{1 - \frac{\int_0^\infty \ln\eta\, e^{-\eta}\,d\eta}{2\ln(\nu_X^+(0)\,T)} - \frac{\int_0^\infty (\ln\eta)^2\,e^{-\eta}\,d\eta}{8(\ln(\nu_X^+(0)\,T))^2} + \dots\Big\} \\ &= \sigma_X\sqrt{2\ln(\nu_X^+(0)\,T)}\,\Big\{1 + \frac{\gamma}{2\ln(\nu_X^+(0)\,T)} - \frac{\frac{\pi^2}{6} + \gamma^2}{8(\ln(\nu_X^+(0)\,T))^2} + \dots\Big\} \end{aligned} \tag{10.52}$$

because $\int_0^\infty \ln\eta\, e^{-\eta}\,d\eta = -\gamma$ and $\int_0^\infty (\ln\eta)^2\,e^{-\eta}\,d\eta = \pi^2/6 + \gamma^2$, where $\gamma = 0.5772...$ denotes Euler's constant. Usually, $\ln(\nu_X^+(0)\,T)$ is sufficiently large to warrant the following approximation

$$\mathrm{E}[M(T)] \approx \sigma_X\Big\{\sqrt{2\ln(\nu_X^+(0)\,T)} + \frac{0.5772...}{\sqrt{2\ln(\nu_X^+(0)\,T)}}\Big\} \tag{10.53}$$

Similarly, it is found that

$$\mathrm{E}[M(T)^2] = \int_0^\infty a(\eta)^2\, e^{-\eta}\, d\eta \approx 2\sigma_X^2 \big\{ \ln(\nu_X^+(0)\, T) + \gamma \big\}. \tag{10.54}$$

From Eqs. (10.54) and (10.53), we may derive the expression for the variance of $M(T)$. It is found that

$$\sigma_{M(T)}^2 = \mathrm{E}[M(T)^2] - \mathrm{E}[M(T)]^2 \approx \frac{\pi^2 \sigma_X^2}{12 \ln(\nu_X^+(0)\, T)}. \tag{10.55}$$

One may note that $\mathrm{E}[M(T)] \to \infty$, while $\sigma_{M(T)} \to 0$ when $T \to \infty$. In Fig. 10.6 it is clearly seen how the mean value of $M(T)$ increases, while the standard deviation decreases with increasing values of T.

Let us conclude this discussion of the extreme value distribution of a stationary Gaussian process by showing that it approaches one of the three asymptotic extreme value distributions from the previous section, viz. the Gumbel distribution. The expression for $F_{M(T)}(a)$ given by Eq. (10.41) can be written as

$$F_{M(T)}(a) = \exp\Big\{ -\exp\Big(-h(a) \Big) \Big\}, \quad (a \to \infty), \tag{10.56}$$

where $h(a) = \frac{a^2}{2\sigma_X^2} + \ln(\nu_X^+(0)\, T)$. Let a_0 denote the solution of the equation $h(a) = 0$. This gives $a_0 = \sigma_X \sqrt{2 \ln(\nu_X^+(0)\, T)}$. It can be verified that the range of values of a where most of the extreme value distribution "lives," cf. Fig. 10.6, satisfies $|a - a_0| \ll a_0$ for $a_0 \to \infty$. Then $h(a) = h(a) - h(a_0) \approx h'(a_0)\,(a - a_0)$ for large values of a_0. Hence, it follows that asymptotically

$$\begin{aligned} F_{M(T)}(a) &\approx \exp\big\{ -\exp\big(-h'(a_0)\,(a - a_0) \big) \big\} \\ &= \exp\Big\{ -\exp\Big(-\frac{a - a_0}{\sigma_0} \Big) \Big\}, \quad (a_0 \to \infty), \end{aligned} \tag{10.57}$$

where $\sigma_0 = \sigma_X / \big(\sqrt{2 \ln(\nu_X^+(0)\, T)}\big)$. This is clearly an extreme value distribution of the Gumbel type. It may be noted that the mean value of this Gumbel distribution is $a_0 + 0.5772\sigma_0$ (Bury, 1975; Melchers, 1999), which agrees with the mean value given in Eq. (10.53). Similarly, the variance of this Gumbel distribution is $\pi^2\sigma_0^2/6$, which coincides with the variance derived in Eq. (10.55).

10.7.1 Example – (Continuation of Examples 9.5.1 and 9.13.1)

For the installation of the equipment module onto the bottom-mounted frame, requirements are imposed concerning the maximum vertical displacement amplitudes of the module due to motions of the crane top. In this example, we model the wave process as stationary and Gaussian. Based on the assumptions introduced previously, the vertical displacement response $U(t)$ will then also be stationary and Gaussian. It is assumed that $U(t)$ is defined so that $\mathrm{E}[U(t)] = 0$.

On the basis of these assumptions, the following installation criterion is now introduced: for a time window of length $T = 3$ hours, we require that

$$\mathrm{E}[M(T)] + 3\sigma_{M(T)} \le 4\,\mathrm{m}. \tag{a}$$

Here, $M(T) = \max\{U(t); 0 \le t \le T\}$.

For simplicity, let us assume that the sea states at the installation site are described by a set of parameterized wave spectra $S_H(\omega;\beta)$, where the value of the parameter β determines the severity of the sea state. It is further assumed that the following results have been established: $\mathrm{Var}[V(t)] = 0.16\,\beta^2$ and $\mathrm{Var}[\dot{V}(t)]/\mathrm{Var}[V(t)] = \beta^{-2}$.

Our task now is to determine the maximum β-value such that the installation criterion is satisfied.

From Example 9.13.1, we know that $U(t) \approx V(t)$ with good accuracy. We can therefore assume that $\mathrm{Var}[U(t)] \approx \mathrm{Var}[V(t)]$ and $\mathrm{Var}[\dot{U}(t)] \approx \mathrm{Var}[\dot{V}(t)]$. To be able to use Eqs. (10.53) and (10.55), we need to calculate the zero upcrossing frequency $\nu_U^+(0)$. Equation (10.23) gives $\nu_U^+(0) = 1/(2\pi\beta)$. Let $Q = \sqrt{2\ln(T/(2\pi\beta))}$. Then, with good accuracy,

$$\mathrm{E}[M(T)] = 0.4\,\beta\left\{Q + \frac{0.5772}{Q}\right\} \tag{b}$$

and

$$3\,\sigma_{M(T)} = 0.4\,\beta\frac{3\pi}{\sqrt{6}\,Q} = 0.4\,\beta\,\frac{3.8476}{Q}. \tag{c}$$

From this, we deduce that the maximum value of β must satisfy the following equation:

$$\beta\left\{Q + \frac{4.4248}{Q}\right\} = 10. \tag{d}$$

This equation cannot be solved explicitly, but an iterative solution is ready at hand. It is seen that Q varies slowly with β. A trial solution $\beta_0 = 1.0$ gives $Q_0 = 3.86$. β_1 is then calculated by solving the equation

$$\beta_1\left\{Q_0 + \frac{4.4248}{Q_0}\right\} = 10. \tag{e}$$

That gives $\beta_1 = 2.0$. Two more iterations lock the solution at $\beta = 2.05$, which determines the maximum sea state for the installation operation.

10.7.2 Example – The Crossing Rate of Transformed Processes

Assume that two stationary and differentiable processes $X(t)$ and $Y(t)$ satisfy the equation

$$Y(t) = h\big[X(t)\big], \tag{a}$$

where $h(\cdot)$ is a given function. The following useful result can then be shown: the upcrossing rate $\nu_Y^+(b)$ of $Y(t)$ is determined by the upcrossing rate $\nu_X^+(a)$ of $X(t)$ by the relation (Naess, 1983; Grigoriu, 1984a),

$$\nu_Y^+(b) = \sum_{j=1}^{n} \nu_X^+(a_j), \tag{b}$$

where $a_1, \ldots, a_n$ denote all possible x-solutions of the equation $b = h(x)$.

Let us apply this result on the example $Y(t) = X(t)^2$, where $X(t)$ is a stationary Gaussian process with mean value zero. It can be shown that $Y(t)$ is also stationary. In this particular case, $h(x) = x^2$ so that the equation $b = h(x)$ has the solutions $a_1 = \sqrt{b}$ and $a_2 = -\sqrt{b}$ ($b \geq 0$). According to Eqs. (b) and (10.22), we obtain that

$$\nu_Y^+(b) = \frac{1}{\pi}\frac{\sigma_{\dot{X}}}{\sigma_X}\exp\left\{-\frac{b}{2\sigma_X^2}\right\}. \tag{c}$$

The extreme value distribution for the $Y(t)$ process then becomes

$$F_{M_Y(T)}(b) = \exp\{-\nu_Y^+(b)\,T\} = \exp\left\{-\nu_Y^+(0)\,T\,\exp\left(-\frac{b}{2\sigma_X^2}\right)\right\}, \tag{d}$$

where

$$\nu_Y^+(0) = \frac{1}{\pi}\frac{\sigma_{\dot{X}}}{\sigma_X}\ (= 2\,\nu_X^+(0)). \tag{e}$$

Note that $F_{M_Y(T)}(b)$ is in fact a Gumbel distribution. Analogously to the Gaussian case, one will find that, for example,

$$\mathrm{E}[M_Y(T)] = 2\sigma_X^2\left\{\ln(\nu_Y^+(0)\,T) + \gamma\right\}. \tag{f}$$

For the similar case of $Z(t) = X(t)\,|X(t)|$, then $\nu_Z^+(b) = \nu_Y^+(|b|)/2$ because in this case the equation $b = h(x) = x|x|$ has only one solution, viz. $a = \mathrm{sign}(b)\sqrt{|b|}$.

10.8 Return Period

Let Z be a random variable, and let

$$p = \mathrm{Prob}\{Z > z\} = 1 - F_Z(z). \tag{10.58}$$

Assume that we can make a series of independent observations of Z. The mean number of observations to the first time the observed (measured) value of Z exceeds z, is called the *return period* for exceedance of z, and we denote it by $\tilde{R}(z)$. It can be shown that

$$\tilde{R}(z) = \frac{1}{p} = \frac{1}{1 - F_Z(z)}. \tag{10.59}$$

This equation can be explained by recognizing that, on average, $1/p$ trials must be conducted before an event of probability p occurs.

Note that $\tilde{R}(z)$ refers to the number of observations and that these are assumed to be statistically independent. To express the return period in terms of time, we need to know about the time interval between the observations. If the observation interval is Δt, the return period specified in terms of time, which we denote by $R(z)$, will be given as

$$R(z) = \Delta t\ \tilde{R}(z). \tag{10.60}$$

The observation interval Δt must be chosen sufficiently long such that the individual observations become approximately independent. Note that $R(\xi_p(T)) = T/(1-p)$, where $\xi_p(T)$ is given by Eq. (10.46) and T is the "observation" interval.

A design load with a probability of 10^{-2} of being exceeded during one year is often used in connection with the design of offshore structures. If we let $X(t)$ denote a relevant load process considered for such a design provision, and let ξ denote the corresponding load level, then $\text{Prob}\{Z > \xi\} = 0.01$, where $Z = \max(X(t); 0 \leq t \leq 1\text{ year})$. The return period for exceedance of ξ then becomes

$$\tilde{R}(z) = \frac{1}{\text{Prob}\{Z > \xi\}} = \frac{1}{0.01} = 100. \tag{10.61}$$

The reference period in this case is one year; therefore, $R(\xi) = 100$ years.

It should be mentioned that time varying loads caused by, for example, ocean waves cannot generally be considered as stationary over an extended period of time. This implies that quantities such as yearly maxima must be calculated by using so-called long-term statistics. This is discussed in Chapter 13.

10.9 Basic Notions of Fatigue Damage

In this section, we give a short introduction to stochastic calculation of fatigue damage, mainly for the purpose of illustrating the application of notions developed in this chapter. A more detailed and extensive discussion of this topic is resumed in Chapter 13.

If a suitably shaped specimen of a metallic material is subjected to harmonic stress cycles of constant amplitude larger than a certain threshold value, which depends on the material, experience has shown that there is often a connection between the stress amplitude and the number of stress cycles to fracture of the material. This connection is usually expressed as

$$N\,S^m = K. \tag{10.62}$$

Here, m and K are material constants. For construction steel, m typically has values in the range of 3 to 5. S denotes the stress range and N denotes the number of stress cycles to fracture.

To calculate fatigue damage for general load time histories, one often applies the so-called Palmgren-Miner's hypothesis: for a load time history of duration T with different stress amplitudes, it is assumed that the accumulated fatigue damage $\tilde{D}(T)$ is given as

$$\tilde{D}(T) = \sum_j \frac{n_j}{N_j}. \tag{10.63}$$

Here, n_j = the number of stress cycles in the time history associated with stress range S_j, or more correctly, with stress ranges between $S_j - \Delta S/2$ and $S_j + \Delta S/2$ for a suitable choice of discretization interval length ΔS. $N_j = K\,S_j^{-m}$ is the number of stress cycles to fracture with stress range S_j. Fracture is assumed to occur when $\tilde{D}(T) = 1$.

Equation (10.63) expresses the assumption that n_j load cycles with stress range S_j consumes a part, which is equal to n_j/N_j, of the "lifetime" of the material (expressed in terms of accumulated damage $\tilde{D}(T) = 1$).

Although many examples show that Palmgren-Miner's hypothesis is not correct, it is still widely used. Even if this hypothesis is not true, it gives a reasonably accurate picture of reality. In addition, Eq. (10.63) is advantageous due to a linear damage accumulation.

Let $X(t)$ denote a stationary, narrow banded process, which may represent the von Mises stress in a section somewhere in a lightly damped structure. Because we are only interested in stress ranges, we assume that $X(t)$ has mean value zero. Let $\tilde{N}(a)\,da$ denote the number of stress cycles with amplitude between a and $a + da$ ($S = 2a$) of $X(t)$ during time T. $\tilde{N}(a)\,da$ is a random variable.

The accumulated damage

$$\tilde{D}(T) = \int_a \frac{\tilde{N}(a)}{N(a)}\,da \tag{10.64}$$

where $N(a) = K(2a)^{-m}$, also becomes a random variable, and its expectation value $D(T) = \mathrm{E}[\tilde{D}(T)]$ is then given as

$$D(T) = \int_0^\infty \frac{\mathrm{E}[\tilde{N}(a)]}{N(a)}\,da. \tag{10.65}$$

Here, $\mathrm{E}[\tilde{N}(a)]\,da$ = expected number of stress cycles with amplitude between a and $a + da$ during time T. Hence,

$$\mathrm{E}[\tilde{N}(a)]\,da = \nu_X^+(0)\,T \cdot f_{X_p}(a)\,da \tag{10.66}$$

because $f_{X_p}(a)\,da$ = the relative number of peaks with amplitude between a and $a + da$ and $\nu_X^+(0)\,T$ = the total number of peaks, which equals the number of stress cycles, during time T. This gives

$$D(T) = \nu_X^+(0)\,T \int_0^\infty \frac{f_{X_p}(a)}{N(a)}\,da = \nu_X^+(0)\,T \int_0^\infty \frac{(2a)^m}{K}\,f_{X_p}(a)\,da. \tag{10.67}$$

For a narrow banded Gaussian process

$$f_{X_p}(a) = \frac{a}{\sigma_X^2}\exp\Big\{-\frac{a^2}{2\sigma_X^2}\Big\}. \tag{10.68}$$

Substituted into Eq. (10.67), this leads to the equation

$$\begin{aligned} D(T) &= \frac{\nu_X^+(0)\,T}{2^{-m}\,K\,\sigma_X^2}\int_0^\infty a^{m+1}\exp\Big\{-\frac{a^2}{2\sigma_X^2}\Big\}\,da \\ &= \nu_X^+(0)\,T\,\frac{(2\sqrt{2}\,\sigma_X)^m}{K}\,\Gamma\Big(1+\frac{m}{2}\Big), \end{aligned} \tag{10.69}$$

where $\Gamma(x)$ denotes the gamma function, which is a standard tabulated function. In particular, $\Gamma(n+1) = n!$ for $n = 0, 1, 2, \ldots$. An estimate of the lifetime can now be derived from Eq. (10.69) by putting $D(T) = 1$. That gives

$$T = \frac{K\,T_z}{(2\sqrt{2}\,\sigma_X)^m\,\Gamma(1+\frac{m}{2})}, \tag{10.70}$$

where $T_z = (\nu_X^+(0))^{-1}$ = the mean zero-crossing period.

11 Statistics for Nonlinear Problems

11.1 Introduction

In this chapter, we discuss two types of nonlinear hydrodynamic forces on marine structures that have been subjected to extensive research since the 1970s. The first kind is related to forces on hydrodynamically slender structures, where viscous effects play a significant role. In the everyday lingo, they are just referred to as dragforces. This term would be recognized by many, not only offshore engineers. But they would perhaps be in the majority when we specialize a little more and introduce the "famous" Morison equation for the total forces on slender structures exerted by ocean waves. This equation was introduced in 1950, and is still extensively used.

The other kind of nonlinear forces that will be discussed, are the so-called second order hydrodynamic forces. They came into focus in the early 1970s due to the exploration of offshore oil fields using moored semisubmersible platforms. For the mooring of the platform, typically a kind of spread mooring system with steel ropes would be used. The consequence of this was that the natural periods of the horizontal motion modes would be of the order of minutes rather than seconds. Because there is virtually no energy in ocean waves at these periods, the only way that such slowly varying motions can occur is the existence of nonlinear hydrodynamic forces. The response pattern of offshore structures in the ocean environment indeed show that it is crucially important to be able to assess the impact of nonlinear hydrodynamic loads for the design of such structures.

11.2 Hydrodynamic Forces on Slender Structures

11.2.1 Morison Equation

An accurate calculation of the hydrodynamic forces on slender structural elements in waves is largely beyond our capabilities. This is due to the complex mixture of viscous and inviscid fluid effects. This situation was, of course, even more conspicuous in 1950, when Morison et al. (1950) proposed an engineering solution to this problem by separating the two fluid phenomena and formulating the famous Morison equation, or Morison formula, for the resulting fluid forces. That the Morison formula is still largely the state-of-the-art formulation of the fluid forces when viscous effects are of importance bears a clear witness to the complexity of the problem, as well as to the fortuitous fact that the Morison formula provides reasonably accurate results, at least from an engineering point of view. Since its introduction in 1950, the amount of literature dealing with the Morison formula is truly formidable. An account of the development during the first three decades is admirably reported by Sarpkaya and Isaacson (1981).

The original formulation of the Morison equation was related to the wave forces on a vertical pile, but their formula has later been adopted to cover a wide spectrum of structural configurations. The general form of the Morison equation for a fixed cylinder can be written as

$$f = k_m \dot{u} + k_d u|u|, \tag{11.1}$$

where f denotes the force per unit length of the cylinder, and u is the incident undisturbed water particle velocity component normal to the cylinder axis. The coefficients k_m and k_d in the particular case of a circular cylinder of diameter D are usually written as

$$k_m = C_m \rho \pi D^2/4, \quad k_d = C_d \rho D/2, \tag{11.2}$$

where ρ denotes the mass density of water, which for offshore applications may be taken as 1,025 kg/m^3. The two dimensionless empirical coefficients C_m and C_d are usually referred to as the inertia coefficient and drag coefficient, respectively. Their values depend on many parameters such as Reynolds number, Keulegan-Carpenter number, surface roughness ratio, etc. (Sarpkaya and Isaacson, 1981; Faltinsen, 1990). Typical values would be $C_m \approx 2$ and $C_d \approx 1$.

The Morison formula was originally proposed for an oscillatory fluid flow with a fixed frequency, and a wavelength assumed to be large compared to the cylinder diameter. Later, the formula was also used to describe the forces in random waves for which the last condition is clearly not satisfied in the high-frequency tail of the wave spectrum. This imposes a need to introduce some form of frequency dependence of the coefficients k_m and k_d, or a suitable frequency cut-off.

An extension of the Morison formula to account for structural flexibility has also been proposed. If the velocity of the structural element is not negligible compared with the maximum water particle velocity at the element, the relative velocity and relative acceleration are introduced, and the proposed modification assumes the form

$$f = k_m(\dot{u} - \ddot{x}) + \rho A \ddot{x} + k_d(u - \dot{x})|u - \dot{x}|, \tag{11.3}$$

where A is the cross-sectional area of the cylinder.

11.2.2 Statistics of Morison-Type Wave Forces

The standard approach to the derivation of the statistical characteristics of the Morison-type forces for random seas is to assume that the wave elevation process $X(t)$ is a stationary Gaussian process. This implies that the water particle velocity

$V_x(t;x,y,z)$ and acceleration $A_x(t;x,y,z)$ also become stationary Gaussian processes, cf. Section 8.2.3. In the following, we assume that the random sea is unidirectional and propagating in the positive x-direction. In the absence of current, a good measure of the relative importance of the drag and inertia terms is

$$\kappa = \frac{k_d\,\sigma_{V_x}^2}{k_m\,\sigma_{A_x}}, \tag{11.4}$$

which assumes a role similar to the Keulegan-Carpenter number for harmonic fluid flow. In the following, we do not restrict ourselves to the zero current case; that is, we may have $\mathrm{E}[V_x] = \overline{V}_x \neq 0$. Also, in this case, the parameter κ assumes an important role. In Eq. (11.4), k_m and k_d should be considered as representative constants. Note that their frequency dependence can be incorporated into the processes $A_x(t;x,y,z)$ and $V_x(t;x,y,z)$, respectively, leading to two new Gaussian processes $\tilde{A}_x(t;x,y,z)$ and $\tilde{V}_x(t;x,y,z)$. The practical consequence of this is that we would write Eq. (11.1) as $F = k_m\tilde{A}_x + k_d\tilde{V}_x|\tilde{V}_x|$, where k_m and k_d are now independent of frequency. This also helps ensure that we may consider σ_{A_x} finite, which we already tacitly did.

To illustrate some of the statistical aspects of Morison-type forces $F_m(t)$ for a random sea, we simplify the notation and write

$$F_m(t) = k_m\,A_x(t) + k_d\,V_x(t)|V_x(t)|, \tag{11.5}$$

For the subsequent analysis, we rewrite Eq. (11.5) as

$$F(t) = W(t) + \kappa U(t)|U(t)|, \tag{11.6}$$

where

$$F(t) = \frac{F_m(t)}{k_m\sigma_{A_x}},\quad \kappa = \frac{k_d\sigma_{V_x}^2}{k_m\sigma_{A_x}},\quad W(t) = \frac{A_x(t)}{\sigma_{A_x}},\quad U(t) = \frac{V_x(t)}{\sigma_{V_x}}, \tag{11.7}$$

The random variables U and W are seen to be independent normal variables, $W \sim \mathrm{N}(0,1)$ and $U \sim \mathrm{N}(\overline{U},1)$, where $\overline{U} = \mathrm{E}[U]$. The drag force part $Q(t) = \kappa U(t)|U(t)|$ of $F(t)$ then has PDF,

$$f_Q(q) = f_U(u)\,\frac{du}{dq} = \frac{1}{2\sqrt{2\pi\kappa|q|}}\exp\left\{-\frac{1}{2}\left(\mathrm{sign}(q)\sqrt{\frac{|q|}{\kappa}} - \overline{U}\right)^2\right\},\quad q \neq 0. \tag{11.8}$$

Note that $f_Q(q)$ has a singularity at the origin, but that $\int_{-\infty}^{\infty} f_Q(q)\,dq$ exists and, of course, equals 1. As expected, the PDF of the drag force is distinctly non-Gaussian with a much higher probability of large values; in fact, a closer look at the PDF of Q reveals that it is almost exponential in the tails. The PDF of the Morison force $F(t)$ can now be calculated by convolution of f_Q with f_W, which is the standard normal density. It is obtained that

$$f_F(q) = \frac{1}{4\pi}\int_{-\infty}^{\infty}\exp\left\{-\frac{(q-y)^2}{2} - \frac{1}{2}\left(\mathrm{sign}(y)\sqrt{\frac{|y|}{\kappa}} - \overline{U}\right)^2\right\}\frac{dy}{\sqrt{\kappa|y|}}. \tag{11.9}$$

The PDF f_F, which is completely determined by the parameters κ and $\overline{U}$, cannot be simply expressed, but the integral can be easily calculated for given values of q, κ, and $\overline{U}$, so there is no need to elaborate further on this formula.

To calculate the statistical moments of $F(t)$, two avenues of approach may be used. Either directly from the PDF of F, or by using Eq. (11.6). We do not derive expressions for the general case here, but rather consider the special case $\mathrm{E}[U] = 0$. Then, clearly, $\mathrm{E}[F] = 0$. The variance of $F(t)$ can now easily be calculated directly from Eq. (11.6) by observing that $\mathrm{Var}[F] = \mathrm{E}[F^2] = \mathrm{E}[W^2] + \kappa^2\mathrm{E}[U^4] = 1 + 3\kappa^2$, because $\mathrm{E}[U^4] = 3$. This gives $\mathrm{Var}[F_m] = (k_m\sigma_{A_x})^2\mathrm{Var}[F] = (k_m\,\sigma_{A_x})^2 + 3(k_d\,\sigma_{V_x}^2)^2$.

For a range of problems it is of particular interest to determine the mean rate of level upcrossings, e.g., for the extreme value distribution. For the Morison-type forces the mean upcrossing rate cannot generally be given in a simple closed form. However, good approximations are available. The most general form seems to have been derived by Vinje (1980). A general method for its derivation was developed by Naess (1983). It is based on using the Rice formula (Rice, 1954; Lin, 1967). We do not go into details about the derivation here, but rather refer the reader to Naess (1983), where it was shown that the exact asymptotic form $\hat{\nu}_F^+(q)$ of the mean rate of level upcrossings of $F(t)$ is given as

$$\hat{\nu}_F^+(q) = \frac{\varrho}{2\pi}\exp\Big\{-\frac{q}{2\kappa} + \overline{U}\Big(\frac{q}{\kappa} - \frac{1}{2\kappa^2}\Big)^{\frac{1}{2}} - \frac{\overline{U}^2}{2} + \frac{1}{8\kappa^2}\Big\}, \tag{11.10}$$

where $\varrho = \sigma_{A_x}/\sigma_{V_x}$ and $q \gg (1+\overline{U})/2\kappa$. However, Vinje (1980) showed that $\hat{\nu}_F^+(q)$ is a remarkably accurate approximation to the true upcrossing rate even for relatively small values of q relative to $(1+\overline{U})/2\kappa$. Special cases of the formula for $\hat{\nu}_F^+(q)$ were derived by Borgman (1972) for $\overline{U} = 0$ and by Moe and Crandall (1977) for $\overline{U} << 1$, see also (Grigoriu, 1984b). If we consider only the drag force $F_d(t) = \kappa U(t)|U(t)|$, we can immediately calculate its mean upcrossing rate $\nu_{F_d}^+(q)$ by using the result of Example 10.7.2. That is, we need to find the u-solutions of the equation $q = \kappa u|u|$. For each q there is only one solution, which is $u = \mathrm{sign}(q)\sqrt{|q|/\kappa}$. This leads to the result

$$\nu_{F_d}^+(q) = \frac{\varrho}{2\pi}\exp\Big\{-\frac{1}{2}\Big(\mathrm{sign}(q)\sqrt{\frac{|q|}{\kappa}} - \overline{U}\Big)^2\Big\}, \tag{11.11}$$

which in the case of zero current assumes the simple form

$$\nu_{F_d}^+(q) = \frac{\varrho}{2\pi}\exp\Big\{-\frac{|q|}{2\kappa}\Big\}. \tag{11.12}$$

A comparison of Eqs. (11.10) and (11.12) shows that in the zero current case there is a constant correction factor $\exp(1/8\kappa^2)$ to account for the presence of the inertia forces.

In the narrow band approximation, the upcrossing rate is often used to provide an approximation for the distribution of peak values, which again may be used to approximate range distributions for calculation of fatigue damage. Assuming the case of zero current, and invoking the results of Section 10.3, the PDF of the peak values F_p of $F(t)$ can be expressed as

$$f_{F_p}(q) = -\frac{1}{\nu_F^+(0)}\frac{d\nu_F^+(q)}{dq}, \tag{11.13}$$

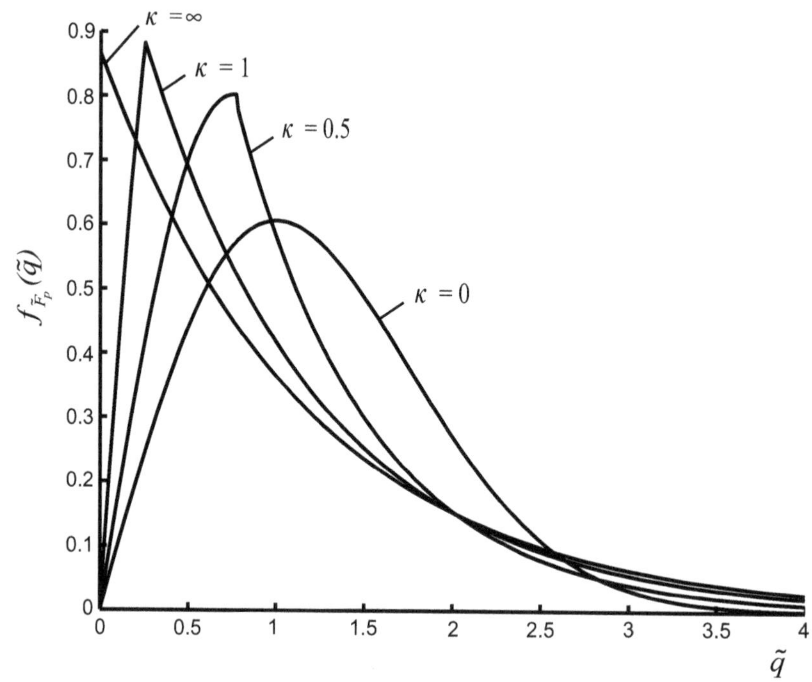

Figure 11.1. Plots of $f_{\tilde{F}_p}(q)$ for $\kappa = 0, 0.5, 1.0, \infty$.

where $\nu_F^+(q)$ denotes the exact mean upcrossing rate of $F(t)$, which is not easily available. The following approximation was derived (Borgman, 1965; Madsen et al., 1986),

$$f_{F_p}(q) = \begin{cases} q \exp(-q^2/2) & : \quad 0 < q < 1/2\kappa \\ \frac{1}{2\kappa} \exp\left\{ -\frac{1}{2\kappa}\left(q - \frac{1}{4\kappa}\right)\right\} & : \quad 1/2\kappa < q \end{cases} \tag{11.14}$$

The PDF is a mixture of a Rayleigh PDF for the smaller arguments with an exponential PDF for the larger arguments. Plots of $f_{\tilde{F}_p}(q)$ for various values of κ are shown in Fig. 11.1, where $\tilde{F}_p = F_p/\sigma_F = F_p/\sqrt{1+3\kappa^2}$.

Brouwers and Verbeek (1983) used Eq. (11.14) together with a narrow-band assumption to approximate the distribution of stress ranges as was done in Section 10.9 by relating the stress response linearly to the wave force in a quasistatic manner. Writing the relation between the number of stress cycles to failure N at stress range S as $N = cS^{-m} \approx \tilde{c}(2F_p/\sigma_F)^{-m}$, we now use Eq. (10.67) to calculate the expected fatigue damage $D(T)$ as

$$D(T) = \frac{\nu_F^+(0)\,T}{\tilde{c}} \int_0^\infty \left(\frac{2q}{\sigma_F}\right)^m f_{F_p}(q)\,dq. \tag{11.15}$$

The integral on the rhs can be expressed in terms of the gamma functions $\gamma(a,x) = \int_0^x t^{a-1}e^{-t}dt$ and $\Gamma(a) = \gamma(a,\infty)$, which are tabulated special functions that are also included as standard functions in commercial software, e.g., Matlab. Using these

Table 11.1. *Values of the ratio $r(m)$*

m	2.5	3	3.5	4	4.5
$r(m)$	2.17	3.14	4.74	7.40	11.91

functions, the following relation is obtained (Madsen et al., 1986):

$$D(T) = \frac{\nu_F^+(0)\,T}{\tilde{c}}\Big(\frac{4}{1+3\kappa^2}\Big)^{m/2}\Big\{2^{m/2}\gamma(1+m/2, 1/8\kappa^2) + (2\kappa)^m e^{1/8\kappa^2}\Big(\Gamma(1+m) - \gamma(1+m, 1/4\kappa^2)\Big)\Big\}. \tag{11.16}$$

Two limiting cases are of particular interest. $\kappa \to 0$ corresponds to the case of dominating inertia forces. This leads to

$$D(T)_{\kappa=0} = \frac{\nu_F^+(0)\,T}{\tilde{c}}(2\sqrt{2})^m\Gamma(1+m/2), \tag{11.17}$$

which agrees with Eq. (10.69). $\kappa \to \infty$ indicates dominating drag forces, and it is found that

$$D(T)_{\kappa=\infty} = \frac{\nu_F^+(0)\,T}{\tilde{c}}\Big(\frac{4}{\sqrt{3}}\Big)^m\Gamma(1+m), \tag{11.18}$$

A comparison of the two limiting forms shows that drag dominance gives the greatest fatigue damage accumulation rate, and the difference increases rapidly with increasing m-parameter. This observation leads one to question the commonly used approach of stochastic linearization (Roberts and Spanos, 1990) for calculating fatigue damage, where typically a stochastic response process is replaced by a Gaussian process. The criterion used is to minimize the mean square deviation. In the context of Morison-type forces, $F(t)$ would be replaced by $F_l(t) = W(t) + \kappa c_l U(t)$, where the constant c_l is determined such that $\mathrm{E}[(F - F_l)^2]$ is minimum. It can be shown that $c_l = \sqrt{8/\pi}$ (Roberts and Spanos, 1990). We do not derive the full expression for the expected fatigue damage for this case here, but rather write down the limiting expression for $\kappa \to \infty$. This is obtained by using the procedure of Section 10.9 because $F_l(t)$ is a Gaussian process. Denoting the expected fatigue damage for $F_l(t)$ by $D_l(T)$, it can be shown that

$$D_l(T)_{\kappa=\infty} = \frac{\nu_F^+(0)\,T}{\tilde{c}}\Big(\frac{8}{\sqrt{3\pi}}\Big)^m\Gamma(1+m/2), \tag{11.19}$$

The ratio $r(m)$ between the two limiting expressions for dominating drag forces becomes

$$r(m) = \frac{D(T)_{\kappa=\infty}}{D_l(T)_{\kappa=\infty}} = \Big(\frac{\sqrt{\pi}}{2}\Big)^m\frac{\Gamma(1+m)}{\Gamma(1+m/2)}. \tag{11.20}$$

By substituting a range of relevant values for the m-parameter, the $r(m)$-values obtained are listed in Table 11.1.

The values of $r(m)$ clearly demonstrate the deficiency of stochastic linearization for the case of drag dominance and quasistatic loading as previously described.

11.3 Nonlinear, Second-Order Forces and Motions

11.3.1 Slow-Drift Response

An important problem in offshore and ocean engineering since the early 1970s is the study of second-order, slowly varying forces and motions of moored floating large-volume structures. This can be attributed in large part to the enormous growth in the drilling activities in search of offshore oil fields, especially in the North Sea. A large amount of this drilling was carried out by semisubmersible drilling rigs, which were moored to the seabed by spread mooring systems. This resulted in dynamic systems with natural periods for the horizontal motion modes such as surge, sway and yaw in the range of typically one to two minutes. Because there is virtually no energy in the ocean waves at those periods, linear system theory by itself would not be sufficient to predict the observed motion response of typical drilling rigs at offshore locations.

This lead to a rapid development of the hydrodynamic theory of second-order, slowly varying forces on floating structures. A common feature to all this theory is the use of a perturbation technique with the wave height as the small parameter. This means that the theory, strictly speaking, is only applicable to small sea states. However, in practice, it is used even for heavy seas. This can be explained by two factors. First, it seems to work reasonably well for such conditions. Second, it is the only practical nonlinear theory available. A detailed description of the basic elements of the second-order hydrodynamic theory is given by Faltinsen (1990).

To simplify the description of the theory, we first restrict ourselves to the case of unidirectional irregular seas modeled as a stationary Gaussian process. Consider the case of the surge response of a semisubmersible platform with the waves propagating in the positive x-direction. Adopting the further simplification that the surge motion is uncoupled from heave and pitch, the total hydrodynamic surge forces $F(t)$ on the platform can be expressed as $F(t) = F_1(t) + F_2(t)$, where the first-order, wave frequency forces $F_1(t)$ are given as

$$F_1(t) = \int_0^\infty k_1(s)X(t-s)\,ds, \tag{11.21}$$

and the second-order, slowly varying forces $F_2(t)$ are given as

$$F_2(t) = \int_0^\infty \int_0^\infty k_2(s_1, s_2)X(t-s_1)X(t-s_2)\,ds_1\,ds_2, \tag{11.22}$$

In these equations, $X(t)$ denotes the ocean surface elevation, which is modeled as a stationary Gaussian process, meaning that we are considering a short-term sea state. This leads us to the representation of the wave elevation process given by Eq. (8.4). Modified to fit the unidirectional seas case, this can be written as

$$X(t) = \Re\Big\{\sum_{j=1}^{n} \sqrt{2\,S_X^+(\omega_j)\,\Delta\omega_j}\,C_j\,e^{i\omega_j t}\Big\} = \sum_{j=-n}^{n} \sqrt{S_X^+(|\,\omega_j|)\,\Delta\omega_j/2}\,C_j\,e^{i\omega_j t}, \tag{11.23}$$

where we have introduced the notation $C_j = C_{-j}^*$, $\omega_j = -\omega_{-j}$ and $\Delta\omega_j = \Delta\omega_{-j}$ for $j = -n, \ldots, -1$. The far rhs expression is very convenient for some theoretical derivations to be given as follows. For convenience, we have also adopted the convention that the summation index, when running from negative to positive integers, avoids zero (if not expressly stated otherwise).

From Eq. (11.21), the function $k_1(s)$ is recognized as an impulse response function describing the linear, time-invariant system connecting the wave process to the first-order surge forces, cf. Section 2.8. In the formulation of the total surge forces, the new and unfamiliar element is the rhs of Eq. (11.22). In analogy with the linear case, the function $k_2(s_1, s_2)$ is called a *quadratic impulse response function*; however, in contrast to $k_1(s)$, it does not have a simple physical interpretation apart from the fact that it characterizes the second-order properties of the surge forces. The formulation of the surge forces as a sum of the two previous integral expressions represents a stochastic Volterra functional expansion of a nonlinear, time-invariant system truncated after the second-order term (Schetzen, 1980). Among early efforts to use this formulation to investigate the statistics of, for example, the surge forces and motions of a floating structure, we mention (Neal, 1974; Vinje, 1983), where further references to the early literature are found.

To perform practical calculations, it is convenient to rewrite the expressions on the rhs of Eqs. (11.21) and (11.22) in terms of the Fourier transforms of the impulse response functions. The reason for this is that general-purpose commercial software for calculating the hydrodynamic forces on submerged objects in waves provides data for these Fourier transforms rather than for the impulse response functions. The Fourier transform of $k_1(s)$ is recognized as the transfer function of a linear, time-invariant system, and it is defined as follows:

$$\hat{K}_1(\omega) = \int_0^\infty k_1(s)\, e^{-\mathrm{i}\omega s}\, ds. \tag{11.24}$$

The analogous Fourier transform of the quadratic impulse response function is called the *quadratic transfer function* (QTF) , and it is defined by the following relation:

$$\hat{K}_2(\omega_1, \omega_2) = \int_0^\infty \int_0^\infty k_2(s_1, s_2)\, e^{-\mathrm{i}(\omega_1 s_1 + \omega_2 s_2)}\, ds_1\, ds_2. \tag{11.25}$$

By substituting the expression for the wave elevation process $X(t)$ given by Eq. (11.23) into Eqs. (11.21) and (11.22), it is a straightforward exercise to show that the surge forces can now be written in the following manner

$$F_1(t) = \sum_{j=-n}^{n} q_j\, C_j\, e^{\mathrm{i}\omega_j t}, \tag{11.26}$$

where

$$q_j = \hat{K}_1(\omega_j) \sqrt{\frac{1}{2}\, S_X^+(|\omega_j|)\Delta\omega_j}\,, \tag{11.27}$$

and

$$F_2(t) = \sum_{i=-n}^{n} \sum_{j=-n}^{n} Q_{ij}\, C_i\, C_j^*\, e^{\mathrm{i}(\omega_i - \omega_j)t}, \tag{11.28}$$

where

$$Q_{ij} = \hat{K}_2(\omega_i, -\omega_j)\,\frac{1}{2}\sqrt{S_X^+(|\omega_i|)\, S_X^+(|\omega_j|)\,\Delta\omega_i\,\Delta\omega_j}\,. \tag{11.29}$$

Equation (11.28) shows that the second-order surge forces will contain frequency components at the sum and difference of pairs of frequencies in the wave spectrum

$S_X(\omega)$. This explains why the second-order wave-induced forces can cause a moored semisubmersible to oscillate. However, there is a practical question related to this. Clearly, from the representation of the wave process expressed by Eq. (11.23), it is realized that the difference frequencies available are $k\,\Delta\omega$, for $k = 1, 2, \ldots$. Hence it is necessary to choose the discretization of the wave spectrum frequency range such that the slowly varying forces are suitably represented. If the natural period in surge is T_s, then $\Delta\omega \ll 2\pi/T_s$.

A particularly important aspect of second-order forces is the mean drift forces $\overline{F}_2 = \mathrm{E}[F_2(t)]$. Because $\mathrm{E}[C_i] = 0$ and $\mathrm{E}[C_i\, C_j^*] = \delta_{ij}$, where $\delta_{ij} = 1$ if $i = j$ and $\delta_{ij} = 0$ if $i \neq j$, it follows from Eq. (11.28) that the mean drift forces are given as follows:

$$\overline{F}_2 = \sum_{i=-n}^{n} Q_{ii} = \sum_{i=-n}^{n} \hat{K}_2(\omega_i, -\omega_i)\,\frac{1}{2}\,S_X^+(|\omega_i|)\,\Delta\omega_i$$

$$= \sum_{i=1}^{n} \hat{K}_2(\omega_i, -\omega_i)\,S_X^+(\omega_i)\,\Delta\omega_i, \tag{11.30}$$

because $\hat{K}_2(\omega_i, -\omega_i)$ is a real, nonnegative number. Physically, $\hat{K}_2(\omega_i, -\omega_i)$ expresses the drift force from a regular wave of circular frequency ω_i, and this is always in the direction of wave propagation.

A thorough discussion of the hydrodynamics of mean drift forces is given by Faltinsen (1990). What makes the mean drift forces of particular practical significance to us here is the fact that the calculation of $\hat{K}_2(\omega_i, -\omega_i)$ is based on the first-order, linear potential, which is much easier to calculate than the second-order potential. This fact combined with the so-called Newman's approximation makes it possible to approximate the full second-order forces by expressions involving only quantities based on the first-order potential (Newman, 1974). Although there are several versions of Newman's approximation for the QTF in use, they all achieve the same goal, simplifying the calculation of the QTF by involving only the first-order potential. Denoting the Newman's approximation to the QTF by $\hat{K}_2^N(\omega_i, -\omega_j)$, the following equation cites one version of this approximation:

$$\hat{K}_2^N(\omega_i, -\omega_j) = 0.5\big(\hat{K}_2(\omega_i, -\omega_i) + \hat{K}_2(\omega_j, -\omega_j)\big), \tag{11.31}$$

There is a simple physical argument to support this kind of approximation. For moored offshore structures only frequency pairs $(\omega_i, -\omega_j)$ close to the diagonal $(\omega_i, -\omega_i)$ are important for the surge response of these structures because the natural period of surge is typically 1 to 2 minutes. Assuming that the QTF is smooth in a narrow banded domain close to the diagonal $(\omega_i, -\omega_i)$, it is clear that Newman's approximation seems to be a reasonable approximation to adopt. In fact, this approximation is widely used in practice for calculating the slowly varying forces on moored offshore structures.

As already pointed out, the second-order forces $F_2(t)$ given by Eq. (11.28) contain both sumfrequency and differencefrequency terms. When the focus is on the slowly varying forces, the sumfrequency terms are clearly of little importance and can be mostly neglected. This is achieved by introducing the slowdrift approximation of the QTF, denoted by $\hat{K}_2^{SD}(\omega_i, -\omega_j)$, which can be expressed as

$$\hat{K}_2^{SD}(\omega_i, -\omega_j) = 0 \text{ when } \omega_i \cdot \omega_j < 0. \tag{11.32}$$

This effectively eliminates all sumfrequency terms in Eq. (11.28).

Even if the forces themselves are rarely the ultimate goal of an analysis, but rather the response induced by these forces, we have chosen to also elaborate on the forces here. The reason is that for many offshore systems, an accurate calculation of the response requires a solution in the time domain, and in such cases, the forces are the crucial input. To produce realizations of the slowly varying surge forces $F_2^{SD}(t)$ on a semisubmersible, one would use the following modified version of Eq. (11.28):

$$F_2^{SD}(t) = \sum_{i=-n}^{n} \sum_{j=-n}^{n} Q_{ij}^{SD} \, C_i \, C_j^* \, e^{i(\omega_i - \omega_j)t}, \tag{11.33}$$

where

$$Q_{ij}^{SD} = \hat{K}_2^{SD}(\omega_i, -\omega_j) \frac{1}{2} \sqrt{S_X^+(|\omega_i|) \, S_X^+(|\omega_j|) \, \Delta\omega_i \, \Delta\omega_j}. \tag{11.34}$$

For a range of offshore structures, reasonably accurate estimates of the response can be obtained by assuming that the equations of motion are linear. To be specific, let us retain the assumption that the surge response of a TLP is decoupled from the other modes of motion and that the equation of surge displacement response can be written as

$$M \ddot{Z}(t) + C \dot{Z}(t) + K Z(t) = F_1(t) + F_2^{SD}(t). \tag{11.35}$$

Here, M denotes the total mass, including added mass in surge; K is the restoring force coefficient due to the tethers; and C is an equivalent damping coefficient.

As we know from Chapter 2, the linear dynamic model expressed by Eq. (11.35) has an impulse response function, $l(s)$, with a corresponding linear transfer function denoted by $\hat{L}(\omega)$. The surge response $Z(t)$ can then be written as

$$Z(t) = \int_0^\infty l(s) F(t-s)\, ds = \int_0^\infty l(s) F_1(t-s)\, ds + \int_0^\infty l(s) F_2^{SD}(t-s)\, ds, \tag{11.36}$$

Hence, it follows that the surge response consists of two components, a first-order, wave frequency part

$$Z_1(t) = \int_0^\infty l(s) F_1(t-s)\, ds, \tag{11.37}$$

and a second-order, slowly varying part

$$Z_2^{SD}(t) = \int_0^\infty l(s) F_2^{SD}(t-s)\, ds. \tag{11.38}$$

By substituting from Eqs. (11.26) and (11.33), the following expressions for $Z_1(t)$ and $Z_2^{SD}(t)$ are obtained:

$$Z_1(t) = \sum_{j=-n}^{n} r_j \, C_j \, e^{i\omega_j t}, \tag{11.39}$$

where $r_j = q_j\,\hat{L}(\omega_j)$, and

$$Z_2^{SD}(t) = \sum_{i=-n}^{n}\sum_{j=-n}^{n} R_{ij}^{SD}\, C_i\, C_j^*\, e^{\mathrm{i}(\omega_i-\omega_j)\,t}\,, \tag{11.40}$$

where $R_{ij}^{SD} = Q_{ij}^{SD}\,\hat{L}(\omega_i - \omega_j)$. The expressions for the two components of the surge response are entirely similar to the corresponding expressions for the forces. In the remaining part of this section, we focus on the response components, but it should be clearly understood that all results will have their counterparts for the force components by making the requisite modifications.

To analyze the statistical properties of the surge response, it is expedient to reformulate the slowly varying component. Introducing the $n \times n$-matrix $\boldsymbol{R} = (R_{ij})$, $i, j = 1, \ldots, n$, it can be demonstrated that $\boldsymbol{R}$ is a Hermitian matrix. By adapting the analysis in (Naess, 1990) to the present formulation, it can now be shown that by solving the eigenvalue problem (assumed nonsingular)

$$\boldsymbol{R}\boldsymbol{v} = \mu\,\boldsymbol{v} \tag{11.41}$$

to find the eigenvalues μ_j and the orthonormal eigenvectors (of dimension n) $\boldsymbol{v}_j$, $j = 1, \ldots, n$, the slowly varying, difference-frequency response can be represented as

$$Z_2^{SD}(t) = \sum_{j=1}^{n} \mu_j\,\big(W_{2j-1}(t)^2 + W_{2j}(t)^2\big). \tag{11.42}$$

Here $W_k(t)$, $k = 1, \ldots, 2n$, are real stationary Gaussian N(0,1)-processes, which can be determined as follows (Naess, 1985c)

$$W_k(t) = \sum_{i=-n}^{n} w_k(\omega_i)\, C_i\, e^{\mathrm{i}\omega_i t} \tag{11.43}$$

where the functions $w_k(\omega_i)$ are defined as

$$w_{2j-1}(\omega_i) = \begin{cases} \frac{1}{\sqrt{2}}\, v_j(\omega_i) & : \ \omega_i > 0 \\ \frac{1}{\sqrt{2}}\, v_j(-\omega_i)^* & : \ \omega_i < 0 \end{cases} \tag{11.44}$$

and

$$w_{2j}(\omega_i) = \begin{cases} -\frac{\mathrm{i}}{\sqrt{2}}\, v_j(\omega_i) & : \ \omega_i > 0 \\ \frac{\mathrm{i}}{\sqrt{2}}\, v_j(-\omega_i)^* & : \ \omega_i < 0 \end{cases} \tag{11.45}$$

where $v_j(\omega_i)$ denotes the i'th component of the eigenvector $\boldsymbol{v}_j$. For each fixed t, the $W_k = W_k(t)$, $k = 1, \ldots, 2n$, becomes a set of independent Gaussian variables. Note that $w_{2j}(\omega_i) = \chi(\omega_i)\, w_{2j-1}(\omega_i)$, where $\chi(\omega) = -i$ for $\omega > 0$, $(\chi(0) = 0)$, and $\chi(\omega) = i$ for $\omega < 0$. $\chi(\omega)$ is recognized as the transfer function of the Hilbert transform. Hence, it follows that that $W_{2j}(t)$ is the Hilbert transform of $W_{2j-1}(t)$. That is, $W_{2j-1}(t)^2 + W_{2j}(t)^2$ is the square of the Hilbert envelope of $W_{2j-1}(t)$ (or $W_{2j}(t)$), pointing to the slowly varying characteristics of $Z_2^{SD}(t)$.

Having achieved the representation of the slowly varying response $Z_2^{SD}(t)$ given by Eq. (11.42), we may proceed to establish a similar representation of the first-order response $Z_1(t)$. The following relation obtains

$$Z_1(t) = \sum_{j=1}^{n} \big(d_{2j-1}\, W_{2j-1}(t) + d_{2j}\, W_{2j}(t) \big), \tag{11.46}$$

where the real expansion coefficients d_k are given by the equation

$$d_{2j-1} - \mathrm{i}\, d_{2j} = \sqrt{2} \sum_{i=1}^{n} r_i\, \nu_j(\omega_i)^* . \tag{11.47}$$

The total surge response process may now be written succinctly as

$$Z(t) = \sum_{k=1}^{2n} \big(\lambda_k\, W_k(t)^2 + d_k W_k(t) \big), \tag{11.48}$$

where the eigenvalue parameters $\lambda_{2j-1} = \lambda_{2j} = \mu_j$. This response process representation facilitates significantly the statistical analysis of the response process $Z(t)$. It may be noted that the PDF of the slow drift response $Z_2^{SD}(t)$ can be calculated in closed form. Denoting the PDF of $Z_2^{SD}(t)$ by f_{SD}, Naess (1986) showed that

$$f_{SD}(z) = \sum_{j=1}^{n} \frac{l_j}{|\mu_j|}\, H\Big(\frac{z}{2\mu_j}\Big) \exp\Big\{ -\frac{z}{2\mu_j} \Big\}, \tag{11.49}$$

where $H(\cdot)$ denotes the Heaviside function defined as $H(x) = 1$ for $x > 0$, $H(0) = 1/2$, and $H(x) = 0$ for $x < 0$, and

$$l_j = \prod_{\substack{k=1 \\ k \neq j}}^{n} \Big(1 - \frac{\mu_k}{\mu_j} \Big)^{-1}. \tag{11.50}$$

Equation (11.49) was derived under the specific condition that the eigenvalues μ_j are simple, that is, that they are all different. If that is not the case, this formula has to be modified, cf. Naess (1990).

Unfortunately, no explicit formula is known for the PDF of the total response $Z(t)$. However, accurate numerical methods are available for calculating this PDF, cf. Naess and Johnsen (1992). These methods are based on first determining the characteristic function of $Z(t)$, which is done by using Eq. (11.48), and then using a numerical technique to calculate the inverse Fourier transform of the characteristic function, which provides the PDF. It is beyond the scope of this book to go into details about these methods. Instead, we present the statistical moments of $Z(t)$, which can be calculated by using Eq. (11.48), cf. Naess (1987). It is the statistical cumulants that are calculated; however, all statistical moments can be expressed in terms of the cumulants, cf. Appendix C. Denoting the cumulant of order m by κ_m, it is obtained that $\kappa_1 = \mathrm{E}[Z(t)] = \mathrm{E}[Z_2^{SD}(t)] = \sum_{j=1}^{2n} \lambda_j = 2\sum_{j=1}^{n} \mu_j$, because $\mathrm{E}[Z_1(t)] = 0$. For $m > 1$,

$$\kappa_m = \sum_{j=1}^{2n} 2^{m-1}\, (m-1)!\, \lambda_j^{m-2}\, \big(\lambda_j^2 + \frac{m}{4}\, d_j^2\big). \tag{11.51}$$

Writing out the first three of these cumulants, which are the ones most frequently used, gives

$$\kappa_2 = \mathrm{Var}[Z(t)] = \sum_{j=1}^{2n} (2\lambda_j^2 + d_j^2) \tag{11.52}$$

and

$$\kappa_3 = \sum_{j=1}^{2n} 2\lambda_j\,(4\lambda_j^2 + 3d_j^2)\,, \qquad \kappa_4 = \sum_{j=1}^{2n} 48\,\lambda_j^2\,(\lambda_j^2 + d_j^2). \tag{11.53}$$

Because Eq. (11.52) is completely general in the sense that it applies to any expression of the form given by Eq. (11.48), it may be deduced that

$$\mathrm{Var}[Z_1(t)] = \sum_{j=1}^{2n} d_j^2\,, \qquad \mathrm{Var}[Z_2^{SD}(t)] = \sum_{j=1}^{2n} 2\lambda_j^2. \tag{11.54}$$

It also follows that $Z_1(t)$ and $Z_2^{SD}(t)$ are uncorrelated random variables for each value of t because $\mathrm{Var}[Z(t)] = \mathrm{Var}[Z_1(t)] + \mathrm{Var}[Z_2^{SD}(t)]$.

κ_3 and κ_4, which are both zero for Gaussian variables, are often used as a measure of the deviation from the Gaussian probability law. $\kappa_3/\kappa^{3/2}$ is referred to as the *coefficient of skewness*, and κ_4/κ_2^2 as the *coefficient of excess* of the underlying probability distribution, cf. Appendix C.

From an applicational point of view, it is often convenient to express key parameters in terms of frequency-domain functions. To this end, the (one-sided) spectral density, say $S_{SD}^{+}(\omega)$, of the stationary process $Z_2^{SD}(t)$ is required. It was derived by Neal (1974), and it can be expressed as

$$S_{SD}^{+}(\omega) = \int_{-\infty}^{\infty} |\hat{H}_2^{SD}(\omega', \omega - \omega')|^2\, S_X^{+}(|\omega'|)\, S_X^{+}(|\omega - \omega'|)\, d\omega', \tag{11.55}$$

where $\hat{H}_2^{SD}(\omega_1, -\omega_2) = \hat{L}(\omega_1 - \omega_2)\,\hat{K}_2^{SD}(\omega_1, -\omega_2)$. From this formula, we may then derive an expression for the variance of $Z_2^{SD}(t)$,

$$\mathrm{Var}[Z_2^{SD}(t)] = \int_0^{\infty} S_{SD}^{+}(\omega)\, d\omega = \int_0^{\infty}\int_0^{\infty} |\hat{H}_2^{SD}(\omega, -\omega')|^2\, S_X^{+}(\omega)\, S_X^{+}(\omega')\, d\omega\, d\omega'. \tag{11.56}$$

The following formula for $\mathrm{E}[Z_2^{SD}(t)]$ is also useful:

$$\mathrm{E}[Z_2^{SD}(t)] = \int_0^{\infty} \hat{H}_2^{SD}(\omega, -\omega)\, S_X^{+}(\omega)\, d\omega. \tag{11.57}$$

Let us end this subsection by a consideration concerning the extreme values of the total response process $Z(t)$. As discussed in Chapter 10, a good approximation to the extreme value distribution is usually obtained if the mean upcrossing rate function of $Z(t)$ can be calculated. Because this involves the joint PDF of $Z(t)$ and its derivative $\dot{Z}(t)$, this is generally difficult. However, accurate numerical methods have been developed for calculating the mean upcrossing rate of $Z(t)$, cf. Naess (2001) and Naess et al. (2006). A detailed discussion of these methods would be beyond the scope of this book. Instead we present a simpler method, which may be used to provide initial estimates. It is based on the assumption that the two random variables $Z(t)$ and $\dot{Z}(t)$ are independent. This implies that the joint PDF of the two variables factorizes. Specifically, $f_{Z\dot{Z}}(\cdot,\cdot) = f_Z(\cdot)\cdot f_{\dot{Z}}(\cdot)$. This property is satisfied exactly in the case of stationary Gaussian response processes. By invoking Rice's formula for the average z-upcrossing rate $\nu_Z^+(z)$ of the total response $Z(t)$, it is found that

$$\nu_{tot}^+(z) = \int_0^\infty s f_{Z\dot{Z}}(z,s)\,ds = \int_0^\infty s f_{\dot{Z}}(s)\,ds \cdot f_Z(z)\,. \tag{11.58}$$

It is now observed that Eq. (11.58) leads to the result

$$\nu_Z^+(z) = \frac{\nu_Z^+(z_{ref})}{f_Z(z_{ref})} f_Z(z)\,, \tag{11.59}$$

where z_{ref} denotes some suitable reference level, typically the mean response level or similar. $f_Z(z)$ can be calculated fairly quickly and with high accuracy (Naess and Johnsen, 1991), and $\nu_Z^+(z_{ref})$ may be estimated by the following "Gaussian" combination formula:

$$\nu_Z^+(z) = \left(\frac{(\nu_{Z_1}^+(0))^2\,\sigma_1^2 + (1/T_e)^2\,\sigma_2^2}{\sigma_1^2+\sigma_2^2}\right)^{1/2}\,, \tag{11.60}$$

where $\sigma_i^2 = Var[Z_i]$, $\nu_{Z_1}^+(0)$ = the mean zero upcrossing rate of the Gaussian component $Z_1(t)$, and $\nu_{Z_2}^+(z_{ref})$ = the mean z_{ref}-upcrossing rate of the slow-drift component $Z_2(t)$. Here, we put $\nu_{Z_2}^+(z_{ref}) = 1/T_e$, where T_e denotes the appropriate slow-drift period. A slightly more involved formula for the mean upcrossing rate than Eq. (11.59) was developed by McWilliam and Langley (1993).

A method for trying to also capture the contribution of viscous effects on the slowly varying response of floating platforms was presented in Donley and Spanos (1990). For other contributions along this line, cf. Spanos and Donley (1991); Quek et al. (1994); Kareem et al. (1995); and Spanos et al. (2003).

11.3.2 Springing Response

When the TLP concept was first evaluated for application as an oil production facility offshore, one of the concerns was the possible excitation of heave, roll, and pitch motions. Example 2.2.5 discusses the natural period in heave of a TLP, and it is pointed out that this would typically have a value in the interval 2 to 4 seconds. Similar values apply to roll and pitch. Even if very little wave energy is present at such periods in sea states of appreciable severity, the concern was that these motions might still be excited by higher-order hydrodynamic forces. This concern was related to a recognition of the fact that the overall damping (including mechanical and hydrodynamic) is very low for these motion modes, so that even small higher order forces may excite resonant response, inducing significant loads in the tethers. These high-frequency tension oscillations in the vertical tethers are often referred to as *springing*, and they may have important implications for fatigue damage.

We saw in the previous subsection that second-order hydrodynamic forces on large-volume structures contain terms oscillating at both the difference and the sum of frequencies occurring in the wave spectrum. In the context of springing response, the sumfrequency terms are of particular interest because they will constitute one of the contributions to the higher-order hydrodynamic forces that may give rise to heave, roll, and pitch motions of a TLP. The importance of the second-order forces relative to other higher-order contributions does not seem to be fully understood even today. It appears that the usual approach is to assume that the second-order, sumfrequency forces constitute the dominating part, at least on a large volume TLP.

To produce time series of the sumfrequency forces on a floating structure, Eq. (11.28) can be used in combination with the sumfrequency approximation of the QTF, denoted by $\hat{K}_2^{SF}(\omega_i, -\omega_j)$, which can be expressed as

$$\hat{K}_2^{SF}(\omega_i, -\omega_j) = 0 \text{ when } \omega_i \cdot \omega_j > 0. \tag{11.61}$$

The sum-frequency approximation removes all difference frequency terms in Eq. (11.28), retaining only the sumfrequency terms. Denoting these forces by $F_2^{SF}(t)$, their representation is given by the following expression:

$$F_2^{SF}(t) = \sum_{i=-n}^{n} \sum_{j=-n}^{n} Q_{ij}^{SF} \, C_i \, C_j^* \, e^{\mathrm{i}(\omega_i - \omega_j)t}, \tag{11.62}$$

where

$$Q_{ij}^{SF} = \hat{K}_2^{SF}(\omega_i, -\omega_j) \frac{1}{2} \sqrt{S_X^+(|\omega_i|) \, S_X^+(|\omega_j|) \, \Delta\omega_i \Delta\omega_j} \,. \tag{11.63}$$

Adopting an equation similar to Eq. (11.35) for the heave response of a TLP, we can carry out an analysis entirely analogous to the one for the surge response in the previous subsection. That is, writing the heave response $Z(t) = Z_1(t) + Z_2^{SF}(t)$, we obtain

$$Z_1(t) = \sum_{j=n}^{n} r_j \, C_j \, e^{\mathrm{i}\omega_j t}, \tag{11.64}$$

where $r_j = q_j \, \hat{L}(\omega_j)$, and

$$Z_2^{SF}(t) = \sum_{i=-n}^{n} \sum_{j=-n}^{n} R_{ij}^{SF} \, C_i \, C_j^* \, e^{\mathrm{i}(\omega_i - \omega_j)t}, \tag{11.65}$$

where $R_{ij}^{SF} = Q_{ij}^{SF} \, \hat{L}(\omega_i - \omega_j)$. In these equations, all transfer functions have, of course, been appropriately modified to fit the heave response case.

For the statistical analysis of the combined first-order, wave frequency response and the second-order, sumfrequency response we use a recipe similar to that used in the previous subsection. By introducing the matrix $\boldsymbol{S} = (R_{ij}^{SF})$, where $i = -n, \ldots, -1$ and $j = 1, \ldots, n$, then the total aggregate of terms R_{ij}^{SF} can be represented by the $2n \times 2n$-matrix

$$\boldsymbol{T} = \begin{pmatrix} \boldsymbol{0} & \boldsymbol{S} \\ \boldsymbol{S}^H & \boldsymbol{0} \end{pmatrix} \tag{11.66}$$

where $\boldsymbol{0}$ is an $n \times n$ zero matrix and $\boldsymbol{S}^H = (\boldsymbol{S}^*)^T$ is the Hermitian conjugate of $\boldsymbol{S}$. The desired representation is obtained by solving the eigenvalue problem

$$\boldsymbol{T}\boldsymbol{\nu} = \lambda \boldsymbol{\nu} \,. \tag{11.67}$$

It is assumed that $\boldsymbol{S}$, and thereby $\boldsymbol{T}$, are nonsingular. Let λ_j and $\boldsymbol{\nu}_j = (\nu_{j,-n}, \ldots, \nu_{j,-1}, \nu_{j,1}, \ldots, \nu_{j,n})^T$, $j = -n, \ldots, -1, 1, \ldots, n$, denote the eigenvalues and orthonormal eigenvectors of $\boldsymbol{T}$. As shown by Naess and Ness (1992), it is appropriate to consider each eigenvector $\boldsymbol{v}_j$ as a function of frequency by writing $\nu_j(\omega_i) = \nu_{j,i}$, where $\nu_{j,i}$ is the i'th component of ν_j. They also showed that the eigenvalues may be ordered so that $\lambda_{-j} = -\lambda_j$ for $j = 1, \ldots, n$, and that $\nu_{-j}(\omega_i) = \chi(\omega_i)\,\nu_j(\omega_i)$, where $\chi(\omega)$ is the transfer function of the Hilbert transform. Using results from Naess (1990), the following representation was then obtained:

$$Z_2^{SF}(t) = \sum_{j=1}^{n} \lambda_j \left(W_j(t)^2 - W_{-j}(t)^2 \right), \tag{11.68}$$

where $W_k(t)$, $k = -n, \ldots, -1, 1, \ldots, n$, are real stationary Gaussian N(0,1)-processes, which can be determined as follows,

$$W_k(t) = \sum_{i=-n}^{n} \nu_k(\omega_i)\, C_i\, e^{i\omega_i t}\,, \tag{11.69}$$

which are zero-mean, stationary real Gaussian processes. Because $\nu_{-j}(\omega_i) = \chi(\omega_i)\cdot \nu_j(\omega_i)$, it follows that $W_{-j}(t)$ is the Hilbert transform of $W_j(t)$, $j = 1, \ldots, n$. From this, it may be deduced that the spectral density of $W_j(t)^2 - W_{-j}(t)^2$ only contains sums of pairs of frequencies occurring in the wave spectrum $S_X^+(\omega)$. This can be demonstrated by writing the representation of $W_j(t)$ as follows:

$$W_j(t) = \sum_{i=1}^{n} U_{j,i}\, \cos(\omega_i t + \Theta_{j,i})\,, \tag{11.70}$$

where $U_{j,i}$ and $\Theta_{j,i}$, $i, j = 1, \ldots, n$, are independent sets of independent random variables, $U_{j,i}$ being Rayleigh distributed and $\Theta_{j,i}$ uniformly distributed, and $\omega_1 < \cdots < \omega_n$ is the initial discretization of the wave spectrum. Using the fact that the Hilbert transform of $\cos(\omega_i t + \Theta_{j,i})$ is $\sin(\omega_i t + \Theta_{j,i})$, then

$$W_{-j}(t) = \sum_{i=1}^{n} U_{j,i}\, \sin(\omega_i t + \Theta_{j,i})\,, \tag{11.71}$$

which gives

$$W_j(t)^2 - W_{-j}(t)^2 = \sum_{i=1}^{n}\sum_{k=1}^{n} U_{j,i} U_{j,k}\, \cos\left((\omega_i + \omega_k)t + \Theta_{j,i} + \Theta_{j,k}\right). \tag{11.72}$$

Hence, it follows that the only frequencies occurring in the representation of $Z_2^{SF}(t)$ given by the rhs of Eq. (11.68) are sums of pairs of frequencies from the initial discretization of the wave spectrum.

The first-order response $Z_1(t)$ can now be represented in terms of the "eigen-processes" $W_k(t)$ (Naess and Ness, 1992):

$$Z_1(t) = \sum_{k=-n}^{n} c_k\, W_k(t), \tag{11.73}$$

where

$$c_k = \sum_{i=-n}^{n} r_i\, \nu_k(\omega_i)^*\,. \tag{11.74}$$

The total response $Z(t) = Z_1(t) + Z_2^{SF}(t)$ can then be written as

$$Z(t) = \sum_{k=-n}^{n} \left(\lambda_k W_k(t)^2 + c_k W_k(t) \right). \tag{11.75}$$

Adapting Eq. (11.51) to this case, we can go on to calculate the statistical moments. It is found that $\kappa_1 = \mathrm{E}[Z(t)] = \mathrm{E}[Z_2^{SF}(t)] = \sum_{k=-n}^{n} \lambda_k = 0$, because obviously $\mathrm{E}[Z_1(t)] = 0$. For $m > 1$,

$$\kappa_m = \sum_{k=-n}^{n} 2^{m-1} (m-1)! \, \lambda_k^{m-2} \left(\lambda_k^2 + \frac{m}{4} c_k^2 \right), \tag{11.76}$$

which immediately leads to the conclusion that $\kappa_{2m'+1} = 0$ for $m' = 1, 2, \ldots$, pointing to the symmetry of the probability distribution of $Z(t)$. Also,

$$\kappa_2 = \mathrm{Var}[Z(t)] = \sum_{k=-n}^{n} (2\lambda_k^2 + c_k^2), \tag{11.77}$$

which again leads to the conclusion that $Z_1(t)$ and $Z_2^{SF}(t)$ are uncorrelated random variables for each t. Likewise, it follows that

$$\kappa_4 = \sum_{k=-n}^{n} 48 \, \lambda_k^2 \left(\lambda_k^2 + c_k^2 \right). \tag{11.78}$$

Even if it is possible to calculate the statistical moments of $Z(t)$ quite easily, no explicit, closed-form expression has been derived for the probability distribution of either $Z(t)$ or $Z_2^{SF}(t)$. However, the numerical method presented by Naess and Johnsen (1991) makes it possible to calculate the PDF with high accuracy. Empirical estimation of the PDF by Monte Carlo simulation is, of course, also available, and offers an attractive alternative due to its conceptual simplicity.

Let us now assume that the goal is to determine the total vertical response $Y(t)$ of a point in a leg of a TLP. To simplify the analysis, it is assumed that the surge, sway, and yaw motions are decoupled from the heave, roll, and pitch motions. This results in a 3DOF system of equations of motion:

$$\boldsymbol{M} \ddot{\boldsymbol{Z}}(t) + \boldsymbol{C} \dot{\boldsymbol{Z}}(t) + \boldsymbol{K} \boldsymbol{Z}(t) = \boldsymbol{F}_1(t) + \boldsymbol{F}_2^{SF}(t). \tag{11.79}$$

Here, $\boldsymbol{M}$ = the total mass matrix, $\boldsymbol{C}$ = the damping matrix and $\boldsymbol{K}$ = the stiffness matrix, and $\boldsymbol{F}_1(t)$ and $\boldsymbol{F}_2^{SF}(t)$ are the requisite first- and second-order excitation forces. $\boldsymbol{Z}(t) = \boldsymbol{Z}_1(t) + \boldsymbol{Z}_2^{SF}(t)$, where $\boldsymbol{Z}_1(t) = (Z_{1,1}(t), Z_{1,2}(t), Z_{1,3}(t))^T$ and $\boldsymbol{Z}_2^{SF}(t) = (Z_{2,1}^{SF}(t), Z_{2,2}^{SF}(t), Z_{2,3}^{SF}(t))^T$, and where the second subindex refers to heave, roll and pitch, respectively. In general, heave, roll, and pitch are coupled, which implies that the 3×3 matrices in Eq. (11.79) are nondiagonal.

Equation (11.79) is associated with a linear transfer function matrix $\hat{\boldsymbol{L}}(\omega) = (\hat{L}_{kl}(\omega))$, $k, l = 1, 2, 3$, given by the expression

$$\hat{\boldsymbol{L}}(\omega) = \left(-\omega^2 \boldsymbol{M} + \mathrm{i}\omega \boldsymbol{C} + \boldsymbol{K} \right)^{-1}. \tag{11.80}$$

Because the relevant motions of the TLP are small compared to the platform dimensions, we may write $Y(t) = Y_1(t) + Y_2^{SF}(t)$, where

$$Y_1(t) = \sum_{k=1}^{3} a_k \, Z_{1,k}(t) = \boldsymbol{a}^T \boldsymbol{Z}_1(t) , \tag{11.81}$$

and

$$Y_2^{SF}(t) = \sum_{k=1}^{3} a_k \, Z_{2,k}^{SF}(t) = \boldsymbol{a}^T \boldsymbol{Z}_2^{SF}(t), \tag{11.82}$$

for a suitable choice of $\boldsymbol{a} = (a_1, a_2, a_3)^T$.

We may now proceed to establish the $\boldsymbol{T}$-matrix associated with the sumfrequency response $Y_2^{SF}(t)$. It is found that the entries R_{ij}^{SF} in this case are given as follows:

$$R_{ij}^{SF} = \frac{1}{2} \sqrt{S_X^{+}(|\omega_i|)\, S_X^{+}(|\omega_j|)\, \Delta\omega_i \Delta\omega_j} \sum_{k=1}^{3} \sum_{l=1}^{3} a_k \hat{L}_{kl}(\omega_i - \omega_j)\, \hat{K}_{2,l}^{SF}(\omega_i, -\omega_j) , \tag{11.83}$$

where $\hat{K}_{2,l}^{SF}(\omega_i, -\omega_j)$ refers to the QTF for the sumfrequency force for motion mode l. The representation and statistical analysis of the response process $Y(t)$ may now proceed in a manner completely analogous to that already described.

11.3.3 Ship Hull Vibrations

In this section, we discuss the response statistics of the vertical vibration modes of a ship hull subjected to random sea waves (Naess, 1996). Only the case of stationary response is considered, and the description is based on a quadratic strip theory developed by Jensen and Pedersen (1979, 1981). This theory makes it possible to calculate the vertical vibration response of the ship hull taking due account up to second order of nonlinearities in the exciting waves, nonvertical ship sides, and nonlinear hydrodynamic forces. The flexibility of the ship is also taken into account by modeling the ship hull as a nonprismatic Timoshenko beam with variable mass and stiffness (Jensen and Pedersen, 1979). On the basis of this theory, it is then possible to express the vibration response as a sum of two components, viz. a linear and a quadratic transformation of the sea surface elevation process. More specifically, the response process can be written as a Volterra series truncated after the quadratic term.

In accordance with the theory developed by Jensen and Pedersen (1979, 1981), the wave-induced bending moment in a ship hull subjected to a long-crested random seaway can be written as a sum of a linear and a quadratic component. Specifically, let $M(t; x)$ denote the bending moment response process at the local position x along the hull, see Fig. 11.2.

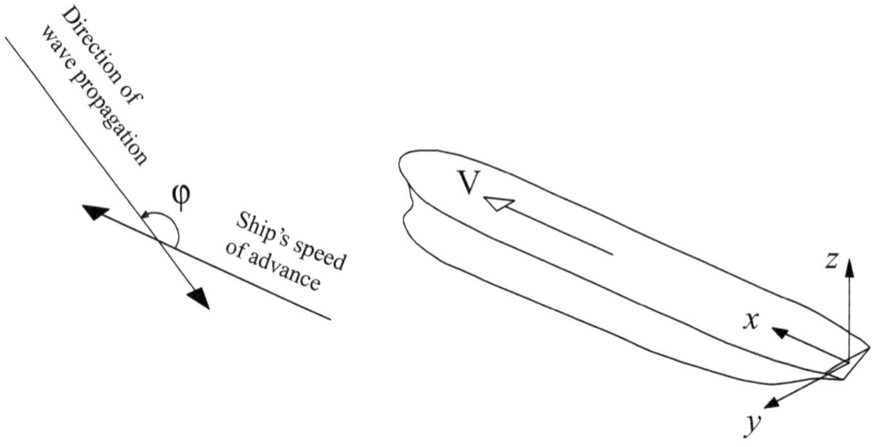

Figure 11.2. Definition of local coordinate system

One may write

$$M(t;x) = M_1(t;x) + M_2(t;x), \tag{11.84}$$

where the subscript 1 signifies the linear component and the subscript 2 refers to the quadratic part. In this chapter, the two components are expressed as,

$$M_1(t;x) = \sum_{j=-n}^{n} q_j B_j e^{i\omega_j t}, \tag{11.85}$$

where

$$q_j = \hat{M}_1(\sigma_j; V, \varphi, x)\left(\frac{1}{2} S_X^+(|\sigma_j|)\Delta\sigma\right)^{1/2}, \tag{11.86}$$

and

$$M_2(t;x) = \sum_{i=-n}^{n} \sum_{j=-n}^{n} Q_{ij} B_i B_j^* e^{i(\omega_i - \omega_j)t}, \tag{11.87}$$

where

$$Q_{ij} = \hat{M}_2(\sigma_i, -\sigma_j; V, \varphi, x)\frac{1}{2}\left[S_X^+(|\sigma_i|) S_X^+(|\sigma_j|)\right]^{1/2} \Delta\sigma. \tag{11.88}$$

The summation index omits zero whenever it runs from negative to positive values. $0 \le \sigma_0 < \sigma_1 < \cdots < \sigma_{n+1}$ is an equidistant discretization of the positive frequency axis of the specified sea state, $\Delta\sigma = (\sigma_{i+1} - \sigma_{i-1})/2$ and $\sigma_{-i} = -\sigma_i$. The assumption of an equidistant discretization is adopted for simplicity and is not necessary. The formulas are easily adapted to cover the situation of nonequidistant discretization. V denotes the speed of the ship, and φ denotes the angle of advance relative to the long-crested seaway, cf. Fig. 11.2. The resulting encounter frequency ω corresponding to a wave frequency σ is assumed to be given by the relation

$$\omega = \omega(\sigma) = \text{sign}(\sigma)\left|\sigma - \frac{\sigma|\sigma|}{g} V \cos\varphi\right| \tag{11.89}$$

where g denotes the gravitational acceleration. This formula ensures that the encounter frequency always has the same sign as the wave frequency, which is convenient for the mathematical treatment where both positive and negative frequencies are needed. $\{B_i\}$, $i = 1, \ldots, n$, is a set of independent, complex Gaussian N(0,1)-variables with independent, identically distributed real and imaginary parts. These variables are extended to negative indices by imposing the relation $B_{-i} = B_i^*$, where * signifies complex conjugation.

The expressions for the linear transfer function (LTF) $\hat{M}_1(\sigma_i) = \hat{M}_1(\sigma_i; V, \varphi, x)$ and the quadratic transfer function (QTF) $\hat{M}_2(\sigma_i, -\sigma_j) = \hat{M}_2(\sigma_i, -\sigma_j; V, \varphi, x)$ are given by Jensen and Pedersen (1979). To facilitate the comparison between the expressions we use here and those in Jensen and Pedersen (1979), the relations between the transfer functions in the case of countering seas, i.e., $\cos\varphi < 0$, $(i, j = 1, \ldots, n)$ are as follows:

$$\hat{M}_1(\sigma_i) = M_i^c + \mathrm{i}\, M_i^s\,, \tag{11.90}$$

$$\hat{M}_2(\sigma_i, -\sigma_j) = 2\{M_{ij}^{c-} + \mathrm{i}\, M_{ij}^{s-}\}\,, \tag{11.91}$$

$$\hat{M}_2(\sigma_i, \sigma_j) = 2\{M_{ij}^{c+} + \mathrm{i}\, M_{ij}^{s+}\}\,. \tag{11.92}$$

The aggregate of terms Q_{ij} indexed by $i, j = -n, \ldots, -1, 1, \ldots, n$ is represented by a $2n \times 2n$ matrix $\boldsymbol{Q} = (Q_{ij})$, which is Hermitian, that is, $\boldsymbol{Q} = \boldsymbol{Q}^H = \boldsymbol{Q}^{*T}$. By invoking the representation theorem proved by Naess (1987, 1990) and expressing it in discretized form, it follows that the eigenvalue problem that has to be solved to obtain the desired representation in this context can be written as

$$\boldsymbol{Q}\boldsymbol{w} = \mu \boldsymbol{w}\,. \tag{11.93}$$

It is assumed that $\boldsymbol{Q}$ is nonsingular. Let μ_α and $\boldsymbol{w}_\alpha = (w_{\alpha,-n}, \ldots, w_{\alpha,-1}, w_{\alpha,1}, \ldots, w_{\alpha,n})^T$, $\alpha = -n, \ldots, -1, 1, \ldots, n$, denote the eigenvalues and orthonormal eigenvectors of $\boldsymbol{Q}$. Without loss of generality, it can also be assumed that $w_{\alpha,-i} = w_{\alpha,i}^*$.

The following decomposition of $\boldsymbol{Q}$ (Naess, 1990) is obtained:

$$Q_{ij} = \sum_{\alpha=-n}^{n} \mu_\alpha w_\alpha(\sigma_i) w_\alpha(\sigma_j)^*\,, \tag{11.94}$$

where $w_\alpha(\sigma_i) = w_{\alpha,i}$.

By substituting for Q_{ij} given by Eq. (11.94) into Eq. (11.87), it is obtained that

$$M_2(t; x) = \sum_{\alpha=-n}^{n} \mu_\alpha W_\alpha(t)^2\,. \tag{11.95}$$

Here

$$W_\alpha(t) = \sum_{i=-n}^{n} w_\alpha(\sigma_i) B_i e^{i\omega_i t}, \tag{11.96}$$

which is a zero-mean, stationary, and real Gaussian process of unit variance. For t fixed, the $W_\alpha = W_\alpha(t)$, $\alpha = -n, \ldots, -1, 1, \ldots, n$ constitute a set of independent variables. Note that the (one-sided) spectral density $S_\alpha^+(\omega)$ of the $W_\alpha(t)$ process, expressed in terms of the encounter frequency ω, is given as follows:

$$S_\alpha^+(\omega_i) = \sum_{\{j\,;\,\omega(\sigma_j)=\omega_i\}} 2\,|w_\alpha(\sigma_j)|^2 \left|\frac{d\omega_i}{d\sigma_j}\right|^{-1}, \quad 0 < \omega_1 < \omega_2 < \cdots < \omega_m. \tag{11.97}$$

To obtain the corresponding representation for the first-order response $M_1(t;x)$, the assumed completeness of the orthonormal set $\{\boldsymbol{w}_\alpha\}$ is invoked. This implies that the following expansion is valid:

$$q_i = \sum_{\alpha=-n}^{n} c_\alpha w_\alpha(\sigma_i), \quad i = -n, \ldots, -1, 1, \ldots, n. \tag{11.98}$$

The orthonormality of the eigenvectors then leads to the expressions ($q_{-i} = q_i^*$, $w_\alpha(\sigma_{-i}) = w_\alpha(\sigma_i)^*$)

$$c_\alpha = \sum_{i=-n}^{n} q_i w_\alpha(\sigma_i)^*. \tag{11.99}$$

By substituting from Eq. (11.98) into Eq. (11.85), $M_1(t;x)$ can be represented as follows:

$$M_1(t;x) = \sum_{\alpha=-n}^{n} c_\alpha W_\alpha(t). \tag{11.100}$$

To get a first idea of how much the probability distribution of the response deviates from a Gaussian law, one may calculate the coefficients of skewness and excess. To this end we write down the expressions for the first four cumulants k_j, $j = 1, \ldots, 4$, of the total response $M(t;x)$. Clearly,

$$\mathrm{E}[M(t;x)] = \mathrm{E}[M_2(t;x)] = k_1 = \sum_{\alpha=-n}^{n} \mu_\alpha. \tag{11.101}$$

Invoking Eq. (11.76), it is obtained that

$$\mathrm{Var}[M(t;x)] = k_2 = \sum_{\alpha=-n}^{n} (2\mu_\alpha^2 + c_\alpha^2), \tag{11.102}$$

$$k_3 = \sum_{\alpha=-n}^{n} (8\mu_\alpha^3 + 6\mu_\alpha c_\alpha^2), \tag{11.103}$$

$$k_4 = \sum_{\alpha=-n}^{n} 48\mu_\alpha^2(\mu_\alpha^2 + c_\alpha^2). \tag{11.104}$$

Referring to Eqs. (11.87) and (11.88), and Eqs. (11.91) and (11.92), the second-order response can be divided into a differencefrequency and a sumfrequency component, that is,

$$M_2(t;x) = M_2^-(t;x) + M_2^+(t;x)\,, \tag{11.105}$$

where

$$M_2^{\pm}(t;x) = \sum_{i=-n}^{n}\sum_{j=-n}^{n} Q_{ij}^{\pm} B_i B_j^* e^{i(\omega_i - \omega_j)t}\,. \tag{11.106}$$

The matrices $\boldsymbol{Q}^{\pm} = \left(Q_{ij}^{\pm}\right)$ are then given as follows:

$$\boldsymbol{Q}^- = \left(Q_{ij}^-\right) = \begin{pmatrix} \boldsymbol{R}^G & \boldsymbol{O} \\ \boldsymbol{O} & \boldsymbol{R} \end{pmatrix}, \tag{11.107}$$

where $\boldsymbol{R} = (R_{ij}), i, j = 1, \ldots, n,$

$$Q_{ij}^- = R_{ij} = \hat{M}_2(\sigma_i, -\sigma_j)\frac{1}{2}\left[S_X^+(\sigma_i)S_X^+(\sigma_j)\right]^{1/2}\Delta\sigma\,; i, j = 1, , 2, \ldots, n\,, \tag{11.108}$$

and

$$\boldsymbol{Q}^+ = \left(Q_{ij}^+\right) = \begin{pmatrix} \boldsymbol{O} & \boldsymbol{S}^H \\ \boldsymbol{S} & \boldsymbol{O} \end{pmatrix}, \tag{11.109}$$

where $\boldsymbol{S} = (S_{ij}), i, j = 1, \ldots, n,$

$$Q_{i,-n+j-1}^+ = S_{ij} = \hat{M}_2(\sigma_i, \sigma_j)\frac{1}{2}\left[S_X^+(\sigma_i)S_X^+(\sigma_j)\right]^{1/2}\Delta\sigma\,; i, j = 1, , 2, \ldots, n\,. \tag{11.110}$$

In Eqs. (11.107) and (11.109), $\boldsymbol{O}$ denotes an $n \times n$ zero matrix and $\boldsymbol{R}^G = \left(\boldsymbol{R}^*\right)^{ST}$, where ST denotes the operation $\boldsymbol{A} = \boldsymbol{B}^{ST}$ if $A_{ij} = B_{n-i+1,n-j+1}$.

11.4 Statistics of Wind Load and Response

The standard engineering approximation to the modeling of the gust wind process $U(t)$ is to assume that it is a zero-mean, stationary Gaussian process for each short-term condition. By adopting the approximation of Eq. (7.16), it also follows that the wind load process becomes a stationary Gaussian process with a spectral density given by, for example, Eq. (7.20) or (7.21). If the along-wind response $X(t)$ of a structure subjected to such a wind load process can be modeled as a linear, time-invariant system with transfer function $H(\omega)$, the response also becomes a stationary Gaussian process with a mean response given as (Eq. (9.7))

$$m_X = \mathrm{E}[X(t)] = H(0)\overline{F}_w = \frac{1}{2}\rho_a C_w A \overline{W}^2\, H(0)\,, \tag{11.111}$$

and a spectral density (Eq. (9.17))

$$S_X^+(\omega) = |H(\omega)|^2 S_F^+(\omega) = 4\frac{\overline{F}_w^2}{\overline{W}^2}\left[\chi\left(\frac{\omega\sqrt{A}}{2\pi\overline{W}}\right)\right]^2 |H(\omega)|^2\, S_U^+(\omega)\,. \tag{11.112}$$

The modeling of wind load and response as described in this section is extensively used in offshore engineering. However, in the case of linear, time-invariant dynamics, the approximation expressed by Eq. (7.16), which is usually quite good, can be avoided by invoking the second-order stochastic analysis described in the previous section, cf. Naess (1987). This would allow us to use the full expression for the wind load given by Eq. (7.15), which is quadratic in the gust wind speed process $U(t)$.

12 Short-Term and Long-Term Extremes

12.1 Introduction

For true probabilistic design of load-carrying members of offshore structures that have to withstand the effects of the environmental forces generated by wind, waves, and current, it would be necessary to provide a simultaneous probabilistic model of all three load-generating sources. As an alternative, a simplified conservative approach may be applied. According to NORSOK N-003 (2007), the 100-year environmental condition can be represented by the 100-year wave and mean wind speed, and the 10-year current.

In the following sections, the focus is on the three commonly used approaches for calculating characteristic values of extreme load effects due to wave forces for ULS design checks, viz.,

- the design wave approach
- the design sea state approach
- the full long-term approach

12.2 Design Wave Approach

For the design of offshore structures where the load effects to be considered are primarily of a quasistatic nature, the extreme load effect for ULS design check can be estimated by the so-called 100-year wave approach. This is understood to mean a wave with a wave height $H^{(100)}$ being exceeded on the average only once every 100 years. Equivalently, $H^{(100)}$ can be interpreted as the wave height that is exceeded during a period of one year with a probability of 10^{-2}, cf. Section 10.8.

Estimation of load effects according to this format would in practice proceed as follows. (1) The design wave height is established on the basis of available data relevant for the offshore location in question. (2) A suitable range of corresponding wave periods, and, if relevant, wave directions, is specified. (3) According to best practice, items 1 and 2 are combined to provide a range of wave profiles for which the corresponding load effects on the structure are established, either by numerical calculations using state-of-the-art computer codes, or in some cases, model tests.

In general, design guidelines applicable in a specific case will specify how the design wave should be chosen. For example, to simplify the calculations if more accurate estimates are not available, NORSOK N-003 (2007) suggested to take $H^{(100)} = 1.9 \cdot H_s$, where H_s denotes the significant wave height corresponding to an annual exceedance probability of 10^{-2}. The same document also recommends varying the corresponding wave period T in the range $\sqrt{6.5 \cdot H^{(100)}} \le T \le \sqrt{11 \cdot H^{(100)}}$, where the left hand side criterion is based on a limiting wave steepness H/λ of 1/10.

12.3 Short-Term Design Approach

When the load effects to be taken into account in the design of load-carrying members are expected to have significant dynamic components, it has been good practice to carry out the design check for load effects established during a 100-year storm of specified duration (e.g., 3 or 6 hours). Traditionally, the notion of a 100-year storm was interpreted to mean a stationary storm condition with a significant wave height not exceeded on the average more than once every 100 years. A corresponding spectral period was then chosen according to some specified rule or table.

In recent years, the environmental contour line approach (Winterstein et al., 1993) has been advocated as a rational basis for choosing the appropriate short-term design storms leading to load and response extremes corresponding to a prescribed return period, e.g., 100 years, or equivalently, a prescribed annual probability of exceedance, which otherwise has to be obtained from a long-term analysis.

Environmental contour line plots are convenient tools for complicated structural dynamic systems where a full long-term response analysis is extremely time consuming. Environmental contour lines make it possible to obtain reasonable long-term extremes by concentrating the short-term considerations to a few sea states in the scatter diagram.

The contour line approach can be applied for an offshore site if the joint PDF for the significant wave height and the spectral peak period is available in the form of a joint model as described by Eqs. (12.16)–(12.18). This joint model must be fitted to the available data given in the form of a scatter diagram such as the one in Table 12.1. As demonstrated by Moan et al. (2005), prediction of extreme values is very sensitive to the amount of environmental data available to represent the long-term variability of the sea states. The fitting of appropriate analytical PDFs ensures a smoothing and can facilitate a reasonably accurate representation of the long-term extreme response.

Contour lines corresponding to a constant annual exceedance probability can be obtained by transforming the joint model to a space consisting of independent, standard Gaussian variables and then using the inverse first order reliability method (IFORM), (see, e.g., Winterstein et al. (1993)). In the standard Gaussian space, the contour line corresponding to an annual exceedance probability of q will be circles with radius $r = \Phi^{-1}(1 - q/2{,}920)$, where Φ denotes the CDF of a standard Gaussian variable, and $2{,}920 = 365 \times 8$ is the number of 3-hour sea states per year. Transforming these circles back to the physical parameter space provides the q-probability contour lines. Approximate contour lines can be obtained by determining the probability density for the point defined by the marginal q-probability significant wave height and the conditional median spectral peak period, and then estimating

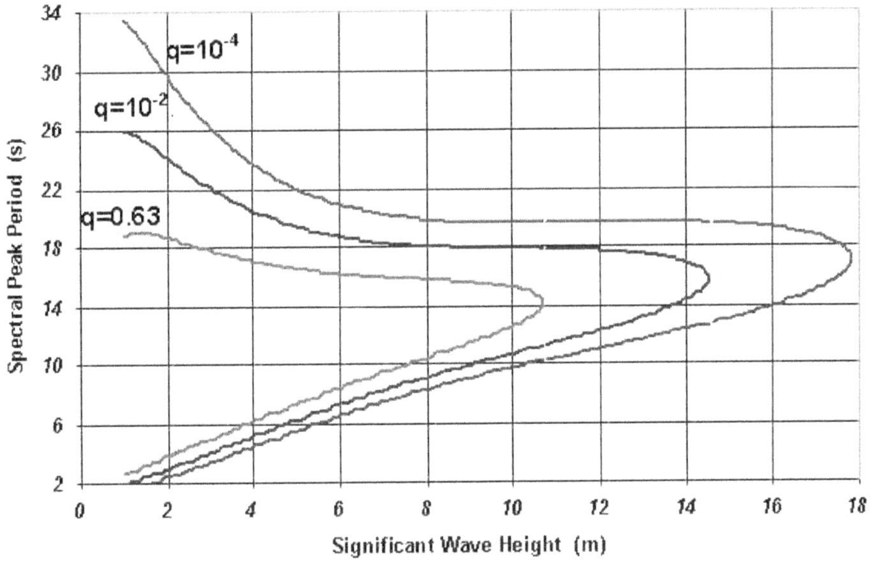

Figure 12.1. Environmental contour line plot for the wave condition in the Statfjord area.

the q-probability contour line by the line of constant probability density. Contour lines based on the joint model discussed in Section 12.4 are plotted in Fig. 12.1, cf. Haver (2002). Even for the most complicated systems, simple methods may often be used to identify the most critical range of the q-probability contour line regarding a prediction of the q-probability response extreme.

The advantage of this method is that analyses of only a few sea states are required. As the most unfavorable sea state along the q-probability contour line is identified, a proper estimate for the q-probability response is taken as the p-fractile of the distribution of the 3-hour extreme response value. It is important to note that the median 3-hour extreme value for this sea state, i.e., $p = 0.50$, will not represent a proper estimate for the q-probability extreme value because this characteristic value will not account for the inherent randomness of the 3-hour extreme value. The fractile level, p, will depend on the aimed exceedance probability target, q, and the degree of nonlinearity of the system. For most practical systems, $p = 0.90$ seems reasonable for $q = 10^{-2}$, while $p = 0.95$ may be more adequate for $q = 10^{-4}$. As an alternative to using a p-fractile above 0.5, the desired load effect may be obtained by multiplying the median or expected maximum value with a factor of 1.2—1.3. Anyway, this simplified contour line method needs to be validated by a full long-term analysis for the relevant type of environmental conditions and load effects.

12.3.1 Short-Term Extreme Values

In the short-term design approach, it is necessary to calculate the extreme loads and responses during a short-term storm condition. In many cases, that would be done by assuming that the stochastic load or response process is stationary and Gaussian, where frequency domain methods can be applied. This case is studied in Chapter 10, and the details are found there. In case the responses cannot be reasonably assumed to be Gaussian, there may be several other options available for determining the relevant extreme values. Often the simplest and most direct approach is to use the Monte Carlo simulation method. This topic is discussed in Chapter 16.

12.4 Long-Term Design Approach

In principle, the most accurate design approach to determine extreme load effects for ULS design checks is based on long-term statistics of response, but it may clearly not be the most economic method from a computational point of view because it may involve response calculations for sea states that contribute little or nothing to the design loads or load effects. Simplified approaches described previously need to be validated by the full long-term approach. Moreover, a full long-term analysis is usually required for the fatigue design as described in Chapter 13.

Clearly, the estimation of the extreme loads or load effects on marine structures subjected to the ocean environment over the design life of the structure must take into account the changing weather conditions. This is done in a consistent manner by invoking the appropriate long-term statistical methods, which are described in this section.

There are basically three different approaches to estimating characteristic long-term extreme values. These methods are based on (1) all peak values, (2) all short-term extremes, or (3) the long-term extreme value. A more detailed description follows, where $X(t)$ denotes a zero-mean stochastic process, for example the wave elevation or a corresponding load effect, that reflects the changing environmental conditions. Therefore, $X(t)$ is a nonstationary process. Let T denote the long-term time duration, e.g., 1 year or a service life of, e.g., 30 years, and let $\tilde{T}$ denote the duration of each short-term weather condition, assuming that $T = N\tilde{T}$, where N is a large integer. The long-term situation is considered to be a sequence of N short-term conditions, where each short-term condition is assumed to be stationary. Significantly, around the world there are different kinds of weather conditions, also at sea. A coarse characterization of sea states:

- extratropical, with slowly varying wave conditions
- tropical, with rare hurricanes that represent very rapidly changing weather conditions

In this section, the discussions and derivations are limited to extratropical conditions.

Let W denote the vector of parameters that describes the short-term environmental condition. W can be considered as a random vector variable. For simplicity, let us assume that $W = (H_s, T_s)$, where H_s is the significant wave height and T_s a suitable spectral period (generic notation). For example, T_s may represent the spectral peak period T_p or the mean zero-crossing period T_z. In principle, the analysis is entirely similar if W contains more parameters, e.g., dominant wave direction, wind speed, etc.

12.4.1 All Peak Values

A peak value of $X(t)$, denoted generically by X_p, is defined here as the maximum value of $X(t)$ between two consecutive zero upcrossings. For each short-term condition, let $F_{X_p|H_sT_s}(\xi|h_s, t_s)$ denote the conditional CDF of the peak value. Battjes (1970) showed that the long-term CDF $F_{X_p}(\xi)$ of the peak value X_p is given as follows:

$$F_{X_p}(\xi) = \frac{1}{\overline{\nu_X^+(0)}} \int_{h_s} \int_{t_s} \nu_X^+(0|h_s, t_s)\, F_{X_p|H_sT_s}(\xi|h_s, t_s)\, f_{H_sT_s}(h_s, t_s)\, dh_s dt_s\,, \qquad (12.1)$$

where $\overline{\nu_X^+(0)}$ denotes the long-term average zero-upcrossing rate given by

$$\overline{\nu_X^+(0)} = \int_{h_s} \int_{t_s} \nu_X^+(0|h_s, t_s)\; f_{H_sT_s}(h_s, t_s)\, dh_s dt_s\,. \qquad (12.2)$$

Here, $\nu_X^+(0|h_s, t_s)$ denotes the average zero-upcrossing rate for the short-term stationary condition characterized by $H_s = h_s$ and $T_s = t_s$.

In practical applications, a commonly adopted statistical distribution for the peak values in a short-term condition is the Rayleigh distribution, that is,

$$F_{X_p|H_sT_s}(\xi|h_s, t_s) = 1 - \exp\Big(- \frac{\xi^2}{2\, \sigma_X(h_s, t_s)^2} \Big)\,. \qquad (12.3)$$

Under the assumption that all peak values can be considered as statistically independent, which may not always be very accurate, the peak value ξ_q with a probability q of being exceeded per year is found by solving the following equation:

$$F_{X_p}(\xi_q) = 1 - \frac{q}{365 \cdot 24 \cdot 3600 \cdot \overline{\nu_X^+(0)}}\,. \qquad (12.4)$$

The short-term duration $\tilde{T}$ does not enter into this analysis. In the long run, the relative frequency of the various sea states is reflected in the PDF $f_{H_sT_s}(h_s, t_s)$, which can be approximated by using an appropriate scatter diagram, if that is available. An example of such a scatter diagram is shown in Table 12.1.

Let the scatter diagram be divided into m intervals for the h_s-values and n intervals for the t_s-values. It may often be an acceptable approximation to assume that $\nu_X^+(0|h_s, t_s) = T_z^{-1} \approx c\, T_s^{-1}$ for a fixed constant c. Equation (12.1) with the Rayleigh approximation for $F_{X_p|H_sT_s}(\xi|h_s, t_s)$ can then be approximately expressed in the following form:

$$F_{X_p}(\xi) \approx \frac{1}{\overline{t_s^{-1}}} \sum_{i=1}^{m} \sum_{j=1}^{n} \Big\{ 1 - \exp\Big(- \frac{\xi^2}{2\, \sigma_X(h_i, t_j)^2} \Big) \Big\} \frac{N_{ij}}{t_j\, N}\,, \qquad (12.5)$$

where

$$\overline{t_s^{-1}} = \sum_{i=1}^{m} \sum_{j=1}^{n} \frac{N_{ij}}{t_j\, N}\,. \qquad (12.6)$$

Here, N_{ij} equals the number of observations in the h_s-interval $(h_i - \Delta h/2, h_i + \Delta h/2)$ and t_s-interval $(t_j - \Delta t/2, t_j + \Delta t/2)$; $i = 1, \ldots, m$, $j = 1, \ldots, n$. $N = \sum_{i=1}^{m} \sum_{j=1}^{n} N_{ij}$ is the total number of observations. Also note that the values for h_s and t_s included on the scatter diagram in Table 12.1 are upper-class limits, that is, $h_i + \Delta h/2$ and $t_j + \Delta t/2$.

Table 12.1. *Scatter diagram northern North Sea, 1973–2001*

	t_p(s)																	
h_s(m)	**3**	**4**	**5**	**6**	**7**	**8**	**9**	**10**	**11**	**12**	**13**	**14**	**15**	**16**	**17**	**18**	**19**	**20**	**> 20**
0.5	18	15	123	113	110	390	260	91	38	42	32	3	19	13	9	1	3	2	7
1.0	16	49	675	433	589	1,442	1,802	959	273	344	125	33	64	29	13	1	7	1	6
1.5	5	32	417	893	1,107	1,486	2,757	1,786	636	731	299	121	92	43	18	10	5	2	13
2.0	1	0	102	741	1,290	1,496	2,575	1,968	780	868	492	200	116	51	31	8	4	4	8
2.5	0	0	9	256	969	1,303	2,045	1,392	803	941	484	181	157	58	23	19	5	1	8
3.0	0	0	1	45	438	1,029	1,702	1,398	705	957	560	218	196	92	40	11	4	2	5
3.5	0	0	1	4	124	650	1,169	1,701	647	865	456	237	162	100	36	12	6	1	5
4.0	0	0	2	0	33	270	780	1,369	573	868	427	193	157	91	51	13	3	0	1
4.5	0	0	0	0	3	90	459	1,017	466	761	380	127	137	86	31	23	6	5	0
5.0	0	0	0	0	0	15	228	647	408	737	354	119	96	50	32	18	2	4	1
5.5	0	0	0	0	0	2	68	337	363	580	283	94	92	31	24	10	6	2	0
6.0	0	0	0	0	0	1	20	166	221	418	307	63	76	24	13	9	4	0	0
6.5	0	0	0	0	0	0	5	50	140	260	257	59	49	20	12	4	2	2	2
7.0	0	0	0	0	0	0	0	23	90	180	193	41	53	20	5	3	3	0	0
7.5	0	0	0	0	0	0	0	6	25	93	121	45	46	17	5	5	0	1	0
8.0	0	0	0	0	0	0	0	3	14	50	84	26	47	11	6	0	1	0	0
8.5	0	0	0	0	0	0	0	0	7	25	45	23	25	20	8	0	0	0	0
9.0	0	0	0	0	0	0	0	1	2	12	30	22	20	19	0	0	0	0	0
9.5	0	0	0	0	0	0	0	0	1	2	20	21	14	7	1	1	0	1	0
10.0	0	0	0	0	0	0	0	0	0	2	5	4	21	6	2	0	0	0	0
10.5	0	0	0	0	0	0	0	0	0	3	4	8	9	12	2	0	0	0	0
11.0	0	0	0	0	0	0	0	0	0	0	2	0	4	3	1	0	1	0	0
11.5	0	0	0	0	0	0	0	0	0	0	2	1	2	3	0	0	0	0	0
12.0	0	0	0	0	0	0	0	0	0	0	0	0	1	2	1	0	0	0	0
12.5	0	0	0	0	0	0	0	0	0	0	0	0	0	1	0	0	0	0	0
13.0	0	0	0	0	0	0	0	0	0	0	0	0	0	0	0	1	0	0	0

Values given for h_s and t_p are upper-class limits.

12.4.2 All Short-Term Extremes

The conditional CDF $F_{\tilde{X}|H_sT_s}(\xi|h_s, t_s)$ of the largest peak value, $\tilde{X}$, during a short-term condition is clearly given by the expression

$$F_{\tilde{X}|H_sT_s}(\xi|h_s, t_s) = \left(F_{X_p|H_sT_s}(\xi|h_s, t_s)\right)^{k^{(st)}}, \tag{12.7}$$

where $k^{(st)} = \nu_X^+(0|h_s, t_s)\,\tilde{T}$ is the number of peak values during the short-term condition specified by $H_s = h_s$ and $T_s = t_s$. The validity of Eq. (12.7) is again based on the assumption that all peak values are independent.

The long-term CDF of the short-term extreme peak values is now approximately obtained as

$$F_{\tilde{X}}(\xi) = \int_{h_s}\int_{t_s} F_{\tilde{X}|H_sT_s}(\xi|h_s, t_s)\, f_{H_sT_s}(h_s, t_s)\, dh_s dt_s\,. \tag{12.8}$$

Although this error is usually not significant, the averaging done in Eq. (12.8) is not quite correct in the sense that it is not a so-called ergodic average (Naess, 1984), which would be the correct approach. To achieve this, Eq. (12.8) has to be modified to read (Krogstad, 1985),

$$F_{\tilde{X}}(\xi) = \exp\left\{\int_{h_s}\int_{t_s} \ln F_{\tilde{X}|H_sT_s}(\xi|h_s, t_s)\, f_{H_sT_s}(h_s, t_s)\, dh_s dt_s\right\}. \tag{12.9}$$

Assuming for illustration that $\tilde{T} = 3$ h, an estimate of the value ξ_q, which has a probability q of being exceeded per year, is in this case determined by the equation

$$F_{\tilde{X}}(\xi_q) = 1 - \frac{q}{365 \cdot 8}. \tag{12.10}$$

If Eq. (12.8) is used, a relation analogous to Eq. (12.5) would be (with $\tilde{T} = 3$ h and $T_z^{-1} \approx c\, T_s^{-1}$)

$$F_{\tilde{X}}(\xi) \approx \sum_{i=1}^{m}\sum_{j=1}^{n}\left\{1 - \exp\left(-\frac{\xi^2}{2\,\sigma_X(h_i, t_j)^2}\right)\right\}^{\frac{60^2\times 3\times c}{t_j}} \frac{N_{ij}}{N}\,. \tag{12.11}$$

12.4.3 Long-Term Extreme Value

The CDF of the extreme value $\hat{X} = \hat{X}(T)$, that is, the global extreme value over a long-term period T, can be expressed as follows (Naess, 1984)

$$F_{\hat{X}}(\xi) = \exp\left\{-T\int_{h_s}\int_{t_s} \nu_X^+(\xi|h_s, t_s)\, f_{H_sT_s}(h_s, t_s)\, dh_s dt_s\right\}, \tag{12.12}$$

where $\nu_X^+(\xi|h_s, t_s)$ denotes the average ξ-upcrossing rate for the short-term stationary situation characterized by $H_s = h_s$ and $T_s = t_s$.

From Eq. (8.28), it follows that for the case of a zero-mean Gaussian process, Eq. (12.12) would read

$$F_{\hat{X}}(\xi) = \exp\left\{-T\int_{h_s}\int_{t_s} \frac{\sigma_{\dot{X}}(h_s, t_s)}{2\,\pi\sigma_X(h_s, t_s)}\exp\left(-\frac{\xi^2}{2\sigma_X(h_s, t_s)^2}\right) f_{H_sT_s}(h_s, t_s)\, dh_s dt_s\right\}, \tag{12.13}$$

where the standard deviations σ_X and $\sigma_{\dot{X}}$ in the long-term situation become functions of the environmental parameters h_s and t_s as indicated.

With $T = 1$ year $= 365 \cdot 24 \cdot 3{,}600$ s, the value ξ_q, which has a probability q of being exceeded per year, is now calculated from the equation

$$F_{\hat{X}}(\xi_q) = 1 - q \, . \tag{12.14}$$

With reference to Table 12.1, Eq. (12.13) can then be expressed as a relation analogous to Eq. (12.5) in the following way (with $T = 1$ year and $T_z^{-1} \approx c\, T_s^{-1}$):

$$F_{\hat{X}}(\xi) \approx \exp\Big\{ - \sum_{i=1}^{m} \sum_{j=1}^{n} \frac{60^2 \times 24 \times 365 \times c}{t_j} \exp\Big(- \frac{\xi^2}{2\,\sigma_X(h_i, t_j)^2} \Big) \frac{N_{ij}}{N} \Big\} . \tag{12.15}$$

For the purpose of estimating extreme load effects, the use of scatter diagrams calls for a certain amount of caution. If the scatter diagram is too coarse, leading to poor resolution in the tail regions, the long-term extreme value estimates may become inaccurate. In such cases, it is recommended to use a properly adapted smooth joint PDF of the parameters characterizing the short-term sea states. For our purposes, the joint PDF of $W = (H_s, T_s)$ is needed. For North Sea applications, the spectral period T_s is often the spectral peak period T_p due to the fact that a commonly adopted spectral model is the JONSWAP spectrum, which is usually parameterized by the significant wave height and the spectral peak period. The marginal distribution of H_s is often modeled as a

- two-parameter Weibull distribution
- combination of a lognormal and a Weibull distribution

The following probabilistic model given by Haver (1980) and Haver and Nyhus (1986) has been frequently adopted as the latter model.

$$f_{H_s}(h_s) = \frac{1}{\sqrt{2\pi}\alpha h_s} \exp\Big\{ - \frac{(\ln h_s - \theta)^2}{2\,\alpha^2} \Big\} \qquad h_s \leq \eta \tag{12.16}$$

and

$$f_{H_s}(h_s) = \frac{\beta}{\rho} \Big(\frac{h_s}{\rho} \Big)^{\beta - 1} \exp\Big\{ - \Big(\frac{h_s}{\rho} \Big)^{\beta} \Big\} \qquad h_s > \eta \tag{12.17}$$

where the value of the transition parameter η separating the lognormal model for the smaller values of H_s from the Weibull model for the larger values, will depend on the geographic location.

This marginal PDF for the significant wave height is complemented by the conditional PDF of the spectral peak period T_p given the value of H_s using a lognormal model:

$$f_{T_p|H_s}(t_p|h_s) = \frac{1}{\sqrt{2\pi}\sigma t_p} \exp\Big\{ - \frac{(\ln t_p - \mu)^2}{2\sigma^2} \Big\} , \tag{12.18}$$

where the parameters μ and σ are assumed to depend on the significant wave height h_s in the following manner:

$$\mu = a_1 + a_2 h_s^{a_3}, \tag{12.19}$$

$$\sigma^2 = b_1 + b_2 \exp(-b_3 h_s), \tag{12.20}$$

for suitably chosen constants a_i and $b_i, i = 1, 2, 3$.

The joint PDF for the environmental parameters is then obtained by multiplying the marginal PDF for the significant wave height with the conditional PDF for the spectral peak period, that is,

$$f_W(w) = f_{H_s T_p}(h_s, t_p) = f_{H_s}(h_s)\, f_{T_p|H_s}(t_p|h_s). \tag{12.21}$$

The following set of parameter values was cited by Haver (2002) for locations in the northern North Sea (Statfjord area): $\alpha = 0.6565$, $\theta = 0.77$, $\eta = 2.90$, $\beta = 2.691$, $\rho = 1.503$, $a_1 = 1.134$, $a_2 = 0.892$, $a_3 = 0.225$, $b_1 = 0.005$, $b_2 = 0.120$, $b_3 = 0.455$.

Based on the earlier work by Jahns and Wheeler (1972) and Haring and Heideman (1978), Tromans and Vanderschuren (1995) proposed an alternative approach to the calculation of the long-term extreme load or load effect. In their approach, the focus is on storm events, similar to what is done in a peaks-over-threshold analysis (Reiss and Thomas, 2001). This approach is particularly relevant for tropical areas with rare hurricanes. Hence, the long-term situation is considered as a sequence of storm events. The method is based on the assumption that the distribution of the storm extreme response value can be approximated by a Gumbel extreme value distribution conditional on the most probable extreme response for that storm. The distribution of the most probable extreme value itself is assumed to follow a generalized Pareto distribution, which is determined by fitting to data. By invoking the rule of total probability, as exemplified by Eq. (12.8), the long-term extreme response value distribution can be calculated.

In this chapter three alternative approaches to estimate the extreme wave-induced response for ULS design checks at a given annual probability (or return period), has been outlined. It is emphasized that a long-term approach is the most accurate approach. The simplified design seastate or wave approaches always need to be validated against the full long-term approach.

The applicability of the simplified approaches depend on the character of the response, especially whether it can be considered quasistatic or dynamic, which response values are relevant and also which accuracy is required; that is, whether the analysis is carried out in pre-engineering or the detailed design phase. In this connection the fact that nonlinear hydrodynamic effects might be present and cause sum- or difference frequency excitation, respectively, should be considered.

A bottom fixed structure with a natural period below 3 seconds, say, could be considered to have a quasi-static behaviour under steady wave loading and a design approach with appropriately chosen wave height and period would be relevant. For structures with natural periods above 3 seconds, the quasistatic approach might still be used in combination with the use of a dynamic amplification factor (DAF) determined by a stochastic analysis for relevant sea states, if the DAF is limited, say, to less than 1.5. The simplified methods would be relevant for early design phases while a stochastic dynamic approach should be used in the detailed design phase.

The motions of floating or compliant platforms subjected to wave loads in general need to be estimated by a stochastic approach (design sea state based, e.g., on the contour line method or the full long-term approach). However, the extreme structural responses in a floating structure, which have a quasistatic character, may be determined by a design wave approach based on an appropriate calibration of the the wave period (length) and height.

12.5 Extreme Values for a Combination of Multiple Stochastic Load Effects in a Short-Term Period

In practical design, it is often necessary to account for the fact that the total load effect is the result of a combination of several stochastic load effect components. If these load effects are determined separately, the challenge is then to provide a rational combination procedure for the estimation of the extreme combined load effect value, which ideally has the prescribed return period of, say, 100 years. In some cases, it is possible to carry out a long-term extreme value analysis as described in the previous section, but some degree of simplification often has to be accepted in practical design work. This situation is exemplified by the case of the total surge response of a TLP in long-crested random seas due to first and second order wave loads. We return to a more thorough discussion of combination of load effects in Chapter 16.

Combination of First- and Second-Order Extreme Response

Let the total horizontal wave-induced response of the TLP structure subjected to a random, long-crested seaway be written as the sum of a linear and a quadratic response component. Specifically, let $Z(t)$ denote the total surge response process of the TLP. Then,

$$Z(t) = \mu_2 + Z_1(t) + Z_2(t), \tag{12.22}$$

where the subscript 1 signifies the linear, first-order, or wave frequency component, while subscript 2 refers to the nonlinear, second-order part of the TLP's surge response. Note that $\mu_2 =$ the mean drift offset, implying that $Z_2(t)$ denotes the zero-mean, slowly varying part of the second-order response. Strictly speaking, the assumed decomposition presupposes that the equation of motion for the surge response is linear, or linearized.

Assume that the quantity to be calculated is the expected largest surge response during a short-term storm condition; that is, $E[\hat{Z}]$, where $\hat{Z} = \hat{Z}(\tilde{T}) = \max\left(Z(t); 0 \le t \le \tilde{T}\right)$ denotes the extreme value of $Z(t)$ during a short-term time interval of length $\tilde{T}$.

Procedures exist for calculating the extreme values of the total response process $Z(t)$ as described in the previous section and based on the level upcrossing rate of $Z(t)$. However, for this case this is not a straightforward matter because it involves extensive calculations of the second-order hydrodynamic coefficients of the TLP for the short-crested seas case, cf. Teigen and Naess (1999).

A more practical approach to estimate the extreme surge response is based on a simplified procedure. For example, API's Recommended Practice document (API, 1993) for combination of wave frequency and low-frequency horizontal excursion of a moored floating structure is based on a specific, simplified combination formula, viz. Turkstra's rule (Turkstra, 1970; Madsen et al., 1986). In this context, it assumes the following form:

$$E[\hat{Z}] = \mu_2 + \max\left\{E[\hat{Z}_1] + \sigma_2,\ \sigma_1 + E[\hat{Z}_2]\right\}, \tag{12.23}$$

where σ_i = standard deviation of $Z_i(t)$. In this equation, σ_i represents an arbitrary point in time value of $Z_i(t)$. By using one standard deviation to represent this value, a small positive correlation effect has been implicitly introduced. A study of the effect of dependence between the load effect components on Turkstra's rule is performed by Naess and Royset (2000).

The main efforts in calculating the expected largest response according to this formula are in the calculation of $E[\hat{Z}_1]$ and $E[\hat{Z}_2]$. The state-of-the-art approach is to assume that the short-term random seaway can be described as a stationary Gaussian process. This implies that the linear response is also a Gaussian process; hence, $E[\hat{Z}_1]$ can be calculated by Eq. (10.53). The slowly varying response $Z_2(t)$ is generally non-Gaussian, which invalidates the use of Eq. (10.53) without further justification. As a first estimate of $E[\hat{Z}_2]$, Naess (1989) proposed the following formula based on a simplified, second-order response representation:

$$E[\hat{Z}_2] = \sigma_2 \ln\left(\frac{\tilde{T}}{T_{SD}}\right), \tag{12.24}$$

where T_{SD} denotes the natural period of the surge motion.

A more accurate estimate of $E[\hat{Z}_2]$ would lead to a result that falls somewhere between the Gaussian estimate provided by Eq. (10.53) and the simplified formula given by Eq. (12.24). One such formula is, cf. Naess (1989),

$$E[\hat{Z}_2] = \sigma_2 \ln\left(\frac{\tilde{T}}{T_{SD}}\left[1 - r^2\left(\frac{\tilde{T}}{T_{SD}}\right)^{-\delta}\right]\right), \tag{12.25}$$

where $r = \exp(-\pi\xi)$, $\delta = (1-r)/(1+r)$ and ξ = damping ratio in the linear or linearized dynamic model. An alternative formula is discussed by Stansberg (1992, 2000).

To finalize the calculation of $E[\hat{Z}]$ as given by Eq. (12.23), μ_2 and σ_2 must be provided. The following formula obtains

$$\mu_2 = \frac{1}{M\,\omega_{SD}^2}\int_0^\infty C(\omega)\, S_X^+(\omega)\, d\omega\,, \tag{12.26}$$

where M = the total mass of the TLP, $\omega_{SD} = 2\pi/T_{SD}$, $C(\omega)$ = the wave drift coefficient = $\hat{K}_2(\omega, -\omega)$, cf. Section 11.3.1; and $S_X^+(\omega)$ = the wave spectrum.

Adopting a Newman's approximation (Newman, 1974) for the calculation of the slowly varying wave forces, see Section 11.3.1, σ_2 can be calculated as follows:

$$\sigma_2^2 = 2\int_0^\infty \left(\int_0^\omega |\hat{L}(\omega-\omega')|^2\, S_X^+(\omega')\, d\omega'\right) C(\omega)^2\, S_X^+(\omega)\, d\omega\,, \tag{12.27}$$

where $\hat{L}$ denotes the linear transfer function for the surge motion, that is,

$$\hat{L}(\omega) = \frac{1}{M(\omega_{SD}^2 - \omega^2 + 2\,\mathrm{i}\,\zeta\,\omega_{SD}\,\omega)}, \tag{12.28}$$

where ζ denotes the damping ratio for the surge motion, and $\mathrm{i} = \sqrt{-1}$.

An alternative combination format that is used extensively in practical design applications is the so-called square root of sum of squares (SRSS) formula. The SRSS equivalent to Eq. (12.23) is the following expression:

$$E[\hat{Z}] = \mu_2 + \left\{E[\hat{Z}_1]^2 + E[\hat{Z}_2]^2 + 2\,\rho_e\,E[\hat{Z}_1]E[\hat{Z}_2]\right\}^{1/2}, \tag{12.29}$$

where the parameter ρ_e denotes a correlation coefficient between first- and second-order wave-induced responses at large response levels. The standard form of the SRSS formula is obtained by neglecting this correlation coefficient, which, in general, cannot be justified for a combination of first- and second-order wave-induced responses. For applications considered here, the correlation coefficient ρ_e can be expected to assume values typically in the interval from 0.2 to 0.4.

13 Dynamic Load Effects for Design Checks

13.1 Introduction

This chapter describes models for loads that have a fundamental variability and their load effects on structures. These load effects are applied in the design and reassessment of structures during operation to ensure adequate structural performance; i.e., serviceability and safety in, say, a period of 20–50 years (ISO 2394, 1998). Serviceability requirements refer to motions, deformations, vibrations, etc., that can hamper the operation, but do not represent a threat to the safety. Safety means the absence of failures and damages, and is ensured by fulfilling requirements to overall stability and ultimate strength, and fatigue life under repetitive loading to avoid ultimate consequences such as fatalities, environmental damage, or property damage. The corresponding so-called limit state criteria are defined by limit states for ultimate failure (ultimate limit state (ULS)) and fatigue failure (fatigue limit state (FLS)), respectively.

ULS criteria for overall stability of bottom-supported structures are based on overturning forces due to wave, current, wind, and stabilizing forces due to permanent and variable payloads. Stability of floating structures is considered in terms of overturning moment by wind only, and uprighting moment due to hydrostatics of the inclined body.

ULS and FLS criteria for relevant components were developed for the relevant failure modes dependent on geometry and load conditions. Permanent and variable payloads, fluid pressure loads, and environmental loads are considered. Environmental loads due to waves, current, wind, and possibly ice and earthquakes are considered. Load and resistance factor design is commonly used for ULS.

Figure 13.1 illustrates the characteristic features of a design check relating to sea loads. Load effects are determined by recognizing the stochastic character of the waves, using hydrodynamic theory to calculate wave forces and structural mechanics to obtain load effects on an appropriate format for the design check.

Engineering decisions, in general, and about safety, in particular, must be made in the presence of uncertainties arising from inherent randomness in many design parameters, imperfect modeling, and lack of experience. Indeed, it is precisely on account of these uncertainties and the potential risks arising therefrom that safety

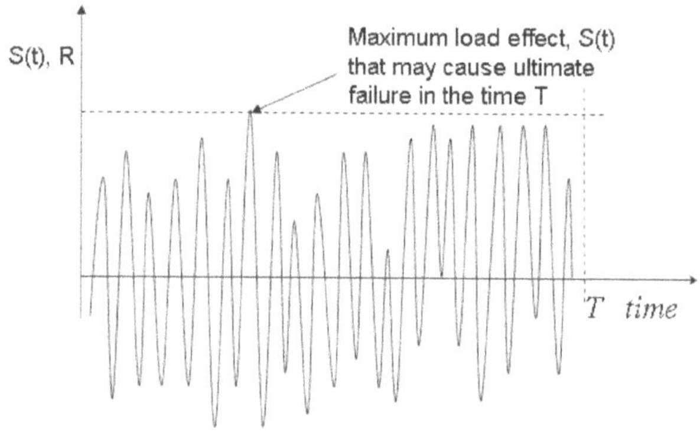

Figure 13.1. Overview of analyses involved in design checks for offshore structures subjected to wave loading.

margins provided by the specification of allowable stresses, resistance and load factors, and the like, are required in design. In the following sections, the probabilistic treatment of uncertainty is therefore expanded to deal with all uncertainties in load effects and resistance that affect structural design. This is the topic of probabilistic design.

Another issue in modern codes is the application of partial safety factors or load and resistance factor design (LRFD) instead of working stress design (WSD). The main advantage of WSD is its simplicity. However, it can lead to designs with less safety than those normally considered adequate, particularly if loads counteract each other. The WSD method is therefore not further discussed in this book.

13.2 Limit State Design Approaches

13.2.1 Ultimate Limit State

To illustrate design against ultimate failure, consider a single structural component subjected to a random load effect process $S(t)$, as shown in Fig. 13.2. The aim of the design process is to choose the dimensions of this component and the material properties (e.g., yield strength) so that the probability of failure during the service life is acceptably small.

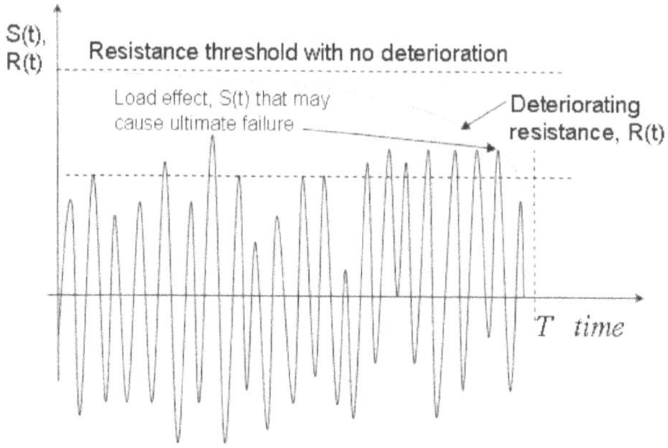

Figure 13.2. Illustration of ultimate overload failure under random loading.

The simplest example of an ultimate limit state design check of a structural component subjected to a single type of load effect due to two sources of loads may be written as

$$R \geq S, \tag{13.1}$$

where R is the resistance, $S = S_1 + S_2$, and S_1 and S_2 are extreme load effects in a reference period of, say, 100 years due to different types of loads, e.g., payload and wave load, respectively. R and S refer to the same physical quantity, e.g., a stress, an axial force or a bending moment.

Because R, S_1, and S_2 are subject to uncertainty and variability, the design format applied is

$$R_c/\gamma_R \geq \gamma_{S1} S_{1c} + \gamma_{S2} S_{2c}, \tag{13.2}$$

where the subindex c refers to a characteristic value, and γ_R, γ_{S1}, and γ_{S2} are resistance and load factors, respectively. The characteristic resistance is obtained by using, e.g., the 5% fractile material strength, while the characteristic load effects due to payload and waves correspond to a specified value and the load with an annual probability of exceedance of 10^{-2}, respectively. Equation (13.2) implies that the failure probability,

$$P_f = \text{Prob}\,(R \leq S)\,, \tag{13.3}$$

for this problem becomes sufficiently small when the load and resistance factors and characteristic values are properly chosen. Ultimate limit state criteria in modern design codes are based on the design format like Eq. (13.2). This approach is denoted semiprobabilistic, while design based on direct calculation of the failure probability is a probabilistic approach. Then, R and S are considered random variables that represent the uncertainties and variability in R and S. In particular, the load effect S needs to reflect the fundamental variability in wave and other environmental loads, but also the model and statistical uncertainty due to lack of knowledge about the load effect processes. Some design codes permit direct probabilistic design for particular situations. Even more important, probabilistic approaches are used to calibrate the semiprobabilistic approaches.

An appreciation of the philosophy underlying such provisions is essential: in the presence of uncertainty, absolute reliability is an unattainable goal. However, probability theory and reliability-based design provide a formal framework for developing criteria for design that ensure that the probability of unfavorable performance corresponds to an acceptably small target value. This aspect is further discussed in Section 13.3.

More general expressions than Eqs. (13.1), (13.2), and (13.3), involving several resistances R_i and corresponding load effects S_{ij} for a given type of load (j), would be required to describe ultimate failure of beams, panels, and shell structures.

Also, when S_{11} and S_{12} represent the vertical bending moment ($i = 1$) for two time-varying loads, such as the stillwater ($j = 1$) and wave loading ($j = 2$) on a ship, the fact that their maxima in a given period do not occur at the same time necessitates particular analysis to determine the maximum of the combined load. By use of stochastic process theory, the characteristic values of the combined loads may be expressed by using $(\psi_{S1} S_{11c}, S_{12c})$ or $(S_{11c}, \psi_{S2} S_{12c})$, where ψ_{Si} are load reduction factors and the individual load effects are still defined separately with respect to the 10^{-2} annual exceedance probability.

13.2.2 Fatigue Limit State

Fatigue is an important consideration for structures in areas with more or less continuous storm loading, especially for dynamically sensitive structures. The fatigue design check is normally based on resistance defined by SN data (e.g., $N = K\,S^{-m}$, where N is the number of cycles to failure under a given stress range S; K, and m are material parameters), that were obtained by laboratory experiments, and the use of Palmgren-Miner's hypothesis of linear cumulative damage, see Section 13.5. If the stress process is narrow banded with a zero mean value, the stress range S can be assumed to be 2 times the stress amplitude. A simple expression for the cumulative damage can then be obtained by assuming that the SN curve is defined by $N\,S^m = K$, and the number $n(S)$ of stress ranges that is experienced by the structure can be obtained from the distribution of stress ranges using, e.g., the two-parameter Weibull distribution. Then,

$$F_S(s) = 1 - \exp\left(-\left\{\frac{s}{A}\right\}^h\right), \tag{13.4}$$

where $A = s_0/\,(\ln N_0)^{1/h}$ and $\text{Prob}\,(S \geq s_0) = 1/N_0$, see Section 13.5.4. The scale parameter, A, is directly related to the extreme response value required for ULS design checks. Guidance on the magnitude of the shape parameter, h, is available for different types of structures. In this way, fatigue loading, at least for initial design and screening to identify the importance of fatigue, can be easily accomplished.

The damage D in a long-term period T with N_T cycles is then, as shown in Section 10.9, given by

$$\begin{aligned} D &= \sum_i \frac{n_i}{N_i} = \sum_i \frac{n(s_i)}{N(s_i)} = \frac{N_T}{K}\sum_i s_i^m\, f_S(s_i)\,\Delta s \approx \frac{N_T}{K}\int_0^\infty s^m\, f_S(s)\,ds \\ &= \frac{N_T}{K} A^m\,\Gamma\left(\frac{m}{h}+1\right) = \frac{N_T\, s_0^m}{K\,(\ln N_0)^{m/h}}\,\Gamma\left(\frac{m}{h}+1\right), \end{aligned} \tag{13.5}$$

where the expression for A is given previously. Further details about the derivation of this equation and its refinement follow later in this chapter.

The fatigue design check may then be written as

$$D_c < \Delta_{\text{all}}, \tag{13.6}$$

where D_c is the characteristic fatigue damage and Δ_{all} is the allowable fatigue damage during service life, including a safety factor. The fatigue design check is normally based on best estimate of stress cycles, i.e., expected values of A and h; while K is taken to be K_c, the characteristic value of K, typically corresponding to 97.7% probability of exceedance (mean minus two standard deviations for a Gaussian distribution), and Δ_{all} is the allowable fatigue damage, usually between 0.1 and 1 for marine structures. Sometimes the fatigue design criteria are formulated in terms of the fatigue design factor (FDF), which corresponds to $\text{FDF} = 1/\Delta_{\text{all}}$.

Until recently, the focus was on achieving serviceability and safety by structural design. However, inspection, maintenance, and repair (IMR) during fabrication and operation are crucial issues in order to maintain safety, especially in connection with deterioration phenomena such as fatigue and corrosion. An adequate treatment of design and IMR issues relating to fatigue requires a probabilistic approach (Moan et al., 1993; Moan, 2005). To some extent these issues are reflected in fatigue design criteria in offshore codes (NORSOK, ISO) where the fatigue design check depends on the consequences of failure and access for inspection, and the allowable cumulative damage, Δ_{all}, therefore varies between 0.1 and 1 to reflect this (NORSOK N-001, 1998). The consequence measure is based on whether the structure fails in a load condition comprising an expected variable load and a 100-year sea load after fatigue failure of the relevant joint.

Slightly different criteria were proposed by API (Karsan, 2005). The difference between the two classes is whether the structure has sufficient residual strength after fatigue failure of the relevant joint to resist expected functional load and 100-year sea loads, and it is thus directly linked to the *accidental collapse limit state* criterion. The fatigue design factor (FDF) varies between 1 and 10. Note that FDF $= 1/\Delta_{all}$. The treatment of both inspection and consequence issues, however, could be improved by taking FDF as a function of a more precise measure of residual strength and an explicit measure of the effect of inspection, including the quality of the inspection.

Normally, FDF $= 1$ is applied for ships, including FPSOs (floating, production, storage and offloading). However, it may be reasonable to apply more restrictive criteria when the consequences of a through-thickness crack may be high. Even if cracks in hulls do not represent an immediate risk of global failure, they can cause oil leaks, which can lead to explosion risk, environmental damage, and reduced oil cargo quality. This fact, together with the large number of potential crack sites in a complex hull with difficult access for inspection, suggests a larger FDF, say, in the range of 2 to 3. This issue is even more important in the case of gas carriers. Liquefied natural gas (LNG) tanks are preferably designed based on the *leak before failure* principle. This means that potential cracks will propagate through the thickness before reaching a critical size. A gas detection system is therefore arranged to detect potential gas leaks. After a through-thickness leak, the tanks will have a sufficient time window (e.g., two weeks) for the vessel to go to port and unload her cargo. The minimum acceptable time for the crack to penetrate the thickness is the maximum value of the planned operational lifetime of the vessel and 20 years. Furthermore, some classification societies would require an FDF of 10 if leak before failure cannot be proven. The consequences of a tendon failure for TLPs are also significant and an FDF of 10 is typically applied.

Ultimate overload and fatigue failure events have been implicitly assumed to be independent. However, if the overload failure is fracture or rupture, it will be dependent on fatigue crack growth; that is, propagation of an initial crack to a crack size that can result in ultimate failure before the fatigue life is exhausted. This failure mode is illustrated in Fig. 13.3 by a (fracture) resistance that is deteriorating until a single high load effect causes failure. Corrosion will have a similar deteriorating effect on the resistance.

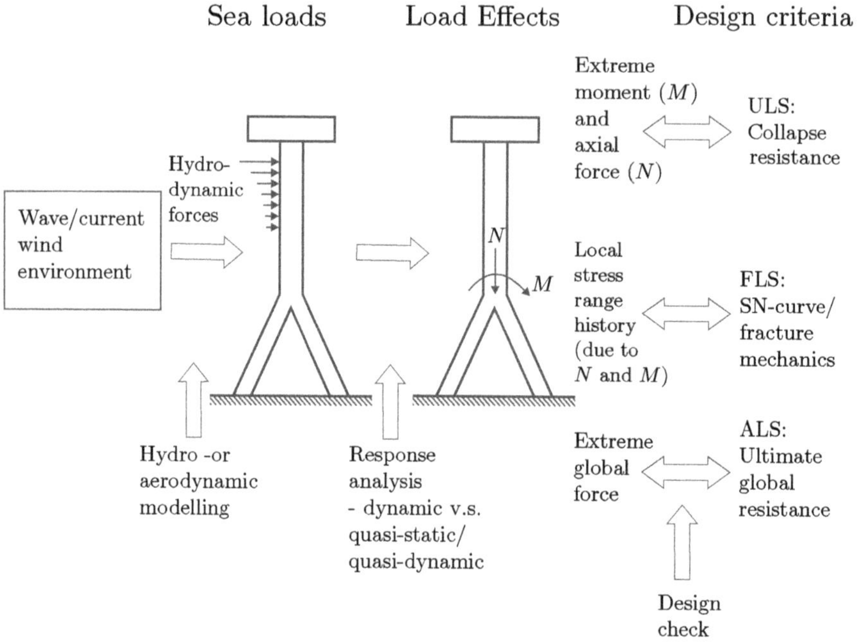

Figure 13.3. Illustration of overload failure following deteriorating resistance due to fatigue crack growth.

13.3 Reliability Framework

13.3.1 Elementary Case

The term *structural reliability* should be considered as having two different meanings – a general one and a mathematical one.

- In the most general sense, the reliability of a structure is its ability to fulfill its design purpose for some specified time under specified conditions.
- In a narrow sense it is the probability that a structure will not attain each specified limit state (ultimate or serviceability) during a specified reference period.

In this context, the focus is on structural reliability in the narrow sense, as the complementary quantity to the *failure probability* P_f defined by Eq. (13.3). Hence, the *reliability*, denoted by P_s, may be determined in terms of R and S for ULS (or D and Δ for FLS) as

$$P_s = 1 - P_f = \text{Prob}\,(R > S)\,. \qquad (13.7)$$

Depending on how R and S are related to time, the failure probability will also be defined with respect to some reference time period, say, one year or the service lifetime. This issue is explored later. First, the basic reliability problem – i.e., the calculation of P_f is examined.

Table 13.1. *Correspondence between values of* β *and* P_f

β	1.0	1.1	1.2	1.3	1.4	1.5	1.6	1.7	1.8	1.9
$P_f \cdot 10^2$	16	14	12	9.7	8.1	6.7	5.5	4.5	3.6	2.9
β	2.0	2.1	2.2	2.3	2.4	2.5	2.6	2.7	2.8	2.9
$P_f \cdot 10^3$	23	18	14	11	8.2	6.2	4.7	3.5	2.6	1.9
β	3.0	3.1	3.2	3.3	3.4	3.5	3.6	3.7	3.8	3.9
$P_f \cdot 10^4$	14	9.7	6.9	4.8	3.4	2.3	1.6	1.1	0.72	0.48
β	4.0	4.1	4.2	4.3	4.4	4.5	4.6	4.7	4.8	4.9
$P_f \cdot 10^6$	32	21	13	8.5	5.4	3.4	2.1	1.3	0.79	0.48

When the independent load effect S and the resistance R for a structural component are described by their PDFs, $f_S(s)$ and $f_R(r)$, respectively, the failure probability P_f is calculated as

$$P_f = \text{Prob}\,(R \leq S) = \int_0^{\infty} F_R(s)\, f_S(s)\, ds = F_M(0), \tag{13.8}$$

where the *safety margin* M is given by $M = R - S$. If the PDF $f_M(m)$ or the CDF $F_M(m)$ is known, P_f may be calculated analytically in a few cases; otherwise, it must be calculated numerically.

The failure probability can be expressed analytically when both R and S are normally distributed, and when both R and S are lognormally distributed. The failure probability can generally be written as

$$P_f = \Phi\,(-\beta)\,, \tag{13.9}$$

where the *reliability index* β is uniquely related to the failure probability P_f by this equation. Table 13.1 shows the corresponding values between P_f and β.

The reliability index β for the two special cases referred to previously can be given explicitly. When both R and S are normal variables, β is obtained as

$$\beta = \beta_N = \frac{\mu_R - \mu_S}{\sqrt{\sigma_R^2 + \sigma_S^2}} = \frac{{}^{\mu_R}/_{\mu_S} - 1}{\sqrt{\left({}^{\mu_R}/_{\mu_S}\right)^2 V_R^2 + V_S^2}}. \tag{13.10}$$

When both R and S are lognormal variables, β is expressed as

$$\beta = \beta_{LN} = \frac{\ln\left[(\mu_R/\mu_S) \cdot \sqrt{(1 + V_S^2)/(1 + V_R^2)}\right]}{\sqrt{\ln\left(1 + V_R^2\right)\left(1 + V_S^2\right)}} \approx \frac{\ln(\mu_R/\mu_S)}{\sqrt{V_R^2 + V_S^2}}. \tag{13.11}$$

It is noted that the failure probability and reliability index has to be referred to a time period (e.g., one year or the service life). If the annual P_f is considered, the load effect S should refer to its annual maximum value.

13.3.2 Example – Implicit Failure Probability in Design Equations

Despite the simplicity of the expression for P_f given by Eq. (13.9) with $\beta = \beta_{LN}$, it has been extensively used in the calibration of design codes, as described later. Here, it is used to illustrate the failure probability implied by Eq. (13.2), with only one load effect, S. Let the random load effect S and resistance R be defined by

$$\mu_S = B_S S_c, \quad B_S \le 1; \quad V_S = 0.15 \text{ to } 0.30, \tag{a}$$

and

$$\mu_R = B_R R_c, \quad B_R \ge 1; \quad V_R = 0.1. \tag{b}$$

The B_S reflects the ratio of the mean load (which refers to an annual maximum if the annual failure probability is to be calculated) and the characteristic load effect (typically, the 100-year value), as well as possible bias in predicting wave load effects.

By using the design equation Eq. (13.2) in combination with the approximate expression of Eq. (13.11), it follows that

$$\beta_{LN} \approx \frac{\ln(\mu_R/\mu_S)}{\sqrt{V_R^2 + V_S^2}} = \frac{\ln(B_R\gamma_R\gamma_S/B_S)}{\sqrt{V_R^2 + V_S^2}}. \tag{c}$$

With $\gamma_R\gamma_S = 1.5$, a typical value $B_S = 0.8$ for wave-induced load effects, $B_R = 1.1$ and $V_R = 0.1$, it is found that β_{LN} is about 2.7 and 3.2 for a V_S of 0.25 and 0.20, respectively. These reliability indices correspond to a P_f of $3.5 \cdot 10^{-4}$ and $7 \cdot 10^{-4}$, respectively. By inspection of the expression for β_{LN}, it is seen that when V_S dominates over V_R, β_{LN} in this case is roughly inversely proportional to V_S, and, hence, very sensitive to the assumed distributions.

13.3.3 Generalization of Reliability Analysis

The methods for calculating the failure probability outlined previously, have resulted in a one-dimensional integral, Eq. (13.8). The failure probability may be expressed as the integral of the probability density function for R and S over the area which corresponds to failure, i.e., $S \ge R$, or, in general,

$$P_f = \text{Prob}\,(g(\mathbf{X}) \le 0) = \int_{g(\mathbf{x}) \le 0} f_{\mathbf{X}}(\mathbf{x})\, d\mathbf{x}, \tag{13.12}$$

where $\mathbf{X} = (X_1, \ldots, X_n)$ is a vector of n random variables used to model the uncertainties identified as relevant for the problem. These random variables are often referred to as basic random variables. $\mathbf{X} = (R, S)$, with $g(R, S) = R - S$, for the case in Eq. (13.3). $f_{\mathbf{X}}(\mathbf{x})$ is the joint PDF of $\mathbf{X}$. The n-dimensional integral is taken over the region where $g(\mathbf{x}) \le 0$, which corresponds to failure. $g(\mathbf{x})$ is generally referred to as the limit state function. There may sometimes be a conflict between modeling accuracy and computational feasibility for reliability problems.

As another example of a limit state, consider fatigue failure formulated by the SN approach. Then, $g(\Delta, D) = \Delta - D$, where Δ is the fatigue damage at failure and

D is given by Eq. (13.5), implying that D depends on several resistance and load effect variables.

An important class of limit states are those for which all variables are treated as time independent, either by neglecting time variations in cases where this is considered acceptable or by transforming time-dependent processes into time-invariant variables (e.g., by using extreme value distributions). The integral of Eq. (13.12) may be calculated by direct integration, simulation, or FORM/SORM methods as described in several textbooks, e.g., Madsen et al. (1986); Melchers (1999); Ang and Tang (2007).

Time-invariant reliability methods have been used extensively to calibrate ultimate strength code checks based on partial safety factors, to comply with a certain target reliability level (Melchers, 1999). This application is centered around current design practice in the sense that the $g(\cdot)$-function can be based on the relevant design equations in an existing code. The main result of the calibration is more consistent safety factors.

13.4 ULS Design Check

13.4.1 Introduction

Ultimate limit states are often governing in design, and they are generally based on extreme load effects. Estimation of relevant sea load effects requires consideration of a variety of sea states and elaborate MDOF hydrodynamic and structural models that are computationally time consuming. Ultimate strength formulations are traditionally based on strength of material formulations and substantiated by extensive test results.

The design criteria and reliability analysis are based on limit state functions. The basic formulation of the limit state for a structural component with one resistance R and a corresponding load effect S can be reformulated by Eq. (13.1) by introducing a limit state function $g(\cdot)$ as follows:

$$g(R, S) = R - S. \tag{13.13}$$

The design criterion, Eq. (13.2), is then given by

$$g(R_d, S_{1d}, S_{2d}) > 0, \tag{13.14}$$

where $R_d = R_c/\gamma_R$, $S_{1d} = \gamma_{S1} S_{1c}$, $S_{2d} = \gamma_{S2} S_{2c}$ are the design values of resistance and load effects, respectively. Similarly, the expression for the failure event is given by $g(R, S_1, S_2) \leq 0$.

As another example of an ultimate limit state involving several basic random variables, consider a steel beam-column subjected to axial force and bending moment. The failure criterion in this case is formulated as

$$\begin{aligned} g(R_1, R_2, R_3, S_{1j}, S_{2j}) &= 1 - \left[\frac{S_{1j}}{R_1} + \frac{S_{2j}}{\left[1 - \frac{S_{1j}}{R_2}\right] R_3} \right] \\ &= 1 - \left[\frac{X_1}{X_2} + \frac{X_3}{\left[1 - \frac{X_1}{X_4}\right] X_5} \right] = g(\mathbf{X}) \leq 0, \end{aligned} \tag{13.15}$$

where S_{ij} and R_i are load effects (for a given type of load j) and resistance terms, respectively. Equation (13.15) is based on a Perry-Robertson approach, with R_1 and R_2 being axial force and bending moment capacity, respectively, and R_3 the Euler load. Many other problems can be formulated by a multiple set of random variables, **X**. If the load effects are due to various load processes (e.g., waves and wind) $S_i = \sum a_{ij} S_{ij}$, the joint occurrence may be accounted for by load reduction factors, ψ_{ij}. In this way, the $g(\cdot)$-function, both for applications in a semiprobabilistic design equation and for reliability calculations, may be expressed by random variables.

Brittle fracture due to overload in welded metal structures has often been treated in a simple manner, that is, by the choice of material quality based on environmental temperature, plate thickness, and not a formal explicit design check analogous to Eq. (13.1). However, by fracture mechanics approaches, formal criteria can be established and treated in a semiprobabilistic or probabilistic manner, see, e.g., Almar-Naess (1985).

If Eq. (13.15) is applied in practical design, the characteristic values of load effects and resistances will be used, while the variables would be considered as random variables in a reliability analysis. The characteristic load effects for ULS check is taken to be loads with a certain low exceedance probability; i.e., an annual exceedance probability of 10^{-2}. To account for the long-term variation of the macroscopic environmental condition, one of the procedures discussed in Section 12.4 should be used. This could be amended by avoiding the use of all sea states in the scatter diagram and focusing on some of the sea states by introducing a measure of contribution to the long-term exceedance probability from each sea state (Farnes, 1990; Videiro and Moan, 1999, 2000).

A strategy, which is sometimes useful in connection with long-term response analysis that involves time-consuming short-term response analyses, is to first perform linearized analyses in the frequency domain to identify the range of sea states that contribute significantly to the extreme value. Then, the complete nonlinear short-term approach is used to determine the expected maximum for relevant sea states to obtain the largest one, which is taken to be the desired extreme value.

The partial safety factors, such as γ_{Si}, are commonly calibrated by structural reliability theory so that a design equation like Eq. (13.14) corresponds to a given target failure probability (Fjeld, 1977; Moses, 1987; Moan, 1995).

In the following Sections 13.4.2 and 13.4.4, the focus is on determining the failure probability in a short-term period. In Section 13.4.3 it is also explained how the failure probability in a long-term period can be dealt with.

13.4.2 Short-Term Reliability for a Single Load Effect

If the resistance R is constant (over time) and the load effect S is caused by a single load process, then the characteristic value of the load effect can be obtained from the distribution of individual maxima or the largest value in a given time period. Similarly, the reliability problem may be based on extreme value statistics to characterize $S_{\max}$, and the failure probability in the short-term period $(0, T)$, in which the loading is assumed to be stationary, can be expressed in the following way:

$$P_f(T) = \text{Prob}\Big(g\big(R, \max_{0 \le t \le T} S(Q(t))\big) \le 0\Big) = \text{Prob}\Big(g\big(R, S_{\max}(T)\big) \leqslant 0\Big), \quad (13.16)$$

where $g(\cdot)$ is the limit state function, R is the "structural resistance," S is the load effect resulting from the load process $Q(t)$.

A second alternative is to apply the upcrossing rate approach. In the case of time-dependent reliability, the main interest lies in the time t_f to first failure. For the simple problem $M(t) = g(R, S(t)) = R - S(t)$, t_f is the first time when $M(t) = 0$. Equivalently, t_f is the time of the first excursion of $M(t)$ from a positive to a negative value, assuming that $M(t)$ is a continuous process. The probability of failure in the period $[0, T]$ may be considered equivalent to the probability that t_f is less than T, which may be expressed by

$$P_f(T) = 1 - \text{Prob}(t_f > T) = 1 - \text{Prob}\left(N(T) = 0 \,\middle|\, M(0) > 0\right) \text{Prob}\left(M(0) > 0\right), \tag{13.17}$$

where $N(t)$ = number of upcrossings in $(0, t)$ (i.e., crossings from the safe to the failure domain) and $M(0) > 0$ signifies that the safety margin $M(t) = R - S(t)$ is in the safe domain at zero time.

In general, the calculation of $P_f(t)$ is a complex task. An approximate solution may be achieved by assuming that $N(t)$ is a Poisson process, i.e., upcrossings of the level R by $S(t)$ are independent with a mean rate of $\nu_S^+(R) = \nu_M^-(0)$ per unit time. It is obtained that

$$\text{Prob}\big(N(T) = 0\big) = \frac{\left(\nu_S^+(R)T\right)^0}{0!} e^{-\nu_S^+(R)T} = e^{-\nu_S^+(R)T}. \tag{13.18}$$

Also, $\text{Prob}\big(M(0) > 0\big) = 1 - P_f(0)$, which equals the probability of no failure at $t = 0$. In the present context, it may be assumed that $P_f(0) = 0$. Then, $P_f(t)$ is given by the following approximate equation:

$$P_f(T) \approx 1 - e^{-\nu_S^+(R)T} \approx \nu_S^+(R)T = \nu_M^-(0)T \tag{13.19}$$

because $\nu_S^+(R)T$ is usually small. The failure probability in the time interval $(0, T)$ may be calculated from Eq. (13.19) when an estimate of $\nu_S^+(R)$ of the actual $S(t)$ has been established.

In the special case that $S(t)$ is a Gaussian process, it follows from Eq. (10.22) that

$$\nu_S^+(R) = \nu_0^+ \exp\left(-\frac{(R - \mu_S)^2}{2\sigma_S^2}\right), \tag{13.20}$$

where $\nu_0^+ = \nu_S^+(0)$.

Hagen and Tvedt (1991) proposed a general procedure for estimating the mean upcrossing rate of a scalar differentiable process $X(t)$, obtained as a parametric sensitivity measure of the probability of an associated parallel system unsafe domain. The method is applicable to Gaussian and non-Gaussian stationary as well as non-stationary random processes, and time-dependent failure surfaces. Other methods are considered by Wen and Chen (1989a,b) and Beck and Melchers (2004).

For a non-Gaussian process, e.g., Weibull process describing the stress amplitude, the upcrossing rate may be approximated by (Marley and Moan, 1992)

$$\nu_S^+(\eta, t) = \nu_0^+ \left\{ -\ln\left(1 - \exp\left[-\left(\frac{\eta(t)}{X_A A_\nu}\right)^h\right]\right)\right\}$$
$$\approx \nu_0^+ \exp\left[-\left(\frac{\eta(t)}{X_A A_\nu}\right)^h\right], \tag{13.21}$$

where A_ν and h are the Weibull parameters of the process, $\eta(t)$ is the time-variant threshold, ν_0^+ is the zero upcrossing rate, and X_A is the model uncertainty, i.e., variability of A_ν.

Various uncertainties other than the fundamental variability of the loading affect the upcrossing rate. For given values of the random variables $\mathbf{X} = (X_1, \ldots, X_n)^T$ that represent those uncertainties, the (conditional) failure probability is determined by the downcrossings of 0 by $M(t; \mathbf{x})$ for $t \geq 0$:

$$P_f(\mathbf{x}) = \text{Prob}\left(\min_{0 \leq t \leq T} M(t|\mathbf{x}) \leq 0\right) = 1 - \exp\left[-\nu_M^-(0; \mathbf{x})\, T\right], \tag{13.22}$$

where the zero-downcrossing rate $\nu_M^-(0; \mathbf{x})$ depends on the value of the parameter vector $\mathbf{x}$.

The total failure probability, considering the uncertainties in $\mathbf{X}$, can then be calculated by the unconditional probability,

$$P_f = \int_{\mathbf{x}} P_f(\mathbf{x})\, f_{\mathbf{X}}(\mathbf{x})\, d\mathbf{x}. \tag{13.23}$$

The integral represents the expected value of $P_f(\mathbf{X})$. Hence, P_f may be calculated as the average value of $P_f(\mathbf{x}_i)$, when the set of outcomes $\mathbf{x}_i$ are generated by Monte Carlo simulation. That is,

$$P_f \approx \frac{1}{N} \sum_{i=1}^{N} P_f(\mathbf{x}_i). \tag{13.24}$$

Instead of using Monte-Carlo simulation to calculate the unconditional P_f, it can be alternatively calculated by the FORM/SORM method developed by Wen and Chen (1987). The problem first needs to be reduced to a time-invariant one. This can be done by applying the following expression

$$P_f = \int_{\mathbf{x}} P_f(\mathbf{x})\, f_{\mathbf{X}}(\mathbf{x})\, d\mathbf{x} = \int_{g(\mathbf{u}, u_{n+1}) \leq 0} f_{\mathbf{U}}(\mathbf{u})\, f_{U_{n+1}}(u_{n+1}) d\mathbf{u}\, du_{n+1}, \tag{13.25}$$

where the (modified) failure function is

$$g(\mathbf{u}, u_{n+1}) = u_{n+1} - \Phi^{-1}\left(P_f(\mathbf{T}^{-1}(\mathbf{u}))\right). \tag{13.26}$$

Here, $\mathbf{U} = (U_1, \ldots, U_n)^T$ is a vector of independent standard normal variables, $\mathbf{U} = \mathbf{T}(\mathbf{X})$ for a suitable transformation $\mathbf{T}$, and U_{n+1} is an auxiliary standard normal variable independent of $\mathbf{U}$.

13.4.3 Long-Term Reliability for a Single Load Effect

For a nonstationary process $M(t;\mathbf{x})$, which could be associated with a long-term period, the basic formula for the failure probability associated with a time interval $(0, T)$ may be written as (Naess, 1984)

$$P_f(\mathbf{x}) = 1 - \exp\left\{-\int_0^T \nu_M^-(0;t,\mathbf{x})\,dt\right\}, \tag{13.27}$$

where the mean downcrossing rate $\nu_M^-(0;t,\mathbf{x})$ now depends on time because the sea states change with time. To be specific, let us assume that the relevant properties of the sea states are captured by the (time-variant) environmental variables $\mathbf{W} = \left(H_s, T_p, \Theta_h, U_c, \Theta_c, U_w, \Theta_w\right)^T$, where H_s = significant wave height, T_p = spectral peak period, Θ_h = mean wave direction, U_c = surface current velocity, Θ_c = current direction, U_w = mean wind speed at a specified reference height above mean sea level, Θ_w = mean wind direction. As discussed by Naess (1984), a prerequisite for deriving the long-term failure probability is to impose an ergodicity assumption on the environmental process $\mathbf{W} = \mathbf{W}(t)$. The following result is then obtained (see Section 12.4.3):

$$P_f(\mathbf{x}) = 1 - \exp\left\{-T\int_{\mathbf{w}} \nu_M^-(0;\mathbf{w},\mathbf{x})\; f_{\mathbf{W}}(\mathbf{w})d\mathbf{w}\right\}, \tag{13.28}$$

where $\nu_M^-(0;\mathbf{w},\mathbf{x})$ denotes the mean zero downcrossing rate of M for the sea state $\mathbf{W} = \mathbf{w}$, and $f_{\mathbf{W}}(\mathbf{w})$ is the PDF of $\mathbf{W}$. The full long-term failure probability, accounting for both environmental variability and parameter uncertainty, is given as (Eq. (13.23)):

$$P_f = \int_{\mathbf{x}} P_f(\mathbf{x})\; f_{\mathbf{X}}(\mathbf{x})\,d\mathbf{x}. \tag{13.29}$$

It is recognized that the failure probability P_f may be calculated by a two-step FORM/SORM method (Wen and Chen, 1987). First, the following calculation is performed for each fixed value of $\mathbf{x}$:

$$\int_{\mathbf{w}} T\nu_M^-(0;\mathbf{w},\mathbf{x})\; f_{\mathbf{W}}(\mathbf{w})d\mathbf{w} = \int_{\tilde{g}(\mathbf{v},v_{l+1})\leq 0} f_{\mathbf{V}}(\mathbf{v})\; f_{V_{l+1}}(v_{l+1})d\mathbf{v}\,dv_{l+1}, \tag{13.30}$$

where

$$\tilde{g}(\mathbf{v}, v_{l+1}) = v_{l+1} - \Phi^{-1}\left(T\nu_M^-(0;\tilde{\mathbf{T}}^{-1}(\mathbf{v}),\mathbf{x})\right). \tag{13.31}$$

Here, $\mathbf{V} = (V_1, \ldots, V_l)^T$ is a vector of independent standard normal variables, $\mathbf{V} = \tilde{\mathbf{T}}(\mathbf{W})$ for a suitable transformation $\tilde{\mathbf{T}}$, and V_{l+1} is an auxiliary standard normal variable independent of $\mathbf{V}$. Finally, Eq. (13.29) is then calculated as shown in Section 13.4.2. An early effort to derive the long-term failure probability can be found in Bjerager et al. (1988), and it is interesting to contrast the discussions in this paper with our derivation here. A derivation of the long-term failure probability more in line with the derivation in this section is given by Schall et al. (2001).

13.4.4 Outcrossing Rate Formulations for Two Load Processes

When the load is a vector process consisting of two or more load effect variables, it is not easy to define a characteristic load effect vector for design. A simple way to deal with this problem is to consider the characteristic values of individual load effect variables and attach a reduction factor to each of them to account for the possible lack of simultaneous occurrence, in principle such that the joint exceedance corresponds to a given exceedance probability. Such an approach then needs to be validated by an outcrossing rate approach, e.g., by using the following limit state function for two variables,

$$g(S_1, S_2) = 1 - \left(\frac{S_1}{R_{u1}}\right)^{\alpha} - \left(\frac{S_2}{R_{u2}}\right)^{\beta}, \tag{13.32}$$

where each load effect variable S_i has two components, due to static gravity and pay loads and dynamic loading: $S_i = S_{i1} + S_{i2}$. α and β are time invariant parameters. No model uncertainties are considered here. To assess the outcrossing rate through the limit state surface, two methods are considered in the following.

The method used by Naess (1983, 1985c) to assess upcrossing rates for Morison force and slow drift response was adopted by Jia and Moan (2009) to establish expressions for the outcrossing rate for the more general case considered herein. By assuming

$$Q(t) = \left(\frac{S_1(t)}{R_{u1}}\right)^{\alpha} + \left(\frac{S_2(t)}{R_{u2}}\right)^{\beta} = (X(t))^{\alpha} + (Y(t))^{\beta}, \tag{13.33}$$

where the load effect processes $S_1(t)$ and $S_2(t)$ are assumed to be jointly stationary Gaussian processes, the mean upcrossing rate is then given by the Rice formula (Lin, 1967):

$$\nu^-(g = 0) = \nu^+(Q = 1) = \int_0^{\infty} s f_{Q\dot{Q}}(1, s)\, ds. \tag{13.34}$$

By the law of marginal probability, and invoking the law of transformation of variables, the following result is obtained

$$f_{Q\dot{Q}}(q, \dot{q}) = \int_{-\infty}^{\infty} f_{QY\dot{Q}}(q, y, \dot{q})\, dy = \int_{-\infty}^{+\infty} f_{XY\dot{Q}}(x(y, q), y, \dot{q}) \left|\frac{\partial x(y, q)}{\partial q}\right| dy, \tag{13.35}$$

where

$$x(y, q) = (q - y^{\beta})^{\frac{1}{\alpha}}, \tag{13.36}$$

$$\left|\frac{\partial x}{\partial q}\right| = \frac{1}{\alpha}(q - y^{\beta})^{\frac{1}{\alpha}-1}, \tag{13.37}$$

$$f_{XY\dot{Q}}(x, y, \dot{q}) = f_{\dot{Q}|X,Y}(\dot{q}|x, y) \cdot f_{XY}(x, y). \tag{13.38}$$

By using Eqs. (13.35) and (13.38) in combination with Eq. (13.34), and considering the end points of the limit state equation,

$$\nu^+ = \int_0^\infty s \left[\int_{y_l}^{y_u} f_{\dot{Q}|X,Y}(s|x(y,1),y) \cdot f_{XY}(x(y,1),y) \left|\frac{\partial x}{\partial q}\right|_{q=1} \cdot dy \right] ds$$
$$= \int_{y_l}^{y_u} f_{XY}(x(y,1),y) \left|\frac{\partial x}{\partial q}\right|_{q=1} \left[\int_0^\infty s f_{\dot{Q}|X,Y}(s|x(y,1),y) ds \right] dy. \quad (13.39)$$

From Eq. (13.33)

$$\dot{Q} = \alpha X^{\alpha-1}\dot{X} + \beta Y^{\beta-1}\dot{Y}. \quad (13.40)$$

The conditional random variable $\{\dot{Q}|X = x, Y = y\}$ is a Gaussian variable. To derive its mean and variance, consider the random vector $W = (X, Y)^T$, $\dot{W} = (\dot{X}, \dot{Y})^T$. Writing the covariance matrix of $(W^T, \dot{W}^T)^T$ as

$$\Sigma = \begin{pmatrix} \Sigma_{11} & \Sigma_{12} \\ \Sigma_{21} & \Sigma_{22} \end{pmatrix}, \quad (13.41)$$

where the Σ_{ij} are 2×2 matrices, the joint PDF of $\dot{W}$ given $W = w = (x, y)^T$ is multivariate normal with mean value $\mu = (\mu_1, \mu_2)^T = \Sigma_{21}\Sigma_{11}^{-1}(w - \mu_W)$ and covariance matrix $\Xi = (\Xi_{ij}) = \Sigma_{22} - \Sigma_{21}\Sigma_{11}^{-1}\Sigma_{12}, i, j = 1, 2$. μ_W is the mean value of W. Hence, the expected value of $\{\dot{Q}|W = w\}$ is given as

$$\tilde{\mu} = \alpha x^{\alpha-1}\mu_1 + \beta y^{\beta-1}\mu_2, \quad (13.42)$$

while the variance is,

$$\tilde{\sigma}^2 = \alpha^2 x^{2(\alpha-1)}\Xi_{11} + \beta^2 y^{2(\beta-1)}\Xi_{22} + \alpha\beta x^{\alpha-1}y^{\beta-1}(\Xi_{12} + \Xi_{21}). \quad (13.43)$$

It can be shown that

$$\int_0^\infty s f_{\dot{Q}|X,Y}(s|x,y) ds = \tilde{\sigma}\phi(\frac{\tilde{\mu}}{\tilde{\sigma}}) + \tilde{\mu}\Phi(\frac{\tilde{\mu}}{\tilde{\sigma}}), \quad (13.44)$$

where ϕ and Φ are the standard normal probability density and cumulative distribution function, respectively. Then, Eq. (13.39) becomes

$$\nu^+ = \int_{y_l}^{y_u} f_{XY}(x(y,1),y) \left|\frac{\partial x}{\partial q}\right|_{q=1} \left[\tilde{\sigma}\phi(\frac{\tilde{\mu}}{\tilde{\sigma}}) + \tilde{\mu}\Phi(\frac{\tilde{\mu}}{\tilde{\sigma}})\right]_{x=x(y,1)} dy. \quad (13.45)$$

In general, a numerical method must be used to perform the integration in Eq. (13.45).

Alternatively, the outcrossing rate may be determined by approximating the failure surface by piecewise linear functions and using analytical results for outcrossing of linear surfaces; see, e.g., Leira (1987). Jia and Moan (2009) compared the two approaches, and both were shown to be efficient and accurate. However, using Eq. (13.45) may sometimes lead to numerical difficulties due to the fact that the effective integration interval can be small. In this case, asymptotic analysis by the Laplace method can be useful. Otherwise, the Monte Carlo-based method described in Section 16.6.6 can be used.

In this section the failure probability in a short-term period is considered. The analysis can be extended to account for both parameter uncertainty and the long-term variation of the sea states as shown in Section 13.4.3 for a single load effect.

13.5 Fatigue Design

13.5.1 Introduction

Fatigue design of offshore structures requires a description of the long-term variation of local stresses due to wave, as well as possible sumfrequency wave actions, variable buoyancy, slamming, and wave- or current-induced vortex shedding. The effect of local (pressure) and global actions should also be accounted for. However, it is noted that the main contribution to fatigue damage is caused by frequently occurring load effects that are of the order of 10% to 20% of the extreme load effects in the service life. Fatigue failures can often be catastrophic; they come without warning and may cause significant property damage and loss of life.

The physical process of fatigue consists of initiation of a crack, stable crack growth, and unstable crack growth (rupture) (see, e.g., Broek (1984); Almar-Naess (1985); Barsom and Rolfe (1999); Dowling (1993)). In welded joints, cracklike defects would exist just after fabrication and make the initiation life relatively short. Once a crack is initiated, it tends to grow in a direction orthogonal to the direction of the oscillatory tensile stresses.

Fatigue is a challenging failure mode to deal with for the following reasons. (1) The fatigue process, especially initiation, is inherently unpredictable. (2) It is difficult to translate laboratory data to in-service conditions. (3) It is difficult to model the load environment and the complex stress states.

A variety of measures are adopted during the design, fabrication, and service life of ships to control fatigue cracks. These include design, material and structural detail requirements, as well as inspection during construction and operation and repair as documented in design codes and classification societies' rules.

13.5.2 SN Approach

As previously discussed, fatigue strength is commonly described by SN curves that have been obtained by laboratory experiments. Fatigue failure can be defined by the occurrence of a visible crack, through thickness crack, or complete loss of load-carrying capacity. Fracture mechanics analysis is used to more accurately assess the different stages of fatigue crack growth, including calculation of residual fatigue life beyond through-thickness crack, which is normally defined as fatigue failure. SN data for fixed and floating platforms are given in design codes and guidelines. Some of the most relevant ones are briefly reviewed by Karsan (2005) and Halkyard (2005).

The SN curves may typically be formulated as

$$N = K\,S^{-m}, \tag{13.46}$$

or

$$N = \begin{cases} KS^{-m} & \text{for} \quad S > S' \\ \infty & \text{for} \quad S \leq S' \end{cases}, \tag{13.47}$$

or

$$N = \begin{cases} K_1 S^{-m_1} & \text{for} \quad S > S' \\ K_2 S^{-m_2} & \text{for} \quad S \leq S' \end{cases}, \tag{13.48}$$

where the point of intersection between the two equations (13.48) is (N', S'), with $N' = K_1(S')^{-m_1} = K_2(S')^{-m_2}$. For structures that are in a corrosive environment and subjected to random loading, the formulation in Eq. (13.48) is the most relevant. However, the simpler version in Eq. (13.47) is conservative if K is chosen equal to K_1 in Eq. (13.48).

Typically, the characteristic SN curve defined by $K = K_c$ (Section 13.2.2) and used in design checks, is taken to be the mean minus two standard deviations, that is,

$$\log_{10} K_c = \mu - 2s, \tag{13.49}$$

where μ and s are the mean value and standard deviation of the parameter $\log_{10} N$. Typically, $\mu \approx 12$ and $s \approx 0.2$. Therefore, $K_c/\mu_K \approx 0.4$.

It is noted that $K = K_{ref}(t_{ref}/t)^{m/4}$, where K_{ref} is the reference parameter in the SN curve and t and t_{ref} is the plate thickness and reference plate thickness, respectively (usually measured in millimeters).

The basic design equation is based on the SN curves and Miner-Palmgren's hypothesis, and may be written as Eq. (13.6). The calculation of the fatigue loading; i.e., local stress ranges, involves estimating stress ranges in various sea states in the long-term period, see, e.g., API (1997), NORSOK N-003 (2007). The stress range used in the SN curve and the corresponding response can be a nominal stress (e.g., as obtained by beam theory) or a hot spot stress. Equation (13.6) may be further refined by applying a semiprobabilistic fracture mechanics (FM) or probabilistic SN or FM approach.

SN curves are traditionally determined by constant amplitude testing with large stress ratios, $R = \sigma_{min}/\sigma_{max} > 0.5$, implying that the stresses σ are in a crack opening mode. Relaxation of residual stresses, external loading with partly compressive stresses, and crack closure effects would make the actual crack growth less than implied by assuming all stress ranges effective and using existing SN curves. However, offshore structures are traditionally designed conservatively by the latter approach. For trading vessels, a correction for the so-called mean stress effect has been introduced, in different ways by various classification bodies (Lotsberg, 2006). Kim and Lotsberg (2004) investigated the effect of residual stresses experimentally by considering fillet welds. Figure 13.4 shows results obtained for a joint subjected to three different load histories. The reduction of residual stresses by applying a large initial load in load case 3 is seen to increase fatigue strength. It is also noted that this effect would vary for different types of joints, depending on the magnitude of residual stresses, etc. Further efforts are needed to clarify this issue.

13.5.3 Fracture Mechanics

As already mentioned, we usually differentiate between initiation, stable growth, and unstable growth (fracture) of cracks. Fracture mechanics provide models for determining the steady growth of a crack and the final fracture. The crack initiation time, however, needs to be based on empirical data.

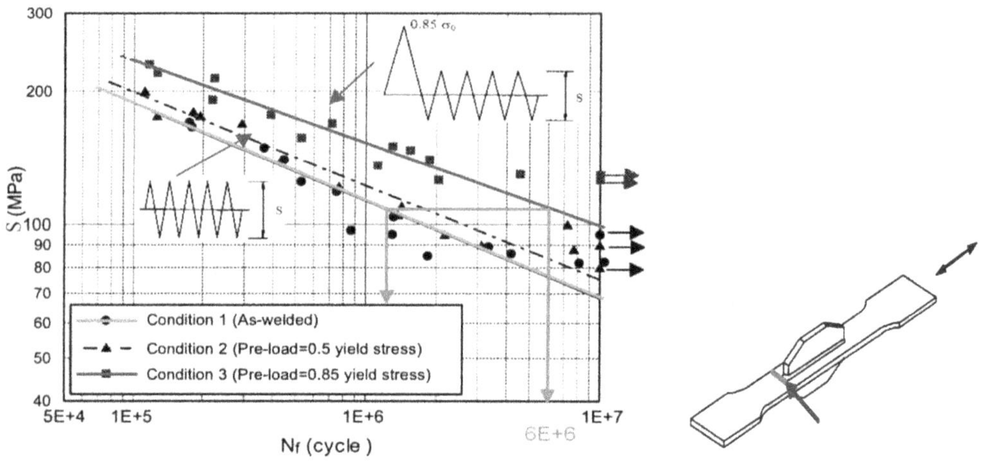

Figure 13.4. Effect of mean stress and stress relaxation (Kim and Lotsberg, 2004).

Despite many efforts to refine the fracture mechanics approach to stable crack propagation, the Paris-Erdogan law is considered the most generally viable model (Almar-Naess, 1985; Maddox, 1991; Sobczyk and Spencer, 1992). In general, it may be formulated as follows:

$$\frac{da}{dN} = \begin{cases} 0 & \text{for} \quad \Delta K \leq \Delta K_{th} \\ C_1 (\Delta K)^{m_1} & \text{for} \quad \Delta K_{th} < \Delta K \leq \Delta K_1 \\ \vdots & \\ C_{n+1} (\Delta K)^{m_{n+1}} & \text{for} \quad \Delta K > \Delta K_n \end{cases} \tag{13.50}$$

where a is crack depth, N is number of cycles, C_i is the crack growth rate parameter for segment i and m_i is its corresponding slope, ΔK_{th} is a threshold of the stress intensity factor range and ΔK_i is the point of intersection of two consecutive segments. ΔK is given as a function of the crack depth a:

$$\Delta K = Y(a) \cdot S\sqrt{\pi a} = (Y_m S_m + Y_b S_b)\sqrt{\pi a}, \tag{13.51}$$

where the one-dimensional compliance function $Y(a)$ is based on a constant aspect ratio a/c; S is the stress range; and the subindices m and b refer to membrane and bending stress, respectively. This one-dimensional model may be extended by generalizing Eq. (13.51) to cover crack growth in a (depth) and in c (length). The definition of the stress intensity factor range ΔK is a crucial aspect in evaluating the fatigue life of a welded joint.

For the particular case where a constant geometry function Y is assumed and $\Delta K_{th} = 0$, the crack growth at time t may be easily integrated resulting in closed form expressions. This procedure is, however, nonconservative in welded joints, e.g., butt joints and fillet welded joints, where the weld toe magnifying effect has important implications. In general, the geometry function, which is dependent on the crack size and shape, may be defined as

$$Y(a) = Y_{plate}(a) \cdot M_k(a) \tag{13.52}$$

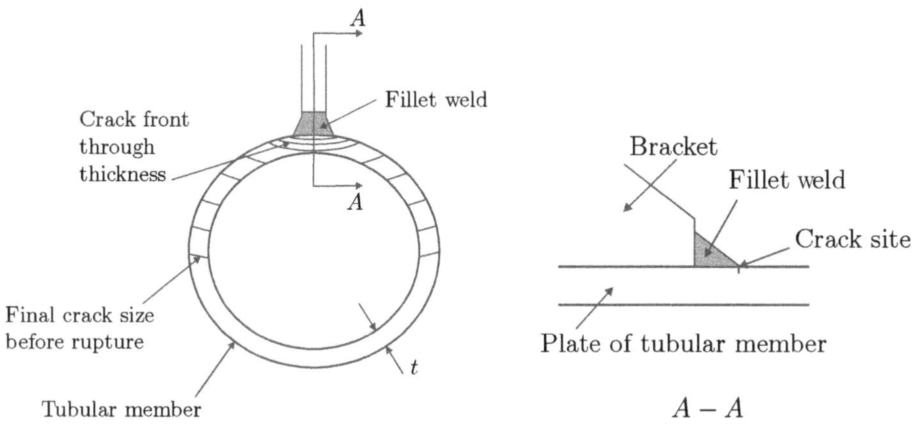

(a) Crack propagation (b) Location of crack at weld toe

Figure 13.5. Crack propagation in two stages – through the thickness and as a through-thickness crack.

where $Y_{plate}(a)$ is the geometry function corresponding to a semielliptical surface crack as proposed by Raju and Newman (1979) from finite element analysis in flat plates subjected to axial and bending remote stresses. $M_k(a)$ is a magnification factor (or correction factor) that depends on the local weld geometry and accounts for the crack size and loading. The cases studied herein considering the $M_k(a)$ effect (joint F) are based on the formulation proposed by Bowness and Lee (1999) and recommended in BS 7910 (1999).

In fracture mechanics analysis under random loading an equivalent constant amplitude loading is applied, see Eq. (13.55). In particular, the fracture mechanics approach may be used to determine the number of cycles N to propagate a crack from the initial size a_i to a critical crack size a_f, which may be equal to the thickness of the plate (Fig. 13.5). Invoking the Paris-Erdogan law, it is seen that

$$N = \int_{a_i}^{a_f} \frac{da}{C(\Delta K)^m}. \tag{13.53}$$

Typically, $a_i = 0.5$ mm (NORSOK N-001, 2000). Parameters C and m depend on environmental conditions, given in, e.g., NORSOK N-004 (1998). A simple example is obtained by putting $Y(a) = 1$ and $\Delta K = \bar{s}\sqrt{\pi a}$. Then,

$$N = \int_{a_i}^{a_f} \frac{da}{C \cdot \bar{s}^m (\pi a)^{m/2}} = \frac{2}{(2-m)\, C \cdot \bar{s}^m \pi^{m/2}} \left(a_f^{-\frac{m}{2}+1} - a_i^{-\frac{m}{2}+1} \right). \tag{13.54}$$

In general, however, numerical integration needs to be applied.

In particular, the fracture mechanics methodology provides a tool for more accurately assessing the different stages of crack growth, especially beyond what is considered fatigue failure (e.g., through-thickness crack in Fig. 13.5) in establishing SN data. Such detailed information about crack propagation is also required to plan inspections and repair. The basis for crack propagation analysis is Paris-Erdogan's law. To account for the mean stress level, the effective stress intensity factor range, ΔK_{eff}, may be introduced as a function of the ratio $R = K_{\min}/K_{\max}$. Many investigators focused their attention on the effective stress intensity factor range (Paris et al., 1999). Before FM formulations can be applied for small crack sizes, they need to be calibrated based on SN data. This is because the initiation of cracks and their initial stages are subjected to uncertainties that are difficult to quantify. This calibration can be performed by a semiprobabilistic or probabilistic (reliability) approach; for example, Moan et al. (1993) and Ayala-Uraga and Moan (2007) applied a probabilistic calibration to various SN curves, using BS 7910 (1999) FM data. The residual strength of ships with cracks makes it possible to use the *leak before break* principle to detect cracks. Even though a formal assessment of the critical crack size in ships, e.g., according to BS 7910 (1999), indicates critical crack sizes of the order of the plate thickness, experiences suggest much larger values (Bach-Gansmo et al., 1987). Further documentation about the residual fatigue life associated with the propagation of large cracks until final fracture was provided by Zhang and Moan (2005).

13.5.4 Fatigue Loading

The local stress for fatigue design needs to be determined with due account of the temporal and spatial variation, see Fig. 13.6. For welded offshore structures the main parameter to represent the variation in time is the stress range. This approach is based on the assumption that tensile residual stresses are always present and that all stress cycles effectively drive the crack. For base material, the mean stress should also be considered as a variable (Almar-Naess, 1985). The spatial stress variation can be accounted for by using a nominal or hot spot stress approach (e.g., Almar-Naess (1985); Karsan (2005); Halkyard (2005)).

Fatigue is caused by dynamic loads such as wind, wave, or traffic, and machinery. For marine structures, the primary cyclic stress that causes fatigue is induced by the external wave loading and calculated based on the environmental conditions. Calculation of the relevant fatigue loading involves a global analysis of loads and their effects in terms of member forces, and a local analysis to determine the hot spot stress, as illustrated in Fig. 13.7.

The hydrodynamic analysis is carried out by Morison's equation for structures consisting of slender members. For large-volume structures, it is important to at least apply 3D theory for validating possible simplified methods. 3D effects may be important because the waves that contribute fatigue loading may be short compared to the size of the structure, implying diffraction and radiation effects. In fatigue design, the total time history of very local stresses is required. Load effects are determined in a hierarchy of analyses, see Fig. 13.8. When dynamic effects are of concern, a model that recognizes the stochastic features of waves is important. Because moderate sea states contribute most to fatigue, linear analyses often suffice. Frequency-domain analyses of wave-induced response in the various sea states can then be efficiently carried out with coarse finite element models of the structure. The structural model used in such analyses at best provides nominal stresses. More refined analyses need to be carried out to obtain the local stresses for fatigue design checks. In some cases, fatigue load effects are influenced by nonlinear effects, and time-domain simulation combined with rainflow counting of stress ranges is necessary. Detailed guidance of load effect analyses is given in design codes, and a review is given by Karsan (2005) and Halkyard (2005).

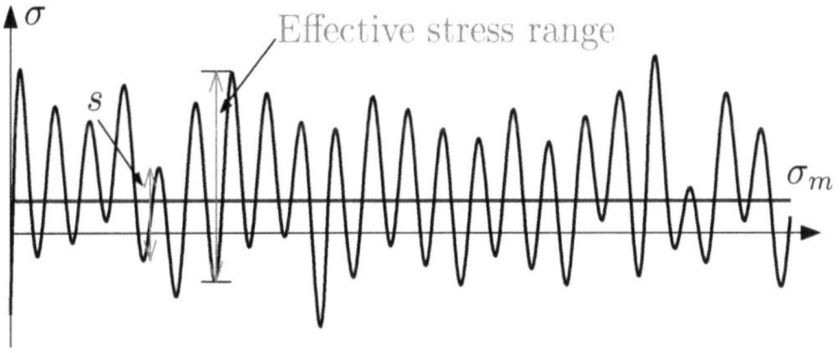

(a) Effective stress range representation

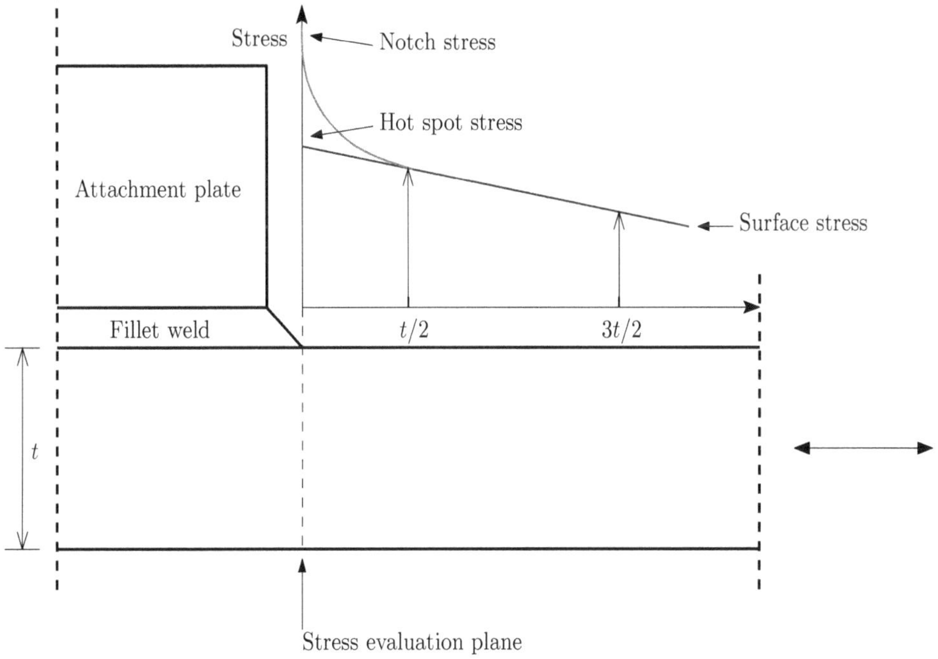

(b) Spatial definition of nominal, hot spot, and notch stress in a plane structure

Figure 13.6. Cyclic stress histories for marine structures.

Fatigue load effects associated with wave-induced loads in a long-term period may be based on alternative approaches – in a hierarchy of procedures with increasing accuracy and complexity. Here, three main alternatives are considered:

- Assume that stress ranges follow a two-parameter Weibull distribution, obtained by estimating s_0 corresponding to an exceedance probability of $1/N_0$, and assume a value of the shape parameter h according to guidance – including the effect of dynamics – mentioned in Section 13.2. Calculation of s_0 and selection of h must obviously be conservative.
- Frequency-domain analysis for each sea state to determine response variance and assume narrow band response, implying Rayleigh distribution of stress

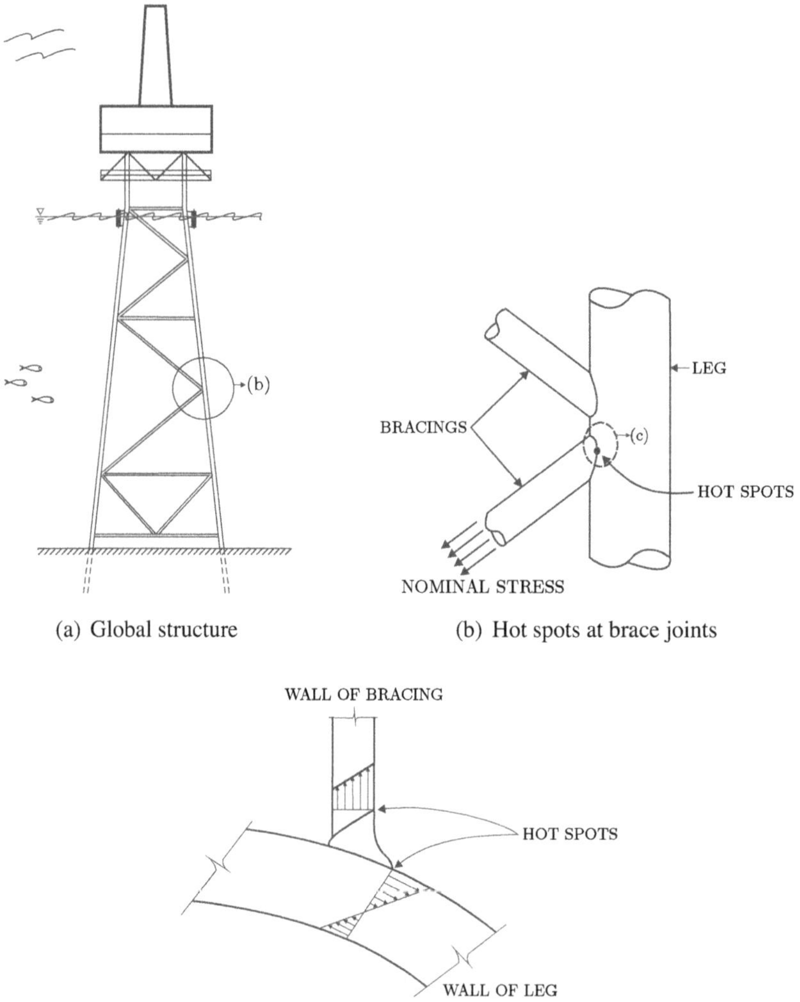

(a) Global structure (b) Hot spots at brace joints

(c) Hot spot stress details

Figure 13.7. Fatigue loading in terms of stress ranges.

ranges. Moderate nonlinearities may be accounted for by determining a quasitransfer function based on time domain analysis or another linearization approach. Factors may be introduced to correct for wideband or non-Gaussian response.

- Time-domain analysis combined with rainflow counting of cycles for all or a representative set of sea states that are found (e.g., by frequency-domain analysis) to contribute most to the fatigue damage. Multiple simulations per sea state are normally required to reduce the statistical uncertainty.

Screening to identify joints with high dynamic stresses and stress concentration, which require more detailed fatigue analyses, may be undertaken by using the first approach. s_0 may be based on the nominal member stress for the extreme event and an appropriate stress concentration factor, and the shape parameter h could be obtained by general guidance.

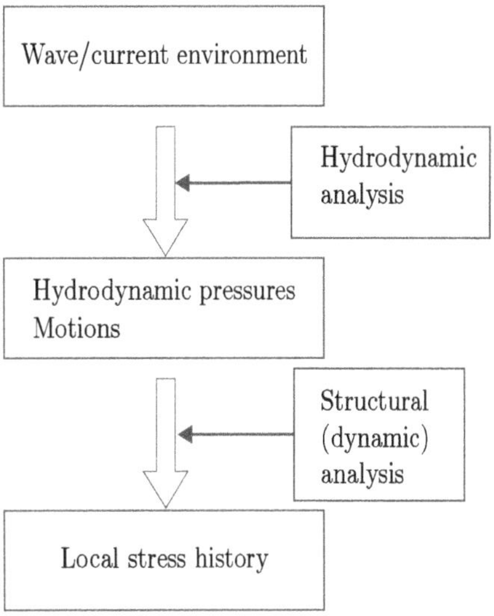

Figure 13.8. Procedure for evaluating load effects.

Detailed fatigue analyses should be performed using conservative deterministic methods or frequency-domain techniques, and in particular situations, by time-domain analysis. Stochastic approaches should be applied for dynamic sensitive structures. For linear systems, frequency-domain techniques are efficient.

More complete time-domain approaches may be necessary in case of strong nonlinearities, e.g., associated with local splash zone behaviour, at least to calibrate simpler methods. Systematic studies carried out for offshore structures in extratropical regions suggest that the long-term statistics of the response variables can be well described by a two-parameter Weibull distribution.

The slowly varying stress due to payloads and the residual fabrication stresses affect the mean stress, see Fig. 13.4. However, if the stress is partly compressive, the effective driving stress range is reduced. This could occur if the mean stress due to external loads is compressive and the welding residual stresses are relaxed. These features were demonstrated in test results by Kim and Lotsberg (2004) for (local) welded joints under constant amplitude loading. Huang and Moan (2007) examined the effect of the stress intensity ratio $R = K_{\min}/K_{\max}$. Rather than using different SN curves, effective stress ranges could be used in this case. In the Joint Tanker Project (JBP, 2005a), the beneficial effect of shakedown induced by local yielding due to external loads is noted. In addition, the effective cyclic stress is considered when the external loads imply compressive stresses. The Joint Tanker Project (JBP, 2005b) uses an effective stress range of the wave-induced stress by $S_{eff} = \sigma_t + 0.6\sigma_c$, where the subindices t and c refer to tensile and compressive stress of the range S, respectively, see, e.g., discussion by Lotsberg (2006). Ayala-Uraga and Moan (2007) proposed an alternative approach based on the work of Huang and Moan (2005).

Obviously, this effect is even more pronounced in case of through-thickness cracks, which may experience compressive residual stresses and wave-induced stresses that are largely compressive, for instance, in ship decks in sagging condition under stillwater and wave loading (Zhang and Moan, 2005).

However, besides wave loads with period equal to the wave period, in the range of 2 to 20 seconds, the presence of certain nonlinear features of the loading may

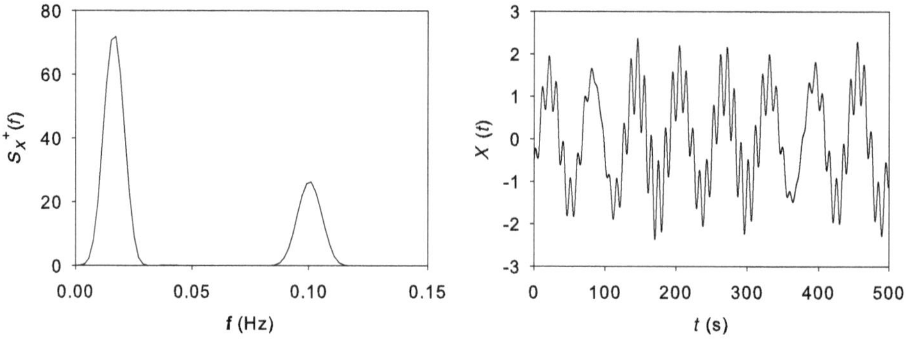

(a) Spectral density function (b) Time series

Figure 13.9. Example of a bimodal Gaussian process.

cause steady-state loads with a period that is a multiple of 1/2, 1/3, ... or 2, 3, ... of the wave period. One such effect is the drag force in Morison's equation, which is nonlinear with respect to water particle velocity, cf. Section 7.2.1.

The drag force in Morison's equation implies superharmonic load components that are normally an order of magnitude smaller than the wave frequency component. However, if the higher-order harmonic load components coincide with the natural frequency, significant amplification occurs, making the high-frequency load effect significant. Slamming and sloshing in tanks can similarly contribute fatigue loading. This means that the stress history comprises a high- and low-frequency response process and becomes wide band, see Figs. 13.9 and 13.10.

Local stresses for fatigue analysis must be determined by shell or solid finite elements and appropriately fit the format of the fatigue resistance. The SN approach may be based on nominal or hot spot stress approaches and SN curves that are consistent with the stress definition. Although methods for tubular joints were established 10-15 years ago (see, e.g., Karsan (2005)), the focus recently has been on procedures for plated structures (e.g., Fricke (2001); Kim and Lotsberg (2004)). Two alternative methods are commonly used for hot spot stress derivation; i.e., based on a linear extrapolation of the stresses to the hot spot from the read out points at a distance $0.5t$ and $1.5t$ from the hot spot or the stress at the readout point $0.5t$ away from the intersection line and multiplied by 1.12.

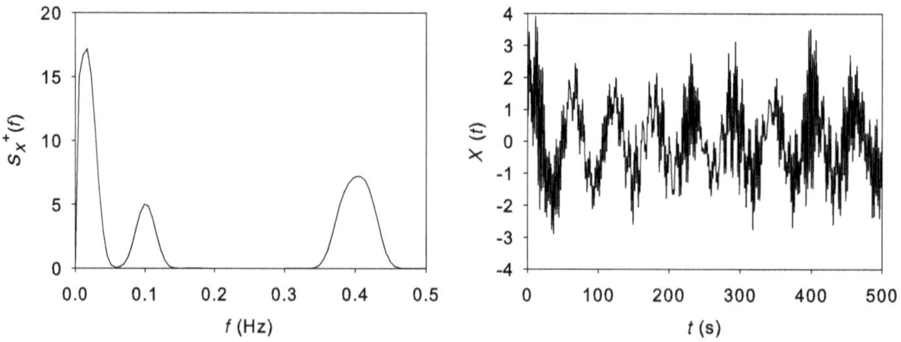

(a) Spectral density function (b) Time series

Figure 13.10. Example of a trimodal Gaussian process.

13.5.5 Fatigue Damage

Equation (13.5) is convenient as a basis for an early screening of fatigue proneness by taking the distribution in Eq. (13.4) to be the long-term distribution of stress ranges and using a simple (conservative) estimate of the extreme response s_0 and by assuming a value for the shape parameter h of the Weibull distribution based on experience. The parameter h depends on the environmental conditions, relative magnitude of drag and inertia forces, and possible dynamic amplification, as discussed by Marshall and Luyties (1982) for jackets and by Odland (1982) for jack-ups. The scale parameter is directly related to the extreme response value required for ultimate strength design checks and can be estimated in connection with determination of extreme response values. In this way, fatigue loading, at least for initial design and screening to identify the importance of fatigue, can be readily accomplished.

Now we are examining the fatigue damage in more detail based on

- fatigue resistance described by SN curves
- fatigue loading described by
  - $f_S(s)$ = probability density function of stress ranges
  - N_T = total number of cycles in the period T, which is estimated by the mean frequency of $S(t)$ times T. For a narrow band stress process with zero mean, $N_T = \nu_S^+(0) \cdot T$ (mean zero upcrossing rate times length of time)

As described in Section 10.9, the expected fatigue damage $D = D(T)$ for the case that the SN curve is given by Eq. (13.46) is

$$D = \int_0^\infty \frac{N_T}{K} s^m f_S(s) ds = \frac{N_T}{K} \mathrm{E}[S^m] = \frac{N_T}{K} \overline{S}^m, \tag{13.55}$$

where N_T is the total number of stress cycles, $\overline{S}$ is defined by $\overline{S} = \left(\mathrm{E}[S^m]\right)^{1/m}$, and E[] denotes the expectation value operator. $\overline{S}$ may be interpreted as a constant amplitude loading that is equivalent to the random loading under the assumption of a single slope SN curve.

For a narrow banded Gaussian process with zero mean and variance σ_X^2, stress ranges follow a Rayleigh distribution, which can be easily determined from the stress amplitude distribution f_A:

$$\begin{aligned} f_S(s) &= f_A(s/2)\frac{da}{ds} = \frac{(s/2)}{\sigma_X^2} \exp\{-\frac{(s/2)^2}{2\sigma_X^2}\}\frac{1}{2} \\ &= \frac{s}{4\sigma_X^2} \exp\{-\frac{s^2}{8\sigma_X^2}\}, \end{aligned} \tag{13.56}$$

which leads to

$$\begin{aligned} D &= \frac{N_T}{K} \mathrm{E}[S^m] = \frac{N_T}{4\sigma_X^2 K} \int_0^\infty s^{m+1} \exp\left\{-\frac{s^2}{8\sigma_X^2}\right\} ds \\ &= \frac{N_T}{K}(2\sqrt{2}\sigma_X)^m \Gamma\left(m/2+1\right). \end{aligned} \tag{13.57}$$

This is a special case of the Weibull distributed stress ranges, see Eq. (13.5). Equation (13.5) may be used to express the cumulative damage in a long-term period τ by applying the equations as obtained in conjunction with the stress range distribution for

- each sea state separately and summing up the contributions to obtain the long-term fatigue damage D.
- the long-term period and determining D directly.

The narrow band linear wave-induced response in a single sea state can be described by a Rayleigh distribution. This corresponds to a Weibull distribution with $h = 2$. In a long-term period τ, the number of (narrow band) cycles associated with sea state no. i is $\tau_i = \tau\, p_i\, \nu_i$, where p_i is the long-term probability and ν_i the number of cycles per time unit of this sea state. Hence, the cumulative damage in τ is

$$D = \sum_i D_i = \sum_i \frac{\tau}{K}\, p_i\, \nu_i\, (2\sqrt{2}\sigma_i)^m\, \Gamma(m/2+1)\, \rho_i, \qquad (13.58)$$

where ρ_i is a correction factor to account for wide band and/or non-Gaussian load effects, as discussed subsequently in Section 13.5.7.

The stress range level that contributes most to D in a long-term period corresponds to the value that yield the maximum of the fatigue damage: dD that is proportional to $f_S(s) \cdot s^m$ (Eq. (13.55)). This stress range is $s = s_0[(m+h-1)/(h \ln N_0)]^{1/h}$ for the Weibull distributed fatigue loading (Eq. (13.4)), implying that fatigue damage is primarily caused by stress ranges that are typically of the order of 10% to 20% of s_0.

The cumulative damage D in a long-term period for the case with a two-slope SN curve (Eq. (13.48)) may be obtained. The damage consists of two contributions, one for each of the two segments of the SN-curve. The cumulative damage for the case when the stress ranges follow a Weibull distribution (Eq. (13.4)) is given by

$$D = \frac{N_T\, A^{m_1}}{K_1}\, \Gamma\left(1+\frac{m_1}{h}; \left(\frac{S'}{A}\right)^h\right) + \frac{N_T\, A^{m_2}}{K_2}\, \gamma\left(1+\frac{m_2}{h}; \left(\frac{S'}{A}\right)^h\right), \qquad (13.59)$$

where $\Gamma(;)$ and $\gamma(;)$ are the complementary incomplete gamma and incomplete gamma functions, respectively.

13.5.6 Stress Ranges from Short-Term Time Histories

Time-domain fatigue analysis results in time series of stress. For narrow band Gaussian response, the cycles are well defined. For more general stress time histories, cycle counting methods need to be applied to all types of response time series, independent of the fundamental properties of the response. In fact, time-domain methods use only the information provided by the series of peaks (local maxima) and valleys (local minima). According to different principles for constructing the effective stress ranges from these identified peaks and valleys, various cycle counting methods are envisaged and their procedures are defined by the American Society of Testing and Materials (ASTM, 1985), including peak counting, range counting, level-crossing counting, and rainflow counting. Of these, rainflow counting is believed to be the best

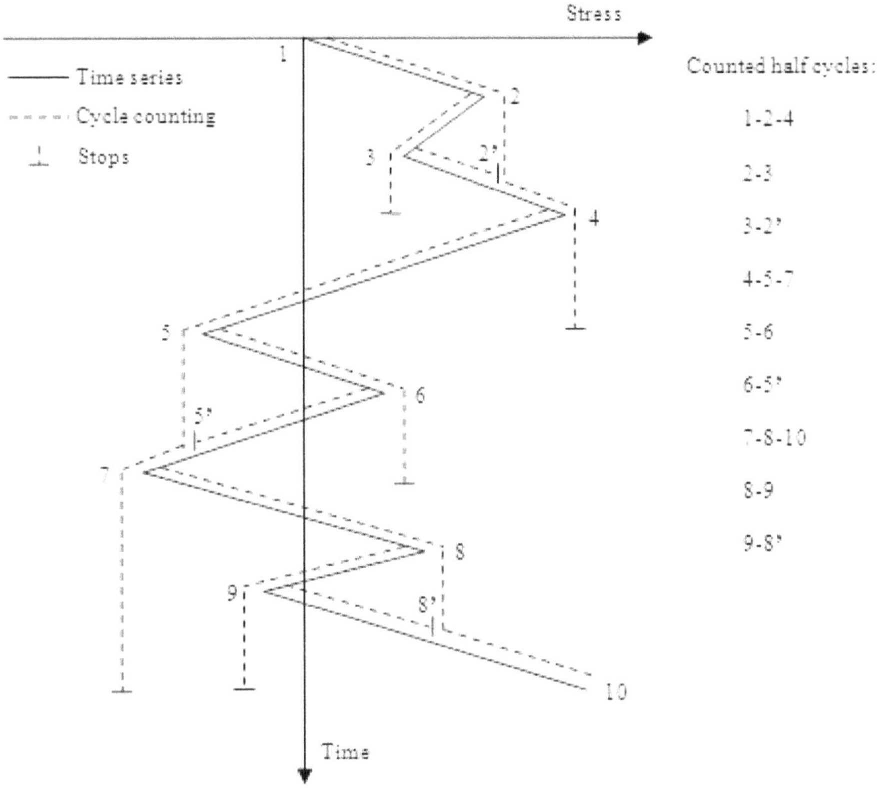

Figure 13.11. Illustration of the rainflow cycle counting method.

method for fatigue damage estimation, cf. Dowling (1972) and Watson and Dabell (1975). The rainflow method was first proposed by Matsuishi and Endo (1968) and it is also known as the pagoda roof method with its original version.

The rainflow counting algorithm is briefly described here. We consider a stress time series of peaks and valleys with the time axis vertically downward, as shown in Fig. 13.11, and in this way the lines connecting peaks and valleys form a series of pagoda roofs. The rainflow method applies the following general rules to produce stress cycles:

- Each rainflow begins at the beginning of the time series and at the inside of every peak and valley.
- Rainflow initiating at a peak (or a valley) drops down until it reaches opposite a peak more positive (or a valley more negative) than the peak (or the valley) from which it started.
- Rainflow also stops when it meets the rainflow from a roof above.
- Rainflow must terminate at the end of the time series.
- The horizontal length of each rainflow is counted as a half-cycle with that stress range.

As shown in Fig. 13.11, the first rainflow starts from the beginning at point 1 as a valley, the second one from the peak 2, the third one from the valley 3, and so on.

The end at 10 would be considered as a peak here. There are 9 half cycles that can be extracted from this time series.

The rainflow initiating at the valley point 1 drops down at 2 and ends at 4 because the following valley 5 has smaller value than the initial point 1. Therefore, a half-cycle of 1-2-4 has been identified. The same rule can be applied to determine the half-cycle of 5-6.

The second rainflow starts from 2 and stops at 3, which leads to a half-cycle of 2-3 because the following peak 4 has larger value than 2. Similarly, the half-cycles of 4-5-7 and 8-9 can be extracted based on the same rule due to the largest peak at 10.

The half-cycles of 3-2', 6-5', and 9-8' are determined because the rainflows started at 3, 6, and 9, respectively, meet the rainflows from roofs above.

The half-cycle of 7-8-10 is formed because the time series ends at 10.

When all 9 half-cycles are identified, the horizontal length of each cycle is used as an effective stress range to calculate the fatigue damage based on, e.g., the linear damage accumulation law.

This is the original proposal of the rainflow cycle counting algorithm (Matsuishi and Endo, 1968). Other versions of the method can be found such as the 3-point algorithm (ASTM 1985), the 4-point algorithm (Amzallag et al., 1994) and the non-recursive definition proposed by Rychlik (1987). These algorithms lead to the same results of fatigue damage. Therefore, no details of other versions are given here.

In general, fatigue damage obtained by the counting methods previously described have the following relationship (see, e.g., Tovo (2002)),

$$D_{RC} \leq D_{RFC} \leq D_{LCC}(= D_{NB}) \leq D_{PC}, \tag{13.60}$$

where D_{RC}, D_{RFC}, D_{LCC}, and D_{PC} represent the fatigue damage estimated by the range counting, rainflow counting, level-crossing counting and peak counting methods, respectively, and D_{NB} denotes the narrow band approximation.

13.5.7 Stress Ranges from Frequency Domain Approaches in Short-Term Periods

As mentioned previously, time-domain analysis with cycle counting methods are extremely time consuming, especially for random fatigue analysis, where a great number of time series of responses must be generated and analyzed. Frequency-domain methods are more computationally efficient. However, the most accurate method, the rainflow cycle counting method, does not have a closed-form solution in the frequency domain for a bi- or trimodal or generally wide band Gaussian process. To propose an accurate frequency-domain method, one must focus on the spectral type and start with particular random processes. The accuracy of various frequency-domain methods for fatigue damage estimation of wide band processes are discussed herein. In particular, the focus is on multimodal processes such as bi- and trimodal processes.

As we know, the shape of the spectral density function of a certain response has a significant effect on the prediction of the induced fatigue damage. Vanmarcke's spectral bandwidth parameter is often used to characterize a random process and is defined as (Vanmarcke, 1972),

$$\delta = \sqrt{1 - \frac{m_1^2}{m_0 m_2}}, \tag{13.61}$$

where m_i is the ith order spectral moment, cf. Section 8.2.4. The value of the bandwidth parameter is between 0 and 1. A process is narrow banded if this parameter is close to 0. Otherwise, it is wide banded. Other bandwidth parameters based on higher spectral moments might also be used.

Often, the load effects are not ideally narrow banded, and their spectra cover a wider range of frequencies. Fatigue damage of wide band processes is difficult to obtain in the frequency domain. Nevertheless, it is clear that damage estimates based on the narrow band assumption are always conservative. For a spectral bandwidth parameter less than 0.3 and 0.5, cf. Eq. (13.61), the error in fatigue damage by using the narrow band approximation is less than 10% and 30%, respectively (Gao and Moan, 2009). Also, the actual wide band fatigue damage is frequently expressed as the narrow band result multiplied with a correction factor. Various methods have been proposed to estimate this factor in terms of the spectral moments.

For a bimodal process consisting of a high-frequency (HF) and a low-frequency (LF) component, where both components are narrow band processes, the rainflow algorithm extracts two types of load cycles: one is the large cycles considering the interaction between the HF and LF components, and the other is the small HF cycles only. Various methods for bimodal fatigue damage estimation under the Gaussian assumption have been proposed by many researchers, e.g., Jiao and Moan (1990), Sakai and Okamura (1995), Fu and Cebon (2000), and Benasciutti and Tovo (2007). For practical use, bimodal fatigue damage could also be expressed in terms of individual damage, as proposed by Lotsberg (2005) and Huang and Moan (2006).

Jiao and Moan (1990) expressed the bimodal fatigue damage as the sum of the HF damage plus the damage induced by an equivalent process of the HF envelope plus the LF process. The number of small cycles is determined by the mean zero upcrossing rate of the HF component, while the number of large cycles is given by that of the equivalent process. They also showed that a closed-form solution could be approximately obtained when assuming that the LF component dominates. The fatigue damage is obtained as

$$D_{JM} = \rho\, D_{NB}. \tag{13.62}$$

Here, D_{NB} is the fatigue damage under the narrow-band assumption (see Eq. (13.57)).

$$\rho = \frac{\nu_{HF}}{\nu_Y}\lambda_{HF}^{\frac{m}{2}} + \frac{\nu_P}{\nu_Y}\left[\lambda_{LF}^{\frac{m}{2}+2}\left(1 - \sqrt{\frac{\lambda_{HF}}{\lambda_{LF}}}\right) + \sqrt{\pi\lambda_{HF}\lambda_{LF}}\,\frac{m\Gamma((m+1)/2)}{\Gamma((m+2)/2)}\right], \tag{13.63}$$

is a correction factor in which $\lambda_{HF} = \sigma_{HF}^2/(\sigma_{HF}^2 + \sigma_{LF}^2)$ and $\lambda_{LF} = \sigma_{LF}^2/(\sigma_{HF}^2 + \sigma_{LF}^2)$ are the normalized variances of the HF and LF components, respectively. $\nu_Y = \sqrt{\lambda_{HF}\nu_{HF}^2 + \lambda_{LF}\nu_{LF}^2}$ and $\nu_P = \sqrt{\lambda_{LF}\lambda_{HF}\nu_{HF}^2\delta_{HF}^2 + \lambda_{LF}^2\nu_{LF}^2}$, in which ν_{HF} and ν_{LF} are the mean zero upcrossing rates of the HF and LF components, respectively,

δ_{HF} is the bandwidth parameter of the HF component and is set to 0.1 here, $\Gamma()$ represents the gamma function and m is the material parameter of the SN curve. This formula is very accurate and is used, e.g., in the DNV offshore standard for fatigue analysis of mooring lines (DNV, 2010a).

Although the single moment method originally proposed by Larsen and Lutes (1991), the formula of Benasciutti and Tovo (2005), the modified Fu and Cebon's method, and Jiao and Moan's method provide quite accurate estimates, the latter gives a damage correction factor that is greater than 1 when the process is close to a narrow band one. However, in such a case, the narrow band approximation is good enough, and no correction factor needs to be applied on the estimation of fatigue damage.

One practical extension of bimodal fatigue is the trimodal fatigue damage estimation, which is relevant, e.g., for a riser system. A riser system attached to a moored vessel could experience vortex-induced vibrations (VIV) with high frequencies in addition to the forces induced by the wave-frequency (WF) and low-frequency (LF) vessel motions. According to the current procedure for fatigue analysis of risers, the WF and LF fatigue damage is estimated, and then the VIV contribution is separately added to obtain the total fatigue damage. In this way, no interaction between the VIV response and the WF and LF response was considered. In principle, the fatigue damage will then be underestimated because of the nonlinear relationship between the fatigue damage and the stress range. Fatigue damage for stress spectra consisting of three frequency components is therefore interesting to study in a way that all interactions can be considered simultaneously. Gao and Moan (2008) extended the work of Jiao and Moan (1990) and proposed an analytical formulation for fatigue damage of ideal trimodal Gaussian processes. Good agreement with the rainflow method was obtained for such wide band processes.

Since the 1980s, frequency-domain methods for general wide band fatigue analysis were proposed by many researchers, e.g., Wirsching and Light (1980), Dirlik (1985), Gall and Hancock (1985), Larsen and Lutes (1991), Naboishikov (1991), Zhao and Baker (1992), and Benasciutti and Tovo (2005). Wide band fatigue damage is normally expressed by an empirical correction term on the narrow band result. The empirical formulas of Dirlik (1985) and Benasciutti and Tovo (2005) have been verified to be the best among empirical approximations of the rainflow counting method. Therefore, the Dirlik and Benasciutti and Tovo formulas are introduced below.

Dirlik (1985) proposed an approximate closed-form solution for the probability density function of effective rainflow stress ranges by using a combination of an exponential and two Rayleigh distributions. A large set of numerical simulations with various spectral shapes have been generated, and a best fit was performed to establish the empirical distribution. Fatigue damage can be obtained by the following closed-form expression:

$$D_{DK} = \frac{\nu_P T}{K}(2\sigma)^m (D_1 Q^m \Gamma(1+m) + (\sqrt{2})^m (D_2 R^m + D_3)\Gamma(1+\frac{m}{2})) \qquad (13.64)$$

where σ is the standard deviation of the random process; ν_P is the mean rate of peaks, $\nu_P = \frac{1}{2\pi}\sqrt{\frac{m_4}{m_2}}$ (where m_i denotes the i'th order spectral moment); T is the duration of interest; K and m are the material parameters of the SN curve; and $\Gamma()$ denotes the gamma function. The parameters D_1, D_2, D_3, Q and R are given as follows:

$$\alpha_2 = \frac{m_2}{\sqrt{m_0 m_4}}, \quad X_m = \frac{m_1}{m_0}\left(\frac{m_2}{m_4}\right)^{\frac{1}{2}}, \quad D_1 = \frac{2(X_m - \alpha_2^2)}{1+\alpha_2^2},$$

$$D_2 = \frac{1-\alpha_2 - D_1 + D_1^2}{1-R}, \quad D_3 = 1 - D_1 - D_2,$$

$$Q = \frac{1.25(\alpha_2 - D_3 - D_2 R)}{D_1}, \quad R = \frac{\alpha_2 - X_m - D_1^2}{1-\alpha_2 - D_1 + D_1^2}. \tag{13.65}$$

Recently, another empirical formula was proposed by Benasciutti and Tovo (2005). As shown by inequality (13.60), the rainflow fatigue damage is bounded by the narrow band approximation and the range counting method. Starting from this point, they proposed a formula of wide-band fatigue damage by using a linear combination of the narrow-band and range counting results that are easy to obtain by closed-form expressions in the frequency domain. The formula is given as the following equation:

$$D_{BT} = bD_{NB} + (1-b)D_{RC} = (b + (1-b)\alpha_2^{m-1})D_{NB}, \tag{13.66}$$

where m is the material parameter of the SN curve, D_{NB} is the fatigue damage under the narrow band assumption (Eq. (13.57)), and the coefficient b is empirically obtained based on extensive numerical simulations:

$$b = \frac{(\alpha_1 - \alpha_2)(1.112(1+\alpha_1\alpha_2 - (\alpha_1+\alpha_2))\exp(2.11\alpha_2) + (\alpha_1 - \alpha_2))}{(\alpha_2 - 1)^2}, \tag{13.67}$$

where

$$\alpha_1 = m_1/\sqrt{m_0 m_2}; \quad \alpha_2 = m_2/\sqrt{m_0 m_4}. \tag{13.68}$$

The formulas by Dirlik (1985) and Benasciutti and Tovo (2005) have been shown to give accurate results in an independent study by Gao and Moan (2008). Moreover, these empirical formulas can be applied with good accuracy to all types of wide and narrow band Gaussian processes.

13.5.8 Fatigue Damage Caused by Non-Gaussian Response in Short-Term Periods

The wave-induced response of marine structures is often non-Gaussian due to the nonlinear environmental forces and/or the nonlinear properties of the structural system itself. Although fatigue damage under non-Gaussian load can be estimated by the time-domain counting methods in the same straightforward manner as for Gaussian processes, they are difficult to obtain in the frequency domain, especially for wide-band processes.

Depending on the structural nonlinearity, non-Gaussianity may cause an increase in fatigue damage estimates as compared with that based on the Gaussian assumption. For instance, fatigue damage induced by the drag force can be several times the one under the Gaussian assumption, see Section 11.2.2. However, for slightly nonlinear responses like the HF springing response of a TLP, the non-Gaussian fatigue damage is only a few percent higher than the Gaussian one as indicated by Naess et al. (1994). Gao and Moan (2007) showed that the non-Gaussian fatigue damage due merely to the LF mooring response is about 1.3 times the Gaussian estimate for the sea state, which contributes the most fatigue damage in a long-term period. Moreover, under particular conditions, the nonlinear fatigue damage is even smaller than that based on the Gaussian assumption of the Morison force, see Madsen et al. (1986) and Winterstein (1988).

Usually, a non-Gaussian process is dealt with by considering it as a monotonic transformation of a Gaussian process. Parametric formulas are often used to define this kind of transformation. For example, a Hermite model was adopted by Winterstein (1985, 1988). Wang and Sun (2005) and Ochi and Ahn (1994) applied a monotonic exponential function, while Sarkani et al. (1994) used a power law model. Nonparametric definition of the transformation was also applied by Rychlik et al. (1997).

Higher-order moments of the non-Gaussian random process such as skewness and kurtosis, are normally involved in the transformation, so the fatigue damage is therefore estimated based on the evaluation of the effect of skewness and kurtosis, see Winterstein (1988) and Wang and Sun (2005). Most studies in this area were based on the assumption that the non-Gaussian process is narrow banded.

If the response nonlinearity can be explicitly expressed and the amplitude distribution can be obtained by a numerical integration, the non-Gaussian fatigue damage might be directly calculated as shown by Gao and Moan (2007), where the LF mooring line tension is obtained based on an approximation proposed in Naess (1986) and the WF tension is expressed as a Morison-type force by using the simplified dynamic model (Larsen and Sandvik, 1990). Finally, the theory of Jiao and Moan (1990) was extended to estimate the total fatigue damage when both the WF and LF line tension are non-Gaussian processes.

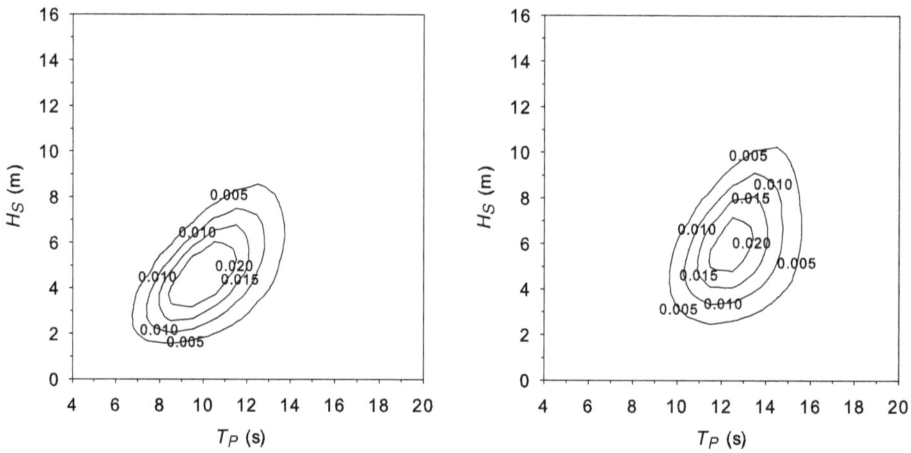

(a) Fatigue damage for a brace in a semi-submersible

(b) Fatigue damage for a hull girder of an FPSO

Figure 13.12. Relative contribution of fatigue damage for a brace in a semi-submersible and the hull girder of an FPSO in the North Sea (Moan et al., 2005).

13.5.9 Long-Term Fatigue Analysis

In principle, fatigue analysis is based on long-term analysis, considering "all" sea states. This may be done by determining the long-term distribution of stress ranges, e.g., in terms of a Weibull distribution, and calculating the damage as described in Section 13.5.5 or by summing up the damage for each sea state as mentioned in Section 13.2.2. This requires extensive efforts and various ways to make the analysis more efficient are particularly important for cases where nonlinear effects require time-domain simulation of load effects (stresses). Figure 13.12 illustrates the fact that the main contribution to the fatigue damage comes from moderate sea states that should therefore be focused on in the analysis.

14 Equations of Motion

14.1 Introduction

The first step in planning response analysis is whether the analysis can be accomplished as a static one or whether a dynamic model must be used. Dynamic analyses are generally necessary in connection with transient loads; otherwise, the results may be significantly conservative or nonconservative. For load processes consisting of several harmonic components, the main criterion is whether the load process contains energy in the range of eigenfrequencies of the system. Figure 14.1 shows an overview of the largest eigenperiod (natural period) of vibration or motion of offshore structures, as well as the relevant range of periods of dynamic loads associated with waves.

14.2 Solution of Equations of Motion

14.2.1 General

The equations of motion for a linear structural system (Section 4.10)

$$\mathbf{M}\ddot{\mathbf{r}} + \mathbf{C}\dot{\mathbf{r}} + \mathbf{K}\mathbf{r} = \mathbf{R}, \tag{14.1}$$

may be solved in the time or frequency domain. The choice of formulation especially depends on:

- The nature of the loading; i.e., whether it is steady state or transient (which often involves response in a wide frequency band).
- Frequency dependence of the dynamic properties (mass, damping, stiffness).
- Nonlinear features of the loading or dynamic properties.

In Chapter 2, solutions of the equation of motion for SDOF systems with different load conditions are described. If the solution method either in time or frequency domain is formulated for the coupled system of equations in Eq. (14.1), the method is denoted as direct.

Alternatively, Eq. (14.1) can first be transformed into an uncoupled system, and then the uncoupled equations can be solved separately as SDOF models. In Section 3.3.10, the modal superposition method is presented. This method can just

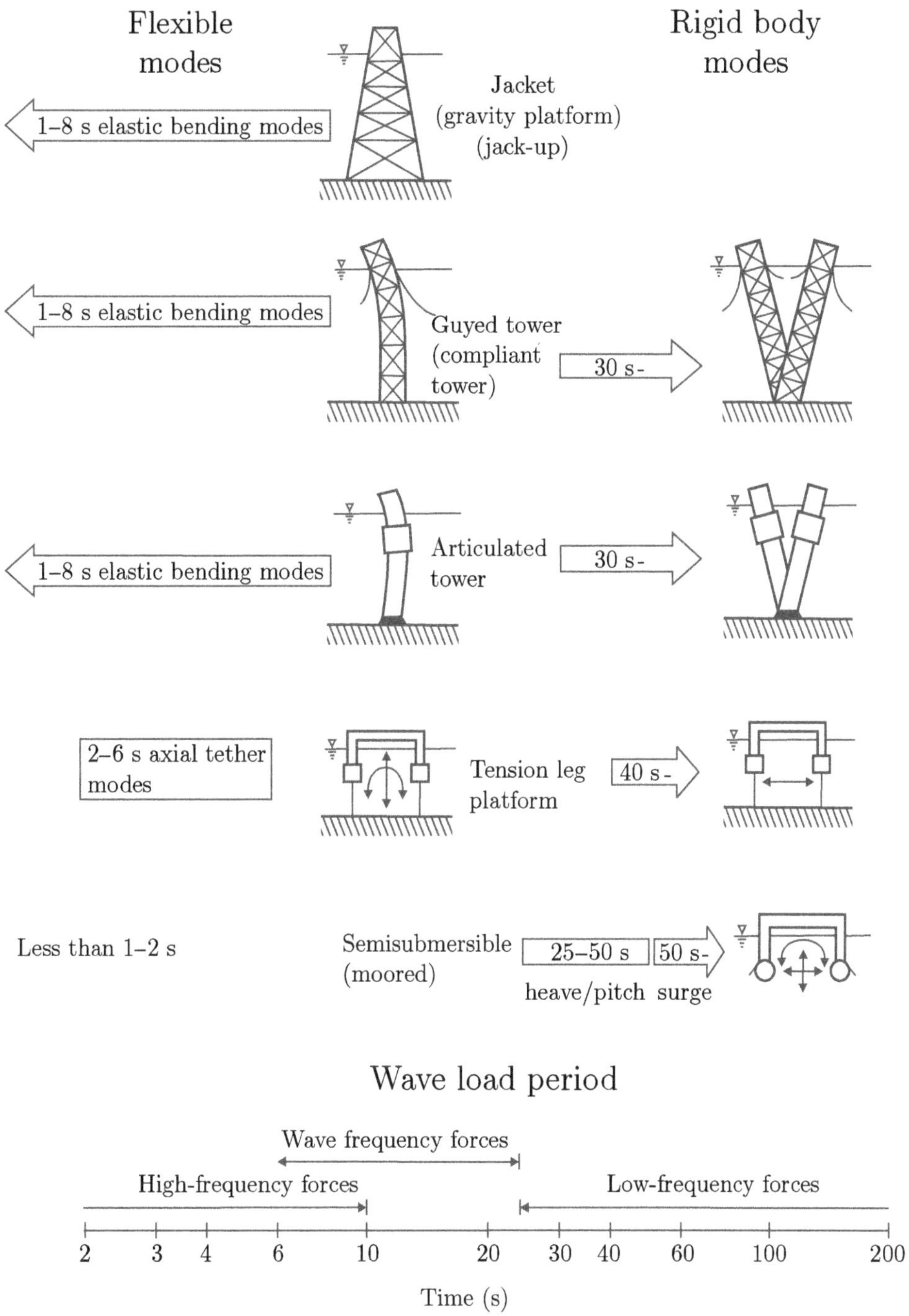

Figure 14.1. Overview of dynamic excitation by wave loads and typical natural periods.

as well be applied to the equations of motion established by the FEM. The modal superposition method consists of the following steps:

- Determination of eigenfrequencies and normal modes by eigenvalue analysis, commonly based on undamped modes.

- Assess damping mechanisms and modeling and the implicit assumption that the modes are also orthogonal with respect to damping.
- Determine the response for each mode by methods described in Chapter 2.
- Superimpose the results for all modes.

Large-scale dynamic analyses should normally be preceded by an estimate of load and natural frequencies. The simplified methods described in Section 3.3.3 may then be applied to estimate the fundamental (lowest) frequency.

Typically, only some of the structural modes will experience dynamic effects. However, in some cases, the quasistatic contribution of higher-order modes may be important. A conventional modal analysis then needs to include many modes. Alternatively, a method that includes the quasistatic solution and the dynamic contributions from the few relevant modes was proposed by Wu and Moan (1996, 2005), and applied to a flexible floating ship structure (floating beam). It is shown that the response may be obtained by superimposing the dynamic effects associated with very few flexible modes.

Frequency-domain methods are suitable for linear systems subjected to stochastic loads when the loading and response can be represented by a few frequency-dependent properties and for nonviscous damping.

After solving the global dynamic equilibrium equation, Eq. (14.1), with respect to $\mathbf{r}$ for a given time instant, the corresponding nodal displacements $\mathbf{v}$ for an element is determined by

$$\mathbf{v}^k = \mathbf{a}^k\mathbf{r}, \tag{14.2}$$

and the stresses within the element no. k are given by analytical (polynomial) functions by the expression

$$\boldsymbol{\sigma}(x, y, z) = \mathbf{D}\boldsymbol{\varepsilon}(x, y, z) = \mathbf{D}\,\mathbf{B}\,\mathbf{v}^k. \tag{14.3}$$

For "one-dimensional structures" such as bars and beams, the stress may alternatively be evaluated based on the nodal forces, $\mathbf{S}^k = \mathbf{k}^k\mathbf{y}^k$, $\mathbf{S}^{ok} = \mathbf{k}^k\mathbf{a}^k\mathbf{r} + \mathbf{S}^{ok}$. This is more accurate than determining the stresses from strains that are derivatives of the displacements, as implied by Eq. (14.3).

14.2.2 Eigenvalue Problem

The eigenmodes and eigenfrequencies are given by Eq. (3.115) when generalized coordinates are used. When the FE method is applied, the eigenvalue problem is given by

$$(\mathbf{K} - \omega^2\mathbf{M})\mathbf{r} = 0. \tag{14.4}$$

Various examples of free vibration problems were presented previously, e.g., in Section 4.10. These problems included 1 or 2 DOFs and were solved by the determinant method, i.e., by requiring that $\det\left[\mathbf{K} - \omega^2\mathbf{M}\right] = 0$. This method works well for systems with a few DOFs. For larger systems, more efficient algorithms need to be used. A variety of methods are available, even in a given computer program for dynamic analysis. Numerical methods include vector iteration, orthogonality transforms, etc. (Craig, 1981; Clough and Penzien, 1993). The methods differ in applicability and efficiency according to matrix sparsity and topology, the number of frequencies desired, and other factors. The former features depend on the type of problem (dimension, element type, use of constant or lumped mass etc.). Usually, only the lowest frequencies are needed.

Some algorithms are sensitive to the form of the stiffness or mass matrix. For instance, a "free," unsupported structure or a multibody, pinned-connected structure will have zero stiffness and eigenfrequencies that are zero for the corresponding modes. In such cases, the actual eigenvalue problem may be replaced by an "eigenvalue" shift to enable computation. The stiffness corresponding to buoyancy for floating structures is relatively small compared with structural stiffness for certain modes. This fact might also cause numerical problems and inaccuracies. Another matter is that zero eigenvalues may occur in some models because a modeling error was made.

Sometimes, when lumped masses are applied and rotational inertia is neglected, some elements on the diagonal of the mass matrix will be zero, implying an infinite frequency. This problem can be dealt with by so-called reduction of the eigenvalue problem, as discussed previously. Another problem associated with eigenvalue solvers is how well they can deal with different modes associated with the same eigenfrequency. A simple test of this feature can be made by using a beam with identical properties about two orthogonal cross-sectional planes, e.g., a cantilever beam with circular cross-section. The choice of method depends on the size of the problem and desired results (number of eigenvalues and eigenvectors, accuracy required, etc.).

For large problems vector iteration such as subspace iteration (Bathe, 1996) is useful for medium-size problems with large bandwidths, or large problems when only the lowest eigenvalues are needed.

Methods to reduce the number of DOFs are briefly described in Sections 4.11 and 4.12.

14.2.3 Frequency Response Method

In Section 2.7, the response of an SDOF linear oscillator to arbitrary loading is described. In this section, this analysis is extended to multi-DOF systems.

It is now assumed that the arbitrary loading in Eq. (14.1) can be transformed into the frequency domain by means of the Fourier transform; that is,

$$\mathbf{X}(\omega) = \frac{1}{2\pi}\int_{-\infty}^{\infty} \mathbf{R}(t)\, e^{-\mathrm{i}\omega t}\, dt, \tag{14.5}$$

so that, inversely,

$$\mathbf{R}(t) = \int_{-\infty}^{\infty} \mathbf{X}(\omega)\, e^{\mathrm{i}\omega t}\, d\omega. \tag{14.6}$$

By an entirely similar procedure to the one used in Chapter 2, the general (steady-state) solution to Eq. (14.1) is obtained by first solving the following equation:

$$\mathbf{M}\ddot{\mathbf{r}} + \mathbf{C}\dot{\mathbf{r}} + \mathbf{K}\mathbf{r} = \mathbf{X}(\omega)\, e^{\mathrm{i}\omega t}, \tag{14.7}$$

which is achieved by assuming $\mathbf{r}(t) = \mathbf{x}(\omega)\, e^{i\omega t}$. Inserting this into Eq. (14.7) shows that $\mathbf{x}(\omega)$ must satisfy the equation

$$\left[-\omega^2\mathbf{M} + i\omega\mathbf{C} + \mathbf{K}\right]\mathbf{x}(\omega) = \mathbf{X}(\omega). \tag{14.8}$$

The bracket on the left-hand side is a complex symmetric and banded matrix, which requires twice as much storage in the computer as the stiffness matrix.

The solution may be expressed as follows:

$$\mathbf{x}(\omega) = \mathbf{H}(\omega)\,\mathbf{X}(\omega), \tag{14.9}$$

where the complex frequency response matrix function $\mathbf{H}(\omega)$ is given as

$$\mathbf{H}(\omega) = \left[-\omega^2\mathbf{M} + i\omega\mathbf{C} + \mathbf{K}\right]^{-1}. \tag{14.10}$$

Finally, the general solution $\mathbf{r}(t)$ of Eq. (14.1) can then be written as

$$\mathbf{r}(t) = \int_{-\infty}^{\infty} \mathbf{H}(\omega)\,\mathbf{X}(\omega)\, e^{i\omega t}\, d\omega. \tag{14.11}$$

Note that according to the convention adopted in this book, cf. Section 9.14, the matrix $\mathbf{H}(\omega) = \left(H_{ij}(\omega)\right)^T$, where $H_{ij}(\omega)$ gives the (complex) response of DOF number j due to a (complex) harmonic load with unit amplitude in DOF number i. Only a limited number of the nodes in the structure carry loads. Hence, only the columns of $\mathbf{H}(\omega)$ that correspond to those degrees of freedom that carry loads are important, not the whole matrix.

It is noted that hysteretic damping in the structure (or soil) is proportional to displacement and in phase with the velocity, and can be modeled by a term $i\mathbf{C}_s\mathbf{r}$, which can be introduced instead of the viscous term $i\omega\mathbf{C}\mathbf{r}$.

The frequency response function for the coupled system may also be expressed by the modal frequency response functions. The modal load for mode no. i, in the frequency domain, is given by

$$\bar{X}_i(\omega) = \boldsymbol{\psi}_i^T(\omega)\mathbf{X}(\omega). \tag{14.12}$$

If the modal frequency response function (for an SDOF) is $\bar{H}_i(\omega)$, then the response in mode i is given by

$$\bar{x}_i(\omega) = \bar{H}_i(\omega)\boldsymbol{\psi}_i^T(\omega)\mathbf{X}(\omega). \tag{14.13}$$

This leads to the following expression for the coupled frequency response:

$$\mathbf{x}(\omega) = \sum\nolimits_{i=1}^{s} \boldsymbol{\psi}_i(\omega)\bar{x}_i(\omega) = \boldsymbol{\psi}(\omega)\bar{\mathbf{H}}(\omega)\boldsymbol{\psi}^T(\omega)\mathbf{X}(\omega), \tag{14.14}$$

where $\boldsymbol{\psi}(\omega) = \left(\boldsymbol{\psi}_1(\omega), \ldots, \boldsymbol{\psi}_s(\omega)\right)$ is the matrix of mode shapes, and

$$\bar{\mathbf{H}}(\omega) = \text{diag}\left\{\bar{H}_1(\omega), \bar{H}_2(\omega), \ldots, \bar{H}_s(\omega)\right\}. \tag{14.15}$$

This implies that

$$\mathbf{H}(\omega) = \boldsymbol{\psi}(\omega)\bar{\mathbf{H}}(\omega)\boldsymbol{\psi}^T(\omega). \tag{14.16}$$

14.2.4 Formulation and Solution of Frequency-Domain Equations

Rigid Floating Structures

By assuming a linear frequency-domain model, it can be shown in a similar way as in Example 9.5.1 that the equations of motion of a rigid floating system (6 DOF) with zero forward speed in sea waves are described by the vector equation:

$$-\omega^2 [\mathbf{M} + \mathbf{A}(\omega)]\,\mathbf{x}(\omega) + i\omega\mathbf{B}(\omega)\mathbf{x}(\omega) + \mathbf{K}\mathbf{x}(\omega) = \mathbf{X}(\omega), \qquad (14.17)$$

where ω is the wave frequency; $\mathbf{M}$ is the rigid-body mass matrix; $\mathbf{A}(\omega)$ is the added mass matrix; $\mathbf{B}(\omega)$ is the potential damping matrix; $\mathbf{K}$ is the restoring coefficient matrix; and $\mathbf{X}(\omega)$ denotes the Fourier transform of the 6D wave loading vector, while $\mathbf{x}(\omega)$ denotes the Fourier transform of the 6D response vector. The solution of these equations is obtained easily and efficiently for the desired frequency resolution.

The wave loading forces are integrated pressure forces obtained from the solution of the diffraction potential over the mean wetted surface of the body. The integrated pressure forces obtained from the solution of the radiation potentials define the radiation forces. The components of the radiation force in phase with acceleration and velocity of the body are then the added mass and the hydrodynamic damping forces, respectively, and are given by

$$\mathbf{X}^r(\omega) = -[-\omega^2\mathbf{A}(\omega)\mathbf{x}(\omega) + i\omega\mathbf{B}(\omega)\mathbf{x}(\omega)]. \qquad (14.18)$$

The restoring forces in Eq. (14.17) (the $\mathbf{K}\,\mathbf{x}(\omega)$ term) are the final form of gravity and buoyancy contributions followed by a Taylor series expansion and keeping up to only the first order. The previous vector quantities (force or response) are defined with respect to the body's equilibrium frame.

Flexible Structures

One of the challenging topics in dynamic response analysis of novel marine structure is the study of dynamic behaviour of flexible floating bodies such as large ships, floating airports, etc., see, e.g., ISSC (2006c), which are examples of very large floating structures (VLFS).

Some of the methods use a mode expansion approach (e.g., Bishop and Price (1979), Newman (1994), Kashiwagi (2000)), also known as the *generalized modes approach*. Hydrodynamic loads corresponding to the radiation are evaluated for unit modal response and are then integrated into the equations of motion where the wave-exciting forces are included. The number of modes required to represent the response depends highly on the problem and can become quite large for VLFS. However, all modes may not contribute to the actual response. Another approach is the so-called direct method (Yago and Endo, 1996; Yasuzawa et al., 1997), where the hydrodynamic boundary value problem (pressure distribution) and the elastic response are solved simultaneously. Therefore, although the direct method avoids the incorporation of any modes into the solution of response, the method is computationally expensive and requires interfacing between the hydrodynamic and structural numerical techniques such as the boundary element method (BEM) and the FEM.

In case of a deformable body with a mode expansion approach, one can rewrite the equations of motion in Eq. (14.17) in a generalized form (Newman, 1994):

$$\left(-\omega^2[\mathbf{M} + \mathbf{A}(\omega)] + i\omega[\mathbf{B}(\omega) + \mathbf{D}] + [\mathbf{K} + \mathbf{G}]\right)\mathbf{x}(\omega) = \mathbf{X}(\omega). \qquad (14.19)$$

Now, the matrices and vectors are correspondingly $(6 + N) \times (6 + N)$ and $(6 + N) \times 1$. N is the number of flexible modes defined in addition to the six rigid-body modes. **G** and **D** are the generalized structural stiffness and the generalized structural damping matrices, respectively. The other coefficients also assume a meaning in a generalized sense.

Multibody Structures

Other types of structures are floating, interconnected, or multibody offshore structures such as aquaculture plants, wave-energy converters, and floating airports, see, e.g., ISSC (2006a,b) and Cruz (2008).

Figure 3.3 shows a simple version of a two-body wave energy converter, consisting of a buoy moving on a guide that is fixed to the semisubmersible platform. The power off-take comes from the relative motion between the two bodies in heave (Hals et al., 2007).

A typical approach is to first evaluate the hydrodynamic loads corresponding to each floating body oscillating in each of the 6 DOFs, then postprocess the radiation and diffraction forces together with other loads such as the power takeoff (PTO) force, and then reduce the DOFs to the original DOFs of the oscillating device by applying the kinematic constraints.

Following this approach and using a software such as WAMIT (2008), no symmetry can be exploited for either the radiation or diffraction problem. Therefore, such an evaluation of hydrodynamic loads and motions is subject to long simulations when the body number increases. In addition, exporting the field particle information such as pressure and wave elevation becomes quite cumbersome because the information passes through the postprocessing stage for each DOF and each body.

Fortunately, for linear hydrodynamic load assessment of multiple floating bodies, one can alternatively use the generalized modes approach to transform the problems to a single body with several DOFs. Such an approach is well known in studying the response of flexible bodies in waves. The geometries in the problem are merged and considered as one. This single body will have "pseudoflexibility" in the sense that different parts of the body (subbodies) are allowed to have different DOFs depending on the problem. In this way, assuming n to be the number of floating bodies, the $6 \times n$ mandatory DOFs for the initial multibody problem reduce to one single body with the original m DOFs where $m << 6 \times n$. Solution by exploiting the geometric/problem symmetry is also a possibility.

14.2.5 Hybrid Frequency- and Time-Domain Models

Formulation

The equations of motion (e.g., Eq. (14.1)) are based on linear theory. Nonlinear effects are most conveniently incorporated by proper time-domain models. One approach may be to formulate the governing equations directly in the time domain. A difficulty then is the accommodation of frequency-dependent properties of, e.g., hydrodynamic added mass and potential damping. An alternative approach is to establish the linear frequency-domain model of the equations of motion (Eq. (14.1)), and then transform this model into the time domain and add nonlinear features. Such a model is referred to as a hybrid frequency- and time-domain model in this context and was initially introduced by Cummins (1962). To date, it has been used successfully in different motion analysis applications, see, e.g., Wu and Moan (1996), Kashiwagi (2004), Taghipour et al. (2009).

To establish this model, Eq. (14.17) is rewritten in the following form:

$$-\omega^2[\mathbf{M}+\mathbf{A}^\infty]\,\mathbf{x}(\omega)+i\omega\mathbf{B}^\infty\mathbf{x}(\omega)+\mathbf{K}\mathbf{x}(\omega)+i\omega\mathbf{C}(\omega)\mathbf{x}(\omega)=\mathbf{X}(\omega), \tag{14.20}$$

where $\mathbf{A}^\infty$ ($\mathbf{B}^\infty$) denotes the high-frequency limit of added mass (wave damping), and

$$\mathbf{C}(\omega)=[\mathbf{B}(\omega)-\mathbf{B}^\infty]+i\omega[\mathbf{A}(\omega)-\mathbf{A}^\infty]. \tag{14.21}$$

Taking the inverse Fourier transform of Eq. (14.20) gives the following vector integro-differential equation,

$$[\mathbf{M}+\mathbf{A}^\infty]\ddot{\mathbf{r}}(t)+\mathbf{B}^\infty\dot{\mathbf{r}}(t)+\int_{-\infty}^{\infty}\mathbf{k}(t-\tau)\dot{\mathbf{r}}(\tau)\,d\tau+\mathbf{K}\mathbf{r}(t)=\mathbf{R}(t). \tag{14.22}$$

The preceding equation is known as the Cummins equation. One can see how the frequency dependency of added mass and damping has resulted in the appearance of a convolution integral. The $\mathbf{k}(t)$ function in the above equation is known as the retardation or memory function. It can be shown that the $\mathbf{k}(t)$ function can be expressed by $\mathbf{C}(\omega)$ in Eq. (14.21) as follows

$$\begin{aligned}\mathbf{k}(t)&=\frac{1}{2\pi}\int_{-\infty}^{\infty}\mathbf{C}(\omega)e^{i\omega t}\,d\omega\\&=\frac{1}{2\pi}\int_{-\infty}^{\infty}\Big([\mathbf{B}(\omega)-\mathbf{B}^\infty]\cos(\omega t)-\omega[\mathbf{A}(\omega)-\mathbf{A}^\infty]\sin(\omega t)\Big)\,d\omega\\&\quad+\frac{i}{2\pi}\int_{-\infty}^{\infty}\Big([\mathbf{B}(\omega)-\mathbf{B}^\infty]\sin(\omega t)+\omega[\mathbf{A}(\omega)-\mathbf{A}^\infty]\cos(\omega t)\Big)\,d\omega,\end{aligned} \tag{14.23}$$

where $\mathbf{k}(t)$ is a real function of time, and thus the imaginary terms in the preceding equations must equal zero. Moreover, $\mathbf{A}(\omega)$ and $\mathbf{B}(\omega)$ are even functions of frequency, which reduces the equation to

$$\mathbf{k}(t)=\frac{1}{\pi}\int_{0}^{\infty}[\mathbf{B}(\omega)-\mathbf{B}^\infty]\cos(\omega t)-\omega[\mathbf{A}(\omega)-\mathbf{A}^\infty]\sin(\omega t)\,d\omega. \tag{14.24}$$

The first term on the rhs of Eq. (14.24) is an even function of time, while the second term is an odd function. In other words,

$$\mathbf{k}(t)=\mathbf{k}_e(t)+\mathbf{k}_o(t). \tag{14.25}$$

One should note that the radiation force is a causal system. There is no force before the structure is oscillating and vice versa. In mathematical terms,

$$\mathbf{k}(t)=\mathbf{0};\quad \text{if } t<0. \tag{14.26}$$

Thus, for $\mathbf{k}(t)$ to represent a causal system in Eq. (14.25), the following relations must be satisfied, $\mathbf{k}_e(t)=\mathbf{k}_o(t)$ if $t\geq 0$; $\mathbf{k}_e(t)=-\mathbf{k}_o(t)$, if $t<0$.

Therefore, Eq. (14.24) reduces to

$$\mathbf{k}(t) = \frac{2}{\pi}\int_0^\infty [\mathbf{B}(\omega) - \mathbf{B}^\infty]\cos(\omega t)\,d\omega$$
$$= -\frac{2}{\pi}\int_0^\infty \omega[\mathbf{A}(\omega) - \mathbf{A}^\infty]\sin(\omega t)\,d\omega. \tag{14.27}$$

Equation (14.27) provides two important expressions that relate added mass and damping to the retardation function. For the sake of numerical convergence, the relation based on the hydrodynamic damping is preferred to the one with added mass.

For floating structures with zero forward speed, the hydrodynamic damping vanishes as the frequency of oscillation approaches infinity, in other words, $\mathbf{B}^\infty = \mathbf{0}$. In practice, the retardation kernel $\mathbf{k}(t)$ tends to vanish for times passing a certain value, say, t_{mem}. Knowing this information and by using causality, one can change Eq. (14.22) to

$$[\mathbf{M} + \mathbf{A}^\infty]\ddot{\mathbf{r}}(t) + \int_0^{t_{mem}} \mathbf{k}(t-\tau)\,\dot{\mathbf{r}}(\tau)\,d\tau + \mathbf{K}\mathbf{r}(t) = \mathbf{R}(t). \tag{14.28}$$

One can now add nonlinearities such as viscous drag forces, “exact body” hydrostatic or Froude-Krylov force, etc., to the rhs of this equation, see, e.g., Hals et al. (2007). Equation (14.28) also provides a suitable means of calculating the linear transient response of marine structures due to moving loads, etc., see, e.g., Kashiwagi (2004), Taghipour et al. (2008).

Replacement of Convolution Integral

In time-domain simulation of random response, the retardation function and the convolution integral must be iterated in time. This is a drawback because integrating the convolution may become computationally demanding, depending on the simulation length, time step, and DOF of the system (ITTC, 2005; ISSC, 2006b). For this reason significant efforts have been devoted toward replacing the convolution terms by other approximately equivalent models. A review and description of these efforts are found in (Taghipour et al., 2008).

These alternative representations can be broadly grouped into the following types:

1. Replacement of the frequency-dependent added mass and damping in Eq. (14.22) by constant coefficients. It is then natural to choose the values at a dominating frequency, such as the spectral mean or peak frequency:

$$\left[\mathbf{M} + \tilde{\mathbf{A}}\right]\ddot{\mathbf{r}}(t) + \tilde{\mathbf{B}}\dot{\mathbf{r}}(t) + \mathbf{K}\mathbf{r}(t) = \mathbf{R}(t). \tag{14.29}$$

2. Replacement of the convolution by a state-space formulation:

$$[\mathbf{M} + \mathbf{A}^\infty]\ddot{\mathbf{r}}(t) + \mathbf{f}^R(t) + \mathbf{K}\mathbf{r}(t) = \mathbf{R}(t), \tag{14.30}$$

$$\dot{\mathbf{z}}(t) = \mathbf{A}'\mathbf{z}(t) + \mathbf{B}'\dot{\mathbf{r}}(t), \tag{14.31}$$

$$\mathbf{f}^R(t) = \mathbf{C}'\mathbf{z}(t) + \mathbf{D}'\dot{\mathbf{r}}(t), \tag{14.32}$$

where the state-space vector $\mathbf{z}(t)$ provides an internal description of the system.

The accuracy of the first, very simple method depends on how frequency sensitive the added mass and damping are for the specific problem. Such an approximation would normally be based on radiation quantities corresponding to the modal (extremum) or the zero crossing frequency of the sea wave spectrum. Such a simple method leads to relatively large errors in modeling the system transient response due to a single-frequency excitation or system steady-state response due to multiple-frequency excitations (Govolato, 1959; Holappa and Falzarano, 1999).

The second approach has been used by numerous researchers in different areas. To the best of the authors' knowledge, state-space models were originally introduced in the field of ocean engineering by Schmiechen (1973), who studied transient ship maneuvering. Later many researchers have proposed different approaches, as reviewed by Taghipour et al. (2008). In the application of the state-space formulation, Eqs. (14.30)–(14.32), the coefficient matrices ($\mathbf{A}', \mathbf{B}', \mathbf{C}', \mathbf{D}'$) must be determined. This may be achieved with different methods. Taghipour et al. (2008) compared the performance of the following identification methods:

1. Impulse response curve fitting
2. Realization theory
3. Regression in the frequency domain

Instead of replacing the convolution term only, a complete state-space model from force to motion may be applied. This approach was originally proposed by Cummins (1962).

The state-space representation discussed in the next section provides an attractive alternative formulation for simulation purposes due to the simple form of its solution, i.e., Eq. (14.44). In addition, state-space models are well suited for the methods of analysis used in automatic control of marine systems (Fossen, 2002).

14.2.6 State-Space Formulation

This section shows how the second-order differential equation may be formulated in terms of a set of first-order differential equations. In this case, no account of frequency-dependent properties in terms of a convolution integral is made.

The general linearized equations of motions for an nDOF system, as expressed by Eq. (14.1), can be recast into the *state variable* form by defining a $2n$ *state vector*, $\mathbf{q}(t)$, as follows:

$$\mathbf{q}(t) = \begin{bmatrix} \mathbf{r}(t) \\ \dot{\mathbf{r}}(t) \end{bmatrix}. \tag{14.33}$$

A first-order matrix equation of motion may then be written as

$$\dot{\mathbf{q}}(t) = \mathbf{G}\mathbf{q}(t) + \mathbf{F}(t), \tag{14.34}$$

where

$$\mathbf{G} = \begin{bmatrix} \mathbf{0} & \mathbf{I} \\ -\mathbf{M}^{-1}\mathbf{K} & -\mathbf{M}^{-1}\mathbf{C} \end{bmatrix} \tag{14.35}$$

and

$$\mathbf{F}(t) = \begin{bmatrix} \mathbf{0} \\ \mathbf{M}^{-1}\mathbf{R}(t) \end{bmatrix}. \tag{14.36}$$

The response to a nonzero forcing **F** can be computed using the equation (Reid, 1983; Roberts and Spanos, 1990)

$$\mathbf{q}(t) = \exp(\mathbf{G}t)\,\mathbf{q}(0) + \int_0^t \exp[\mathbf{G}(t-\tau)]\mathbf{F}(\tau)\,d\tau. \tag{14.37}$$

If the system is initially at rest at $t = 0$, then $\mathbf{q}(0) = 0$, and Eq. (14.37) reduces to

$$\mathbf{q}(t) = \int_0^t \exp[\mathbf{G}(t-\tau)]\,\mathbf{F}(\tau)\,d\tau, \tag{14.38}$$

which can also be expressed as

$$\mathbf{q}(t) = \int_0^t \exp[\mathbf{G}\upsilon]\,\mathbf{F}(t-\upsilon)\,d\upsilon. \tag{14.39}$$

Equation (14.38) or (14.39) is simply the convolution integral relating $\mathbf{q}(t)$ to $\mathbf{F}(t)$; hence, $\exp(\mathbf{G}\upsilon)$ is the appropriate impulse response matrix for those input and output vectors. If the transition matrix is partitioned as

$$\exp(\mathbf{G}\upsilon) = \begin{bmatrix} \mathbf{a}(\upsilon) & \mathbf{b}(\upsilon) \\ \mathbf{c}(\upsilon) & \mathbf{d}(\upsilon) \end{bmatrix}, \tag{14.40}$$

where the submatrices are all $n \times n$, it follows from Eqs. (14.36), (14.39) and (14.40) that

$$\mathbf{r}(t) = \int_0^t \mathbf{b}(\upsilon)\mathbf{M}^{-1}\mathbf{R}(t-\upsilon)\,d\upsilon = \int_0^t \mathbf{h}(t)\mathbf{R}(t-\upsilon)\,d\upsilon, \tag{14.41}$$

where $\mathbf{h}(t)$ is the impulse response matrix relating **r** to **R**.

To facilitate the computation of $\exp(\mathbf{G}t)$, and hence $\mathbf{h}(t)$, it is convenient to first compute the eigenvalues $\lambda_1, \lambda_2, \ldots, \lambda_{2n}$ of the $2n \times 2n$ matrix **G**. Assuming that

$$\mathbf{T} = [\mathbf{d}_1, \mathbf{d}_2, \ldots, \mathbf{d}_{2n}] \tag{14.42}$$

is the $2n \times 2n$ *complex modal matrix* (see, e.g., Fang and Wang (1986); Faravelli et al. (1988)) whose columns are the eigenvectors, or *complex modes*, of **G**, and

$$\boldsymbol{\eta} = \mathrm{diag}(\lambda_1, \lambda_2, \ldots, \lambda_{2n}) \tag{14.43}$$

is the $2n \times 2n$ diagonal matrix containing the eigenvalues, it is obtained that

$$\exp(\mathbf{G}t) = \mathbf{T}\,\mathrm{diag}(\exp(\lambda_1 t), \exp(\lambda_2 t), \ldots, \exp(\lambda_{2n} t))\mathbf{T}^{-1}. \tag{14.44}$$

Once the eigenvalues and the eigenvectors of **G** are obtained, Eq. (14.44) makes it possible to easily determine the elements of the impulse response matrix.

15 Numerical Solution Techniques

15.1 Introduction

We have seen that when the equations of motion of an offshore facility can be modeled as linear, there are sometimes analytical methods available that allow us to calculate the statistical properties of the response. In general, this possibility is lost as soon as nonlinear elements enter the dynamic model. Under such circumstances, the most general and practical option available to obtain statistical information about the response process is to perform a Monte Carlo simulation, which consists of three steps:

1. A sample of load time histories is generated.
2. For each load time history, the equations of motion are solved numerically by a time-stepping method to produce a corresponding sample of response time histories.
3. For the sample of response time histories, statistical techniques are used to derive estimates of the requested statistics. The desired accuracy of these estimates will determine the necessary sample size.

The first step is discussed in the previous section. Here, we discuss the second step. To simplify this discussion, it is expedient to start with the 1DOF case. The equation of motion is then written in the form

$$m\ddot{u}(t) + c\dot{u}(t) + f(u(t), \dot{u}(t)) = p(t), \tag{15.1}$$

subject to the initial conditions

$$u(0) = u_0, \quad \dot{u}(0) = \dot{u}_0, \tag{15.2}$$

The nonlinear function $f(u, \dot{u})$ may contain both nonlinear damping terms and nonlinear restoring force terms. $p(t)$ is an external (deterministic) applied force.

The numerical solution of Eq. (15.1) is achieved by a recursive integration procedure in time. The usual approach is to choose a time increment Δt, and then define a discretization of time by $t_j = j\Delta t$, $j = 0, 1, 2, \ldots$. The numerical procedures described allow us to calculate the values $\ddot{u}_{i+1} = \ddot{u}(t_{i+1})$, $\dot{u}_{i+1} = \dot{u}(t_{i+1})$ and $u_{i+1} = u(t_{i+1})$ from the value $p_{i+1} = p(t_{i+1})$ and the calculated values at the previous times. There is a large literature on solution methods that would apply to the equations of motion relevant to us. We limit ourselves to a brief introduction to the two classes of methods you are most likely to meet or, in fact, use yourself: the Newmark methods and the Runge-Kutta methods.

15.2 Newmark Methods for SDOF Models

The Newmark methods (Newmark, 1959) are based on the following set of equations, where the first one is the equation of dynamic equilibrium:

$$\begin{aligned}\ddot{u}_{i+1} &= m^{-1}\big(-c\dot{u}_{i+1} - f(u_{i+1}, \dot{u}_{i+1}) + p_{i+1}\big),\\ \dot{u}_{i+1} &= \dot{u}_i + (1-\gamma)\Delta t\ddot{u}_i + \gamma\Delta t\ddot{u}_{i+1},\\ u_{i+1} &= u_i + \Delta t\dot{u}_i + (0.5-\beta)\Delta t^2\ddot{u}_i + \beta\Delta t^2\ddot{u}_{i+1}.\end{aligned} \tag{15.3}$$

When applied successively with $i = 0, 1, 2, \ldots,$ the time-stepping procedure for the solution, as implied by these equations, provides the desired response at all times $t_1, t_2, \ldots$. The known initial conditions u_0 and $\dot{u}_0$, combined with $\ddot{u}_0 = m^{-1}\big(-c\dot{u}_0 - f(u_0, \dot{u}_0) + p_0\big)$, provide the necessary information to start the procedure. The parameters β and γ define the variation of acceleration over a time step and determine the important properties of stability and accuracy of the method. Typical values are $1/6 \le \beta \le 1/2$ and $\gamma = 1/2$, and it may be noted that the two special cases $\beta = 1/2$ and $\beta = 1/6$, both with $\gamma = 1/2$, correspond to the assumptions of constant average acceleration between time points and linear variation of acceleration between time points. The Newmark methods also include the central difference method and the Houbolt method.

15.2.1 Linear Models

A quick look at Eq. (15.3) reveals an apparent complication to the solution of the three equations, and that is the presence of the term $\ddot{u}_{i+1}$ on the rhs of the last two equations. In general, this necessitates an iteration process at each time t_i in order to achieve a solution of the equations. This solution method is therefore called *implicit*. However, for a linear system, which is equivalent to $f(u, \dot{u}) = ku$, it is possible to modify Newmark's original formulation to obtain a solution method of Eq. (15.3) without this iteration process, that is, an explicit method. For this purpose, it is expedient to reformulate Eq. (15.3) in terms of incremental quantities $\Delta u_i = u_{i+1} - u_i$, with entirely similar definitions for the other variables. This is not necessary for the case of linear equations, but is particularly convenient for the nonlinear case. The last two equations in Eq. (15.3) can be rewritten as

$$\begin{aligned}\Delta\dot{u}_i &= \Delta t\ddot{u}_i + \gamma\Delta t\Delta\ddot{u}_i,\\ \Delta u_i &= \Delta t\dot{u}_i + \frac{\Delta t^2}{2}\ddot{u}_i + \beta\Delta t^2\Delta\ddot{u}_i.\end{aligned} \tag{15.4}$$

From the second of these equations, it is found that

$$\Delta\ddot{u}_i = \frac{1}{\beta\Delta t^2}\Delta u_i - \frac{1}{\beta\Delta t}\dot{u}_i - \frac{1}{2\beta}\ddot{u}_i. \tag{15.5}$$

Substituting Eq. (15.5) into the first equation in Eq. (15.4) gives

$$\Delta \dot{u}_i = \frac{\gamma}{\beta \Delta t} \Delta u_i - \frac{\gamma}{\beta} \dot{u}_i + \Delta t \left(1 - \frac{\gamma}{2\beta}\right) \ddot{u}_i. \tag{15.6}$$

The next step is to substitute Eqs. (15.5) and (15.6) into the incremental form of the equation of motion, which in the linear case assumes the form

$$m\Delta \ddot{u}_i + c\Delta \dot{u}_i + k\Delta u_i = \Delta p_i. \tag{15.7}$$

This substitution leads to the result

$$\hat{k}\Delta u_i = \Delta \hat{p}_i, \tag{15.8}$$

where

$$\hat{k} = k + \frac{\gamma}{\beta \Delta t} c + \frac{1}{\beta \Delta t^2} m \tag{15.9}$$

and

$$\Delta \hat{p}_i = \Delta p_i + \left(\frac{1}{\beta \Delta t} m + \frac{\gamma}{\beta} c\right) \dot{u}_i + \left(\frac{1}{2\beta} m + \Delta t \left(\frac{\gamma}{2\beta} - 1\right) c\right) \ddot{u}_i. \tag{15.10}$$

From these equations, it is seen that we have the information we need to take the solution one step further from time t_i to t_{i+1}. First, Eq. (15.8) is solved to determine Δu_i. Then, Eq. (15.6) can be solved for $\Delta \dot{u}_i$, and, finally, $\Delta \ddot{u}_i$ is found from Eq. (15.7) or (15.5).

A closer look at Eq. (15.8) reveals that the parameter γ only occurs in combination with the damping parameter c and that γ controls the numerical damping in the solution algorithm. There are three different cases:

$\gamma = \frac{1}{2}$, no numerical damping

$\gamma > \frac{1}{2}$, positive numerical damping

$\gamma < \frac{1}{2}$, negative numerical damping

As already pointed out, the usual choice is $\gamma = 1/2$, but it may sometimes be desirable to introduce some numerical damping by choosing γ slightly larger.

Numerical procedures that lead to bounded solutions as long as the time step remains shorter than some constant are called conditionally stable. If bounded solutions are obtained whatever the size of the time step, the procedure is called *unconditionally stable.* Note that these definitions say nothing about the accuracy of the solutions. In fact, in most cases, the desired accuracy will impose much stricter limitations on the size of time steps than mere stability.

The Newmark methods are unconditionally stable if

$$\gamma \geq \frac{1}{2}, \quad \text{and} \quad \beta \geq \frac{1}{4}\left(\gamma + \frac{1}{2}\right)^2. \tag{15.11}$$

For smaller values of β, the method becomes conditionally stable. The criterion for stability can be expressed as

$$\frac{\Delta t}{T_e} \leq \frac{1}{2\pi} \frac{1}{\sqrt{\frac{1}{4}\left(\gamma + \frac{1}{2}\right)^2 - \beta}}, \tag{15.12}$$

where $T_e = 2\pi\sqrt{m/k}$ is the undamped natural period of the linear system.

From these criteria, it follows that the average acceleration method, which has $\beta = 1/4$ and $\gamma = 1/2$, is unconditionally stable, that is, stable for any Δt. However, requirements on the accuracy of the solution will strongly limit the size of Δt. For the linear acceleration method with $\beta = 1/6$ and $\gamma = 1/2$, the stability requirement is that

$$\frac{\Delta t}{T_e} \leq 0.551. \tag{15.13}$$

However, requirements on accuracy will generally lead to a significant reduction of this number.

Several methods have been proposed to improve on the Newmark methods for linear problems, notably, Wilson's θ-method, Hilbers' α-method and the Bossak-Newmark method. Except for Wilson's method, we do not discuss these methods any further here, but the interested reader may consult Wood (1984) and Bathe (1996) for detailed descriptions.

15.2.2 Nonlinear Models

The modifications to the Newmark solution procedures for linear problems, as described previously, that are needed for solving nonlinear problems are directly connected to the formulation of the incremental form of the equation of motion, that is, Eq. (15.1). More precisely, we must express the increment of the nonlinear term, that is, $\Delta f_i = f(u_{i+1}, \dot{u}_{i+1}) - f(u_i, \dot{u}_i)$, in a form suitable for numerical solution. If the function $f(\cdot, \cdot)$ is continuously differentiable, a possible choice is to use a first-order Taylor expansion, that is,

$$\Delta f_i \approx k_i \Delta u_i + c_i \Delta \dot{u}_i. \tag{15.14}$$

Here, $k_i = k^T(u_i, \dot{u}_i)$, where $k^T(x, y) = \partial f(x, y)/\partial x$, and $c_i = c^T(u_i, \dot{u}_i)$, where $c^T(x, y) = \partial f(x, y)/\partial y$. By this approximation, a tangential stiffness k_i and a tangential damping c_i are introduced, and the incremental form of the equation of motion becomes

$$m\Delta \ddot{u}_i + (c + c_i)\Delta \dot{u}_i + k_i \Delta u_i = \Delta p_i. \tag{15.15}$$

For this approximation, we may therefore use exactly the same solution procedure that we used for the linear model expressed by Eq. (15.7) because this procedure did not depend on c and k being independent of the time step t_i.

Unfortunately, this approximation can lead to unacceptably inaccurate solutions when used with a constant time step Δt. Significant errors may occur for two reasons: (1) tangential stiffness and damping were used instead of secant values, which would require knowledge of u_{i+1} and $\dot{u}_{i+1}$, or (2) sudden changes in the value of the nonlinear term $f(u, \dot{u})$ are not accurately captured with constant Δt.

These errors can be minimized by using an iterative solution method for each time step. The key equation in the Newmark solution procedure was Eq. (15.8). Modified to cover the nonlinear case, we get

$$\hat{k}_i \Delta u_i = \Delta \hat{p}_i, \tag{15.16}$$

where

$$\hat{k}_i = k_i + \frac{\gamma}{\beta \Delta t}(c + c_i) + \frac{1}{\beta \Delta t^2} m \tag{15.17}$$

and

$$\Delta \hat{p}_i = \Delta p_i + \left(\frac{1}{\beta \Delta t} m + \frac{\gamma}{\beta}(c + c_i) \right) \dot{u}_i + \left(\frac{1}{2\beta} m + \Delta t \left(\frac{\gamma}{2\beta} - 1 \right)(c + c_i) \right) \ddot{u}_i. \tag{15.18}$$

A nonlinear tangent stiffness k_i as a function of u_i and $\dot{u}_i$ leads to a nonlinear $\hat{k}_i$. For a static analysis, $\hat{k}_i = k_i$, and the nonlinearity of $\hat{k}_i$ is exactly the same as that of k_i. However, for dynamic analyses, it is often the case that the constant term $m/(\beta \Delta t^2)$ in Eq. (15.17) is much larger than k_i for typical values of Δt. The consequence is that $\hat{k}_i$ may become only weakly nonlinear even if k_i is strongly nonlinear.

The iteration procedure for time t_i starts by solving Eq. (15.16) to determine $\Delta u_i^{(1)}$, and then $\Delta \dot{u}_i^{(1)}$ and $\Delta \ddot{u}_i^{(1)}$ as previously described. To this solution is associated $\Delta f_i^{(1)} = f(u_i^{(1)}, \dot{u}_i^{(1)}) - f(u_i, \dot{u}_i)$. If $\Delta u_i^{(1)}$ is a sufficiently accurate solution,

$$m \Delta \ddot{u}_i^{(1)} + c \Delta \dot{u}_i^{(1)} + \Delta f_i^{(1)} = \Delta p_i, \tag{15.19}$$

within a preassigned error tolerance. If this is not the case, a corrected load increment is introduced:

$$\Delta p_i^{(1)} = \Delta p_i - \left(m \Delta \ddot{u}_i^{(1)} + c \Delta \dot{u}_i^{(1)} + \Delta f_i^{(1)} \right), \tag{15.20}$$

which leads to a correction $\Delta \hat{p}_i^{(1)}$ given by

$$\begin{aligned} \Delta \hat{p}_i^{(1)} &= \Delta p_i^{(1)} + \left(\frac{1}{\beta \Delta t} m + \frac{\gamma}{\beta}(c + c_i^{(1)}) \right) \dot{u}_i^{(1)} \\ &+ \left(\frac{1}{2\beta} m + \Delta t \left(\frac{\gamma}{2\beta} - 1 \right)(c + c_i^{(1)}) \right) \ddot{u}_i^{(1)}, \end{aligned} \tag{15.21}$$

where $c_i^{(1)} = c^T(u_i^{(1)}, \dot{u}_i^{(1)})$, $u_i^{(1)} = u_i + \Delta u_i^{(1)}$, $\dot{u}_i^{(1)} = \dot{u}_i + \Delta \dot{u}_i^{(1)}$ and $\ddot{u}_i^{(1)} = \ddot{u}_i + \Delta \ddot{u}_i^{(1)}$. Similarly,

$$\hat{k}_i^{(1)} = k_i^{(1)} + \frac{\gamma}{\beta \Delta t}(c + c_i^{(1)}) + \frac{1}{\beta \Delta t^2} m, \tag{15.22}$$

where $k_i^{(1)} = k^T(u_i^{(1)}, \dot{u}_i^{(1)})$. A correction to the first iteration solution $u_i^{(1)}$ is now obtained by solving the equation

$$\hat{k}_i^{(1)} \Delta u_i^{(2)} = \Delta \hat{p}_i^{(1)}, \tag{15.23}$$

and then $\Delta \dot{u}_i^{(2)}$ and $\Delta \ddot{u}_i^{(2)}$ are determined as before. The associated nonlinear increment is $\Delta f_i^{(2)} = f(u_i^{(2)}, \dot{u}_i^{(2)}) - f(u_i^{(1)}, \dot{u}_i^{(1)})$. The next step is to determine whether $\Delta u_i^{(2)}$ satisfies the equation,

$$m \Delta \ddot{u}_i^{(2)} + c \Delta \dot{u}_i^{(2)} + \Delta f_i^{(2)} = \Delta p_i^{(1)}, \tag{15.24}$$

within the preassigned error bound. If not, a corrected load increment is again introduced:

$$\Delta p_i^{(2)} = \Delta p_i^{(1)} - \left(m\Delta\ddot{u}_i^{(2)} + c\Delta\dot{u}_i^{(2)} + \Delta f_i^{(2)}\right), \tag{15.25}$$

which leads to a correction $\Delta\hat{p}_i^{(2)}$ exactly as above, and it is seen how this iteration process can be continued to give a sequence of corrections $\Delta u_i^{(1)}, \Delta u_i^{(2)}, \ldots$. The iteration process is stopped as soon as the equation

$$m\Delta\ddot{u}_i^{(k)} + c\Delta\dot{u}_i^{(k)} + \Delta f_i^{(k)} = \Delta p_i^{(k-1)} \tag{15.26}$$

is satisfied with the required accuracy, which implies that $\Delta p_i^{(k)}$ is negligible. By reference to Eqs. (15.20) and (15.25), it is seen that the following set of equations are satisfied ($\Delta p_i^{(0)} = \Delta p_i$):

$$m\Delta\ddot{u}_i^{(j)} + c\Delta\dot{u}_i^{(j)} + \Delta f_i^{(j)} = \Delta p_i^{(j-1)} - \Delta p_i^{(j)}, \quad j = 1, 2, \ldots, k. \tag{15.27}$$

Adding these equations, and neglecting $\Delta p_i^{(k)}$, we get the desired result

$$m\Delta\ddot{u}_i^{[k]} + c\Delta\dot{u}_i^{[k]} + \Delta f_i^{[k]} = \Delta p_i, \tag{15.28}$$

where $\Delta\dot{u}_i^{[k]} = \sum_{j=1}^{k} \Delta\dot{u}_i^{(j)}$, $\Delta\ddot{u}_i^{[k]} = \sum_{j=1}^{k} \Delta\ddot{u}_i^{(j)}$ and $\Delta f_i^{[k]} = \sum_{j=1}^{k} \Delta f_i^{(j)} = f(u_{i+1}, \dot{u}_{i+1}) - f(u_i, \dot{u}_i)$ because $u_{i+1} = u_i + \Delta u_i^{[k]} = u_i + \sum_{j=1}^{k} \Delta u_i^{(j)}$. The result of the iteration process is that Eq. (15.28) is satisfied to the accuracy we require, and the increments we use to bring us one step further, that is, from t_i to t_{i+1}, in the recursive solution of the equation of motion are then $\Delta u_i^{[k]}$, $\Delta\dot{u}_i^{[k]}$ and $\Delta\ddot{u}_i^{[k]}$.

In the preceding iterative solution process, we update the tangential stiffness and damping terms at each iteration, and the resulting iteration algorithm is therefore a Newton-Raphson method. This updating requires additional calculations and is therefore often skipped. The resulting iteration algorithm, without updating, is generally referred to as a modified Newton-Raphson method. The original Newton-Raphson method converges more rapidly than the modified version, but the latter method is faster, and the reduction in CPU time to reach a converged solution may be significant, especially for MDOF systems. For more detailed discussions, the reader may consult Hughes (1987), Argyris and Mlejnek (1991), or Bathe (1996).

15.3 Newmark Methods for MDOF Models

When we move from an SDOF model to an MDOF model, the question about the stability of solutions becomes less transparent and is, in fact, a critical issue for MDOF models. Although conditionally stable procedures can be used effectively for large MDOF models, it is generally necessary to use unconditionally stable methods for nonlinear response analysis of such models.

15.3.1 Linear Models

The equation of motion for a linear MDOF model can be written as

$$\boldsymbol{m}\ddot{\boldsymbol{u}}(t) + \boldsymbol{c}\dot{\boldsymbol{u}}(t) + \boldsymbol{k}\boldsymbol{u}(t) = \boldsymbol{p}(t), \tag{15.29}$$

where $\boldsymbol{m}$, $\boldsymbol{c}$, and $\boldsymbol{k}$ are $n \times n$ matrices, and n = number of DOF of the model. $\boldsymbol{p}(t) = (p_1(t, \ldots, p_n(t))^\top$ denotes the external load, and $\boldsymbol{u} = (u_1, \ldots, u_n)^\top$ is the response vector.

If n is moderate to small, it may be appropriate to solve Eq. (15.29) in its given form. However, for large n, it may be expedient to seek an approximate solution expressed in terms of the first few natural vibration modes $\boldsymbol{\phi}_j$ of the undamped model, cf. Chapter 3. Thus, the nodal displacements of the system are approximated by a linear combination of the first m modes as

$$\boldsymbol{u} = \sum_{j=1}^{m} \boldsymbol{\phi}_j q_j(t) = \boldsymbol{\Phi} \boldsymbol{q}(t), \tag{15.30}$$

where $q_j(t)$ denotes the generalized coordinates, $\boldsymbol{q}(t) = (q_1(t), \ldots, q_m(t))^\top$, and $\boldsymbol{\Phi} = (\boldsymbol{\phi}_1, \ldots, \boldsymbol{\phi}_m)$ is an $m \times m$ matrix with the mode vectors as columns. The choice of m must be based on an assessment of how many modes will actually contribute to the response of the structural model. Using this reduction technique, the equation of motion can be expressed as

$$\boldsymbol{M}\ddot{\boldsymbol{q}}(t) + \boldsymbol{C}\dot{\boldsymbol{q}}(t) + \boldsymbol{K}\boldsymbol{q} = \boldsymbol{P}(t), \tag{15.31}$$

where $\boldsymbol{M} = \boldsymbol{\Phi}^\top \boldsymbol{m} \boldsymbol{\Phi}$ and $\boldsymbol{K} = \boldsymbol{\Phi}^\top \boldsymbol{k} \boldsymbol{\Phi}$ are diagonal matrices, $\boldsymbol{C} = \boldsymbol{\Phi}^\top \boldsymbol{c} \boldsymbol{\Phi}$, and $\boldsymbol{P}(t) = \boldsymbol{\Phi}^\top \boldsymbol{p}(t)$. If $m \ll n$, it may be advantageous to calculate the solution of Eq. (15.31) instead of Eq. (15.29). The computational savings may be considerable despite the necessity to calculate the first m modes.

In general, the damping matrix $\boldsymbol{C}$ is not diagonal. If it is, the system of equations in Eq. (15.29) decouples and we get m separate modal equations, as discussed in Chapter 3. Each equation can then be solved separately by using the numerical solution techniques for SDOF models, as previously discussed. The subsequent discussion therefore applies to the case of nondiagonal $\boldsymbol{C}$. It can be shown that conditionally stable numerical methods can be used to solve Eq. (15.31). What is required is that $\Delta t / T_m$ is small enough to ensure an accurate solution for the last undamped modal response. Here, T_m denotes the undamped natural period of this mode. Clearly, this requirement will also ensure accurate solutions also for all lower modes.

The steps in Newmark solution procedure for Eq. (15.31) are similar to those for the SDOF case. These steps are therefore given without further comments as follows:

Initial Calculations

1. $q_{j,0} = \frac{\boldsymbol{\phi}_j^\top \boldsymbol{m} \boldsymbol{u}_0}{\boldsymbol{\phi}_j^\top \boldsymbol{m} \boldsymbol{\phi}_j}, \quad \dot{q}_{j,0} = \frac{\boldsymbol{\phi}_j^\top \boldsymbol{m} \dot{\boldsymbol{u}}_0}{\boldsymbol{\phi}_j^\top \boldsymbol{m} \boldsymbol{\phi}_j}, \boldsymbol{q}_0 = (q_{1,0}, \ldots, q_{m,0})^\top, \quad \dot{\boldsymbol{q}}_0 = (\dot{q}_{1,0}, \ldots, \dot{q}_{m,0})^\top$
2. $\boldsymbol{P}_0 = \boldsymbol{\Phi}^\top \boldsymbol{p}_0$
3. Solve $\boldsymbol{M}\ddot{\boldsymbol{q}}_0 = \boldsymbol{P}_0 - \boldsymbol{C}\dot{\boldsymbol{q}}_0 - \boldsymbol{K}\boldsymbol{q}_0 \Rightarrow \ddot{\boldsymbol{q}}_0$
4. Select Δt
5. $\hat{\boldsymbol{K}} = \boldsymbol{K} + \frac{\gamma}{\beta \Delta t}\boldsymbol{C} + \frac{1}{\beta \Delta t^2}\boldsymbol{M}$
6. $\boldsymbol{a} = \frac{1}{\beta \Delta t}\boldsymbol{M} + \frac{\gamma}{\beta}\boldsymbol{C}, \quad \boldsymbol{b} = \frac{1}{2\beta}\boldsymbol{M} + \Delta t\left(\frac{\gamma}{2\beta} - 1\right)\boldsymbol{C}$

Calculations for Each Time Step

1. $\boldsymbol{P}_i = \boldsymbol{\Phi}^\top \boldsymbol{p}_i$
2. $\Delta\hat{\boldsymbol{P}}_i = \Delta\boldsymbol{P}_i + \boldsymbol{a}\dot{\boldsymbol{q}}_i + \boldsymbol{b}\ddot{\boldsymbol{q}}_i$
3. Solve: $\hat{\boldsymbol{K}}\Delta\boldsymbol{q}_i = \Delta\hat{\boldsymbol{P}}_i, \Rightarrow \ \Delta\boldsymbol{q}_i$
4. $\Delta\dot{\boldsymbol{q}}_i = \frac{\gamma}{\beta\Delta t}\Delta\boldsymbol{q}_i - \frac{\gamma}{\beta}\dot{\boldsymbol{q}}_i - \Delta t\left(\frac{\gamma}{2\beta} - 1\right)\ddot{\boldsymbol{q}}_i$
5. $\Delta\ddot{\boldsymbol{q}}_i = \frac{1}{\beta\Delta t^2}\Delta\boldsymbol{q}_i - \frac{1}{\beta\Delta t}\dot{\boldsymbol{q}}_i - \frac{1}{2\beta}\ddot{\boldsymbol{q}}_i$
6. $\boldsymbol{q}_{i+1} = \boldsymbol{q}_i + \Delta\boldsymbol{q}_i, \ \dot{\boldsymbol{q}}_{i+1} = \dot{\boldsymbol{q}}_i + \Delta\dot{\boldsymbol{q}}_i, \ \ddot{\boldsymbol{q}}_{i+1} = \ddot{\boldsymbol{q}}_i + \Delta\ddot{\boldsymbol{q}}_i$
7. $\boldsymbol{u}_{i+1} = \boldsymbol{\Phi}\boldsymbol{q}_{i+1}$

As for the SDOF case, the most common Newmark methods for MDOF models are also the average acceleration method ($\gamma = 1/2$, $\beta = 1/4$), which is unconditionally stable, and the linear acceleration method ($\gamma = 1/2$, $\beta = 1/6$), which is conditionally stable.

15.3.2 Nonlinear Models

The techniques of dimension reduction discussed previously for linear MDOF models cannot generally be used for nonlinear models. Also, conditionally stable solution procedures should be avoided if there is uncertainty concerning their applicability. We have therefore chosen to present only the Newmark average acceleration method here. And, as commonly done, it is combined with the modified Newton-Raphson iteration method. That is, the tangential stiffness and damping terms are not updated during the iteration phase. For MDOF models, this updating can be a rather time-consuming part of the solution procedure.

The equation of motion for the nonlinear MDOF model is the following:

$$\boldsymbol{m}\ddot{\boldsymbol{u}}(t) + \boldsymbol{c}\dot{\boldsymbol{u}}(t) + \boldsymbol{f}\big(\boldsymbol{u}(t), \dot{\boldsymbol{u}}(t)\big) = \boldsymbol{p}(t), \tag{15.32}$$

where $\boldsymbol{f}(\cdot,\cdot) = \big(f^{kl}(\cdot,\cdot)\big), k, l = 1, \ldots, n$, is a nonlinear matrix function, and the other terms are defined as in Eq. (15.29). As in the SDOF case, we need to express the increment of the nonlinear term in a form suitable for numerical solution, that is, we write

$$\Delta\boldsymbol{f}_i = \boldsymbol{f}(\boldsymbol{u}_{i+1}, \dot{\boldsymbol{u}}_{i+1}) - \boldsymbol{f}(\boldsymbol{u}_i, \dot{\boldsymbol{u}}_i) \approx \boldsymbol{k}_i\Delta\boldsymbol{u}_i + \boldsymbol{c}_i\Delta\dot{\boldsymbol{u}}_i, \tag{15.33}$$

which defines the tangential stiffness matrix $\boldsymbol{k}_i$ and the tangential damping matrix $\boldsymbol{c}_i$ at time step i. The entries in these matrices can be expressed in a manner entirely analogous with the SDOF case. The incremental form of the equation of motion can then be written as

$$\boldsymbol{m}\Delta\ddot{\boldsymbol{u}}_i + (\boldsymbol{c} + \boldsymbol{c}_i)\Delta\dot{\boldsymbol{u}}_i + \boldsymbol{k}_i\Delta\boldsymbol{u}_i = \Delta\boldsymbol{p}_i. \tag{15.34}$$

The numerical solution of the equation of motion with the Newmark average acceleration method can be represented by the following steps:

Initial Calculations

1. Solve $\boldsymbol{m}\ddot{\boldsymbol{u}}_0 = \boldsymbol{p}_0 - \boldsymbol{c}\dot{\boldsymbol{u}}_0 - \boldsymbol{f}(\boldsymbol{u}_0, \dot{\boldsymbol{u}}_0) \ \Rightarrow \ \ddot{\boldsymbol{u}}_0$.
2. Select Δt.

Calculations for Each Time Step i

1. Determine $\boldsymbol{k}_i$ and $\boldsymbol{c}_i$.
2. $\Delta\hat{\boldsymbol{p}}_i = \Delta\boldsymbol{p}_i + \big((4/\Delta t)\boldsymbol{m} + 2(\boldsymbol{c}+\boldsymbol{c}_i)\big)\dot{\boldsymbol{u}}_i + 2\boldsymbol{m}\ddot{\boldsymbol{u}}_i$
3. $\hat{\boldsymbol{k}}_i = \boldsymbol{k}_i + \frac{2}{\Delta t}(\boldsymbol{c}+\boldsymbol{c}_i) + \frac{4}{\Delta t^2}\boldsymbol{m}$
4. Solve for $\Delta\boldsymbol{u}_i$, $\Delta\dot{\boldsymbol{u}}_i$, and $\Delta\ddot{\boldsymbol{u}}_i$ using the modified Newton-Raphson method .
5. $\boldsymbol{u}_{i+1} = \boldsymbol{u}_i + \Delta\boldsymbol{u}_i$, $\dot{\boldsymbol{u}}_{i+1} = \dot{\boldsymbol{u}}_i + \Delta\dot{\boldsymbol{u}}_i$, $\ddot{\boldsymbol{u}}_{i+1} = \ddot{\boldsymbol{u}}_i + \Delta\ddot{\boldsymbol{u}}_i$

Modified Newton-Raphson Method at Time Step i for $j = 0, 1, \ldots$

1. $\boldsymbol{u}_i^{(0)} = \boldsymbol{u}_i$, $\Delta\hat{\boldsymbol{p}}_i^{(0)} = \Delta\hat{\boldsymbol{p}}_i$
2. Solve $\hat{\boldsymbol{k}}_i \Delta\boldsymbol{u}_i = \Delta\hat{\boldsymbol{p}}_i^{(j)} \Rightarrow \Delta\boldsymbol{u}_i^{(j+1)}$.
3. $\Delta\dot{\boldsymbol{u}}_i^{(j+1)} = (2/\Delta t)\Delta\boldsymbol{u}_i^{(j+1)} - 2\dot{\boldsymbol{u}}_i^{(j)}$
4. $\Delta\ddot{\boldsymbol{u}}_i^{(j+1)} = (4/\Delta t^2)\Delta\boldsymbol{u}_i^{(j+1)} - (4/\Delta t)\dot{\boldsymbol{u}}_i^{(j)} - 2\ddot{\boldsymbol{u}}_i^{(j)}$
5. $\boldsymbol{u}_i^{(j+1)} = \boldsymbol{u}_i^{(j)} + \Delta\boldsymbol{u}_i^{(j+1)}$, $\dot{\boldsymbol{u}}_i^{(j+1)} = \dot{\boldsymbol{u}}_i^{(j)} + \Delta\dot{\boldsymbol{u}}_i^{(j+1)}$, $\ddot{\boldsymbol{u}}_i^{(j+1)} = \ddot{\boldsymbol{u}}_i^{(j)} + \Delta\ddot{\boldsymbol{u}}_i^{(j+1)}$
6. $\Delta\boldsymbol{f}_i^{(j+1)} = \boldsymbol{f}(\boldsymbol{u}_i^{(j+1)}, \dot{\boldsymbol{u}}_i^{(j+1)}) - \boldsymbol{f}(\boldsymbol{u}_i^{(j)}, \dot{\boldsymbol{u}}_i^{(j)})$
7. $\Delta\boldsymbol{p}_i^{(j+1)} = \Delta\boldsymbol{p}_i^{(j)} - \boldsymbol{m}\Delta\ddot{\boldsymbol{u}}_i^{(j+1)} - \boldsymbol{c}\Delta\dot{\boldsymbol{u}}_i^{(j+1)} - \Delta\boldsymbol{f}_i^{(j+1)}$
8. $\Delta\hat{\boldsymbol{p}}_i^{(j+1)} = \Delta\boldsymbol{p}_i^{(j+1)} + \big((4/\Delta t)\boldsymbol{m} + 2(\boldsymbol{c}+\boldsymbol{c}_i)\big)\dot{\boldsymbol{u}}_i^{(j+1)} + 2\boldsymbol{m}\ddot{\boldsymbol{u}}_i^{(j+1)}$
9. If $\Delta\boldsymbol{p}_i^{(j+1)}$ is nonnegligible, go to step 2. Otherwise, go to step 10.
10. If $\Delta\boldsymbol{p}_i^{(k)}$ is negligible, define $\Delta\boldsymbol{u}_i = \sum_{j=1}^{k}\Delta\boldsymbol{u}_i^{(j)}$, $\Delta\dot{\boldsymbol{u}}_i = \sum_{j=1}^{k}\Delta\dot{\boldsymbol{u}}_i^{(j)}$, $\Delta\ddot{\boldsymbol{u}}_i = \sum_{j=1}^{k}\Delta\ddot{\boldsymbol{u}}_i^{(j)}$.

One weakness associated with the Newmark average acceleration method is that it does not provide any numerical damping. This may be a disadvantage. A method that does offer numerical damping is the Wilson θ-method. It is described in detail by Bathe (1996).

So far, we have not discussed at any length methods for verifying that the numerically obtained solution is, in fact, an accurate representation of the target solution. Instead of entering a detailed description of the various methods to ensure this, we encourage the interested reader to consult the literature. The main reason behind the decision not to go into too much detail on this aspect is that you are most likely to resort to the following practical approach to this problem: the most common, if not overly sophisticated, method for verifying the accuracy of a numerical solution is to start with a reasonable value for Δt, and then repeat the calculations with half that time step. If the solutions do not agree within an acceptable error margin, another solution is calculated with half the time step of the previous solution. This is continued until sufficient agreement is obtained.

15.4 Runge-Kutta Methods

This is a family of both explicit and implicit one-step methods. The most popular of these is the 4th order explicit method, which is the one we discuss here and call the RK4 method. One should note that the RK4 method, like all explicit Runge-Kutta methods, is only conditionally stable. The formulation of the method is exactly the same for SDOF and MDOF models, and for this purpose, and economy of notation, we introduce the state-space vector $\boldsymbol{r}(t) = \big(u_1(t), \ldots, u_n(t), \dot{u}_1(t), \ldots, \dot{u}_n(t)\big)^\top$ for an

nDOF model. The equations of motion are then written in state-space form as

$$\dot{\boldsymbol{r}}(t) = \boldsymbol{g}(\boldsymbol{r}(t), t), \quad \boldsymbol{r}(0) = \boldsymbol{r}_0. \tag{15.35}$$

The RK4 method for this problem can now be expressed as the following recursive scheme, where $\boldsymbol{r}_i = \boldsymbol{r}(t_i)$:

$$\boldsymbol{r}_{i+1} = \boldsymbol{r}_i + \frac{\Delta t}{6}(\boldsymbol{g}_{1i} + 2\boldsymbol{g}_{2i} + 2\boldsymbol{g}_{3i} + \boldsymbol{g}_{4i}), \tag{15.36}$$

where

$$\begin{aligned}
\boldsymbol{g}_{1i} &= \boldsymbol{g}(\boldsymbol{r}_i, t_i), \\
\boldsymbol{g}_{2i} &= \boldsymbol{g}\Big(\boldsymbol{r}_i + \frac{\Delta t}{2}\boldsymbol{g}_{1i}, t_i + \frac{\Delta t}{2}\Big), \\
\boldsymbol{g}_{3i} &= \boldsymbol{g}\Big(\boldsymbol{r}_i + \frac{\Delta t}{2}\boldsymbol{g}_{2i}, t_i + \frac{\Delta t}{2}\Big), \\
\boldsymbol{g}_{4i} &= \boldsymbol{g}(\boldsymbol{r}_i + \Delta t\boldsymbol{g}_{3i}, t_i + \Delta t).
\end{aligned} \tag{15.37}$$

Thus, the next value $\boldsymbol{r}_{i+1}$ is calculated as the sum of the present value $\boldsymbol{r}_i$ and an estimated increment $\Delta\boldsymbol{r}_i$ over the time interval. It is seen that the calculation of the increment $\Delta\boldsymbol{r}_i$ is based on a weighted average of "slopes" putting more emphasis on the midpoint values:

1. $\boldsymbol{g}_{1i}$ is the "slope" at the start of the interval,
2. $\boldsymbol{g}_{2i}$ is the "slope" at the midpoint of the interval, using $\boldsymbol{g}_{1i}$ to calculate the value of $\mathbf{r}$ at this point by Euler's method,
3. $\boldsymbol{g}_{3i}$ is still the "slope" at the midpoint, but $\boldsymbol{g}_{2i}$ is now used to calculate the value of $\mathbf{r}$,
4. $\boldsymbol{g}_{4i}$ is the "slope" at the end of the interval, using $\boldsymbol{g}_{3i}$ to calculate the value of $\mathbf{r}$ at the endpoint.

The total accumulated error of the RK4 method is of the order Δt^4, which is reflected in the name of the method. Several standard software packages, such as Matlab, contain RK4 modules. The RK4 method may sometimes be a good alternative to the Newmark methods. However, it seems to be competitive only when the characteristic time constants associated with the dynamic system, and its excitation is much larger than the time increment Δt. It is therefore not a good alternative for analysis of systems with white noise excitations. In such cases, an unconditionally stable Newmark-type method would be a better choice. The average acceleration method ($\beta = 1/4, \gamma = 1/2$) or Wilson's θ-method are then viable procedures.

16 Monte Carlo Methods and Extreme Value Estimation

16.1 Introduction

The last decade has seen a dramatic increase in the use of Monte Carlo methods for solving stochastic engineering problems. There are primarily two reasons for this increase. First, the computational power available today, even for a laptop computer, is formidable and steadily increasing. Second, the versatility of Monte Carlo methods make them very attractive as a way of obtaining solutions to stochastic problems. The drawback of Monte Carlo methods for a range of problems has been that the required numerical calculations may take days, weeks or even months to do. But this situation is changing, some numerical problems that required several days of computer time for their solution just a few years ago can now be solved in minutes or hours. This has really opened the door for the use of Monte Carlo-based methods for solving a wide array of stochastic engineering problems. In this chapter the focus is on adapting Monte Carlo methods for estimation of extreme values of stochastic processes encountered in various engineering disciplines.

16.2 Simulation of Stationary Stochastic Processes

The approach to the simulation of stationary stochastic processes favored in this book is the spectral representation method. The main reason for this is its simplicity and transparency for practical applications. This choice has, in fact, already been implemented at several places previously in this book. Taken together, they constitute a fairly complete description on the level needed here of how to simulate a stationary process.

The procedure is described in Section 6.2, where Eq. (6.8) is the key relation. Useful background information for this is contained in Examples 5.7.2 and 5.8.2. Combining this with Example 6.3.1 provides a hands-on guide on how to simulate realizations of a stationary Gaussian process. Then, in Section 8.2.1 on random wave processes, all expressions for wave elevation, wave particle velocity, or wave particle acceleration are given as spectral representations (Shinozuka and Deodatis, 1991), which are then immediately applicable for simulation purposes.

The use of the fast Fourier transform (FFT) technique can substantially speed up the production of realizations of stationary Gaussian processes, cf. Shinozuka (1974) and Newland (1991).

16.3 Monte Carlo Simulation of Load and Response

The linear or linearized equations of motion for marine structures considered in this book and their solutions are discussed in Chapter 14. For such equations, the comments of the previous section apply. An example of a nonlinear dynamic model that is often adopted for an offshore structure can be written in the following form:

$$\boldsymbol{M}\ddot{\boldsymbol{X}}(t) + \boldsymbol{C}\dot{\boldsymbol{X}}(t) + \boldsymbol{f}\big(\boldsymbol{X}(t), \dot{\boldsymbol{X}}(t), t\big) = \boldsymbol{P}(t), \tag{16.1}$$

where $\boldsymbol{M}$ and $\boldsymbol{C}$ are suitable mass and damping matrices, respectively; $\boldsymbol{P}(t) = (P_1(t), \ldots, P_n(t))'$ denotes a stochastic loading process; $\boldsymbol{f}(\cdot,\cdot) = \big(f^{kl}(\cdot,\cdot)\big)$, $k, l = 1, \ldots, n$, is a nonlinear matrix function; and $\boldsymbol{X}(t) = (X_1(t), \ldots, X_n(t))'$ is the corresponding response process.

The Monte Carlo method applied to such a system would consist of generating a statistical sample of specified size N, say, of response time histories by first generating a sample of time histories of the same size of the loading process, and then solve Eq. (16.1) for each of the load time histories in the sample using the methods of Chapter 15. When the desired sample of response time histories is produced, the statistical analysis of the response may then proceed as discussed in the remaining part of this chapter.

16.4 Sample Statistics of Simulated Response

An important element for graphic representation of sampled data from a statistical population for which the underlying distribution function is unknown is the so-called plotting position formula. It is based on the notion of order statistics.

Let us start by assuming that X is a continuous random variable with a CDF $F(x)$ and a PDF $f(x)$. The given sample of independent observations $x_1, x_2, \ldots, x_n$ is now ordered in an increasing sequence $x_{(1)} \le x_{(2)} \le \ldots \le x_{(n)}$. The random variable $X_{(m)}$ corresponding to $x_{(m)}$ is called the mth order statistic, $m = 1, \ldots, n$. The PDF $f_m(x)$ of $X_{(m)}$ follows from the observation that $X_{(m)} = x$ implies the event that there are $m-1$ outcomes of X with values less than (or equal to) x, and $n-m$ outcomes with values greater than x. According to the binomial distribution, the probability of this event equals $n!/\big((m-1)!(n-m)!\big)\, F(x)^{m-1}\big(1-F(x)\big)^{n-m}$. Hence, it is obtained that

$$f_m(x) = \frac{n!}{(m-1)!(n-m)!} F(x)^{m-1}\big(1-F(x)\big)^{n-m} f(x). \tag{16.2}$$

We now want to calculate $\mathrm{E}[F(X_{(m)})]$. This is given by

$$\begin{aligned}\mathrm{E}[F(X_{(m)})] &= \int_x F(x) f_m(x)dx = m\binom{n}{m}\int_x F(x)^m\big(1-F(x)\big)^{n-m} f(x)dx \\ &= m\binom{n}{m}\int_0^1 F^m(1-F)^{n-m}dF = m\binom{n}{m}\frac{m!(n-m)!}{(n+1)!} = \frac{m}{n+1}.\end{aligned} \tag{16.3}$$

Similarly, the variance of $F(X_{(m)})$ is calculated to be

$$\begin{aligned}\mathrm{Var}[F(X_{(m)}] &= \mathrm{E}[F(X_{(m)})^2] - \mathrm{E}[F(X_{(m)})]^2 \\ &= \frac{m(m+1)}{(n+1)(n+2)} - \frac{m^2}{(n+1)^2} = \frac{m(n+1-m)}{(n+1)^2(n+2)}.\end{aligned} \tag{16.4}$$

The results of Eqs. (16.3) and (16.4) are useful because they provide a means of plotting the sample of observations $x_1, x_2, \ldots, x_n$ in order to estimate the CFD $F(x)$ empirically. Equation (16.3) states that the expected value of the distribution function evaluated at the observation of order m is equal to $m/(n+1)$. This result suggests that an optimal plotting strategy is obtained by plotting the points $\big(x_{(m)}, m/(n+1)\big)$, $m = 1, \ldots, n$. Equation (16.4) provides information on the variance of $F(X_{(m)})$, that is, the ordinate of the plotting point. Due to the symmetry of the expression, it attains its maximum $1/[4(n+2)]$ at the median and decreases symmetrically to $n/[(n+1)^2(n+2)]$ toward the ends of the interval $[0, 1]$. Because the distribution function itself is not known, these results are known as distribution-free results.

We complement the preceding results by calculating the probability that the mth observation is not exceeded by a future observation. Now, the conditional probability that a single observation will not exceed $X_{(m)}$ given that $X_{(m)} = x$ is equal to $F(x)$. The corresponding unconditional probability p_m, say, is then obtained by using the law of total probability, which gives exactly the same result as given by Eq. (16.3), that is,

$$p_m = \int_x F(x) f_m(x) dx = \frac{m}{n+1}. \tag{16.5}$$

This result shows that a new observation of the continuous random variable X has equal probability of assuming a value in any of the $n+1$ intervals defined by the previous n observations. This lends further support to the optimality of the plotting position formula previously discussed.

It may be pointed out that several alternative plotting position formulas for specific classes of distributions have been suggested over the years, which primarily aim at correcting for sample bias. We do not enter a discussion of this, but rather refer the reader to the literature.

A useful diagnostic tool to check the accuracy of an assumed statistical distribution F for the observed data is obtained by comparing the fitted distribution $\hat{F}$ with the data on a quantile plot (QQ-plot) or a probability plot (PP-plot). Assuming that $\hat{F}$ is strictly increasing and continuous, the QQ-plot is obtained by comparing the ordered data with the corresponding quantiles of the fitted distribution by plotting

$$\Big(\hat{F}^{-1}\big(\frac{m}{n+1}\big), x_{(m)}\Big). \tag{16.6}$$

The name *QQ-plot* derives from the fact that both $\hat{F}^{-1}(\frac{m}{n+1})$ and $x_{(m)}$ are estimates of the $m/(n+1)$th quantile of F. If F is a good choice for the distribution of the data, the QQ-plot should be close to the straight line of slope 1 passing through the origin. Alternatively, the fit of $\hat{F}$ to the data can be checked by the PP-plot, which is obtained by plotting the points

$$\left(\hat{F}(x_{(m)}), \frac{m}{n+1}\right). \tag{16.7}$$

A good fit is again demonstrated if the plotted graph is approximately a straight line. The main difference between the two plots is that the QQ-plot gives a more clear impression of the fit of the tail data, which may be of particular significance for extreme value statistics.

It is worth noting that for a range of distribution functions QQ-plots can be constructed without having to estimate distribution parameters. This typically applies to distributions characterized by a scale and a location parameter, e.g., the normal distribution. In such cases the intercept of the line fitted to the QQ-plot would represent location, while the slope represents scale. For example, a QQ-plot for a normal distribution can be achieved by plotting $\left(\Phi^{-1}(m/(n+1)), x_{(m)}\right)$.

16.5 Latin Hypercube Sampling

Latin hypercube sampling (LHS) (McKay et al., 1979) is a method for effectively reducing the sample size for Monte Carlo simulations of stochastic response processes that depend on many random parameters. In cases where the parameters of a dynamic model, such as mass, damping, and stiffness are modeled as random variables, the response process will then also depend on these random variables. Specifically, let us assume that the model depends on the random parameters $Y_1, \ldots, Y_m$. To highlight the dependence on these parameters, a random response process $X(t)$ of this model may then be written as $X(t) = X(t; Y_1, \ldots, Y_m)$. Because the external loading is often modeled as a stochastic process, the response $X(t; y_1, \ldots, y_m)$ becomes a stochastic process for each sample $y_1, \ldots, y_m$ of the random parameters. If m is not small, then the number of samples needed to provide good sample statistics for the response process may become quite large if no consideration is made on how to effectively represent the statistical variability of the random parameters. LHS is a very good and simple method for this purpose.

LHS starts by selecting k different values from each of the m random variables in the following manner. The interval (0, 1) is divided into k equally long intervals $I_j = ((j-1)/k, j/k)$, $j = 1, \ldots, k$. Let $F_i(y)$ denote the distribution function of Y_i, $i = 1, \ldots, m$. For each i, k independent outcomes $u_1, \ldots, u_k$ of the random number U, which is uniformly distributed on (0,1), are produced. The resulting ordered sample for Y_i is $y_{i,1} < \ldots < y_{i,k}$ where $y_{i,j} = F_i^{-1}((j-1)/k + u_j/k)$. Note that $(j-1)/k + u_j/k$ is nothing but an outcome of a random number uniformly distributed on the interval $((j-1)/k, j/k)$. A Latin hypercube (LH) presample for Y_i is then obtained as $y_{i,r_1}, \ldots, y_{i,r_k}$, where $r_1, \ldots, r_k$ is a random reordering of $1, \ldots, k$. Finally, an LH sample for $Y_1, \ldots, Y_m$, which will also be of size k, is now represented by the $m \times k$ array or matrix (y_{i,r_j}), where each column is an element in the LH sample.

We tacitly assume that the random parameters are independent. If this is not the case, LHS can also deal with correlated parameters. Standard statistical software packages usually offer LHS as an optional sampling technique.

16.6 Estimation of Extreme Response

The commonly held view that estimation of extreme values is generally prohibitive in terms of computer time is a truth in contention. In fact, with present-day computational power, it is possible to perform simulations on a scale that allows estimation of extreme structural response for a range of problems. Admittedly, it is not difficult to describe a dynamic model for which the required simulation time for direct estimation of extreme response would be beyond any acceptable bounds, but the systems for which this is the case are steadily diminishing in step with the development of increasingly powerful computers. This being the case, it is good reason to believe that Monte Carlo simulation-based methods will become much more frequently used even for extreme value estimation than what is the case today. It is therefore considered appropriate to discuss such methods here. Note that the methods described in this section are equally applicable to measured response time histories, obtained, for instance, from model tests of a floating structure. For simplicity, we only discuss the case of positive extreme values. The necessary modifications to deal with other situations are usually obvious.

16.6.1 Peaks-Over-Threshold Method

Generalized Pareto Distribution

The peaks-over-threshold (POT) method is based on what is called the generalized Pareto (GP) distribution in the following manner. according to Pickands (1975), asymptotically, the excess values above a high level will follow a GP distribution if and only if the parent distribution belongs to the domain of attraction of one of the extreme value distributions. The assumption of a Poisson process model for the exceedance times combined with GP distributed excesses can be shown to lead to the generalized extreme value (GEV) distribution for corresponding extremes. The expression for the GP distribution is

$$P(y) = P(y; a, c) = \text{Prob}(Y \le y) = 1 - \left(1 + c\frac{y}{a}\right)_+^{-1/c}. \tag{16.8}$$

Here, $a > 0$ is a scale parameter and $c(-\infty < c < \infty)$ determines the shape of the distribution. $(z)_+ = \max(0, z)$.

The asymptotic result previously referred to implies that Eq. (16.8) can be used to represent the conditional cumulative distribution function of the excess $Y = X - u$ of the observed variate X over the threshold u, given that $X > u$ for u sufficiently large (Pickands, 1975). The cases $c > 0$, $c = 0$, and $c < 0$ correspond to Fréchet (Type II), Gumbel (Type I), and reverse Weibull (Type III) domains of attraction, respectively. For $c = 0$, the expression between the parentheses in Eq. (16.8) is understood in a limiting sense as the exponential function $\exp(-y/a)$.

Generalized Extreme Value Distribution

When the analysis is based on the maximum observation during specified periods of time such as one year, the usual procedure is to assume that the resulting set of data is independent and identically distributed (iid) and to follow a GEV distribution for maxima with CDF:

$$G(x) = G(x; a, b, c) = \exp\left\{-\left(1 + c\frac{x-b}{a}\right)_+^{-1/c}\right\}. \tag{16.9}$$

Here, $a > 0$ and b are scale and location parameters, respectively. c is a parameter determining the shape of the distribution. The cases $c > 0$, $c = 0$, and $c < 0$ correspond to Fréchet (Type II), Gumbel (Type I), and reverse Weibull (Type III) extreme value distributions, respectively. The special case $c = 0$ is again interpreted as a limiting case, that is,

$$G(x; a, b, 0) = \exp\left\{-e^{-(x-b)/a}\right\}, \tag{16.10}$$

which is a standard expression for the Gumbel distribution.

Similarly as for the distribution of exceedances for high levels discussed in the previous section, the assumption that yearly maxima follow a GEV distribution is also based on a general asymptotic result, see, e.g., Leadbetter et al. (1983). Of particular interest here is the observation that for a particular process, the c-parameter in the GP distribution representing the asymptotic distribution of exceedances is the same as the c-parameter in the corresponding GEV distribution for the same process, explaining the use of the same symbol in the two cases. It is important to note, however, that this statement is valid only asymptotically. The practical significance of this is that, in general, the c-parameter estimated on the basis of a finite set of data rarely equals that of the correct asymptotic extreme value distribution. Also relevant in this context is the observation made by Fisher and Tippet (1928) that the best fit to a finite set of extreme value data from a stationary Gaussian process is provided by a Type III distribution and not by the asymptotically correct Type I distribution.

Return Periods

The return period R of a given wind speed, in years, is defined as the inverse of the probability that the specified wind speed will be exceeded in any one year. If λ denotes the mean crossing rate of the threshold u per year (i.e., the average number of data points above the threshold u per year), the return period R of the value of X corresponding to the level $x_R = u + y$ is given by the relation

$$R = \frac{1}{\lambda \text{Prob}(X > x_R)} = \frac{1}{\lambda \text{Prob}(Y > y)}. \tag{16.11}$$

Hence, it follows that

$$\text{Prob}(Y \leq y) = 1 - 1/(\lambda R). \tag{16.12}$$

Invoking Eq. (16.8) for $c \neq 0$ leads to the result

$$x_R = u - a[1 - (\lambda R)^c]/c. \tag{16.13}$$

Similarly, for $c = 0$, it is found that

$$x_R = u + a \ln(\lambda R), \tag{16.14}$$

where u is the threshold used in the estimation of c and a.

de Haan Estimators

Let n denote the total number of data points, while the number of observations above the threshold value u is denoted by k. The threshold u then represents the $(k+1)$th highest data point(s). An estimate for λ is $\hat{\lambda} = k/n_{yrs}$, where n_{yrs} denotes the length of the record in years. The highest, second highest, ..., kth highest, $(k+1)$th highest variates are denoted by $X_n^*, X_{n-1}^*, \ldots, X_{n-k+1}^*, X_{n-k}^* = u$, respectively.

The parameter estimators proposed by de Haan (1994) are based on the following two quantities:

$$H_{k,n} = \frac{1}{k}\sum_{i=0}^{k-1}\{\ln(X_{n-i}^*) - \ln(X_{n-k}^*)\} \tag{16.15}$$

and

$$H_{k,n}^{(2)} = \frac{1}{k}\sum_{i=0}^{k-1}\{\ln(X_{n-i}^*) - \ln(X_{n-k}^*)\}^2. \tag{16.16}$$

Estimators for a and c are then given by the relations

$$\hat{a} = \rho\, X_{n-k}^* H_{k,n} = \rho\, u H_{k,n} \tag{16.17}$$

and

$$\hat{c} = H_{k,n} + 1 - \frac{1}{2}\left\{1 - \frac{(H_{k,n})^2}{H_{k,n}^{(2)}}\right\}^{-1} \tag{16.18}$$

where $\rho = 1$ if $\hat{c} \geq 0$, while $\rho = 1 - \hat{c}$ if $\hat{c} < 0$.

Subject to general conditions on the underlying probability law, de Haan (1994) showed that $\hat{a} \to a$ and $\hat{c} \to c$ as $n \to \infty$ (in probability).

Closely related to the de Haan estimators are the Hill estimators. Their application to the problem of estimating extreme wind speeds was investigated by Naess and Clausen (1999). Their conclusion was that the Hill estimators lead to results that are quite similar to those provided by the de Haan estimators. Because the Hill estimators require considerably higher numerical efforts than the de Haan estimators, we have therefore chosen to exclude the Hill estimators from the present discussion. The interested reader is referred to Naess and Clausen (1999) for details.

Moment Estimators

In terms of the the mean value $\mathrm{E}(Y)$ and the standard deviation $s(Y)$ of the exceedance variate Y, it can be shown that (Hosking and Wallis, 1987)

$$a = \frac{1}{2}\mathrm{E}(Y)\{1 + [\mathrm{E}(Y)/s(Y)]^2\} \tag{16.19}$$

and

$$c = \frac{1}{2}\{1 - [\mathrm{E}(Y)/s(Y)]^2\}. \tag{16.20}$$

Hence, empirical estimates of the first two moments of Y provide estimates of a and c. We refer to the resulting estimators as the moment estimators for a and c.

Maximum Likelihood Estimators

The maximum likelihood estimators (MLEs) are often preferred due to their asymptotic efficiency. Let $y_1, \ldots, y_k$ denote the observed sample of exceedances above the threshold u from an observed sample of peak values $x_1, \ldots, x_n$. The log-likelihood function $\ell(a, c|y_1, \ldots, y_k)$ for the sample $y_1, \ldots, y_k$ is given by

$$\ell(a, c|y_1, \ldots, y_k) = -k \ln a - \left(\frac{1}{c} + 1\right) \sum_{i=1}^{k} \ln\left(1 + c\frac{y_i}{a}\right), \tag{16.21}$$

provided $(1 + cy_i/a) > 0$ for $i = 1, \ldots, k$. If $c = 0$, the log-likelihood assumes the form

$$\ell(a, 0|y_1, \ldots, y_k) = -k \ln a - \frac{1}{a} \sum_{i=1}^{k} y_i. \tag{16.22}$$

The MLE $\hat{a}$ and $\hat{c}$ are obtained by maximizing $\ell(a, c|y_1, \ldots, y_k)$ with respect to a and c. These values are found by numerical methods, except for the special case $c = 0$, for which a simple, closed-form solution exists for $\hat{a}$. It is given as $\hat{a} = \sum_{i=1}^{k} y_i/k$.

Estimating Confidence Intervals by Bootstrapping

The bootstrapping method is a statistical technique of fairly recent origin that can be used for estimating confidence intervals on statistical estimates on the basis of a limited sample (Efron and Tibshirani, 1993; Davison and Hinkley, 1997). It is based on resampling from a distribution determined by the available sample of data. Despite the fact that the name of the method alludes to lifting oneself up by the bootstraps (Baron von Munchausen), the method appears to be quite effective for the specific purpose of estimating confidence bands. For convenience, a brief discussion of some basic features of the bootstrapping method is provided here.

Assume that $\boldsymbol{u} = (u_1, u_2, \ldots, u_n)$ is a sample or vector consisting of n independent observations of a random variable U. There are in principle two available options for obtaining bootstrap estimates. One is the nonparametric approach, where a purely empirical distribution function (EDF) is established for U on the basis of the observed data by allocating a probability of $1/n$ to each of the observed data points. The other is the parametric bootstrap, which is obtained by assuming that U has a specified CDF $F_U(u; \theta) = \text{Prob}(U \le u)$, where θ denotes a vector of unknown parameters, which determine the CDF. These parameters are then estimated from the observed data $\boldsymbol{u}$, giving $\hat{\theta}$, and $F_U(u; \hat{\theta})$ is adopted as the CDF of U.

In this section on the POT method, only the nonparametric bootstrap is used. In the next section on the Gumbel method, the parametric bootstrap is applied. The goal of both methods is to estimate some statistical quantity R given by the unknown CDF on the basis of the observed sample $\boldsymbol{u}$. Let $\hat{R}$ denote the estimate of R based on the given sample. The nonparametric bootstrapping technique for estimating confidence intervals on the point estimator $\hat{R}$ is based on resampling (with replacement) from the EDF provided by the observed sample $\boldsymbol{u}$.

This is done as follows: the EDF gives rise to an empirical random variable U^*. l bootstrap samples $\boldsymbol{u}^*_j$, $j = 1, \ldots, l$, with n independent observations of U^* in each sample are now generated. Each sample $\boldsymbol{u}^*_j$ gives rise to an estimate R^*_j of R.

A simple estimator for confidence intervals on $\hat{R}$ is derived by calculating the sample standard deviation s_R^*:

$$s_R^* = \sqrt{\frac{1}{l-1}\sum_{j=1}^{l}(R_j^* - \bar{R}^*)^2}, \tag{16.23}$$

where $\bar{R}^* = (1/l)\sum_{j=1}^{l} R_j^*$. An approximate confidence interval at level $1-q$ is then obtained as

$$(\hat{R} - w_{q/2}s_R^*, \hat{R} + w_{q/2}s_R^*), \tag{16.24}$$

where $w_{q/2}$ denotes the $100(1-q/2)\%$ standard normal fractile. To get stable results for this estimator, usually 20-30 bootstrap samples are sufficient. To avoid making the assumption that the bootstrap estimates are normally distributed, the true distribution may be approximated by generating a large number of bootstrap samples, usually several thousand are needed, especially for small values of q. If l samples were generated, the R_j^* are rearranged in increasing order. A $100(1-q)\%$ confidence interval is then

$$(R_L^*, R_M^*) \tag{16.25}$$

where $L = [ql/2]$ and $M = [(1-q/2)l]$ ($[a]$ means the integer part of a). Such estimates may be further improved as described by Davison and Hinkley (1997). However, we do not enter such details here.

16.6.2 Gumbel Method

It was pointed out in Chapter 10 that for response processes relevant for marine structures, the appropriate extreme value distribution would almost always be the Gumbel distribution. Therefore, let us assume that this is indeed the case for the response process $X(t)$, which can be simulated by a suitable procedure. Now, assume that N independent response time histories, each of duration T, have been simulated for a given environmental condition. For the Gumbel method, the extreme response is then identified for each time series. These extreme value data are assumed to be Gumbel distributed, and plotting the obtained data set of extreme values using a Gumbel probability plot should then ideally result in a straight line. In practice, one cannot expect this to happen, but on the premise that the data follow a Gumbel distribution, a straight line can be fitted to the data. A popular method for fitting this straight line is the method of moments. That is, writing the Gumbel distribution of the extreme value $M(T)$ as

$$\text{Prob}(M(T) \leq \xi) = \exp\{-\exp(-a(\xi - b))\}, \tag{16.26}$$

it is known that the parameters $a > 0$ and b are related to the mean value m_M and standard deviation σ_M of $M(T)$ as follows: $b = m_M - 0.57722a^{-1}$ and $a = 1.28255/\sigma_M$ (Bury, 1975). The estimates of m_M and σ_M obtained from the available sample therefore provide estimates of a and b, which leads to the fitted Gumbel distribution by the moment method.

Typically, a specified fractile value of the fitted Gumbel distribution is then extracted and used in a design consideration. To be specific, let us assume that the requested fractile value is the $100(1-\alpha)\%$ fractile, where α is usually a small number, for example $\alpha = 0.1$. To quantify the uncertainty associated with the obtained $100(1-\alpha)\%$ fractile value based on a sample of size N, the 95% confidence interval of this value is often used. A good estimate of this confidence interval can be obtained by using a parametric bootstrapping method (Efron and Tibshirani, 1993; Davison and Hinkley, 1997), cf. Section 16.6.1. In our context, this simply means that the initial sample of N extreme values is assumed to have been generated from an underlying Gumbel distribution, whose parameters are, of course, unknown. If this Gumbel distribution had been known, it could have been used to generate a large number of (independent) samples of size N. For each sample, a new Gumbel distribution would be fitted and the corresponding $100(1-\alpha)\%$ fractile value identified. If the number of samples had been large enough, an accurate estimate of the 95% confidence interval on the $100(1-\alpha)\%$ fractile value based on a sample of size N could be found. Because the true parameter values of the underlying Gumbel distribution are unknown, they are replaced by the estimated values obtained from the initial sample. This fitted Gumbel distribution is then used as previously described to provide an approximate 95% confidence interval. Note that the assumption that the initial N extreme values are actually generated with good approximation from a Gumbel distribution cannot be easily verified, which is a drawback of this method. Compared with the POT method, the Gumbel method would also seem to use much less of the information available in the data. This may explain why the POT method has become increasingly popular over the past years, but the Gumbel method is still widely used in practice.

16.6.3 Naess-Gaidai Method

We know from Eq. (10.30) that a good approximation of the CDF of the extreme value $M(T)$ is obtained from the formula

$$\text{Prob}(M(T) \leq \xi) = \exp\{-\nu^+(\xi)T\}, \tag{16.27}$$

where $\nu^+(\xi)$ denotes the mean upcrossing rate of $X(t)$.

The method we are about to discuss here relies on this particular approximation. This implies that the mean upcrossing rate needs to be estimated from the simulated time series. Assuming the so-called ergodic mean value property, it is obtained that

$$\nu^+(\xi) = \lim_{T\to\infty} \frac{1}{T} n^+(\xi; T), \tag{16.28}$$

where $n^+(\xi; T)$ denotes a realization of $N^+(\xi; T)$; that is, $n^+(\xi; T)$ denotes the counted number of upcrossings during time T from a particular simulated time history. In practice, k time histories of a suitable length T_0, say, are provided by simulation. The appropriate ergodic mean value estimate of $\nu^+(\xi)$ is then

$$\hat{\nu}^+(\xi) = \frac{1}{kT_0} \sum_{j=1}^{k} n_j^+(\xi; T_0), \tag{16.29}$$

where $n_j^+(\xi; T_0)$ denotes the counted number of upcrossings of the level ξ from time history no. j. This will often be the chosen approach to the estimation of the mean upcrossing rate.

For a suitable number k, e.g., $k \geq 20-30$, and provided that T_0 is sufficiently large, a fair approximation of the 95% confidence interval ($CI_{0.95}$) for the value $\nu^+(\xi)$ can be obtained as

$$CI_{0.95}(\xi) = \left(\hat{\nu}^+(\xi) - 1.96\frac{\hat{s}(\xi)}{\sqrt{k}},\ \hat{\nu}^+(\xi) + 1.96\frac{\hat{s}(\xi)}{\sqrt{k}}\right), \tag{16.30}$$

where the empirical standard deviation $\hat{s}(\xi)$ is given as

$$\hat{s}(\xi)^2 = \frac{1}{k-1}\sum_{j=1}^{k}\left(\frac{n_j^+(\xi;T_0)}{T_0} - \hat{\nu}^+(\xi)\right)^2. \tag{16.31}$$

Note that k and T_0 may not necessarily be the number and length of the actually simulated response time series. Rather, they were chosen to optimize the estimate of Eq. (16.31). If, initially, $\tilde{k}$ time series of length $\tilde{T}$ are simulated, then $k = \tilde{k}k_0$ and $\tilde{T} = k_0 T_0$. That is, each initial time series of length $\tilde{T}$ was divided into k_0 time series of length T_0, assuming, of course, that $\tilde{T}$ is large enough to allow for this in an acceptable way. The consistency of the estimates obtained by Eq. (16.31) can be checked by the observation that $\mathrm{Var}[N^+(\xi;t)] = \nu^+(\xi)t$ when $N^+(\xi;t)$ becomes a Poisson random variable, which by assumption occurs for large values of ξ. This leads to the equation

$$s(\xi)^2 = \frac{1}{k}\mathrm{Var}\left[\sum_{j=1}^{k}\frac{N_j^+(\xi;T_0)}{T_0}\right] = \frac{\nu^+(\xi)}{T_0}, \tag{16.32}$$

where $\{N_1^+(\xi;T_0),\ldots,N_k^+(\xi;T_0)\}$ denotes a random sample with a possible outcome $\{n_1^+(\xi;T_0),\ldots,n_k^+(\xi;T_0)\}$. Hence, $\hat{s}(\xi)^2/k \approx \hat{\nu}^+(\xi)/(kT_0)$. Because this last relation is consistent with the adopted assumptions (for large ξ), it could have been used as the empirical estimate of the variance in the first place. It is also insensitive to the blocking of data discussed previously because $kT_0 = \tilde{k}\tilde{T}$. However, the accuracy of this approach may be poor for small to moderate values of ξ, where the Poisson assumption about the upcrossing events may fail. In contrast, the advantage of Eq. (16.31) is that it does not rely on any specific assumptions about the statistical distributions involved.

The idea underlying the development of the new approach (Naess and Gaidai, 2008) described here is based on the observation that for dynamic models relevant for marine structures, the mean ξ-upcrossing rate as a function of the level ξ is highly regular in a particular way. As is shown later, the mean upcrossing rate tail, say, for $\xi \geq \xi_0$, behaves similarly to $\exp\{-a(\xi-b)^c\}$ $(\xi \geq \xi_0)$, where $a > 0, b \leq \xi_0$, and $c > 0$ are suitable constants. Hence, as discussed in detail by Naess and Gaidai (2008), it may be assumed that

$$\nu^+(\xi) \approx q(\xi)\exp\{-a(\xi-b)^c\},\ \xi \geq \xi_0, \tag{16.33}$$

where the function $q(\xi)$ is slowly varying compared with the exponential function $\exp\{-a(\xi-b)^c\}$. Equation (16.33) can be rewritten as

$$\ln\left|\ln\left(\nu^+(\xi)/q(\xi)\right)\right| \approx c\ln(\xi-b) + \ln a,\ \xi \geq \xi_0. \tag{16.34}$$

It follows that by plotting $\ln\left|\ln\left(\nu^+(\xi)/q(\xi)\right)\right|$ versus $\ln(\xi-b)$, it is expected that an almost perfectly linear tail behavior will be obtained. Now, as it turns out, the function $q(\xi)$ can be largely considered as a constant q, say, for tail values of ξ. This suggests using a method for identifying the parameters q and b by optimizing the linear fit in the tail. When this is achieved, the corresponding values of a and c can then be extracted from the plot. This is discussed at some length in Naess and Gaidai (2008). A plot of $\ln \left| \ln \left(\nu^+(\xi)/q \right) \right|$ versus $\ln(\xi - b)$ for optimal parameters b and q will be referred to as an optimal transformed plot.

An alternative method for optimizing the fit to the data is described. Although it involves simultaneous optimization with respect to all four parameters, it is straightforward and quite robust. Specifically, it is suggested to do the optimization on the log level by minimizing the following mean square error function with respect to all four arguments:

$$F(q, a, b, c) = \sum_{j=1}^{N} w_j \left| \ln \hat{\nu}^+(\xi_j) - \ln q + a(\xi_j - b)^c \right|^2, \tag{16.35}$$

where w_j denotes a weight factor that puts more emphasis on the more reliable data points. The choice of weight factor is to some extent arbitrary. Previously, $w_j = \left(\ln C^+(\xi_j) - \ln C^-(\xi_j) \right)^{-\theta}$, with $\theta = 1$ and 2, was used in combination with a Levenberg-Marquardt least squares optimization method (Gill et al., 1981). This usually works well provided reasonable, initial values for the parameters are chosen. For the applications in this chapter, $\theta = 2$ has been used. Note that the form of w_j puts some restriction on the use of the data. Usually, there is a level ξ_j beyond which w_j is no longer defined. Hence, the summation in Eq. (16.35) has to stop before that happens. Also, the data should be preconditioned by establishing the tail marker ξ_0 in a sensible way.

It is realized by scrutinizing Eq. (16.35) that if b and c are fixed, the optimization problem reduces to a standard weighted linear regression problem. That is, with both b and c fixed, the optimal values of a and $\ln q$ are found using closed-form weighted linear regression formulas in terms of w_j, $y_j = \ln \hat{\nu}^+(\xi_j)$, and $x_j = (\xi_j - b)^c$. In that light, it can also be concluded that the best linear unbiased estimators (BLUEs) are obtained for $w_j = \sigma_{yj}^{-2}$, where $\sigma_{yj}^2 = \mathrm{Var}[y_j]$ (empirical) (Draper and Smith, 1998; Montgomery et al., 2002). Unfortunately, this is not a practical weight factor for this kind of problem because the summation in Eq. (16.35) would typically have to stop at undesirably small values of ξ_j.

It is obtained that the optimal values of a and q are given by the relations

$$a^*(b, c) = -\frac{\sum_{j=1}^{N} w_j (x_j - \overline{x})(y_j - \overline{y})}{\sum_{j=1}^{N} w_j (x_j - \overline{x})^2} \tag{16.36}$$

and

$$\ln q^*(b, c) = \overline{y} + a^*(b, c)\overline{x}, \tag{16.37}$$

where $\overline{x} = \sum_{j=1}^{N} w_j x_j / \sum_{j=1}^{N} w_j$, and $\overline{y} = \sum_{j=1}^{N} w_j y_j / \sum_{j=1}^{N} w_j$.

To calculate the final optimal set of parameters, the Levenberg-Marquardt method can now be used on the function $\tilde{F}(b, c) = F(q^*(b, c), a^*(b, c), b, c)$ to find the optimal values b^* and c^*, and Eqs. (16.36) and (16.37) are then used to calculate the corresponding a^* and q^*.

For estimation of a confidence interval for a predicted value of the upcrossing rate function provided by the optimal curve, the empirical confidence band is reanchored to the optimal fitted curve. That is, each of the empirical $CI_{0.95}$ is centered on the optimal curve, thus creating a 95% confidence band around this curve. Under the premise that the specified class of parametric curves fully describes the behavior of the upcrossing rate function in the tail, parametric curves are fitted as previously described to the boundaries of the reanchored confidence band. These curves are used to determine a confidence interval of the predicted extreme value, which can be interpreted as a 95% confidence level measure of the uncertainty of the prediction made (Naess and Gaidai, 2009). However, although the procedure discussed here provides a simple and quick measure of this uncertainty, to obtain a more precise estimate of a confidence interval, a bootstrapping method would be recommended, cf. Section 16.6.1.

As a final point, it was observed that the predicted value is not very sensitive to the choice of ξ_0, provided it is chosen with some care.

In offshore applications, it is quite common to assume that the observed extreme value response data do follow a Gumbel distribution, cf. Section 16.6.2. The problem with this approach is that classical extreme value theory cannot be used to decide to what extent the asymptotic distribution is actually valid for a given set of extreme value data. Note that the asymptotic Gumbel distribution given by Eq. (16.26) corresponds to an asymptotic upcrossing rate that is purely exponential, that is, with $c = 1$ in Eq. (16.33). Hence, by adopting a much more general class of functions, with the purely exponential functions as a subclass, to represent the upcrossing rate, as done here, the ability to capture subasymptotic behavior is greatly enhanced. By this, the necessity to adopt a strictly asymptotic extreme value distribution of questionable validity is avoided.

Note that the so-called Weibull method for extreme value prediction, which is based on the assumption that the local peak values follow a three parameter Weibull distribution is, in fact, basically a prejudiced version of the Naess-Gaidai method in the sense that the parameter q is a priori given the value 1.

In cases where the approximation implied by Eq. (16.27) may be questioned, the ACER method developed by Naess and Gaidai (2009) should be applied. This method is capable of providing, in a statistical sense, a representation of the exact extreme value distribution under study. Thus, the ACER method makes it possible to investigate the level of accuracy involved in adopting Eq. (16.27) for prediction purposes. A brief description of the ACER method is given in the next section.

16.6.4 The Average Conditional Exceedance Rate Method

This method, referred to as the ACER method for short, is derived for a discretely sampled response process. To fix ideas, consider a stochastic process $Z(t)$, which was observed over a time interval $(0, T)$. Assume that values $X_1, \ldots, X_N$, which were derived from the observed process, are allocated to the discrete times $t_1, \ldots, t_N$ in $(0, T)$. This could be simply the observed values of $Z(t)$ at each t_j, $j = 1, \ldots, N$, or it could be average values or peak values over smaller time intervals centered at the t_js. Our goal now is to accurately determine the distribution function of the extreme value $M_N = \max\{X_j; j = 1, \ldots, N\}$. Specifically, we want to estimate $P(\eta) = \text{Prob}(M_N \le \eta)$ accurately for large values of η.

From the definition of M_N, it follows that

$$P(\eta) = \text{Prob}(M_N \le \eta) = \text{Prob}\{X_1 \le \eta, \ldots, X_N \le \eta\}. \tag{16.38}$$

The joint distribution function on the rhs of this equation cannot in general be estimated directly from the data. However, by introducing a cascade of conditioning approximations $P_k(\eta)$ of $P(\eta)$, where $P_k(\eta) \to P(\eta)$ as k increases, this problem can be solved in practice in a very efficient manner (Naess and Gaidai, 2009). In can be shown that for $N >> 1$ and $k = 1, 2, \ldots,$

$$P_k(\eta) \approx \exp\Big(-\sum_{j=k}^{N} \alpha_{kj}(\eta)\Big), \tag{16.39}$$

where

$$\alpha_{kj}(\eta) = \text{Prob}\{X_j > \eta | X_{j-1} \le \eta, \ldots, X_{j-k+1} \le \eta\}. \tag{16.40}$$

For the empirical estimation of the requisite quantities in the $P_k(\eta)$, it is expedient to introduce the concept of average conditional exceedance rates (ACER) as follows:

$$\varepsilon_k(\eta) = \frac{1}{N-k+1} \sum_{j=k}^{N} \alpha_{kj}(\eta), \ k = 1, 2, \ldots. \tag{16.41}$$

In practice, there are typically two scenarios for the underlying process $Z(t)$. We may consider it to be a stationary process, or even an ergodic process. The other alternative is to view $Z(t)$ as a process that depends on certain parameters whose variation in time may be modeled as an ergodic process in its own right. For each set of values of the parameters, the premise is that $Z(t)$ can be modeled as an ergodic process. This would be the scenario that can be used to model long-term statistics (Naess, 1984).

For both scenarios, the empirical estimation of the ACER $\varepsilon_k(\eta)$ proceeds in a completely analogous way by counting the total number of favorable incidents, that is, exceedances conditional on the requisite number of preceding nonexceedances, for the total data time series, and then finally dividing by $N - k + 1 \approx N$. This can be shown for the long-term situation by using a similar analysis as in Naess (1984).

A few more details on the numerical estimation of $\varepsilon_k(\eta)$ for $k \ge 2$ are useful. We start by introducing the following random functions:

$$A_{kj}(\eta) = \mathbf{1}\{X_j > \eta, X_{j-1} \le \eta, \ldots, X_{j-k+1} \le \eta\},$$
$$j = k, \ldots, N, \ k = 2, 3, \ldots \tag{16.42}$$

and

$$B_{kj}(\eta) = \mathbf{1}\{X_{j-1} \le \eta, \ldots, X_{j-k+1} \le \eta\},$$
$$j = k, \ldots, N, \ k = 2, \ldots, \tag{16.43}$$

where $\mathbf{1}\{\mathcal{A}\}$ denotes the indicator function of some event $\mathcal{A}$. Then,

$$\alpha_{kj}(\eta) = \frac{\mathrm{E}[A_{kj}(\eta)]}{\mathrm{E}[B_{jk}(\eta)]}, \; j = k, \ldots, N, \; k = 2, \ldots, \tag{16.44}$$

where $\mathrm{E}[\cdot]$ denotes the expectation operator. Assuming an ergodic process, then obviously $\varepsilon_k(\eta) = \alpha_{kk}(\eta) = \ldots = \alpha_{kN}(\eta)$, and it may be assumed that for the time series at hand,

$$\varepsilon_k(\eta) = \lim_{N\to\infty} \frac{\sum_{j=k}^{N} a_{kj}(\eta)}{\sum_{j=k}^{N} b_{kj}(\eta)}, \tag{16.45}$$

where $a_{kj}(\eta)$ and $b_{kj}(\eta)$ are the realized values of $A_{kj}(\eta)$ and $B_{kj}(\eta)$, respectively, for the observed time series.

Clearly, $\lim_{\eta\to\infty} \sum_{j=k}^{N} b_{jk}(\eta) = N - k + 1 \approx N$. Hence, $\lim_{\eta\to\infty} \tilde{\varepsilon}_k(\eta)/\varepsilon_k(\eta) = 1$, where

$$\tilde{\varepsilon}_k(\eta) = \lim_{N\to\infty} \frac{\sum_{j=k}^{N} a_{kj}(\eta)}{N - k + 1}. \tag{16.46}$$

In the following, we use $\tilde{\varepsilon}_k(\eta)$ instead of $\varepsilon_k(\eta)$ for $k \geq 2$, whenever convenient. The advantage of using the modified ACER function $\tilde{\varepsilon}_k(\eta)$ for $k \geq 2$ is that it is easier to use for non-stationary or long-term statistics than $\varepsilon_k(\eta)$. Because our focus is on the values of the ACER at the extreme levels, we may use any function that provides correct estimates of the ACER function at the extreme levels.

For both stationary and nonstationary time series, the sample estimate of $\tilde{\varepsilon}_k(\eta)$ would be

$$\hat{\varepsilon}_k(\eta) = \frac{1}{R} \sum_{r=1}^{R} \hat{\varepsilon}_k^{(r)}(\eta), \tag{16.47}$$

where R is the number of realizations (samples), and

$$\hat{\varepsilon}_k^{(r)}(\eta) = \frac{\sum_{j=k}^{N} a_{kj}^{(r)}(\eta)}{N - k + 1}, \tag{16.48}$$

where the index (r) refers to realization no. r.

It is of interest to note what events are actually counted for the calculation of the various $\hat{\varepsilon}_k(\eta)$, $k \geq 2$. Let us start with $\hat{\varepsilon}_2(\eta)$. It follows from the definition of $\tilde{\varepsilon}_2(\eta)$ that $\tilde{\varepsilon}_2(\eta)(N-1)$ can be interpreted as the expected number of exceedances above the level η satisfying the condition that an exceedance is counted only if it is immediately preceded by a nonexceedance. A reinterpretation of this is that $\hat{\varepsilon}_2(\eta)(N-1)$ equals the average number of clumps of exceedances above η for the realizations considered, where a clump of exceedances is defined as a maximum number of consecutive exceedances above η. In general, $\hat{\varepsilon}_k(\eta)(N-1)$ then equals the average number of clumps of exceedances above η separated by at least $k-1$ nonexceedances. This means that all consecutive clumps of exceedances separated by less than $k-1$ nonexceedances will be joined to form one larger clump.

Now, let us look at the problem of estimating a confidence interval for $\varepsilon_k(\eta)$. The sample standard deviation $\hat{s}_k(\eta)$ can be estimated by the standard formula

$$\hat{s}_k(\eta)^2 = \frac{1}{R-1}\sum_{r=1}^{R}\left(\hat{\varepsilon}_k^{(r)}(\eta) - \hat{\varepsilon}_k(\eta)\right)^2. \tag{16.49}$$

Assuming that realizations are independent, for a suitable number R, e.g., $R \geq 20$, Eq. (16.49) leads to a fairly good approximation of the 95% confidence interval (CI) for the value $\varepsilon_k(\eta)$ given as

$$\mathrm{CI}(\eta) = \hat{\varepsilon}_k(\eta) \pm 1.96\hat{s}_k(\eta)/\sqrt{R}. \tag{16.50}$$

As for the Naess-Gaidai method, the approach to extreme value prediction presented here is linked to an assumption about the sampled time series to be used as a basis for prediction. This assumption derives from an underlying premise concerning the relevant asymptotic extreme value distribution, which is assumed here to be of Gumbel type, in agreement with common modeling assumptions and empirical data. The implication of this on the possible subasymptotic functional forms of $\varepsilon_k(\eta)$ cannot easily be decided. However, again using the asymptotic form as a guide, we assume that the behavior of the mean exceedance rate in the tail is dominated by a function of the form $\exp\{-a(\eta - b)^c\}$ $(\eta \geq \eta_0 \geq b)$, where a, b, and c are suitable constants, and η_0 is an appropriately chosen tail level. Hence, it will be assumed that

$$\varepsilon_k(\eta) \approx q_k(\eta)\exp\{-a_k(\eta - b_k)^{c_k}\}, \quad \eta \geq \eta_0, \tag{16.51}$$

where the function $q_k(\eta)$ is slowly varying compared with the exponential function $\exp\{-a_k(\eta - b_k)^{c_k}\}$ and a_k, b_k, and c_k are suitable constants, that will generally be dependent on k. Note that the value $q_k = c_k = 1$ corresponds to the assumed asymptotic form.

In general, $q_k(\eta)$ is not constant, but its variation in the tail region is usually sufficiently slow to allow for its replacement by a constant for a suitable choice of the tail marker η_0. A similar approach was successfully used for mean upcrossing rate estimation for extreme value analysis of the response processes related to different dynamic systems, cf. the previous discussion of the Naess-Gaidai method (Naess et al., 2007; Naess and Gaidai, 2008).

The optimal tail parameters may be obtained in an entirely similar manner as for the Naess-Gaidai method. For construction of a confidence interval for a predicted, deep tail extreme value given by a particular ACER function as provided by the fitted parametric curve, one can follow a procedure analogous to that described for the Naess-Gaidai method.

16.6.5 A Comparison of Methods

In this subsection, we compare the performance of the Naess-Gaidai method with that of the Gumbel method. This is done for the particular case of the horizontal deck response of a jacket structure installed on the Kvitebjørn field in the North Sea. For this kind of response process, the Naess-Gaidai approach is very accurate

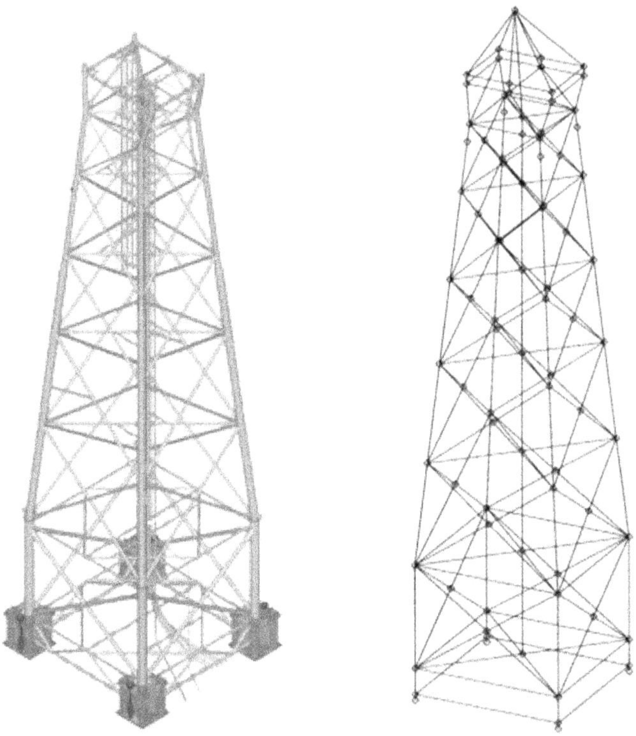

Figure 16.1. Left: Sketch of the Kvitebjørn platform with the superstructure removed. Right: Computer model of the Kvitebjørn platform (Karunakaran et al., 2001).

because a plot of the ACER functions shows that there are no dependence effects that need to be accounted for.

Figure 16.1 depicts the Kvitebjørn jacket platform with the superstructure removed together with the corresponding three-dimensional computer model used for the Monte Carlo simulations, see Naess et al. (2007) for details.

For the simulations discussed here, two long-crested sea states described by a JONSWAP wave spectrum as listed in Table 16.1 were used. Twenty independent response time histories, each of 3 hours' duration, were simulated for each sea state. For the Gumbel method, the extreme horizontal deck response in the wave direction is identified for each time series. These extreme value data are then assumed to be Gumbel distributed, and plotting each data set as a Gumbel probability plot results in Figs. 16.2 and 16.3. Specifically, the observed 3-hour extremes M_k are plotted versus $-\ln\big(\ln(21/k)\big)$, for $k = 1, \ldots, 20$. The fitted straight line in each figure, which represents the fitted Gumbel distribution, is based on the moment estimation method, cf. Section 16.6.2.

Table 16.1. *Representative sea states*

H_s (m)	T_p (s)
12.0	12.0
14.7	16.5

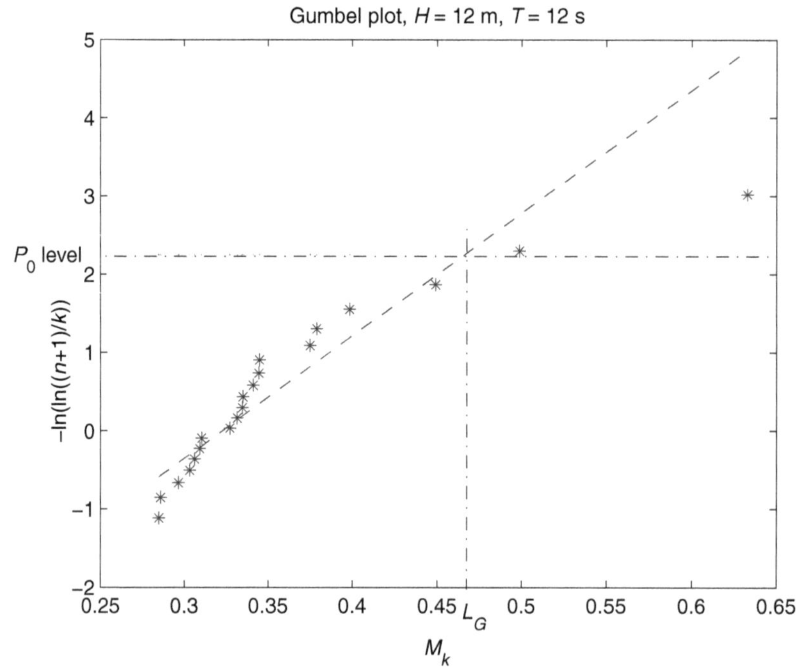

Figure 16.2. Empirical Gumbel plot of the 20 simulated 3-hour extremes for the sea state $H_s = 12$ m and $T_p = 12$ s together with the fitted Gumbel distribution (− − −).

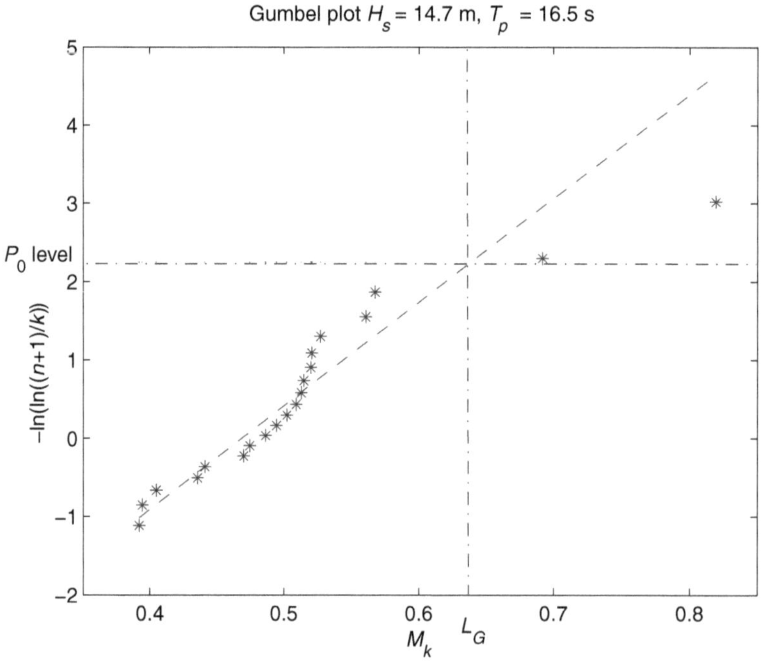

Figure 16.3. Empirical Gumbel plot of the 20 simulated 3-hour extremes for the sea state $H_s = 14.7$ m and $T_p = 16.5$ s together with the fitted Gumbel distribution (− − −).

Table 16.2. *90% Fractile values of fitted Gumbel distributions*

L_G (m)	$CI_{0.95}$
0.466	(0.396, 0.540)
0.632	(0.558, 0.726)

The 90% fractile value L_G of the fitted Gumbel distribution is identified and shown in each figure. Table 16.2 lists the obtained 90% fractile values. To quantify the uncertainty associated with the obtained 90% fractile value based on a sample of size 20, the 95% confidence interval ($CI_{0.95}$) of this value is used. A good estimate of this confidence interval can be obtained by using a parametric bootstrapping method (Efron and Tibshirani, 1993; Davison and Hinkley, 1997). In our context, this simply means that the initial sample of 20 extreme values is assumed to have been generated from an underlying Gumbel distribution, whose parameters are, of course, unknown. If this Gumbel distribution had been known, it could have been used to generate many samples of size 20. For each sample, a new Gumbel distribution would be fitted, and the corresponding 90% fractile value identified. If the number of samples had been large enough, an accurate estimate of the 95% confidence interval on the 90% fractile value based on a sample of size 20 could be found. Because the true parameter values of the underlying Gumbel distribution are unknown, they are replaced by the estimated values obtained from the initial sample. This fitted Gumbel distribution is then used as previously described to provide an approximate 95% confidence interval. Note that the assumption that the initial 20 extreme values are actually generated from a Gumbel distribution is quite accurate in this case, as discussed later.

Invoking the parametric bootstrap, the 95% confidence interval is estimated for each case based on 100,000 bootstrap samples. The obtained results are listed in Table 16.2. This way of estimating the 90% fractile value of the 3-hour extreme value distribution is referred to as the Gumbel method. The empirical PDFs obtained for the predicted 90% fractile values with the $CI_{0.95}$ indicated are shown in Figs. 16.4 and 16.5.

Let us now compare the results provided by the Gumbel method previously discussed, with the results obtained by the Naess-Gaidai method. Using the general transformation approach sketched in Section 16.6.3, $\ln|\ln(\nu^+(\xi)/q)|$ is plotted versus $\ln(\xi - b)$, and the results shown in Figs. 16.6 and 16.7 are obtained. To dimensionalize the response in Figs. 16.6 and 16.7, one should multiply the nondimensionalized response ξ/σ (given on the horizontal axis) with the response standard deviation σ. For the two sea states from Table 16.1, σ is 0.047 meters and 0.068 meters. Using a Levenberg-Marquardt least squares optimization method leads to the results shown in Figs. 16.8 and 16.9.

As shown in Figs. 16.10 and 16.11, when the mean upcrossing rate is plotted on a logarithmic scale, the tails are closely linear. This means that the associated extreme value distribution can be expected to be similar to a Gumbel distribution, which would correspond to exactly linear tails, cf. Eq. (10.30). Although it may not be obvious that the data plotted in Figs. 16.2 and 16.3 come from a distribution

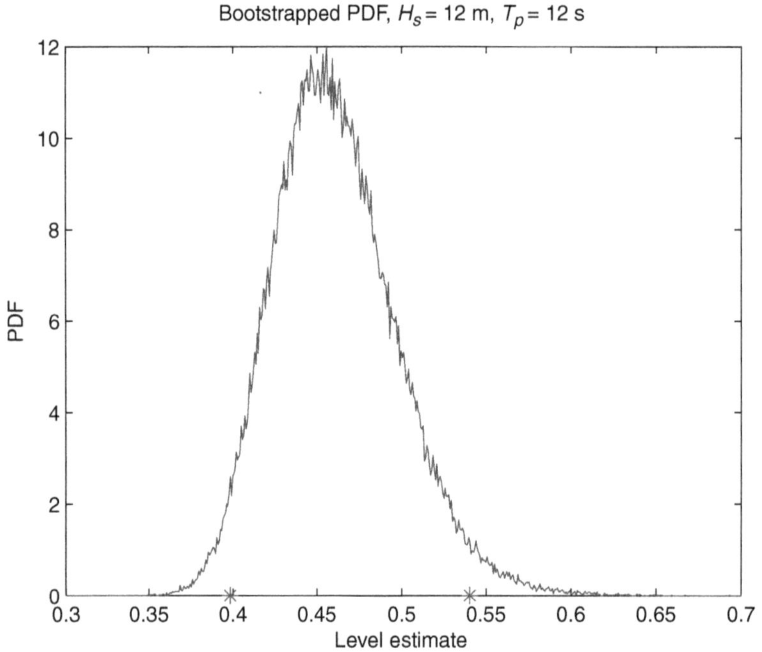

Figure 16.4. Empirical PDF of the predicted 90% fractile value based on sample of size 20 for the sea state with $H_s = 12$ m, $T_p = 12$ s. The $*$ indicates the limits of $CI_{0.95}$.

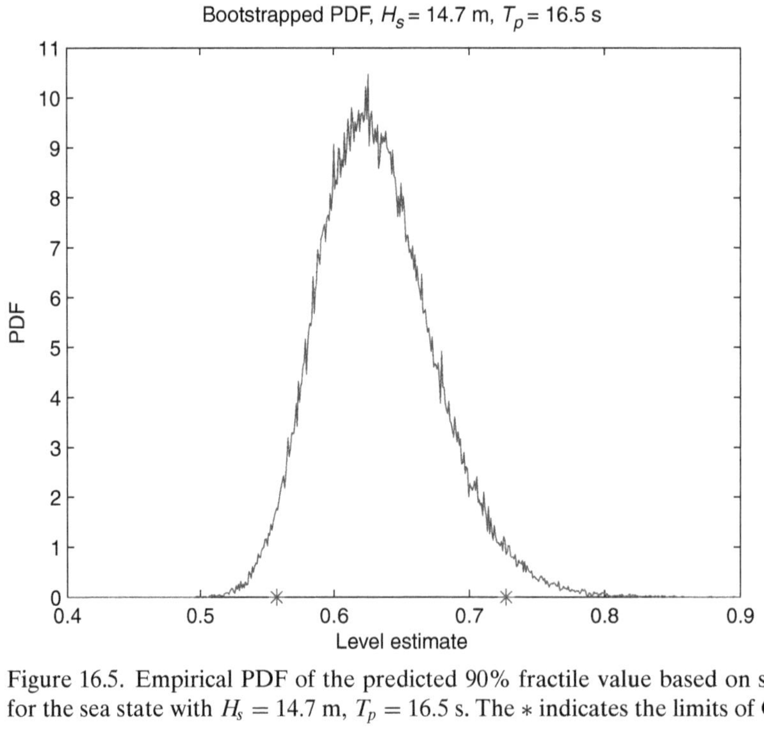

Figure 16.5. Empirical PDF of the predicted 90% fractile value based on sample of size 20 for the sea state with $H_s = 14.7$ m, $T_p = 16.5$ s. The $*$ indicates the limits of $CI_{0.95}$.

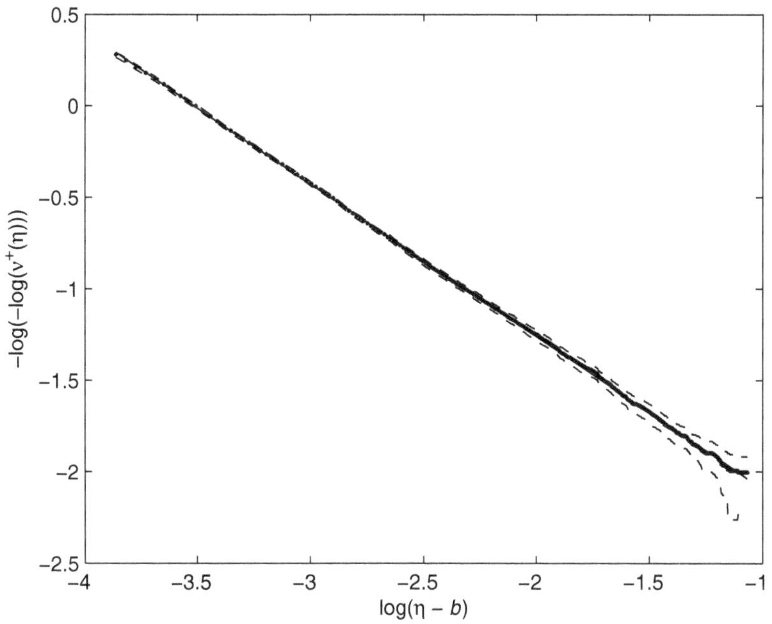

Figure 16.6. Optimal transformed plot of the empirical upcrossing rate with the 95% confidence band (−−) for the sea state with $H_s = 12$ m, $T_p = 12$ s, $\sigma = 0.047$ m. $*$: Monte Carlo; —— : linear fit, $q = 0.116$, $b = 0.03\sigma$.

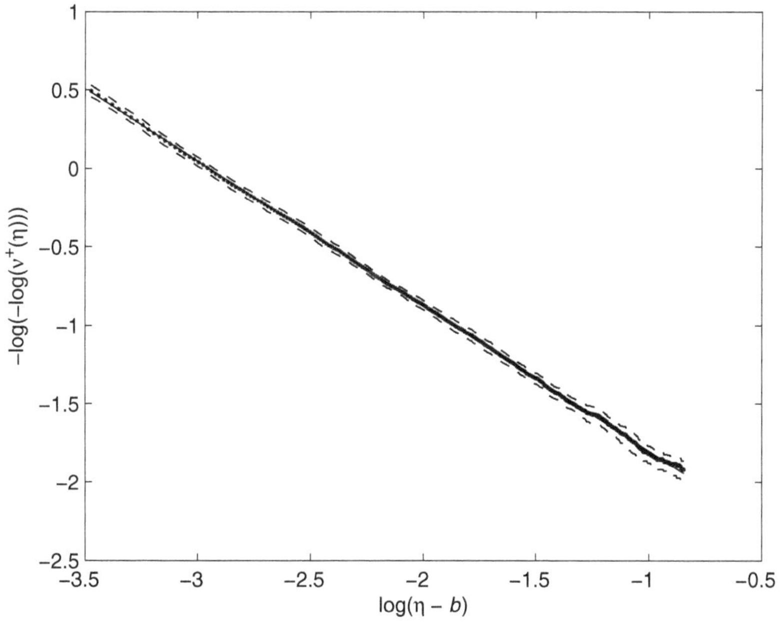

Figure 16.7. Optimal transformed plot of the empirical upcrossing rate with the 95% confidence band (−−) for the sea state with $H_s = 14.7$ m, $T_p = 16.5$ s, $\sigma = 0.068$ m. $*$: Monte Carlo; —— : linear fit, $q = 0.063$, $b = 0.07\sigma$.

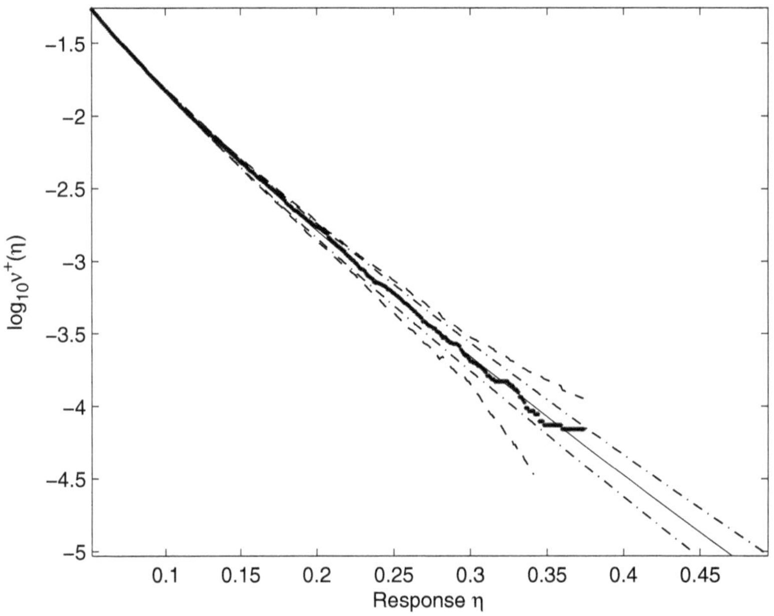

Figure 16.8. Log plot of the empirical and fitted upcrossing rate with the reanchored 95% empirical confidence band (−−) and fitted confidence band (− ·) for the sea state with H_s = 12 m, T_p = 12 s. 90% fractile estimate = 0.471, with $\mathrm{CI}_{0.95}$ = (0.449, 0.496).

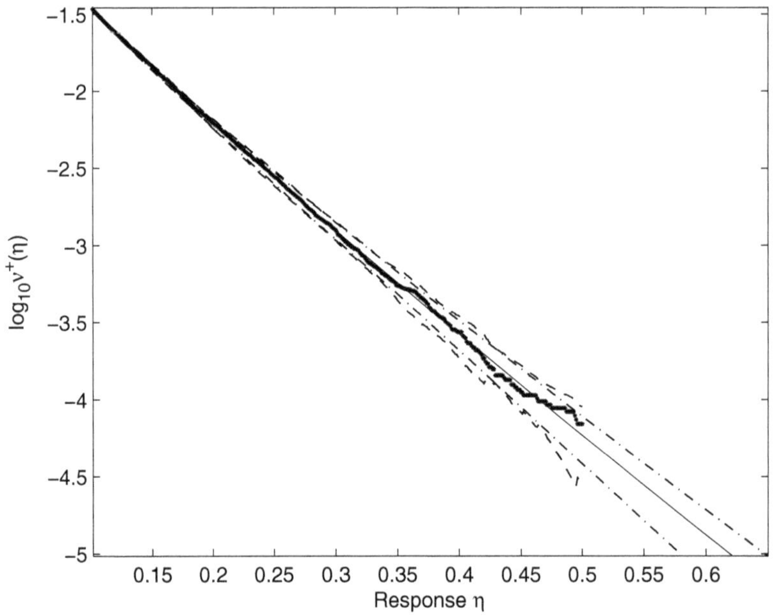

Figure 16.9. Log plot of the empirical and fitted upcrossing rate with the reanchored 95% empirical confidence band (−−) and fitted confidence band (− ·) for the sea state with H_s = 14.7 m, T_p = 16.5 s. 90% fractile estimate = 0.624, with 95% $\mathrm{CI}_{0.95}$ = (0.584, 0.653).

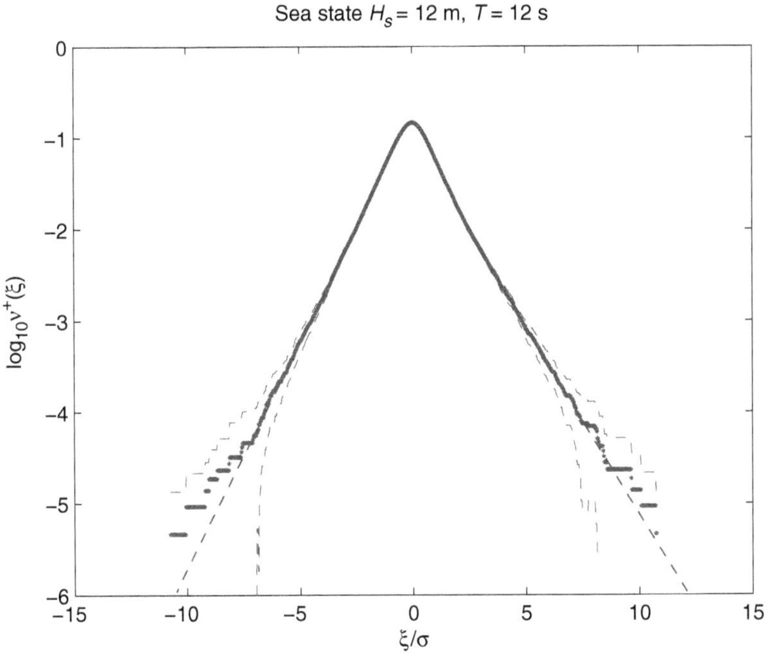

Figure 16.10. Mean upcrossing rate statistics along with 95% confidence bands (−−) for the sea state with $H_s = 12$ m, $T_p = 12$ s, $\sigma = 0.047$ m. $*$: Monte Carlo; − − − : linear fit.

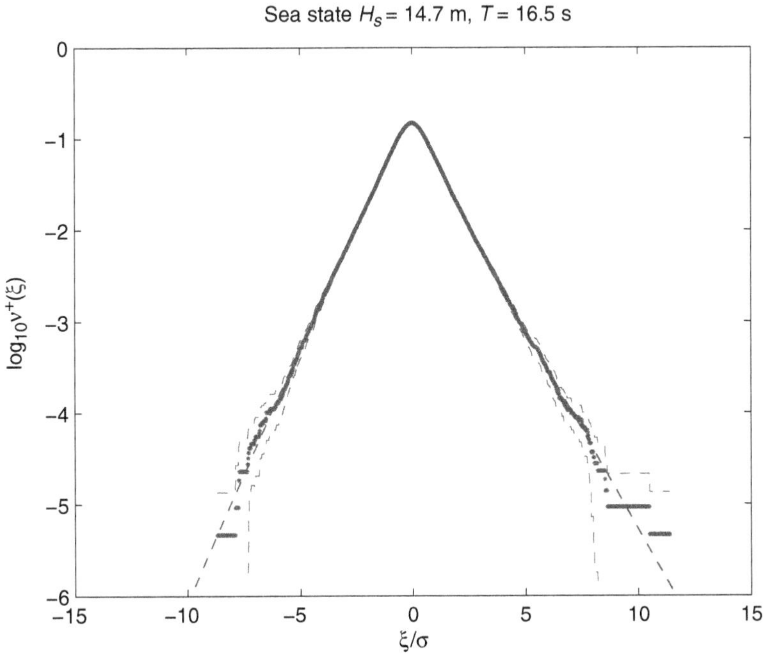

Figure 16.11. Mean upcrossing rate statistics along with 95% confidence bands (−−) for the sea state with $H_s = 14.7$ m, $T_p = 16.5$ s, $\sigma = 0.068$ m. $*$: Monte Carlo; − − − : linear fit.

Table 16.3. *90% Fractile values by the Naess-Gaidai method*

L_{CR} (m)	$\mathrm{CI}_{0.95}$
0.471	(0.449, 0.496)
0.624	(0.584, 0.653)

very close to a Gumbel distribution, the similar to linear exponential decay of the crossing rate strongly supports this assumption. This is yet another indication of the usefulness of the mean crossing rate function.

Aiming at $T = 3$-hour extreme response prediction, one needs upcrossing rates down to about 10^{-4}–10^{-6}. Accurate estimates based on direct Monte Carlo simulation down to this order are expensive in terms of CPU time for the considered structure. It is therefore convenient when accurately estimated upcrossing rates down to about 10^{-3} can be used as a basis for extrapolation down to the appropriate response level ξ (with $\nu^+(\xi) \approx 10^{-5}$), as illustrated in Figs. 16.8–16.9.

Returning now to the specific prediction of the 90% fractile of the 3-hour extreme value distribution, L_{CR}, Figs. 16.8 and 16.9 lead to the estimates listed in Table 16.3. The estimated 95% confidence intervals are given in Table 16.3, and indicated in the figures. They are significantly smaller than those obtained by the Gumbel method. The prediction accuracy is thus significantly higher for the proposed method. However, it is also observed that there is good agreement between the L_{CR}-values and the L_G-values, which is to be expected because the exact extreme value distribution is very close to a Gumbel distribution.

16.6.6 Combination of Multiple Stochastic Load Effects

A prominent problem in the design of structures subjected to random loads is to find methods for the combination of resulting load effects at high and extreme response levels. In codified design, this is usually implemented as linear combination rules of specified characteristic values of the individual load effects (Madsen et al., 1986; Melchers, 1999). For nonlinear dynamic structures, the precision level of such procedures would seem highly questionable. One of the reasons for adopting such simplified procedures is the complexity of the task to accurately predict the extreme value statistics of the combined load effects, even in the case of linear combinations. Over the years, several simplified procedures have been suggested for the linear combination of load effects, most notably the Ferry Borges-Castanheta method (Ditlevsen and Madsen, 1996), Turkstra's rule (Turkstra, 1970; Madsen et al., 1986), the load coincidence method (Wen, 1990; Melchers, 1999), the SRSS method (Wen and Pearce, 1981; Wen, 1990), and the point crossing approximation method (Larrabee and Cornell, 1981; Madsen et al., 1986). A main shortcoming of these combination procedures is that they apply only to the case of independent load effect components. A method for lifting this restriction from the point crossing approximation is proposed by Toro (1984). An effort to extend Turkstra's rule to dependent processes is described by Naess and Royset (2000). In this section, we illustrate the use of the Naess-Gaidai method for stochastic load effect combination problems.

The general formulation of the load effect combination problem to be studied here is the following:

$$H(t) = h[X_1(t), \ldots, X_N(t)], \tag{16.52}$$

where the stochastic load effect component processes $X_1(t), \ldots, X_N(t)$ are combined according to a specified deterministic function h to produce the load effect combination process $H(t)$. The component processes may derive from a vector solution process of a dynamic model for the structural response of an offshore platform to random waves. They may often be modeled as stationary stochastic processes, but that is not a requirement for the application of the Naess-Gaidai method.

The typical problem to answer concerning the load effect combination process $H(t)$ is to determine the probability of exceeding a critical threshold h_c during a specified time interval T. Let us call this the failure probability and denote it by $p_f = p_f(T)$. Defining $M(T) = \max\{H(t) : 0 \leq t \leq T\}$, it is realized that the goal is to calculate

$$p_f = 1 - \text{Prob}(M(T) \leq h_f). \tag{16.53}$$

In many practical applications, the structure of the process $H(t)$ is quite involved and the dimension N can be high. This makes a direct analytical approach virtually impossible. In such cases, Monte Carlo simulations of some sort would seem to be the most attractive way to provide estimates of the failure probability.

Two different load effect combination examples are used for illustration purposes: von Mises yielding stress and linear combination of non-Gaussian load effects. In both cases, the load components are correlated stochastic processes. A Newmark integration method was used to produce accurate response time series, cf. Chapter 15.

The load effect components $X_k(t)$ are modeled as stationary processes, being the response of Duffing-type systems to the same stationary Gaussian white noise excitation $W(t)$ with $\text{E}[W(t)W(t+\tau)] = \delta(\tau)$, where $\delta(\cdot)$ denotes the Dirac delta function. That is,

$$\ddot{X}_k + 2\zeta_k\omega_k\dot{X}_k + \omega_k^2(X_k + \varepsilon X_k^3) = W(t)/m_k, \tag{16.54}$$

with specific damping constants ζ_k and resonance frequencies $\omega_k = 2\pi/T_k$, and m_k represent masses, $k = 1, \ldots, N$. In this chapter N is taken to be 3.

First, the linear case, i.e., $\varepsilon = 0$ in Eq. (16.54), is considered. In this case, the (two-sided) PSD of the process $X_k(t)$ is given as $S_k(\omega) = |A_k(\omega)|^2$, where

$$A_k(\omega) = \frac{1}{\sqrt{2\pi}m_k(-\omega^2 + 2i\zeta_k\omega_k\omega + \omega_k^2)}, \tag{16.55}$$

with $k = 1, 2, 3$. The correlation coefficient $\rho_{kl} = E[X_k(t)X_l(t)]/\sigma_k\sigma_l$ between $X_k(t)$ and $X_l(t)$ is given by

$$\rho_{kl} = \int_{-\infty}^{\infty} A_k(\omega)A_l^*(\omega)d\omega/\sigma_k\sigma_l. \tag{16.56}$$

Here, $\sigma_k^2 = \text{Var}[X_k(t)] = \int_{-\infty}^{\infty} S_k(\omega)d\omega$ is the variance of $X_k(t)$, $k = 1, 2, 3$.

Table 16.4. *Model parameters*

k	ζ_k	T_k (s)	m
1	0.04	1.8	1
2	0.04	2.0	1
3	0.04	2.2	1

von Mises Stress Combination

Let σ_x, σ_y, σ_z be axial stresses and τ_{xy}, τ_{xz}, τ_{yz} be shear stresses in a structural element. According to the von Mises yield criterion (Madsen et al., 1986), yielding occurs if

$$(\sigma_x - \sigma_y)^2 + (\sigma_x - \sigma_z)^2 + (\sigma_y - \sigma_z)^2 + 6(\tau_{xy}^2 + \tau_{xz}^2 + \tau_{yz}^2) \geq 2\sigma_Y^2, \qquad (16.57)$$

where σ_Y is the yield stress. In many cases, in practice, several stress components are zero. In the analysis carried out here, only xy-plane stresses are encountered, meaning that $\sigma_z = \tau_{xz} = \tau_{yz} = 0$. According to the von Mises criterion, yielding then occurs if

$$(\sigma_x - \sigma_y)^2 + \sigma_x^2 + \sigma_y^2 + 6\tau_{xy}^2 \geq 2\sigma_Y^2. \qquad (16.58)$$

The load effect vector in three-dimensional space is introduced as

$$(X_1(t), X_2(t), X_3(t)) \propto (\sigma_x(t), \sigma_y(t), \tau_{xy}(t)), \qquad (16.59)$$

where the components $X_k(t)$ are determined by Eq. (16.54).

The von Mises criterion (Eq. (16.58)) states that yielding (failure) occurs when

$$H_{vM}(t) = \left((X_1 - X_2)^2 + X_1^2 + X_2^2 + 6X_3^2\right)^{1/2} \geq h_f, \qquad (16.60)$$

where $h_f \propto \sqrt{2}\sigma_Y$.

Because in the linear case ($\varepsilon = 0$ in Eq. (16.54)) the process $H_{vM}(t)^2$ becomes a quadratic expression in stationary Gaussian processes, the saddle point method described by Naess and Karlsen (2004) is applicable. Because this method is very accurate, it provides an opportunity to check the accuracy and efficiency of the Naess-Gaidai method. Table 16.4 lists the damping values and resonance periods used in Eq. (16.54) for the load effect processes $X_k(t)$.

For the linear case ($\varepsilon = 0$), it is straightforward to calculate the standard deviations and correlation coefficients for the X_k: $\sigma_1 = 0.38$, $\sigma_2 = 0.45$, $\sigma_3 = 0.52$, $\rho_{12} = 0.36$, $\rho_{13} = 0.14$, $\rho_{23} = 0.41$. There is significant correlation between some of the load effect components. For each example, 20 time series of length approximately 0.92 hours each were simulated. The total computation time for each example was less than a minute, including simulation time and optimization.

The log plot presented in Fig. 16.12 shows the optimized fitted parametric curve for the case of linear dynamics, i.e., $\varepsilon = 0$, together with the confidence band generated by the full range of allowed parametric curves that stay within the reanchored empirical 95% confidence band. For illustration purposes, we indicate the predicted value given by $\nu^+(\xi) = 10^{-6}$, which corresponds to the 99% fractile value of a 3-hour extreme value distribution. Predicted value = 6.838 = 10.033σ, with 95% CI = (6.603, 7.123).

In Fig. 16.13, it is demonstrated that when the fitted parametric curve shown in Fig. 16.12 is replotted as an optimal transformed plot, which yields a straight line,

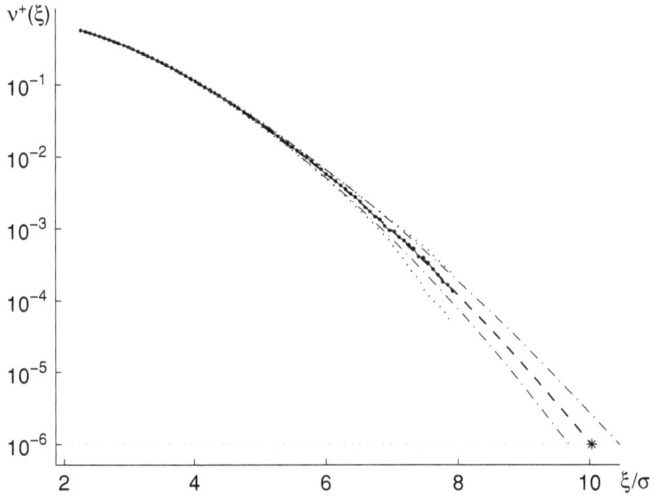

Figure 16.12. Log plot of the mean upcrossing rate of $H_{vM}(t)$: Monte Carlo (•), reanchored empirical 95% confidence band (- -), fitted curve (– –), fitted confidence band (– ·). Linear case: $\varepsilon = 0$. $a = 0.795$, $b = 0.996$, $c = 1.607$, $\ln q = -0.249$, $\sigma = 0.682$.

the empirical curve is also largely indistinguishable from a straight line, verifying the validity of our assumption about the representation of the upcrossing rate function. The results from saddle point calculations, which are practically exact, are also plotted in Fig. 16.13, and they are seen to agree very well with the extrapolated straight line results.

Figure 16.14 shows the optimized fitted parametric curve for the case of nonlinear dynamics, i.e., $\varepsilon = 1$, together with the confidence band as generated previously. Predicted 99% fractile value of 3-hour extreme = 4.290 = 8.038 σ, 95% confidence interval = (4.205, 4.358). Figure 16.15 presents the mean upcrossing rate function of

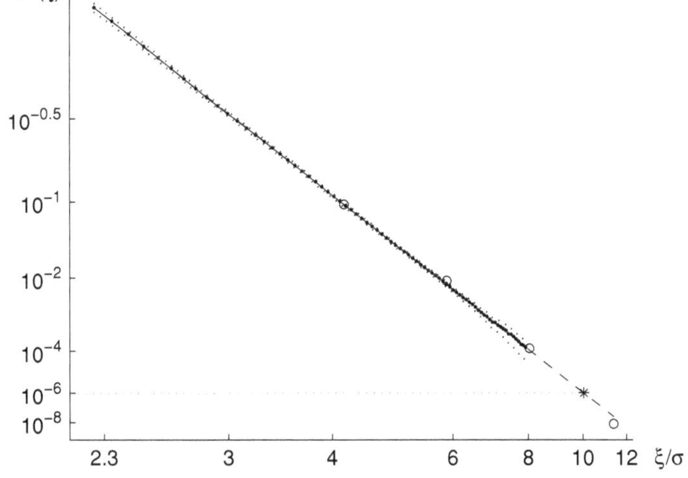

Figure 16.13. Optimal transformed plot of the mean upcrossing rate of $H_{vM}(t)$: Monte Carlo (•), empirical confidence band (- -), saddle point results (o), linear extrapolation (– –). Linear case: $\varepsilon = 0$. $b = 0.996$, $\ln q = -0.249$, $\sigma = 0.682$.

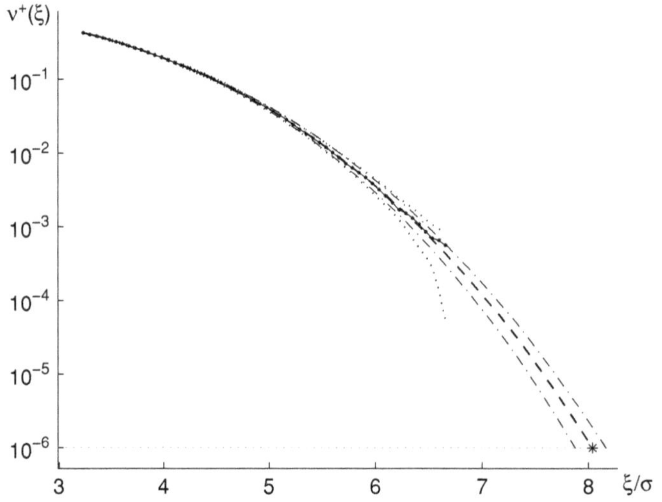

Figure 16.14. Log plot of the mean upcrossing rate of $H_{vM}(t)$: Monte Carlo (•), reanchored empirical 95% confidence band (- -), fitted curve (– –), fitted confidence band (– ·). Nonlinear case: $\varepsilon = 1.0$. $a = 0.082$, $b = -0.382$, $c = 3.334$, $\ln q = 0.124$, $\sigma = 0.534$.

$H_{vM}(t)$ plotted on the optimal transformed scale. It is observed that the assumption is again fully verified.

Linear Combination of Non-Gaussian Load Effects

To get a flexible model that also provides a convenient way of investigating the effect of statistical dependence between load components, we use an example from Naess and Royset (2000). The input load components are again assumed to be given by Eq. (16.54), with the same parameters as in the previous case study, cf. Table 16.4.

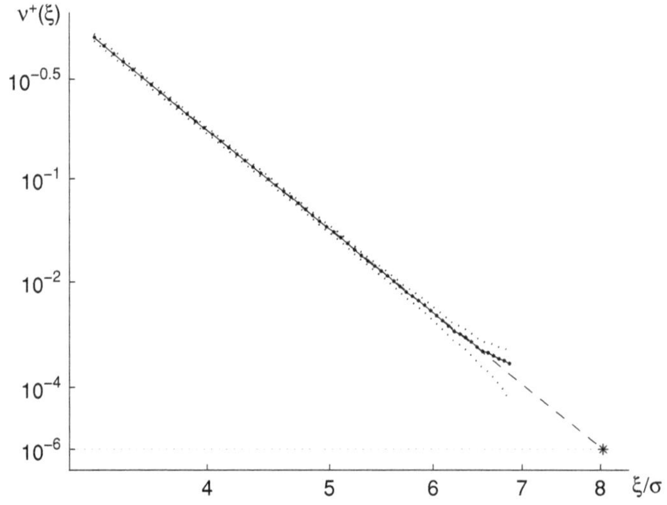

Figure 16.15. Optimal transformed plot of the mean upcrossing rate of $H_{vM}(t)$: Monte Carlo (•), empirical confidence band (- -), linear extrapolation (–). Nonlinear case: $\varepsilon = 1.0$. $b = -0.382$, $\ln q = 0.124$, $\sigma = 0.534$.

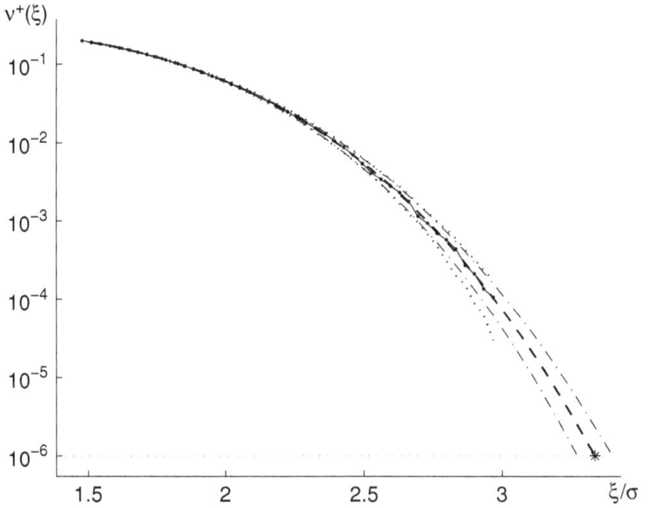

Figure 16.16. Log plot of the mean upcrossing rate of $H_{lc}(t)$: Monte Carlo (•), empirical confidence band (- -), fitted curve (– –), fitted confidence band (– ·). Linear case: $\varepsilon = 0$. $a = 0.147$, $b = 0.129$, $c = 3.742$, $\ln q = -1.103$, $\sigma = 1.026$, $\alpha = -0.5$.

As an example of non-Gaussian load effect component processes, we consider memoryless transformations of the input processes $X_j(t)$ provided by Eq. (16.54). In this section, we assume accordingly that

$$H_{lc}(t) = \sum_{k=1}^{3} Z_k(t), \tag{16.61}$$

with

$$Z_k(t) = X_k(t)|X_k(t)|^{\alpha}, \quad -1 < \alpha < 1. \tag{16.62}$$

We choose two α-values here, −0.5 and 0.5. The number of terms in the sum (16.61) is chosen to be three, but it can be any positive integer because it does not matter much for the Monte Carlo simulation.

Figure 16.16 shows the optimized fitted parametric curve for $\alpha = -0.5$ and the case of linear dynamics, i.e., $\varepsilon = 0$, together with the confidence band generated by the allowed parametric curves. Predicted 99% fractile value of 3 hour extreme $= 3.423 = 3.335\,\sigma$, 95% confidence interval $= (3.358, 3.488)$. In Fig. 16.17 it is demonstrated that when the fitted parametric curve shown in Fig. 16.16 is replotted as an optimal transformed plot, the empirical curve is largely indistinguishable from a straight line, supporting the assumption about the representation of the upcrossing rate function.

Figure 16.18 shows the optimized fitted parametric curve for $\alpha = 0.5$ and the case of linear dynamics, together with the confidence band generated by the allowed parametric curves. Predicted 99% fractile value of 3 hour extreme $= 5.150 = 7.687\,\sigma$, 95% confidence interval $= (4.951, 5.372)$. In Fig. 16.19 it is demonstrated that when the fitted parametric curve shown in Fig. 16.18 is replotted as an optimal transformed plot, the empirical curve is largely indistinguishable from a straight line, again supporting the assumption about the representation of the upcrossing rate function.

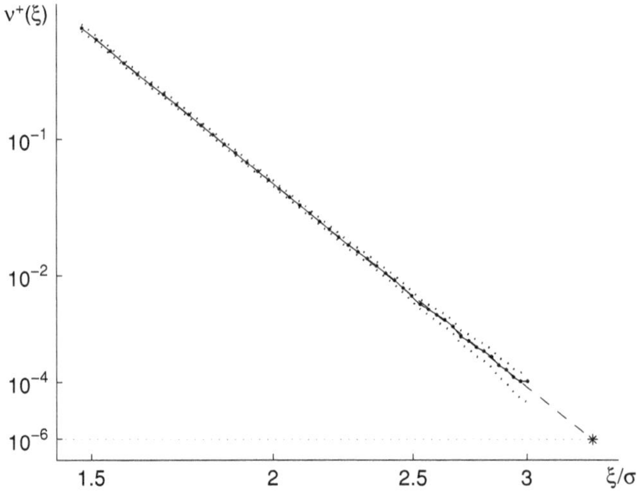

Figure 16.17. Optimal transformed plot of the mean upcrossing rate of $H_{lc}(t)$: Monte Carlo (•), confidence band (- -), linear extrapolation (–). Linear case: $\varepsilon = 0$. $b = 0.129$, $\ln q = -1.103$, $\sigma = 1.026$, $\alpha = -0.5$.

Figures 16.20 and 16.21 show the log plot and the optimal transformed plots for the case $\alpha = -0.5$, under nonlinear dynamics with $\varepsilon = 1$. Predicted 99% fractile value of the 3-hour extreme $= 3.043 = 3.175\,\sigma$, 95% confidence interval $= (2.996, 3.089)$. By way of comment to the rather high optimal value for c that was found in this particular case, it may be of interest to observe that the function $F(q^*(c), a^*(c), b^*(c), c)$ is almost constant for a range of values from about $c = 4$ to 7, where the calculations stopped. From the transformed optimal plot, the assumption is also still verified for this case.

Figures 16.22 and 16.23 show the log plot and the optimal transformed plots for the case $\alpha = 0.5$, under nonlinear dynamics with $\varepsilon = 1$. Predicted 99% fractile

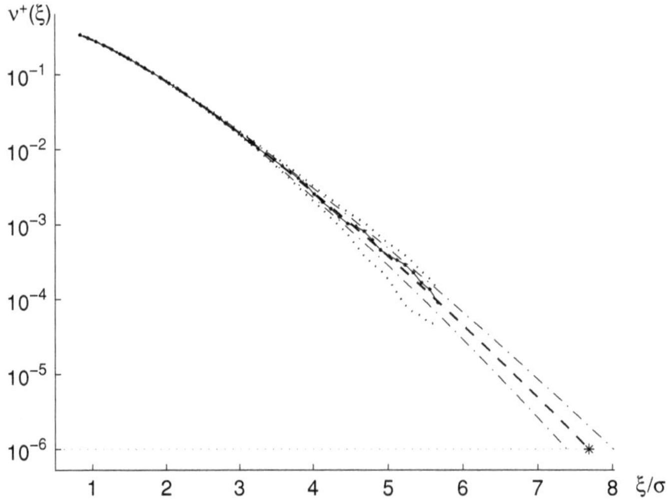

Figure 16.18. Log plot of the mean upcrossing rate of $H_{lc}(t)$: Monte Carlo (•), empirical confidence band (- -), fitted curve (– –), fitted confidence band (– ·). Linear case: $\varepsilon = 0$. $a = 1.771$, $b = 0.449$, $c = 1.280$, $\ln q = -0.980$, $\sigma = 0.670$, $\alpha = 0.5$.

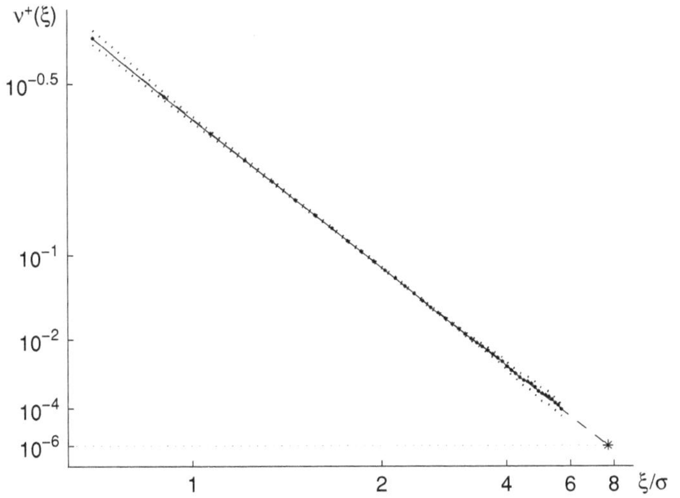

Figure 16.19. Optimal transformed plot of the mean upcrossing rate of $H_{lc}(t)$: Monte Carlo (•), confidence band (- -), linear extrapolation (–). Linear case: $\varepsilon = 0$. $b = 0.449$, $c = 1.280$, $\ln q = -0.980$, $\sigma = 0.670$, $\alpha = 0.5$.

value of 3-hour extreme $= 3.061 = 6.1465\ \sigma$, 95% confidence interval $= (2.968, 3.121)$. From the transformed optimal plot, it is again seen that the assumption is fully verified also for this case.

We also extract 20 maxima M_k, $k = 1, \ldots, 20$, one from each realization in order to view it on a Gumbel plot. The latter is a plot of M_k versus $-\ln(\ln \frac{20+1}{k})$, $k = 1, \ldots, 20$.

A 95% confidence interval for the Gumbel estimate of the response level L_{90} not being exceeded during time $T = 500 \max(T_1, T_2, T_3)$ (cf. Table 16.4), with

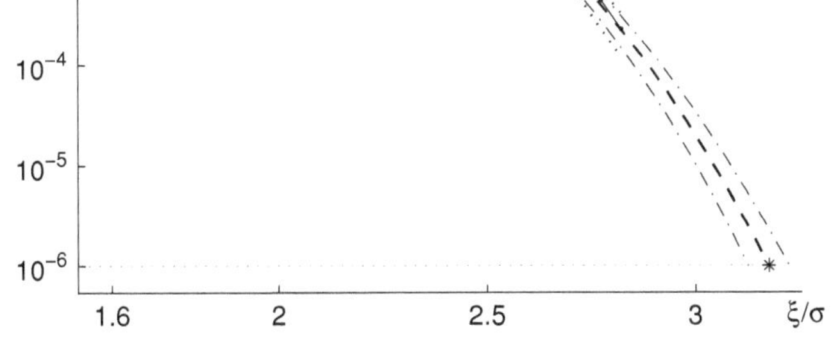

Figure 16.20. Log plot of the mean upcrossing rate of $H_{lc}(t)$: Monte Carlo (•), empirical confidence band (- -), fitted curve (– –), fitted confidence band (– ·). Nonlinear case: $\varepsilon = 1$. $a = 0.014$, $b = -0.456$, $c = 5.423$, $\ln q = -1.048$, $\sigma = 0.958$, $\alpha = -0.5$.

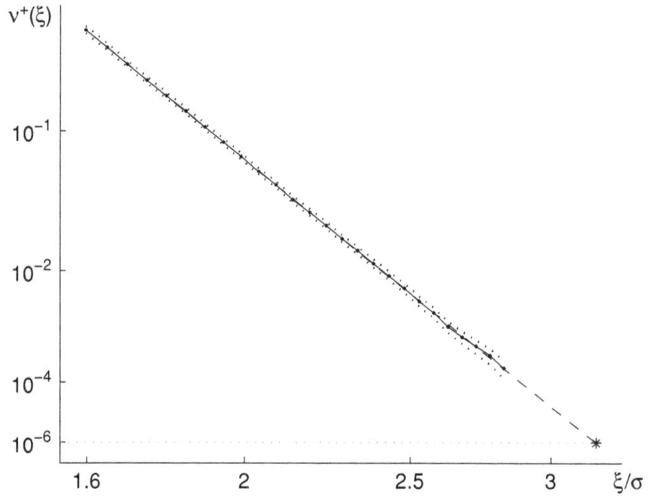

Figure 16.21. Optimal transformed plot of the mean upcrossing rate of $H_{lc}(t)$: Monte Carlo (•), confidence band (- -), linear extrapolation (–). Nonlinear case: $\varepsilon = 1$. $b = -0.456$, $\ln q = -1.048$, $\sigma = 0.958$, $\alpha = -0.5$.

probability 90% based on a sample of size 20 can be obtained by the MC technique and with parametric bootstrapping from the fitted Gumbel distribution (Davison and Hinkley, 1997). One hundred thousand bootstrap samples where used to estimate the PDF of the 90% fractile, and from this PDF, the desired confidence interval was extracted. Figures 16.24 and 16.25 present the L_{90} estimates by the Naess-Gaidai and Gumbel methods for the nonlinear system $\varepsilon = 1$, $\alpha = 0.5$. Fig. 16.26 presents the parametrically bootstrapped PDF of L_{90} from the Gumbel distribution.

The estimate of L_{90} from the Naess-Gaidai method is 3.70 with 95% confidence interval (3.65, 3.75). For the Gumbel method, the estimate of L_{90} is 3.55 with 95% confidence interval (3.23, 3.85).

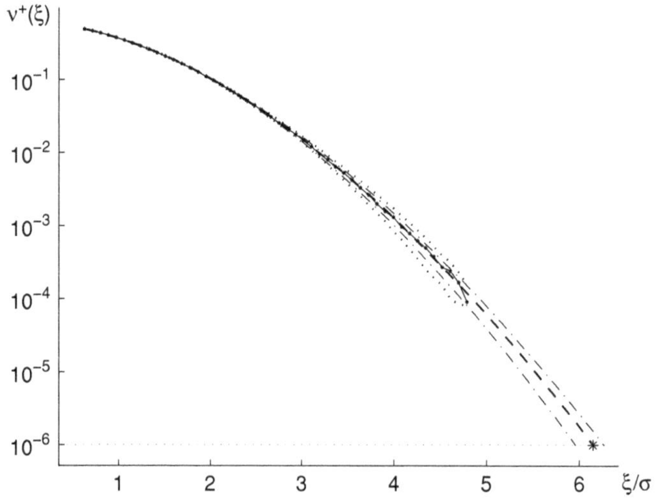

Figure 16.22. Log plot of the mean upcrossing rate of $H_{lc}(t)$: Monte Carlo (•); empirical confidence band (- -); fitted curve (– –); fitted confidence band (– ·). Nonlinear case: $\varepsilon = 1$. $a = 1.872$, $b = 0.030$, $c = 1.767$, $\ln q = -0.527$. $\sigma = 0.498$, $\alpha = 0.5$.

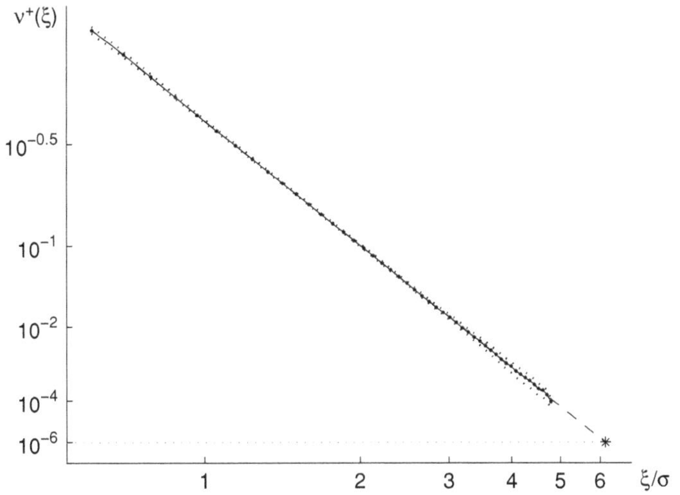

Figure 16.23. Optimal transformed plot of the mean upcrossing rate of $H_{lc}(t)$: Monte Carlo (•), confidence band (- -), linear extrapolation (–). Nonlinear case: $\varepsilon = 1$. $b = 0.030$, $\ln q = -0.527$. $\sigma = 0.498$, $\alpha = 0.5$.

16.6.7 Total Surge Response of a TLP

We now return to the problem of predicting the total surge response of a TLP structure that is briefly discussed in Section 12.5. With the tools developed in this chapter, we are in a position to solve this problem in a more satisfactory manner. The presented material is largely taken from Naess et al. (2007).

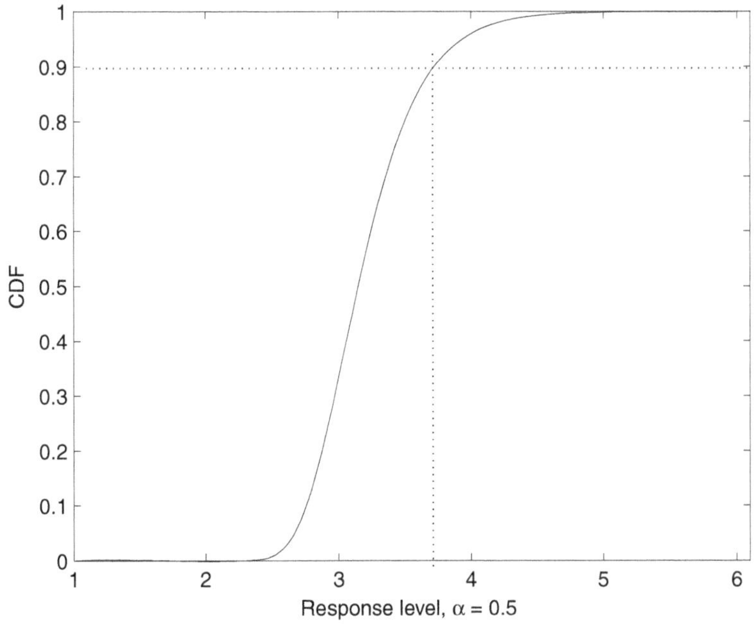

Figure 16.24. CDF by the Naess-Gaidai method; nonlinear system, $\varepsilon = 1$, $\alpha = 0.5$.

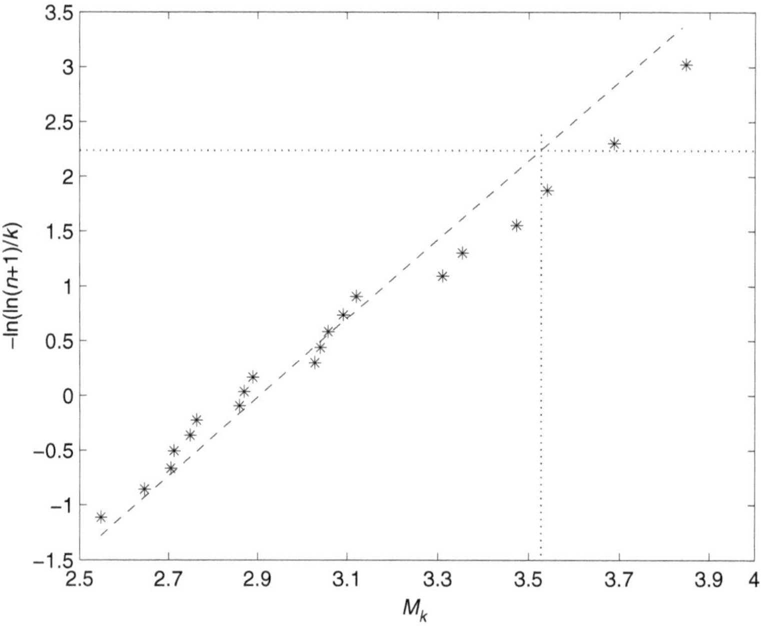

Figure 16.25. Gumbel plot of 20 maxima; nonlinear system, $\varepsilon = 1$, $\alpha = 0.5$.

The equations of motion for a floating, rigid-body TLP structure subjected to environmental forces such as wind, waves and current would generally be written as

$$\boldsymbol{M}\ddot{\boldsymbol{Z}}(t) + \boldsymbol{H}(\boldsymbol{Z}(t), \dot{\boldsymbol{Z}}(t), t) = \boldsymbol{F}(t). \tag{16.63}$$

Here, $\boldsymbol{M}$ denotes a generalized 6×6 mass matrix, $\boldsymbol{Z} = (Z_1, \ldots, Z_6)^T$ = the structure's response vector, while $\boldsymbol{H}$ denotes a nonlinear vector function. $\boldsymbol{F}(t)$ denotes

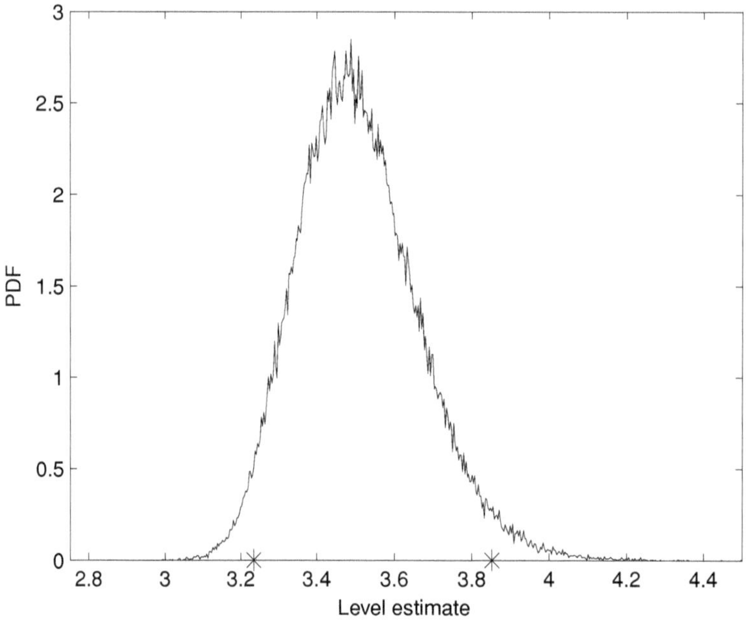

Figure 16.26. Bootstrapped PDF; nonlinear system, $\varepsilon = 1$, $\alpha = 0.5$.

a stochastic loading process, which in general also depends on the response of the structure. This point is further discussed later.

Because the main purpose of this section is to illustrate the versatility and accuracy of the proposed method, we chose to discuss a simplified SDOF model for the surge response of the TLP in random waves. Except for interaction effects between different motion modes, the SDOF model allows for the introduction of most of the relevant nonlinear effects that may influence the response of the TLP. Hence, the following SDOF equation of motion is studied

$$M\ddot{Z}(t) + H(Z(t), \dot{Z}(t), t) = F(t), \tag{16.64}$$

where $Z = Z(t)$ denotes the surge response of the TLP; M is the mass of the platform, including added mass; and H is a nonlinear function to be specified. As discussed in Section 11.3.1, the hydrodynamic loading process $F(t)$ is assumed to consist of two components: a linear, first-order (wave frequency) term $F_1(t)$, and a nonlinear, slowly varying, second-order term $F_2(t)$.

To set up the proper dynamic model for the surge response $Z(t)$, it is necessary to take into account the fact that hydrodynamic loading on a floating body depends on the motions of the body. In case of the slow drift motions of the TLP, it is of some importance to take into account the dependence of the slow drift force $F_2(t)$ on the slowly varying surge velocity $\dot{Z}_2(t)$. Because we have a nonlinear dynamic model, we have to introduce a definition of the slow drift response $Z_2(t)$. A suitable definition would seem to be the following: The slow-drift response $Z_2(t)$ is obtained from the total response $Z(t)$ by a low-pass filter that removes all wave frequency components. In practice, this may be achieved by a running mean operator used iteratively.

To account for the dependence of the slow drift force on the slowly varying velocity, it is appropriate to write $F_2(t, \dot{Z}_2(t))$ rather than $F_2(t)$. However, $F_2(t, \dot{Z}_2(t))$ is not directly available to us, but only $F_2(t) \equiv F_2(t, 0)$ given by Eq. (11.28). Because, in the context of slow drift motions, $\dot{Z}_2(t)$ is small, the following approximation is adopted:

$$F_2(t, \dot{Z}_2(t)) \approx F_2(t, 0) + \frac{\partial F_2(t, 0)}{\partial \dot{Z}_2} \dot{Z}_2(t). \tag{16.65}$$

It is shown by Naess and Johnsen (1993) that for a TLP structure $\frac{\partial F_2(t,0)}{\partial \dot{Z}_2} \approx -cF_2(t, 0) \equiv -cF_2(t)$ for a suitable constant $c > 0$ may to some extent serve as a useful approximation to capture qualitatively the time-variant damping effect, which is the result of the expansion in Eq. (16.65).

The first dynamic model adopted for $Z(t)$ is now the following,

$$\tilde{M}\ddot{Z} + C\dot{Z} + K(Z + \varepsilon Z^3) = F_1(t) + F_2(t, \dot{Z}_2) \approx F(t) - cF_2(t)\dot{Z}_2. \tag{16.66}$$

Here, $\tilde{M} = M + \tilde{m}$, where $\tilde{m}$ is an appropriately chosen (constant) added mass. C, K, and ε are suitably chosen positive constants. This equation is rewritten in the equivalent form

$$\ddot{Z} + 2\omega_e\zeta\dot{Z} + 2\omega_e\tilde{c}F_2(t)\dot{Z}_2 + \omega_e^2(Z + \varepsilon Z^3) = \frac{F(t)}{\tilde{M}}, \tag{16.67}$$

where $\omega_e^2 = K/\tilde{M}$, $\zeta = C/(2\omega_e\tilde{M})$, and $\tilde{c} = c/(2\omega_e\tilde{M})$.

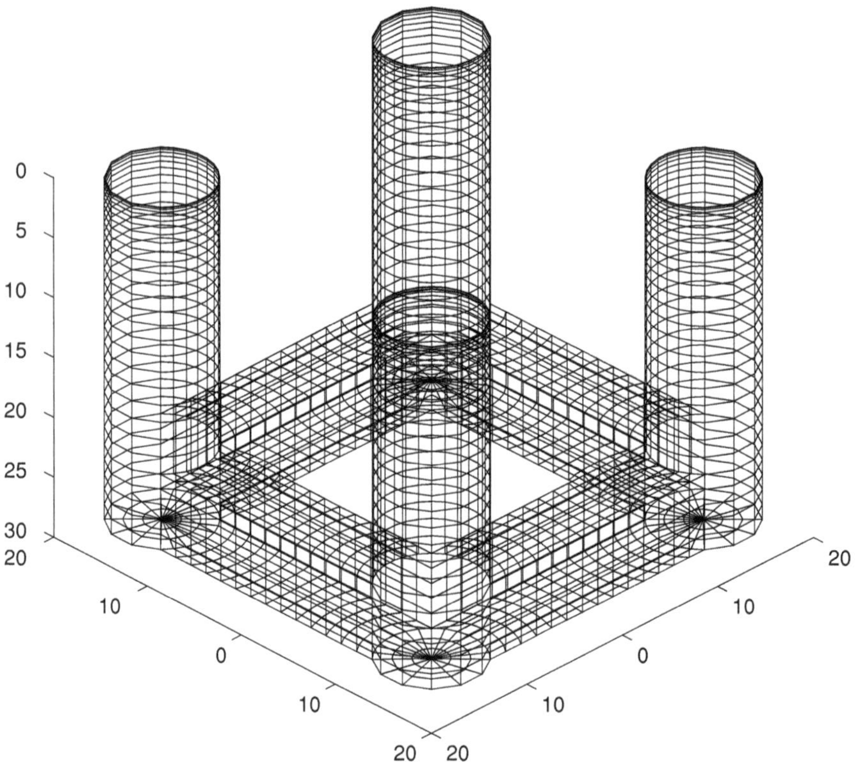

Figure 16.27. Sketch of submerged part of TLP. Units in meters

Thus, the dynamic system is nonlinear with a Duffing-type hardening stiffness nonlinearity and time-varying damping. Because for the TLP the relative damping coefficient ζ is usually small, the contribution from the time-varying term is nonnegligible, especially for severe seas for which the slow drift response is significant. The third-order term in the restoring force, generally referred to as the *set down* effect, is caused by the fact that the tethers will induce the TLP to act like an inverted pendulum. Note that the set down effect will also have an influence on the hydrodynamic loading process, which depends not only on $\dot{Z}$, but also on Z. Even if this dependence could have been taken into account, it was neglected here because it is of minor importance.

For the numerical simulations, a particular model of a TLP is considered, and the corresponding LTF and QTF are computed using the second-order diffraction program (WAMIT, 2008). For simplicity, unidirectional seas are used, meaning that the directional argument β is skipped. This simplification should have no effect on the conclusions based on the comparison of accuracy. The combined first-order and second-order slowly varying surge deck motion is studied applying the single-degree-of-freedom model. The TLP particulars are detailed in Table 16.5, and the subsurface part of the structure is shown in Fig. 16.27.

The values in Table 16.5 are used to obtain the second-order response. This means that for the second-order response, we used a simplified version of Eq. (16.66), where mass $\tilde{M}$, stiffness K, and damping coefficient C are frequency independent,

Table 16.5. *Particulars of the TLP*

Column diameter D (m)	10.0
Eigenperiod surge T_e (s)	128.8
Relative damping ζ	0.05
Total mass (incl. added mass) $\tilde{M}$ (kg)	$1.5 \cdot 10^7$

which is a good approximation for the slow drift motion. The time-invariant damping part ζ is considered to be 5%.

Two versions of Eq. (16.67) are used. The first version is a linear, time-invariant model obtained by putting $\tilde{c} = \varepsilon = 0$. The second version is the fully nonlinear model where the parameter $\tilde{c}$ in Eq. (16.67) is chosen such that $\mathrm{Var}[2\omega_e \tilde{c} F_2(t) \dot{Z}_2(t)]$ is about 10% of $\mathrm{Var}[2\omega_e \zeta \dot{Z}(t)]$. The parameter ε is estimated from the condition that $0.2Z(t) \geq \varepsilon Z^3(t)$ when $Z(t) \leq 6\sigma_Z$, i.e., even in the extreme response region, stiffness hardening contributes not more than 20% relative to the linear part for severe sea, which lead to $\varepsilon = 1.36 \cdot 10^{-4}$. Finally, the following approximate values were found: $\tilde{c} = 30/(\tilde{M}g)$, $g = 9.81 \mathrm{m/s}^2$ for moderate seas, and $\tilde{c} = 90/(\tilde{M}g)$ for severe seas. The adopted parameter values are largely arbitrary, but the choices made seem to provide a reasonable model for the chosen TLP structure.

To get an accurate representation of the response process, there is a specific requirement that must be observed. Because the damping ratio is only 5%, the frequency resolution $\Delta\omega$ must secure a sufficient number of frequency values over the resonance peak. This will ensure that the second-order, difference-frequency response component captures the TLP surge dynamics with sufficient accuracy. It is commonly required that there are at least 5 discrete frequencies over the frequency range where $|\hat{L}(\omega)|^2$ is equal to or higher than half of the resonance peak height $\max(|\hat{L}(\omega)|^2)$, where

$$\hat{L}(\omega) = \left(-\omega^2 + 2i\zeta\omega_e\omega + \omega_e^2 \right)^{-1}. \tag{16.68}$$

For the surge force QTF $\hat{K}_2(\omega, \omega')$, a suitable initial frequency grid must be chosen for which the values of the force QTF are calculated. The calculation of the force QTF is generally the most time-consuming part of the numerical analysis. Therefore, the initial grid is usually rather coarse to avoid excessive computer time. For the calculations at hand, the discrete frequency range was the following: $\omega_1 = 2\pi/30.0, \ldots, \omega_n = 2\pi/4.0$ (rad/s), $n = 30$. This necessitates the use of an interpolation procedure to be able to provide values of the QTF on a much finer grid than the initial one to comply with the requirement of sufficient frequency resolution to capture the dynamics of slow drift motion. In this chapter cubic spline interpolation is used. In the particular case considered here, the resolution requirement led to the choice $\Delta\omega = 0.0018$ rad/s and $L = 760$ interpolated discrete frequencies.

Each random stationary sea state is specified by a JONSWAP spectrum, which is given as follows:

$$S_\eta(\omega) = \frac{\alpha g^2}{\omega^5} \exp\left\{ -\frac{5}{4}\left(\frac{\omega_p}{\omega}\right)^4 + \ln\gamma \exp\left[-\frac{1}{2\sigma^2}\left(\frac{\omega}{\omega_p} - 1\right)^2 \right] \right\} \tag{16.69}$$

Table 16.6. *Representative sea states, along with response standard deviations and linear and nonlinear TLP*

H_s (m)	T_p (s)	σ_Z (m), lin.	σ_Z (m), nonlin.
10.0	11	9.3	8.2
15.0	17	6.4	6.0

where $g = 9.81$ ms^{-2}, ω_p denotes the peak frequency in rad/s, and γ and σ are parameters affecting the spectral shape. $\sigma = 0.07$ when $\omega \leq \omega_p$, and $\sigma = 0.09$ when $\omega > \omega_p$. The parameter γ is chosen to be equal to 3.3. The parameter α is determined from the following empirical relationship:

$$\alpha = 5.06\left(\frac{H_s}{T_p^2}\right)^2 (1 - 0.287 \ln \gamma), \tag{16.70}$$

where H_s denotes the significant wave height and $T_p = 2\pi/\omega_p$ is the spectral peak wave period. For this analysis, two short-term JONSWAP sea states were used - moderate and severe. Table 16.6 presents the sea state parameters, along with the corresponding response standard deviations.

Figure 16.28 shows the LTF for the wave exciting force amplitude, while Fig. 16.29 depicts the spline interpolated QTF. Because the QTF is complex valued, only its absolute value is plotted.

For the two chosen sea states, Figs. 16.30 to 16.33 present the corresponding response tail crossing rates obtained by Monte Carlo simulation for the linear system given by putting $\varepsilon = 0$ and $\tilde{c} = 0$ in Eq. (16.67). Figures 16.30 and 16.32 show

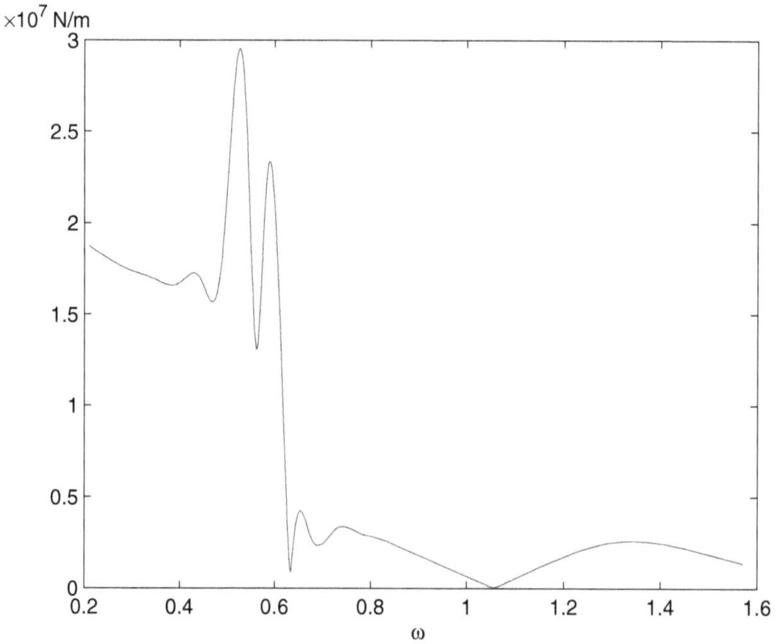

Figure 16.28. Wave exciting force amplitude, surge LTF.

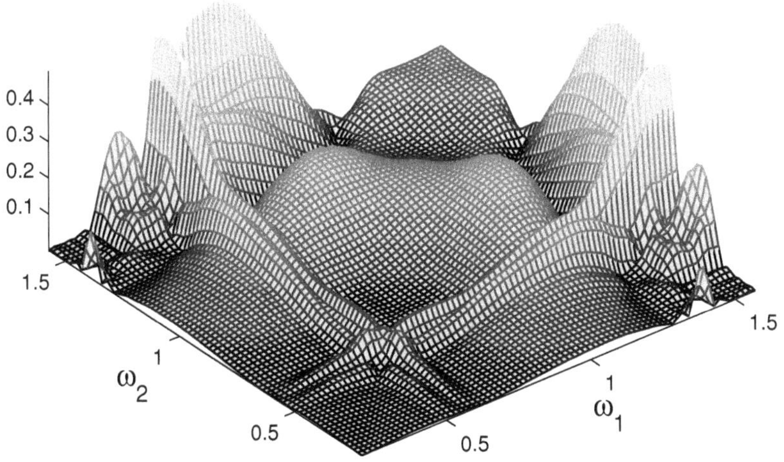

Figure 16.29. Wave exciting force amplitude, surge QTF.

the results obtained from 1,000 realizations, each of duration 1 hour. The crossing rate plots are done on the transformed scale, see Eq. (16.34). Extreme response prediction based on Eq. (16.27) will typically involve crossing rates of the orders $10^{-6} - 10^{-7}$, but to illustrate the achieved accuracy the extrapolated results at the crossing rate level 10^{-10} are highlighted. Figures 16.30 and 16.32 also show the highly accurate results obtained by using a saddle point integration technique (Naess et al., 2006). These results cannot be distinguished from those obtained by linear extrapolation of the mean upcrossing rate function provided by Monte Carlo simulations.

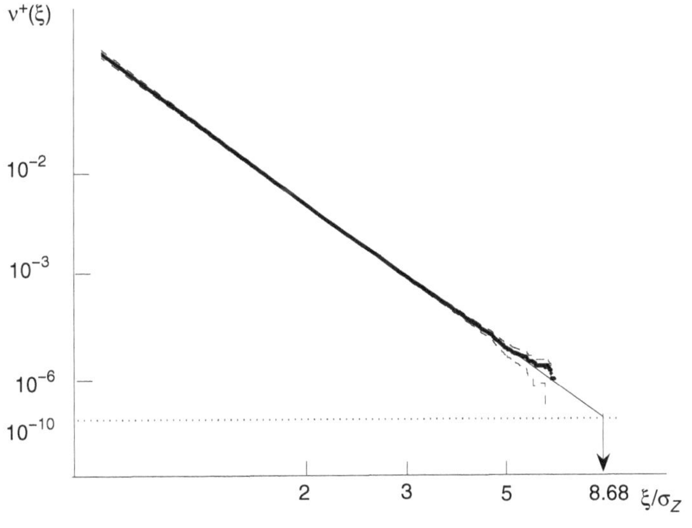

Figure 16.30. Empirical crossing rates by Monte Carlo simulation ($*$) with 95% confidence bands (– –) based on 1,000 hours of response time histories for the case of linear dynamics ($\tilde{c} = \varepsilon = 0$). Saddle point integration results (——) coincide with the optimized linear fit with $b = 0.75\sigma_Z$, $q = 0.0205$, where $\sigma_Z = 9.3$ m (see Table 16.6). Sea state with $H_s = 10$ m, $T_p = 11$ s.

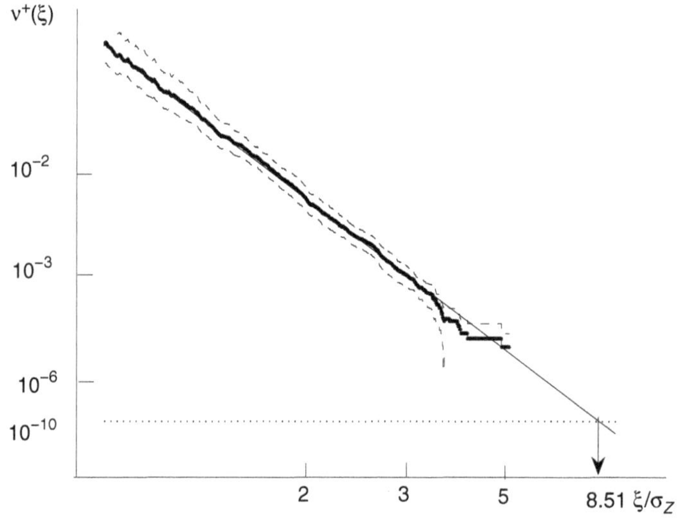

Figure 16.31. Empirical crossing rates by Monte Carlo simulation ($*$) with 95% confidence bands (--) based on 25 hours of response time histories; optimized linear fit (——) with $b = 0.75\sigma_Z$, $q = 0.020$, where $\sigma_Z = 9.3$ m (see Table 16.6), for the case of linear dynamics ($\tilde{c} = \varepsilon = 0$). Sea state with $H_s = 10$ m, $T_p = 11$ s.

Hence, linear extrapolation can be done over several orders of magnitude with high accuracy.

To illustrate the fact that good accuracy can be obtained with much shorter time simulation records, Figs. 16.31 and 16.33 show the results of using only 25 hours of

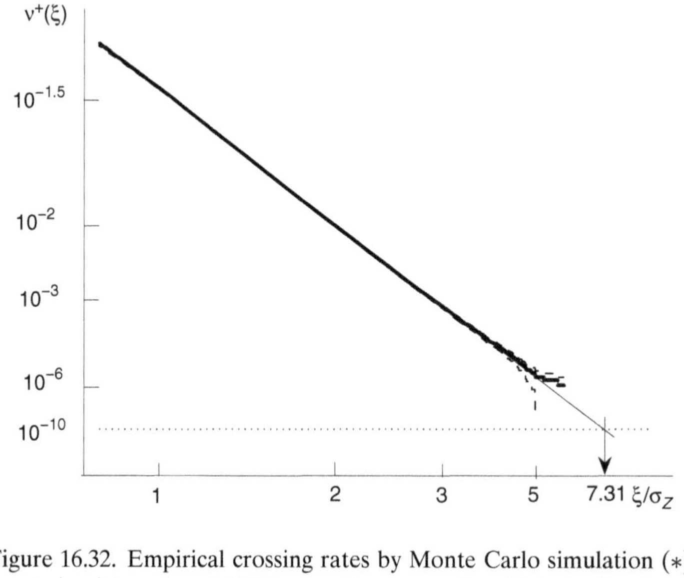

Figure 16.32. Empirical crossing rates by Monte Carlo simulation ($*$) with 95% confidence bands (--) based on 1,000 hours of response time histories for the case of linear dynamics ($\tilde{c} = \varepsilon = 0$). Saddle point integration results (——) coincide with the optimized linear fit with $b = 0.39\sigma_Z$, $q = 0.042$, where $\sigma_Z = 6.4$ m (see Table 16.6). Sea state with $H_s = 15$ m, $T_p = 17$ s.

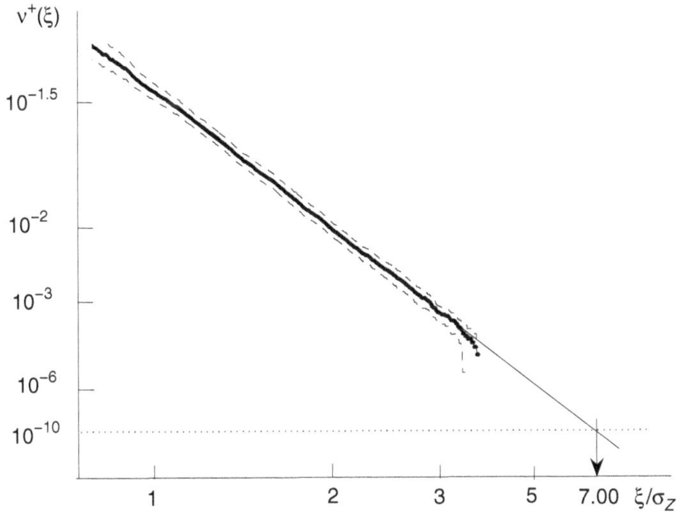

Figure 16.33. Empirical crossing rates by Monte Carlo simulation ($*$) with 95% confidence bands (– –) based on 25 hours of response time histories for the case of linear dynamics ($\tilde{c} = \varepsilon = 0$); optimized linear fit (——) with $b = 0.39\sigma_Z$, $q = 0.042$, where $\sigma_Z = 6.4$ m (see Table 16.6). Sea state with $H_s = 15$ m, $T_p = 17$ s.

simulated response time histories, which required about 10 minutes on a standard desktop PC.

Figures 16.34–16.37 present response tail crossing rates for the two sea states obtained by Monte Carlo simulation for the fully nonlinear model given by Eq. (16.67). In this case the Monte Carlo simulation results are the only results available for verification of the extrapolation method for the nonlinear model, but the experience from the linear case indicates that the results obtained from Figs. 16.34

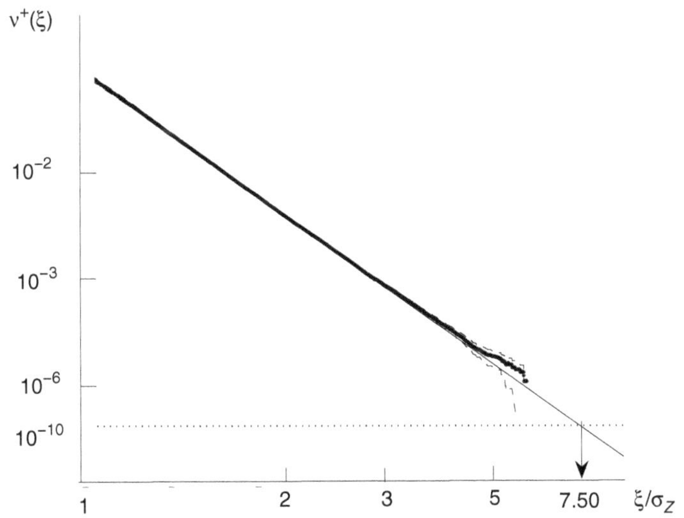

Figure 16.34. Empirical crossing rates by Monte Carlo simulation ($*$) with 95% confidence bands (– –) based on 1,000 hours of response time histories for the fully nonlinear model; optimized linear fit (——) with $b = 0.54\sigma_Z$, $q = 0.0248$, where $\sigma_Z = 8.2$ m (see Table 16.6). Sea state with $H_s = 10$ m, $T_p = 11$ s.

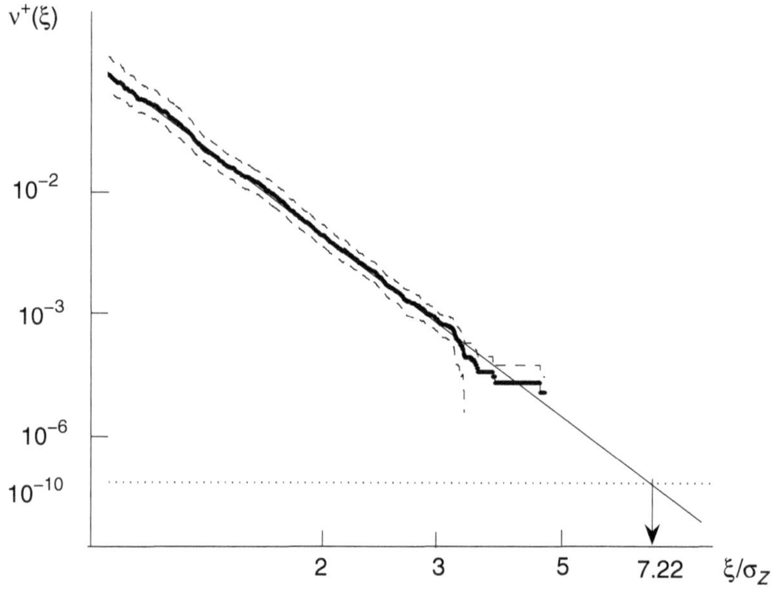

Figure 16.35. Empirical crossing rates by Monte Carlo simulation (∗) with 95% confidence bands (−−) based on 25 hours of response time histories for the fully nonlinear model; optimized linear fit (——) with $b = 0.54\sigma_Z$, $q = 0.0234$, where $\sigma_Z = 8.2$ m (see Table 16.6). Sea state with $H_s = 10$ m, $T_p = 11$ s.

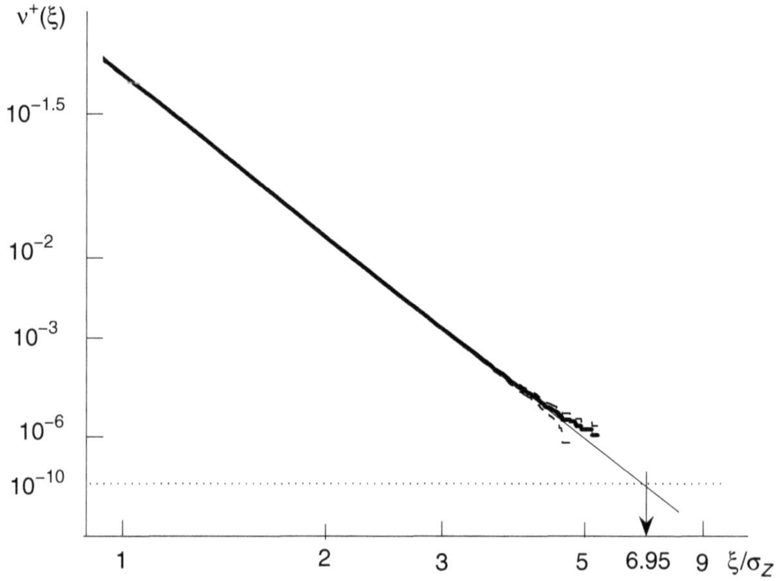

Figure 16.36. Empirical crossing rates by Monte Carlo simulation (∗) with 95% confidence bands (−−) based on 1000 hours of response time histories for the fully nonlinear model; optimized linear fit (——) with $b = 0.39\sigma_Z$, $q = 0.045$, where $\sigma_Z = 6.0$ m (see Table 16.6). Sea state with $H_s = 15$ m, $T_p = 17$ s.

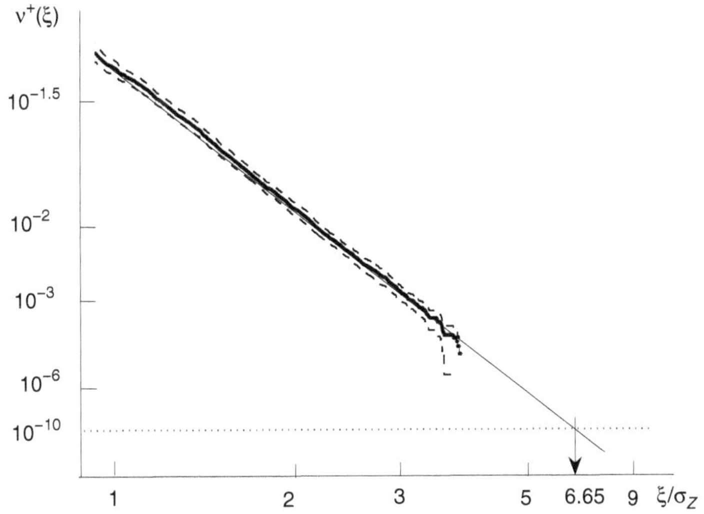

Figure 16.37. Empirical crossing rates by Monte Carlo simulation ($*$) with 95% confidence bands (– –) based on 25 hours of response time histories for the fully nonlinear model; optimized linear fit (——) with $b = 0.39\sigma_Z$, $q = 0.045$, where $\sigma_Z = 6.0$ m (see Table 16.6). Sea state with $H_s = 15$ m, $T_p = 17$ s.

and 16.36 are very accurate. Again, the crossing rate plots are done on the transformed scale, see Eq. (16.34), and for illustration purposes the extrapolated results at the crossing rate level 10^{-10} are highlighted. These results can be compared with the corresponding results obtained from only 25 h of simulated response time histories shown in Figs. 16.35 and 16.37. The agreement is again good.

APPENDIX A

Integrals

In connection with white noise excitation, we have to calculate integrals of the type

$$I = \int_{-\infty}^{\infty} |H(\omega)|^2 \, d\omega \,. \tag{A.1}$$

For a massless vibration system, $H(\omega)$ is generally given as

$$H(\omega) = \frac{B_0}{A_0 + \mathrm{i}\omega \, A_1}. \tag{A.2}$$

It can be shown that the integral I is then given as follows:

$$I = \frac{\pi \, B_0^2}{A_0 \, A_1}. \tag{A.3}$$

Similarly, for a vibration system with mass, the general expression for $H(\omega)$ is

$$H(\omega) = \frac{B_0 + \mathrm{i}\omega \, B_1}{A_0 + \mathrm{i}\omega \, A_1 - \omega^2 \, A_2}, \tag{A.4}$$

and I will be given as

$$I = \frac{\pi (A_0 \, B_1^2 + A_2 \, B_0^2)}{A_0 \, A_1 \, A_2}. \tag{A.5}$$

The following particular case is also relevant for this book:

$$H(\omega) = \frac{B_0 + \mathrm{i}\omega \, B_1 - \omega^2 \, B_2}{A_0 + \mathrm{i}\omega \, A_1 - \omega^2 \, A_2 - \mathrm{i}\omega^3 \, A_3 + \omega^4 \, A_4}, \tag{A.6}$$

and then I is given as

$$I = \frac{\pi \big(- A_0 A_1 A_4 \, B_2^2 + A_0 A_3 A_4 (2 B_0 B_2 - B_1^2) + A_4 B_0^2 (A_1 A_4 - A_2 A_3) \big)}{A_0 A_4 \big(A_0 A_3^2 + A_1^2 A_4 - A_1 A_2 A_3 \big)}. \tag{A.7}$$

These results are adapted from Newland (1991), where more general expressions are also found.

APPENDIX B

Poisson Process

In this appendix we show the validity of Eq. (10.29). The derivation of the formula for the mean rate of upcrossings of a given level, as it is expressed in Eq. (10.9), was based on assumptions about the probability of the number of upcrossings during short time intervals. Using similar notation as in Section 6.2, that is, $p_n = p_n(\Delta t) = \text{Prob}\{N^+(a, \Delta t) = n\}$, it was assumed in particular that $p_n/p_1 \to 0$, when $\Delta t \to 0$ for $n \geq 2$. In other words, the probability of occurrence of two or more upcrossings during a short time interval can be neglected compared with the probability of one upcrossing during the same time interval. To simplify the notation somewhat, we write $\lambda = \nu_X^+(a)$. According to Eq. (10.9), we can therefore assume that for sufficiently small Δt

$$p_1(\Delta t) = \lambda \Delta t \tag{B.1}$$

and

$$p_n(\Delta t) = 0\,, \quad n \geq 2. \tag{B.2}$$

Because $\sum_{n=0}^{\infty} p_n = 1$, it follows that

$$p_0(\Delta t) = 1 - \lambda \Delta t. \tag{B.3}$$

To proceed, we now need to use the assumption that the upcrossings of the level a are independent events. This implies that the number of upcrossings in a given time interval is statistically independent of the number of upcrossings in another, nonoverlapping time interval. This may be used to derive the following equation:

$$\begin{aligned} p_0(t + \Delta t) &= \text{Prob}\{\text{No upcrossings in } (0, t + \Delta t)\} \\ &= \text{Prob}\{[\text{No upcrossings in } (0, t)] \text{ and } [\text{No upcrossings in } (t, t + \Delta t)]\} \\ &= p_0(t)\, p_0(\Delta t) \end{aligned} \tag{B.4}$$

Equation (B.4) together with Eq. (B.3) gives

$$\frac{p_0(t + \Delta t) - p_0(t)}{\Delta t} = -\lambda\, p_0(t). \tag{B.5}$$

Strictly speaking, this equation is only approximately correct. However, on the basis of the assumptions we make, we know that the approximation becomes better the smaller Δt becomes. This leads to the differential equation

$$\frac{dp_0(t)}{dt} = -\lambda\, p_0(t), \tag{B.6}$$

which has the solution $p_0(t) = C\, \exp(-\lambda t)$, where C is a constant. Clearly $p_0(0) = 1$. This gives $C = 1$. The solution is therefore

$$p_0(t) = e^{-\lambda t}, \tag{B.7}$$

which corresponds to Eq. (10.29).

While we are at it, let us also derive the expression for $p_n(t)$. It is realized that $n \geq 1$ upcrossings in the interval $(0, t + \Delta t)$ can occur as follows: { n upcrossings in $(0, t)$ and 0 upcrossings in $(t, t + \Delta t)$ } or { $n - 1$ upcrossings in $(0, t)$ and 1 upcrossing in $(t, t + \Delta t)$ }, etc. This can be expressed by the equation

$$p_n(t + \Delta t) = \sum_{i=0}^{n} p_{n-i}(t)\, p_i(\Delta t) = p_n(t)\, p_0(\Delta t) + p_{n-1}(t)\, p_1(\Delta t). \tag{B.8}$$

The last equality follows from Eq. (B.2).

Similarly to the preceding derivation, this leads to the differential equations

$$\frac{dp_n(t)}{dt} = -\lambda\, p_n(t) + \lambda\, p_{n-1}(t), \quad n = 1, 2, \ldots \tag{B.9}$$

with the initial conditions $p_n(0) = 0\ (n \geq 1)$. These equations can be solved in several ways. One way is to introduce the auxiliary functions $u_n(t)$, $n = 0, 1, \ldots$ defined by $p_n(t) = e^{-\lambda t}\, u_n(t)$. When this is substituted into Eq. (B.9), it leads to the equations

$$\frac{du_n(t)}{dt} = \lambda\, u_{n-1}(t), \quad n = 1, 2, \ldots, \tag{B.10}$$

with initial conditions $u_n(0) = 0$, $n = 1, 2, \ldots$. In particular,

$$\frac{du_1(t)}{dt} = \lambda\, u_0(t) = \lambda, \tag{B.11}$$

which gives $u_1(t) = \lambda\, t$, because $u_1(0) = 0$. Successive solution of Eq. (B.10) gives

$$u_n(t) = \frac{(\lambda\, t)^n}{n!}, \tag{B.12}$$

and thereby

$$p_n(t) = \frac{(\lambda\, t)^n}{n!}\, e^{-\lambda t}, \tag{B.13}$$

which holds for $n = 0, 1, 2, \ldots$ $(0! = 1)$.

The Poisson process is a frequently used model for phenomena characterized by events that occur approximately independent of each other. Subject to certain conditions, the Poisson process can be used to model the stream of telephone calls through a telephone exchange or the stream of cars passing through a road crossing.

APPENDIX C

Statistical Moments and Cumulants

The statistical moments μ_k, $k = 1, 2, \dots$, of a random variable X is defined as

$$\mu_k = \mathrm{E}[X^k], \tag{C.1}$$

provided $\mathrm{E}[|X|^k] < \infty$. Note that $\mu_1 = m_X$ according to previous notation. Clearly, if $\mathrm{E}[|X|^k] < \infty$, then $\mathrm{E}[|X|^r] < \infty$ for $r = 1, \dots, k$. Hence, the existence of a statistical moment μ_k implies the existence of all lower-order moments.

The central moments μ'_k, $k = 1, 2, \dots$, are defined as

$$\mu'_k = \mathrm{E}[(X - m_X)^k]. \tag{C.2}$$

If μ_k exists, so does μ'_k, and vice versa. Also, the existence of μ'_k implies the existence of all lower-order central moments.

It is realized that each μ'_k can be expressed in terms of μ_j for $j \leq k$. In particular, it can be shown that

$$\begin{aligned} \mu'_2 &= \mu_2 - \mu_1^2, \\ \mu'_3 &= \mu_3 - 3\mu_1\mu_2 + 2\mu_1^3, \\ \mu'_4 &= \mu_4 - 4\mu_1\mu_3 + 6\mu_1^2\mu_2 - 3\mu_1^4, \end{aligned} \tag{C.3}$$

The reverse relations will be

$$\begin{aligned} \mu_2 &= \mu'_2 + \mu_1^2, \\ \mu_3 &= \mu'_3 + 3\mu_1\mu'_2 + \mu_1^3, \\ \mu_4 &= \mu'_4 + 4\mu_1\mu'_3 + 6\mu_1^2\mu'_2 + \mu_1^4, \end{aligned} \tag{C.4}$$

For these results and many more, the reader may consult Kendall and Stuart (1976).

To every (real) random variable X, we may associate a characteristic function ϕ_X. It is defined as

$$\phi_X(t) = \mathrm{E}[\exp(\mathrm{i}tX)]. \tag{C.5}$$

Because $|\exp(\mathrm{i}tX)| \leq 1$ for every real number t, it follows that $\phi(t)$ exists for such t. If $\mathrm{E}[|X|^k] < \infty$, then $\phi(t)$ can be differentiated k times at $t = 0$, and

$$\mu_k = (-\mathrm{i})^k \frac{d^k \phi(t)}{dt^k}|_{t=0} . \tag{C.6}$$

If the moments of every order exist, which holds true if X is a normal random variable, then

$$\phi_X(t) = 1 + \sum_{k=1}^{\infty} \frac{(\mathrm{i}t)^k}{k!} \mu_k , \tag{C.7}$$

provided the sum converges. From the equality $\ln(1+z) = \sum_{j=1}^{\infty} (-1)^{j+1} z^j / j$, it follows that

$$\ln \phi_X(t) = \Big(\sum_{k=1}^{\infty} \frac{(\mathrm{i}t)^k}{k!} \mu_k \Big) - \frac{1}{2} \Big(\sum_{k=1}^{\infty} \frac{(\mathrm{i}t)^k}{k!} \mu_k \Big)^2 + \ldots . \tag{C.8}$$

Ordering the terms of this expansion according to increasing powers of $\mathrm{i}t$, it is obtained that

$$\ln \phi_X(t) = \sum_{k=1}^{\infty} \frac{(\mathrm{i}t)^k}{k!} \kappa_k . \tag{C.9}$$

The expansion coefficients κ_k are called the *cumulants* of the random variable X. Because they can be obtained from Eq. (C.8) by differentiation, that is,

$$\kappa_k = (-\mathrm{i})^k \frac{d^k \ln \phi_X(t)}{dt^k}|_{t=0} , \tag{C.10}$$

$\ln \phi_X(t)$ is often referred to as the *cumulant generating function.*

The relationships between the moments and the cumulants can be derived by using Eqs. (C.8) and (C.9) and equating the coefficients of like powers of $\mathrm{i}t$. This gives (Cramer, 1946; Kendall and Stuart, 1976)

$$\begin{aligned} \kappa_1 &= \mu_1 , \\ \kappa_2 &= \mu_2 - \mu_1^2 , \\ \kappa_3 &= \mu_3 - 3\mu_1\mu_2 + 2\mu_1^3 , \\ \kappa_4 &= \mu_4 - 3\mu_2^2 - 4\mu_1\mu_3 + 12\mu_1^2\mu_2 - 6\mu_1^4 , \end{aligned} \tag{C.11}$$

and

$$\begin{aligned} \mu_1 &= \kappa_1 , \\ \mu_2 &= \kappa_2 + \kappa_1^2 , \\ \mu_3 &= \kappa_3 + 3\kappa_1\kappa_2 + \kappa_1^3 , \\ \mu_4 &= \kappa_4 + 3\kappa_2^2 + 4\kappa_1\kappa_3 + 6\kappa_1^2\kappa_2 + \kappa_1^4 . \end{aligned} \tag{C.12}$$

The relationships between the first central moments and the cumulants are particularly simple:

$$\begin{aligned}
\kappa_1 &= \mu_1 \,, \\
\kappa_2 &= \mu_2' \,, \\
\kappa_3 &= \mu_3' \,, \\
\kappa_4 &= \mu_4' - 3\mu_2'^2 \,.
\end{aligned} \tag{C.13}$$

As measures of deviation from the normal distribution function, the coefficients of skewness (γ_3) and kurtosis (γ_4) are sometimes used. They are defined as $\gamma_3 = \kappa_3/\kappa_2^{3/2}$ and $\gamma_4 = 3 + \kappa_4/\kappa_2^2$. The coefficient of excess is defined as $\gamma_4 - 3$. For a normal random variable, $\gamma_3 = 0$ and $\gamma_4 = 3$.

References

Adams, A. J. and N. D. P. Barltrop (1991). *Dynamics of Fixed Marine Structures* (Third Edition). London: Butterworth-Heineman Ltd.

Almar-Naess, A. (1985). *Fatigue handbook: offshore steel structures*. Trondheim, Norway: Tapir Publishers.

Amzallag, C., J. P. Gerey, J. L. Robert, and J. Bahuaud (1994). Standardization of the rainflow counting method for fatigue analysis. *International Journal of Fatigue 16*, 287–293.

Ang, A. and W. Tang (2007). *Probability Concepts in Engineering* (Second Edition). Chichester, UK: John Wiley and Sons, Ltd.

API (1993). *Recommended Practice for Design, Analysis and Maintenance of Moorings for Floating Production Systems* (API-RP2FP1, 1st Edition). Washington, DC: American Petroleum Institute.

API (1997). *Recommended Practice for Planning, Designing and Constructing Fixed Offshore Platforms* (API RP2A-WSD). Washington, DC: American Petroleum Institute.

API (2000). *Recommended Practice for Planning, Designing and Constructing Fixed Offshore Platforms – Working Stress Design* (API-RP2A, 21st Edition). Washington, DC: American Petroleum Institute.

Argyris, J. and H.-P. Mlejnek (1991). *Dynamics of Structures*. Amsterdam, The Netherlands: Elsevier Science Publishers B.V.

ASTM (1985). *Standard Practices for Cycle Counting in Fatigue Analysis* (Designation E 1049-85). West Conshohocken, PA 19428: American Society for Testing and Materials.

Ayala-Uraga, E. and T. Moan (2007). Time-variant reliability assessment of FPSO hull girder with long cracks. *Journal of Offshore Mechanics and Arctic Engineering 129*(1), 81–89.

Azadi, M. E. (1998). *Analysis of Static and Dynamic Pile-Soil-Jacket Behaviour*. MTA Report 121, Norwegian University of Science and Technology, Trondheim, Norway.

Bach-Gansmo, O., C. A. Carlsen, and T. Moan (1987). Fatigue Assessment of Hull Girder for Ship Type Floating Production Vessels. In *Proceedings of Conference on Mobile Offshore Units*. London, UK: City University.

Barsom, J. M. and S. T. Rolfe (1999). *Fracture and Fatigue Control in Structures: Applications of Fracture Mechanics* (Third Edition). West Conshohocken, PA: ASTM.

Batchelor, G. K. (1953). *The Theory of Homogeneous Turbulence*. London: Cambridge University Press.

Bathe, K.-J. (1996). *Finite Element Procedures*. Englewood Cliffs, NJ: Prentice Hall PTR.

Battjes, J. A. (1970). *Long Term Wave Height Distributions of Seven Stations around the British Isles* (Report A.44). Godalming, UK: National Institute of Oceanography.

Beck, A. T. and R. E. Melchers (2004). On the ensemble crossing rate approach to time variant reliability analysis of uncertain structures. *Probabilistic Engineering Mechanics 19*, 9–19.

Benasciutti, D. and R. Tovo (2005). Spectral methods for lifetime prediction under wide-band stationary random processes. *International Journal of Fatigue 27*, 867–877.
Benasciutti, D. and R. Tovo (2007). On fatigue damage assessment in bimodal random processes. *International Journal of Fatigue 29*(2), 232–244.
Benjamin, J. R. and C. A. Cornell (1978). *Probability, Statistics and Decision for Civil Engineers*. New York, NY: McGraw-Hill, Ltd.
Biggs, J. M. (1964). *Introduction to Structural Dynamics*. New York, NY: McGraw-Hill, Ltd.
Bishop, R. E. D. and W. Price (1979). *Hydroelasticity of Ships*. London, UK: Cambridge University Press.
Bjerager, P., R. Løseth, S. R. Winterstein, and C. A. Cornell (1988). Reliability Method for Marine Structures under Multiple Environmental Load Processes. In *Proceedings 5th International Conference on the Behavior of Offshore Structures (BOSS)*, pp. 1239–1251. Trondheim: Norwegian Institute of Technology (NTH).
Boccotti, P. (1982). Relations between characteristic sea wave parameters. *Journal of Geophysical Research 87*, 4267.
Borgman, L. E. (1965). Wave Forces on Piling for Narrow-Band Spectra. *Journal of the Waterways and Harbors Division, ASCE 91*, 65–90.
Borgman, L. E. (1972). Statistical models for ocean waves and wave forces. In V. T. Chow (ed.), *Advances in Hydroscience*, Volume 8. New York, NY: Academic Press.
Bowness, D. and M. M. K. Lee (1999). Weld toe magnification factors for semi-elliptical cracks in T-butt joints. Technical report, Health Safety Executive.
Broek, D. (1984). *Elementary Engineering Fracture Mechanics*. The Hague, The Netherlands: Martinus Nijhoff.
Brouwers, J. J. H. and P. H. J. Verbeek (1983). Expected fatigue damage and expected extreme response for Morison-type wave loading. *Applied Ocean Research 5*(5), 129–133.
Brown, T. G. and M. Määttänen (2009). Comparison of Kemi-I and Confederation Bridge cone ice load measurement results. *Cold regions Science and Technology 55*(1), 3–13.
BS 7910 (1999). *"Guidance on Methods for Assessing the Acceptability of Flaws in Fusion Welded Structures"*. London, UK: British Standards.
Burton, T., N. Jenkins, D. Sharp, and E. Bossanyi (2011). *Wind Energy Handbook* (Second Edition). Chichester, UK: John Wiley and Sons, Ltd.
Bury, K. V. (1975). *Statistical Models in Applied Sciences*. New York, NY: John Wiley & Sons, Inc.
Chakrabarti, S. (2005). *Handbook of Offshore Engineering*. Elsevier Ltd.
Chopra, A. K. (2001). *Dynamics of Structures: Theory and Applications to Earthquake Engineering* (Second Edition). Upper Saddle River, NJ: Prentice Hall, Inc.
Clauss, G., E. Lehmann, and C. Ostergaard (1991). *Offshore Structures, Vol. 1*. Berlin, Germany: Springer-Verlag.
Clough, R. W. and J. Penzien (1993). *Dynamics of Structures*. New York, NY: McGrawHill.
Cook, R. D., D. S. Malkus, M. E. Plesha, and R. J. Witt (2002). *Concepts and Applications of Finite Element Analysis* (Fourth Edition). New York, NY: John Wiley & Sons Ltd.
Craig, R. R. (1981). *Structural Dynamics: An Introduction to Computer Methods*. New York, NY: John Wiley & Sons Ltd.
Craig, R. R. (1987). A review of time-domain and frequency-domain component-mode synthesis methods. *International Journal of Analytical and Experimental Modal Analysis 2*(2), 57–72.
Cramer, H. (1946). *Mathematical Methods of Statistics*. Princeton, NJ: Princeton University Press.
Cruz, J. (2008). *Ocean Wave Energy*. Berlin: Springer-Verlag.
Cummins, W. E. (1962). The impulse response function and ship motions. *Schifftechnik 47*(9), 101–109.
Davenport, A. G. (1977). The Prediction of the Response of Structures to Gusty Wind. In I. Holand, D. Kavlie, G. Moe, and R. Sigbjørnsson (eds.), *Safety of Structures under Dynamic Loading*, pp. 257–284. Trondheim, Norway: Tapir A/S.

Davison, A. C. and D. V. Hinkley (1997). *Bootstrap Methods and their Applications*. London, UK: Cambridge University Press.

de Haan, L. (1994). Extreme Value Statistics. In J. Galambos, J. A. Lechner, and E. Simiu (eds.), *Extreme Value Theory and Applications*. Dordrecht, The Netherlands: Kluwer Academic Publishers.

Demirbilek, Z. (1989). *Tension Leg Platforms – A State of the Art Review* (ASCE Task Group on Compliant Offshore Structures). New York, NY: ASCE.

Dickens, J. M., J. M. Nakagawa, and M. J. Wittbrodt (1997). A critique of mode acceleration and modal truncation augmentation methods for modal response analysis. *Computers and Structures 62*(6), 985–998.

Dirlik, T. (1985). *Application of computers in fatigue analysis*. PhD Thesis, University of Warwick, Warwick, UK.

Ditlevsen, O. and H. O. Madsen (1996). *Structural Reliability Methods*. Chichester: John Wiley & Sons, Inc.

DNV (2010a). *DNV-OS-E301: Offshore Standard – Position Mooring*. Oslo, Norway: Det Norske Veritas.

DNV (2010b). *DNV-RP-C205: Environmental Conditions and Environmental Loads*. Oslo, Norway: Det Norske Veritas.

Donley, M. G. and P. D. Spanos (1990). *Dynamic Analysis of Non-Linear Structures by the Method of Statistical Quadratization*, Volume 57 of *Lecture Notes in Engineering*. Berlin, Germany: Springer-Verlag.

Dowling, N. E. (1972). Fatigue failure predictions for complicated stress-strain histories. *Journal of Materials 7*, 71–87.

Dowling, N. E. (1993). *Mechanical Behavior of Materials*. Englewood Cliffs, NJ: Prentice Hall.

Dowrick, D. J. (1977). *Earthquake Resistant Design*. Chichester: John Wiley & Sons, Inc.

Draper, N. R. and H. Smith (1998). *Applied Regression Analysis*. New York, NY: Wiley-Interscience.

Dyrbye, C. and S. O. Hansen (1997). *Wind Loads on Structures*. Chichester, UK: John Wiley & Sons, Inc.

Edwards, C. H. and D. E. Penney (2002). *Calculus* (Sixth Edition). Upper Saddle River, NJ: Prentice Hall, Inc.

Efron, B. and R. J. Tibshirani (1993). *An Introduction to the Bootstrap*. New York, NY: Chapman and Hall.

Faltinsen, O. M. (1990). *Sea Loads on Ships and Offshore Structures*. Cambridge, UK: Cambridge University Press.

Fang, T. and Z. N. Wang (1986). Mean Square Response to Band-Limited White Noise Excitation. *AIAA Journal, 0001-1452 24*(5), 860–862.

Faravelli, L., F. Casciati, and M. P. Singh (1988). Stochastic Equivalent Linearization Algorithms and Their Applicability to Hysteretic Systems. *Meccanica 23*(2).

Farnes, K. A. (1990). *Long-term Statistics of Response in Non-linear Marine Structures*. MTA-report 1990:74, Division of Marine Structures, NTH (NTNU), Trondheim, Norway.

Felton, L. P. and R. B. Nelson (1997). *Matrix Structural Analysis*. New York: John Wiley and Sons, Ltd.

Fish, J. and T. A. Belytschko (2007). *A First Course in Finite Elements*. Chichester, UK: John Wiley & Sons Ltd.

Fisher, R. A. and L. H. C. Tippet (1928). Limiting forms of the frequency distribution of the largest or smallest member of a sample. *Proc. Camb. Phil. Soc. 24*, 180–190.

Fjeld, S. (1977). Reliability of Offshore Structures. In *Proceedings of the Offshore Technology Conference*, Number OTC 3027. Houston, TX.

Forristall, G. Z. (1978). On the statistical distribution of wave heights in a storm. *Journal of Geophysical Research 83*(C5), 2353–2358.

Forristall, G. Z. (2000). Wave Crest Distributions: Observations and Second-Order Theory. *Journal of Physical Oceanography 30*, 1931–1943.

Fossen, T. I. (2002). *Marine Control Systems: Guidance, Navigation and Control of Ships, Rigs and Underwater Vehicles*. Trondheim, Norway: Marine Cybernetics AS.

Fricke, W. (2001). Recommended Hot Spot Analysis Procedure for Structural Details of FPSO and Ships Based on Round-robin FE Analyses. In *Proceedings 11th International Offshore and Polar Engineering Conference*, Stavanger, Norway. ISOPE.

Fu, T. T. and D. Cebon (2000). Predicting fatigue lives for bi-modal stress spectral densities. *International Journal of Fatigue 22*(1), 11–21.

Gall, D. S. and J. W. Hancock (1985). Fatigue crack growth under narrow and broad band stationary loading. Technical report, University of Glasgow, Glasgow, Scotland.

Gao, Z. and T. Moan (2007). Fatigue damage under combined high and low frequency Gaussian load processes considering a two-slope SN curve. In *Proceedings 10th International Conference on Applications of Statistics and Probability in Civil Engineering*, Tokyo, Japan. ICASP.

Gao, Z. and T. Moan (2008). Frequency-domain fatigue analysis of wide-band stationary Gaussian processes using a trimodal spectral formulation. *International Journal of Fatigue 30*(10-11), 1944–1955.

Gao, Z. and T. Moan (2009). Accuracy of the narrow-band approximation of stationary wide-band Gaussian processes for extreme value and fatigue analysis. In H. Furuta, D. Frangopol, and M. Shinozuka (eds.), *Proceedings 10th International Conference on Structural Safety and Reliability (ICOSSAR'09)*, pp. 997–1004. IASSAR: CRC Press.

Gill, P., W. Murray, and M. H. Wright (1981). *Practical Optimization*. London, UK: Academic Press.

Govolato, P. (1959). A study of the transient pitching oscillations of a ship. *Journal of Ship Research 2*(4), 22–30.

Gravesen, H., S. L. Sorensen, P. Volund, A. Barker, and G. Timco (2005). Ice loading on Danish wind turbines: Part 2. Analyses of dynamic model test results. *Cold regions Science and Technology 41*(1), 25–47.

Grigoriu, M. (1984a). Crossings of Non-Gaussian Translation Processes. *Journal of Engineering Mechanics, ASCE 110*(4), 610–620.

Grigoriu, M. (1984b). Extremes of Wave Forces. *Journal of Engineering Mechanics, ASCE 110*(12), 1731–1742.

Guyan, R. J. (1965). Reduction of stiffness and mass matrices. *AIAA Journal 3*(2), 380.

Hagen, Ø. and L. Tvedt (1991). Vector process out-crossings as parallel system sensitivity measure. *Journal of Engineering Mechanics, ASCE 117*(10), 2201–2220.

Halkyard, J. E. (2005). Floating Offshore Platform Design. In S. K. Chakrabarti (ed.), *Handbook of Offshore Engineering*, Chapter 7. Amsterdam: Elsevier.

Hals, J., R. Taghipour, and T. Moan (2007). Dynamics of a Force Compensated Two-Body Wave Energy Converter in Heave with Hydraulic Power Take-Off Subject to Phase Control. In *Proceedings of the 7th European Wave and Tidal Energy Conference*, Porto, Portugal.

Hansteen, O. E. and K. Bell (1988). On the accuracy of mode superposition analysis in structural dynamics. *Earthquake Engineering and Structural Dynamics 7*(5), 405–411.

Haring, R. E. and J. C. Heideman (1978). Gulf of Mexico Rare Wave Return Periods. In *Proceedings of the Offshore Technology Conference*, Number OTC 3230. Houston, TX.

Haring, R. E., A. R. Osborne, and L. P. Spencer (1976). Extreme wave parameters based on continental shelf storm wave records. In *Proceedings of the 15th Conference on Coastal Engineering*. Honolulu, Hawaii: ASCE.

Hasselmann, K., T. P. Barnett, E. Bouws, H. Carlson, D. E. Cartwright, K. Eake, J. A. Euring, A. Gicnapp, D. E. Hasselmann, P. Kruseman, A. Meerburg, P. Mullen, D. J. Olbers, K. Richren, W. Sell, and H. Walden (1973). Measurement of wind wave growth and swell decay during the Joint North Sea Wave Project (JONSWAP). *Deutschen Hydrographischen Zeitschrift Reihe A 8*(12), 95.

Haver, S. (1980). *Analysis of Uncertainties related to the Stochastic Modelling of Ocean Waves*. Report UR-80-09, Division of Marine Structures, NTH (NTNU), Trondheim, Norway.

Haver, S. (2002). On the prediction of extreme wave crest heights. In *Proceedings of 7th International Workshop On Wave Hindcasting and Forecasting*. Banff, Canada: Meteorological Service of Canada, Environment Canada.

Haver, S. and K. A. Nyhus (1986). A wave climate description for long term response calculations. In *Proceedings 9th International Conference on Offshore Mechanics and Arctic Engineering*. New York: ASME.

Hellan, Ø. (1995). *Nonlinear Pushover and Cyclic Analyses in Ultimate Limit State Design and Reassessment of Tubular Steel Offshore Structures*. MTA Report 108, Norwegian University of Science and Technology, Trondheim, Norway.

Hoen, C., T. Moan, and S. Remseth (1993). System identification of structures exposed to environmental loads. In *Proceedings of EURODYN'93*, Volume 2, pp. 835–844. A.A. Balkema Publishers, Rotterdam, Netherlands.

Holappa, K. and J. Falzarano (1999). Application of extended state space to nonlinear ship rolling. *Journal of Ocean Engineering 26*, 227–240.

Hosking, J. R. M. and J. R. Wallis (1987). Parameter and Quantile Estimation for the Generalized Pareto Distribution. *Technometrics 29*, 339–349.

Housner, G. W. and P. C. Jennings (1982). Earthquake Design Criteria. Technical report, California Institute of Technology, Los Angeles, CA.

Huang, W. and T. Moan (2005). Combination of global still-water and wave load effects for reliability-based design of floating production storage and offloading (FPSO) vessels. *Applied Ocean Research 27*(3), 127–141.

Huang, W. and T. Moan (2006). Fatigue Under Combined High and Low Frequency Loads. In *Proceedings 25th International Conference on Offshore Mechanics and Arctic Engineering*, Number OMAE2006-92247, Hamburg, Germany. ASME.

Huang, X. and T. Moan (2007). Improved Modeling of the Effect of R-ratio on Crack Growth Rate. *International Journal of Fatigue 29*(4), 591–602.

Hughes, T. J. R. (1987). *The Finite Element Method*. Englewood Cliffs, NJ: Prentice-Hall.

IEC (1999). *Wind turbine generator systems – Part 1: Safety requirements* (International standard 61400-1). Washington, DC: International Electrotechnical Commission.

Irvine, H. M. (1988). *Structural Dynamics for the Practicing Engineer*. London, UK: Unwin Hyman, Ltd.

ISO 19902 (2007). *Petroleum and natural gas industries – Fixed steel offshore structures*. Geneva, Switzerland: International Organization for Standardization.

ISO 2394 (1998). *General Principles on Reliability for Structures*. London, UK: International Standardization Organization.

ISSC (2006a). Report of Specialist Task Committee I.2, Loads. In P. Frieze and R. Shenoi (eds.), *Proceedings of the 16th International Ship and Offshore Structures Congress*, pp. 85–172. Elsevier, Southampton.

ISSC (2006b). Report of Specialist Task Committee II.1, Quasi-static Response. In P. Frieze and R. Shenoi (eds.), *Proceedings of the 16th International Ship and Offshore Structures Congress*, pp. 173–257. Elsevier, Southampton.

ISSC (2006c). Report of Specialist Task Committee VI.2, Very Large Floating Structures. In P. Frieze and R. Shenoi (eds.), *Proceedings of the 16th International Ship and Offshore Structures Congress*, pp. 391–442. Elsevier, Southampton.

ITTC (2005). The Seakeeping Committee: Final Reports and Recommendations to the 24th ITTC. Technical report, International Towing Tank Organization, Edinburgh, UK.

Jahns, H. O. and J. D. Wheeler (1972). Long Term Wave Probabilities Based on Hindcasting of Severe Storms. In *Proceedings of the Offshore Technology Conference*, Number OTC 1590. Houston, TX.

JBP (2005a). Common Structural Rules for Bulk Carriers. Technical report, Joint Tanker Project, London, UK.

JBP (2005b). Common Structural Rules for Double Hull Oil Tankers. Technical Report Sect. 9, Design verification, Joint Tanker Project, London, UK.

Jensen, J. J. and P. T. Pedersen (1979). Wave-Induced Bending Moments in Ships – a Quadratic Theory. *RINA, Supplementary Papers 121*, 151–165.

Jensen, J. J. and P. T. Pedersen (1981). Bending Moments and Shear Forces in Ships Sailing in Irregular Seas. *Journal of Ship Research 24*(4), 243–251.

Jia, H. and T. Moan (2009). Comparative reliability analysis of ships under vector load processes. *Journal of Marine Science and Technology 14*, 485–498.

Jiao, G. and T. Moan (1990). Probabilistic Analysis of Fatigue due to Gaussian Load Processes. *Probabilistic Engineering Mechanics 5*(2), 76–83.

Jordaan, I. J. (2001). Mechanics of ice-structure interaction. *Engineering Fracture Mechanics 68*, 1923–1960.

Kareem, A., J. Zhao, and M. A. Tognarelli (1995). Surge response statistics of tension leg platforms under wind and wave loads: a statistical quadratization approach. *Probabilistic Engineering Mechanics 10*, 225–240.

Kärnä, T., Y. Qu, and W. Kühnlein (2004). New Spectral Method for Modeling Dynamic Ice Actions. In *Proceedings 23rd International Conference on Offshore Mechanics and Arctic Engineering*, pp. OMAE2004–51360. New York: ASME.

Karsan, D. I. (2005). Fixed Offshore Platform Design. In S. K. Chakrabarti (ed.), *Handbook of Offshore Engineering*, Chapter 6. Amsterdam: Elsevier.

Karunakaran, D., M. Baerheim, and B. J. Leira (1997). Measured and Simulated Dynamic Response of a Jacket Platform. In *Proceedings of the 16th International Conference on Offshore Mechanics and Arctic Engineering*, Volume II, Porto, Portugal, pp. 157–164.

Karunakaran, D., S. Haver, M. Baerheim, and N. Spidsoe (2001). Dynamic behaviour of Kvitebjørn jacket in the North Sea. In *Proceedings of the 20th International Conference on Offshore Mechanics and Arctic Engineering*, pp. OMAE2001/OFT1184. Rio de Janeiro, Brasil: ASME.

Kashiwagi, M. (2000). A time-domain mode expansion method for calculating transient elastic responses of a pontoon-type VLFS. *Journal of Marine Science and Technology 5*, 89–100.

Kashiwagi, M. (2004). Transient response of a VLFS during landing and take-off of an airplane. *Journal of Marine Science and Technology 9*, 14–23.

Keinonen, A. (1977). *An Analytical Method for Calculating The Pure Ridge Resistance Encountered by Ships in First-year Ice Ridges.* PhD Thesis, Ship Hydrodynamics Laboratory, Helsinki University of Technology, Helsinki, Finland.

Kendall, M. and A. Stuart (1976). *The Advanced Theory of Statistics* (Fourth Edition). London: Charles Griffin & Comp. Ltd.

Kim, W. S. and I. Lotsberg (2004). Fatigue Test Data for Welded Connections in Ship Shaped Structures. In *Specialty Symposium on FPSO Integrity*, Number OMAE-FPSO'04-0018, Houston, TX. ASME.

Kreyszig, E. (1993). *Advanced Engineering Mathematics* (Seventh Edition). New York, NY: John Wiley & Sons, Inc.

Krogstad, H. E. (1985). Height and period distributions of extreme waves. *Applied Ocean Research 7*(3), 158–165.

Krokstad, J. R., C. T. Stansberg, A. Nestegård, and T. Marthinsen (1996). A new non-slender load approach verified against experiments. In *Proceedings of the 15th International Conference on Offshore Mechanics and Arctic Engineering*. Florence, Italy: ASME.

Larrabee, R. D. and C. A. Cornell (1979). Upcrossing rate solutions for load combinations. *Journal of Structural Engineering, ASCE 105*(1), 125–132.

Larrabee, R. D. and C. A. Cornell (1981). Combination of various load processes. *Journal of the Structural Division, ASCE 107*, 223–239.

Larsen, C. E. and L. D. Lutes (1991). Prediction of fatigue life of offshore structures by the single-moment spectral method. *Probabilistic Engineering Mechanics 6*(2), 96–108.

Larsen, K. and P. C. Sandvik (1990). Efficient methods for the calculation of dynamic mooring line tension. In *Proceedings 1st European Offshore Mechanics Symposium*, Trondheim, Norway.

Larsen, L. M. (1981). The influence of bandwidth on the distribution of heights of sea waves. *Journal of Geophysical Research 86*, 4299.

Leadbetter, M. R., G. Lindgren, and H. Rootzen (1983). *Extremes and Related Properties of Random Sequences and Processes*. New York, NY: Springer-Verlag.

Leira, B. J. (1987). *Gaussian Vector-Processes for Reliability Analysis involving Wave-induced Load Effects*. Report UR-87-57, Division of Marine Structures, NTH (NTNU), Trondheim, Norway.

Leung, A. Y. T. (1988). A simple dynamic substructure method. *Earthquake Engineering and Structural Dynamics 16*(6), 827–837.

Lin, Y. K. (1967). *Probabilistic Theory of Structural Dynamics*. New York, NY: MCGraw-Hill Inc.

Liu, X., G. Li, R. Oberlies, and Q. Yue (2009). Research on short term dynamic ice cases for dynamic analysis of ice-resonant jacket platform in the Bohai Gulf. *Marine Structures 22*(3), 457–479.

Livesley, R. K. (1983). *Finite elements: an introduction for engineers*. London, UK: Cambridge University Press.

Longuet-Higgins, M. S. (1952). On the statistical distribution of the heights of sea waves. *Journal of Marine Research 11*, 245.

Longuet-Higgins, M. S. (1980). On the distribution of the heights of sea waves: Some effects of nonlinearity and finite bandwidth. *Annals of Statistics 85*, 1519.

Lotsberg, I. (2005). Background for revision of DNV-RP-C203 fatigue analysis of offshore steel structures. In *Proceedings 24th International Conference on Offshore Mechanics and Arctic Engineering*, Number OMAE2005-67549, Halkidiki, Greece. ASME.

Lotsberg, I. (2006). Assessment of fatigue capacity in the new bulk carrier and tanker rules. *Journal of Marine Structures 19*(1), 83–96.

Lutes, L. D. and S. Sarkani (2004). *Random Vibrations – Analysis of Structural and Mechanical Systems*. Oxford, UK: Elsevier Butterworth-Heineman.

Määttänen, M. (1996). Ice failure and ice loads on a conical structure – Kemi-1 cone full-scale ice force measurement data analysis. In *Proceedings of 13th International Symposium on Ice*, Number Vol. 1, pp. 8–17. IAHR, Beijing, China.

Maddox, N. R. (1974). On the number of modes necessary for accurate response and resulting forces in dynamic analysis. *Journal of Applied Mechanics, ASME 42*, 516–517.

Maddox, S. J. (1991). *Fatigue Strength of Welded Structures* (Second Edition). Cambridge, UK: Woodhead Publishing.

Madsen, H. O., S. Krenk, and N. C. Lind (1986). *Methods of Structural Safety*. New Jersey: Prentice-Hall Inc.

Manwell, J. F., J. G. McGowan, and A. L. Rogers (2002). *Wind Energy Explained: Theory, Design and Application*. New Yok: John Wiley & Sons, Inc.

Marley, M. J. and T. Moan (1992). Time variant formulation for fatigue reliability. In *Proceedings of 11th International Conference on Offshore Mechanics and Arctic Engineering*, Paper No. OMAE92-1203, Alberta, Canada. ASME.

Marshall, P. W. and W. H. Luyties (1982). Allowable Stresses for Fatigue Design. In *Proceedings BOSS '82*, Volume 2, New York, NY, pp. 2–25. McGraw-Hill.

Masterson, D. M., R. M. W. Frederking, B. Wright, T. Kärnä, and W. P. Maddock (2007). A revised ice pressure-area curve. In *Proceedings of the Nineteenth International Conference on Port and Ocean Engineering under Arctic Conditions*, Vol. 1, pp. 305–314. Dalian University of Technology Press.

Matsuishi, M. and T. Endo (1968). Fatigue of Metals Subjected to Varying Stress. *Japan Society of Mechanical Engineering, Fukuoka, Japan* , 37–40.

McGuire, W., R. H. Gallagher, and R. D. Ziemian (2000). *Matrix Structural Analysis* (Second Edition). New York: John Wiley and Sons, Ltd.

McKay, M. D., W. J. Conover, and R. J. Beckman (1979). A Comparison of Three Methods for Selecting Values of Input Variables in the Analysis of Output from a Computer Code. *Technometrics 21*, 239–245.

McWilliam, S. and R. S. Langley (1993). Extreme values of first and second order wave induced vessel motions. *Applied Ocean Research 15*(3), 169–181.

Melchers, R. E. (1999). *Structural Reliability Analysis and Prediction* (Second Edition). West Sussex, UK: John Wiley & Sons, ISBN 0471987719.

Mo, O. (1983). *Stochastic Time Domain Analysis of Slender Offshore Structures*. Report UR-83-33, Norwegian University of Science and Technology, Trondheim, Norway.
Mo, O. and T. Moan (1984). Environmental Load Effect Analysis of Guyed Towers. In *Proceedings of the Third OMAE Conference*, Houston, TX.
Moan, T. (1995). Safety Level across Different Types of Structural Forms and Material – Implicit in Codes for Offshore Structures. Technical Report STF70 A95210, prepared for ISO/TC250/SC7, Trondheim, Norway.
Moan, T. (2003). *Finite Element Modelling and Analysis of Marine Structures* (Lecture Notes, UK-03-98). Trondheim, Norway: Department of Marine Technology, NTNU.
Moan, T. (2005). Reliability-based Management of Inspection, Maintenance and Repair of Offshore Structures. *Journal of Structural and Infrastructural Engineering 1*(1), 33–62.
Moan, T., Z. Gao, and E. Ayala-Uraga (2005). Uncertainty of Wave-induced Response of Marine Structures Due to Long Term Variation of Extratropical Wave Conditions. *Journal of Marine Structures 18*(4), 359–382.
Moan, T., G. Hovde, and A. Blanker (1993). Reliability-based Fatigue Design Criteria for Offshore Structures Considering the Effect of Inspection and Repair. In *Proc. 25th OTC*, Volume 2 of *Offshore Technology Conference*, Houston, pp. 591–599. OTC 7189.
Moan, T., K. Syvertsen, and S. Haver (1976). Stochastic Dynamic Response Analysis of Gravity Platforms. Technical Report SK/M33, NTH(NTNU), Trondheim, Norway.
Moan, T., K. Syvertsen, and S. Haver (1977). Stochastic Dynamic Response Analysis of Gravity Platforms. In *Proceedings of the 2nd STAR Symposium* (Invited paper). San Francisco, CA, pp. 119–146. SNAME.
Moe, G. and S. H. Crandall (1977). Extremes of Morison-type wave loading on a single pile. *Journal of Mechanical Design, ASME 100*, 100–104.
Montgomery, D. C., E. A. Peck, and G. G. Vining (2002). *Introduction to Linear Regression Analysis*. Amsterdam, The Netherlands: Elsevier Science Publishers B. V.
Morison, J. R., M. P. O'Brien, J. W. Johnson, and S. A. Shaaf (1950). The Force Exerted by Surface Waves on Piles. *Petroleum Transactions, AIME 189*, 149–154.
Moses, F. (1987). *Load and Resistance Factor Design – Recalibration LRFD* (Report, API PRAC 8722). Dallas, TX: American Petroleum Institute.
Muhonen, A. (1996). *Evaluation of three ice-structure interaction models*. PhD Thesis, Faculty of Mechanical Engineering, Helsinki University of Technology, Helsinki, Finland.
Munkeby, T. (1996). The Heidrun TLP and Concept Development for Deep Water. In *Proceedings 6th International Offshore and Polar Engineering Conference*, Volume 1, pp. 1–11. Los Angeles, CA: ISOPE.
Naboishikov, S. (1991). On the distribution of local extremes, ranges and means of Gaussian processes. In *Proceedings 4th IFIP WG 7.5 Conference*, Munich, Germany, pp. 305–312.
Nadim, F. and R. Dahlberg (1996). Numerical Modelling of Cyclic Pile Capacity in Clay. In *Proceedings OTC Conference*, Volume 1, pp. 347–356. ASME, Houston, TX.
Naess, A. (1983). Prediction of extremes of Morison type loading – an example of a general method. *Ocean Engineering 10*(5), 313–324.
Naess, A. (1984). On the long-term statistics of extremes. *Applied Ocean Research 6*(4), 227–228.
Naess, A. (1985a). The joint crossing frequency of stochastic processes and its application to wave theory. *Applied Ocean Research 7*(1), 35–50.
Naess, A. (1985b). On the distribution of crest to trough wave heights. *Ocean Engineering 12*(3), 221–234.
Naess, A. (1985c). Statistical analysis of second-order response of marine structures. *Journal of Ship Research 29*(4), 270–284.
Naess, A. (1986). The statistical distribution of second-order slowly-varying forces and motions. *Applied Ocean Research 8*(2), 110–118.
Naess, A. (1987). The Response Statistics of Non-Linear Second-Order Transformations to Gaussian Loads. *Journal of Sound and Vibration 115*(1), 103–129.
Naess, A. (1989). Prediction of extremes of combined first-order and slow-drift motions of offshore structures. *Applied Ocean Research 11*(2), 100–110.

Naess, A. (1990). Statistical analysis of nonlinear, second-order forces and motions of offshore structures in short-crested random seas. *Probabilistic Engineering Mechanics 5*(4), 192–203.

Naess, A. (1996). A Second-Order Theory for the Response Statistics of Wave Induced Ship Hull Vibrations in Random Seas. *Probabilistic Engineering Mechanics 9*, 389–408.

Naess, A. (2001). Crossing rate statistics of quadratic transformations of Gaussian processes. *Probabilistic Engineering Mechanics 16*(3), 209–217.

Naess, A., A. J. Berstad, and L. A. Moen (1994). Stochastic fatigue analysis of the tethers of a tension leg platform. In P. D. Spanos (ed.), *Proceedings 2nd International Conference on Computational Stochastic Mechanics, Athens*. Rotterdam: Balkema.

Naess, A. and P. H. Clausen (1999). Statistical Extrapolation and the Peaks Over Threshold Method. OMAE–99–6422. In *Proceedings 18th International Conference on Offshore Mechanics and Arctic Engineering*. New York: ASME.

Naess, A. and O. Gaidai (2008). Monte Carlo methods for estimating the extreme response of dynamical systems. *Journal of Engineering Mechanics, ASCE 134*(8), 628–636.

Naess, A. and O. Gaidai (2009). Estimation of extreme values from sampled time series. *Structural Safety 31*, 325–334.

Naess, A., O. Gaidai, and S. Haver (2007). Efficient estimation of extreme response of drag dominated offshore structures by Monte Carlo simulation. *Ocean Engineering 34*(16), 2188–2197.

Naess, A., O. Gaidai, and P. S. Teigen (2007). Extreme response prediction for nonlinear floating offshore structures by Monte Carlo simulation. *Applied Ocean Research 29*(4), 221–230.

Naess, A. and J. M. Johnsen (1991). Response Statistics of Nonlinear Dynamic Systems by Path Integration. In F. C. N. Bellomo (ed.), *Proceedings IUTAM Symposium on Nonlinear Stochastic Dynamics*, pp. 401–414. Berlin: Springer-Verlag.

Naess, A. and J. M. Johnsen (1992). An Efficient Numerical Method for Calculating the Statistical Distribution of Combined First-Order and Wave-Drift Response. *Journal of Offshore Mechanics and Arctic Engineering 114*(3), 195–204.

Naess, A. and J. M. Johnsen (1993). Response statistics of nonlinear, compliant offshore structures by the path integral solution method. *Probabilistic Engineering Mechanics 8*(2), 91–106.

Naess, A. and H. C. Karlsen (2004). Numerical calculation of the level crossing rate of second order stochastic Volterra systems. *Probabilistic Engineering Mechanics 19*(2), 155–160.

Naess, A., H. C. Karlsen, and P. S. Teigen (2006). Numerical methods for calculating the crossing rate of high and extreme response levels of compliant offshore structures subjected to random waves. *Applied Ocean Research 28*(1), 1–8.

Naess, A. and G. Ness (1992). Second-Order Sum Frequency Response Statistics of Tethered Platforms in Random Waves. *Applied Ocean Research 14*(1), 23–32.

Naess, A. and J. O. Royset (2000). Extensions of Turkstra's rule and their application to combination of dependent load effects. *Structural Safety 22*, 129–143.

Naess, A., C. T. Stansberg, and O. Batsevych (2012). Prediction of Extreme Tether Tension for a TLP by the AUR and ACER Methods. *Journal of Offshore Mechanics and Arctic Engineering 134*(2), 021103(1–9).

Neal, E. (1974). Second order hydrodynamic forces due to stochastic excitation. In *Proceedings 10th ONR Symposium*. Cambridge, Mass.

Newland, D. E. (1991). *An Introduction to Random Vibrations and Spectral Analysis* (Second Edition). London, UK: Longman.

Newman, J. N. (1974). Second order slowly varying forces on vessels in irregular waves. In *Symposium on Dynamics of marine vehicles and structures in waves*. London, UK.

Newman, J. N. (1994). Wave effects on deformable bodies. *Applied Ocean Research 16*(1), 47–59.

Newmark, N. M. (1959). A Method of Computation for Structural Dynamics. *Journal of the Engineering Mechanics Division, ASCE 85*, 67–94.

Nielsen, F. G. (2007). *Lecture Notes on Marine Operations*. Trondheim, Norway: Department of Marine Technology, Norwegian University of Science and Technology.

Noor, A. K. (1994). Recent advances and applications of reduction methods. *Applied Mechanics Reviews, ASME 47*(5), 125–146.

NORSOK N-001 (1998). Structural Design. Technical Report NORSOK N-001, Rev. 3, 16 pages, Norwegian Technology Standards, Oslo, Norway.

NORSOK N-001 (2000). *Structural Design* (N-001). Oslo, Norway: Norwegian Technology Standards Institution.

NORSOK N-003 (2007). *Actions and Action Effects* (N-003, Rev. 2). Oslo, Norway: Norwegian Technology Standards Institution.

NORSOK N-004 (1998). *Steel Structures* (N-004). Oslo, Norway: Norwegian Technology Standards Institution.

Ochi, M. K. (1998). *Ocean Waves – The Stochastic Approach.* Cambridge, UK: Cambridge University Press.

Ochi, M. K. and K. Ahn (1994). Probability distribution applicable to non-Gaussian random processes. *Probabilistic Engineering Mechanics 9*, 255–264.

Odland, J. (1982). Response and strength analysis of jack-up platforms. *Norwegian Maritime Research 10*(4), 225.

Ogilvie, T. F. (1963). First- and second-order forces on a cylinder submerged under a free surface. *Journal of Fluid Mechanics 16*, 451–472.

Papoulis, A. (1965). *Probability, Random Variables and Stochastic Processes.* New York, NY: MCGraw-Hill Inc.

Paris, P. C., H. Tada, and J. K. Donald (1999). Service Load Fatigue Damage – A Historical Perspective. *International Journal of Fatigue 21*, S35–46.

Phillips, O. M. (1958). The Equilibrium Range in the Spectrum of Wind Generated Waves. *Journal of Fluid Mechanics 4*(4), 426–434.

Phillips, O. M. (1984). Spectral and Statistical Properties of the Equilibrium Range in Wind Generated Gravity Waves. *Journal of Fluid Mechanics 156*, 505–531.

Pickands, J. (1975). Statistical Interference Using Order Statistics. *Annals of Statistics 3*, 119–131.

Quek, S. T., X. M. Li, and C. G. Koh (1994). Stochastic response of jack-up platform by the method of statistical quadratization. *Applied Ocean Research 16*(2), 113–122.

Raju, I. S. and J. C. Newman (1979). Stress Intensity Factors for a Wide Range of Semi-Elliptical Surface Cracks in Finite-Thickness Plates. *Engineering Fracture Mechanics 11*(4), 817–829.

Reid, J. G. (1983). *Linear System Fundamentals.* New York, NY: McGraw-Hill.

Reiss, R.-D. and M. Thomas (2001). *Statistical Analysis of Extreme Values.* Basel, Switzerland: Birkhauser Verlag.

Rice, S. O. (1954). Mathematical Analysis of Random Noise. In N. Wax (ed.), *Selected Papers on Noise and Stochastic Processes*, pp. 133–294. New York, NY: Dover Publications, Inc.

Roberts, J. B. and P. D. Spanos (1990). *Random Vibration and Statistical Linearization.* Chichester, UK: John Wiley & Sons.

Rodriguez, G., C. G. Soares, M. Pacheco, and E. Perez-Martell (2002). Wave Height Distribution in Mixed Sea States. *Journal of Offshore Mechanics and Arctic Engineering 124*(1), 34–40.

Rychlik, I. (1987). A new definition of the rainflow cycle counting method. *International Journal of Fatigue 9*(2), 119–121.

Rychlik, I., P. Johannesson, and M. R. Leadbetter (1997). Modelling and statistical analysis of ocean wave data using transformed Gaussian processes. *Marine Structures 10*, 13–47.

Sack, R. L. (1984). *Structural Analysis.* New York: McGraw-Hill Book Company.

Sakai, S. and H. Okamura (1995). On the distribution of rainflow range for Gaussian random processes with bimodal PSD. *JSME Int J Ser A 38*, 440–445.

Sanderson, T. J. O. (1988). *Ice Mechanics and Risks to Offshore Structures.* London: Graham and Trotman, Ltd.

Sarkani, S., D. P. Kihl, and J. E. Beach (1994). Fatigue of welded joints under narrowband non-Gaussian loadings. *Probabilistic Engineering Mechanics 9*, 179–190.

Sarpkaya, T. and M. Isaacson (1981). *Mechanics of Wave Forces on Offshore Structures*. New York, NY: Van Nostrand Reinhold Comp.

Schall, G., M. H. Faber, and R. Rackwitz (2001). Ergodicity assumption for sea states in the reliability estimation of offshore structures. *Journal of Offshore Mechanics and Arctic Engineering 123*(3), 241–246.

Schetzen, M. (1980). *The Volterra and Wiener Theories of Nonlinear Systems*. New York, NY: John Wiley & Sons, Inc.

Schmiechen, M. (1973). On State Space Models and their Application to Hydromechanic Systems. Technical Report NAUT 5002, University of Tokyo, Tokyo, Japan.

Shinozuka, M. (1974). Digital Simulation of Random Processes in Engineering Mechanics with the Aid of FFT Techniques. In *Stochastic Problems in Mechanics*, pp. 277–286. Waterloo, Ontario, Canada: University of Waterloo Press.

Shinozuka, M. and G. Deodatis (1991). Simulation of Stochastic Processes by Spectral Representation. *Applied Mechanics Review 44*(4), 191–203.

Shinozuka, M. and C.-M. Jan (1972). Digital Simulation of Random Processes and Its Application. *Journal of Sound and Vibration 25*(1), 111–128.

Shyu, W.-H., Z.-D. Ma, and G. M. Hulbert (1997). A new component mode synthesis method: Quasi-static mode compensation. *Finite Elements in Analysis and Design 24*(4), 271–281.

Simiu, E. and R. H. Scanlan (1996). *Wind Effects on Structures* (Third Edition). New York, NY: John Wiley & Sons.

SNAME (1988). *Principles of Naval Architecture – Stability and Strength*, Volume 1. New York, NY: Society of Naval Architects and Marine Engineers.

Sobczyk, K. and B. F. Spencer (1992). *Random Fatigue: From Data to Theory*. Boston, USA: Academic Press, Inc.

Soong, T. T. and M. Grigoriu (1993). *Random Vibration of Mechanical and Structural Systems*. Englewood Cliffs, NJ: PTR Prentice Hall.

Søreide, T. H. and J. Amdahl (1994). *USFOS – A Computer Program for Ultimate Strength Analysis of Framed Offshore Structures* (Theory Manual: Report STF71 A86049). Trondheim, Norway: SINTEF Structural Engineering.

Spanos, P. D. and M. G. Donley (1991). Equivalent statistical quadratization for nonlinear systems. *Journal of Engineering Mechanics, ASCE 117*(6), 1289–1310.

Spanos, P. D., G. Failla, and M. D. Paola (2003). Spectral approach to equivalent statistical quadratization and cubicization methods for nonlinear oscillators. *Journal of Engineering Mechanics, ASCE 129*(1), 31–42.

Stansberg, C. T. (1992). Model Scale Experiments on Extreme Slow-Drift Motions in Irregular Waves. In *Proceedings of the BOSS-92 Conference, Vol. 2*, pp. 1207–1222. London, UK: BPP Technical Services, Ltd.

Stansberg, C. T. (2000). Prediction of Extreme Slow-Drift Amplitudes. In *Proceedings of the 19th International Conference on Offshore Mechanics and Arctic Engineering*, Number OMAE00-6135. New Orleans, USA: ASME.

St. Denis, M. and W. J. Pierson (1953). On the motions of ships in confused seas. *Transactions of SNAME 61*, 280–357.

Stewart, G. (1992). Non-Linear Structural Dynamics by Pseudo-Force Influence Method, Part II: Application to Offshore Platform Collapse. In *Proceedings of ISOPE'92*. San Francisco, CA: ISOPE.

Storhaug, G. (2007). *Experimental investigation of wave induced vibrations and their effect on the fatigue loading of ships*. Theses at NTNU, 2007:133, Norwegian University of Science and Technology, Trondheim, Norway.

Sundararajan, C. and D. V. Reddy (1973). Stochastic analysis of ice structure interaction. In *Proceedings 2nd International Conference on Port and Ocean Engineering under Arctic Conditions (POAC'73)*, pp. 345–353. Reykjavik, Iceland.

Taghipour, R., T. Perez, and T. Moan (2008). Hybrid Frequency-Time Domain Models for Dynamic Response Analysis Of Marine Structures. *Ocean Engineering 35*, 685–705.

Taghipour, R., T. Perez, and T. Moan (2009). Time-Domain Hydroelastic Analysis of a Flexible Marine Structure Using State-Space Models. *International Journal of Offshore Mechanics and Arctic Engineering 131*(1), 011603.

Teigen, P. and A. Naess (1999). Stochastic Response Analysis of Deepwater Structures in Short-Crested Random Waves. *Journal of Offshore Mechanics and Arctic Engineering, ASME 121*, 181–186.

Toro, G. R. (1984). *Probabilistic Analysis of Combined Dynamic Responses*. Report no. 65, Stanford University, Palo Alto, CA.

Torsethaugen, K. and S. Haver (2004). Simplified Double Peak Spectral Model for Ocean Waves. In *Proceedings of the 14th International Offshore and Polar Engineering Conference*, Number Paper No. 2004-JSC-193. Montain View, CA: ISOPE.

Tovo, R. (2002). Cycle distribution and fatigue damage under broad-band random loading. *International Journal of Fatigue 24*, 1137–1147.

Tromans, P. S. and L. Vanderschuren (1995). Response Based Design Conditions in the North Sea: Application of a New Method. In *Proceedings of the Offshore Technology Conference*, Number OTC 7683. Houston, TX.

Turkstra, C. J. (1970). *Theory of Structural Safety*. Ontario, Canada: SM Study No. 2, Solid Mechanics Division, University of Waterloo.

UBC (1988). *Handbook to the Uniform Building Code: An Illustrative Commentary*. Whittier, CA 90601: International Conference of Building Officials.

Van der Hoven, I. (1957). Power spectrum of horizontal wind speed in the frequency range from 0.0007 to 900 cycles per hour. *Journal of Meteorology 14*, 160–164.

Vanmarcke, E. H. (1972). Properties of spectral moments with applications to random vibration. *Journal of the Engineering Mechanics Division, ASCE 98*, 425–426.

Videiro, P. M. and T. Moan (1999). Efficient evaluation of long-term distributions. In *Proceedings of the 18th International Conference on Offshore Mechanics and Arctic Engineering*, St. Johns, Canada: ASME.

Videiro, P. M. and T. Moan (2000). Reliability analysis of offshore structures under multiple long-term wave load effects. In *Proceedings of the 8th International Conference on Applications of Statistics and Probability (ICASP8)*, pp. 1165–1173. CERRA: A.A. Balkema, Rotterdam.

Vinje, T. (1980). Statistical distributions of hydrodynamic forces on objects in current and waves. *Norwegian Maritime Research 8*(2), 20–26.

Vinje, T. (1983). On the statistical distribution of second-order forces and motions. *International Shipbuilding Progress 30*, 58–68.

Vugts, J. H. (1970). *The Hydrodynamic Forces and Ship Motions in Waves*. PhD Thesis, Technical University, Delft, The Netherlands.

WAMIT (2008). *WAMIT Inc.* www.wamit.com.

Wang, X. and J. Q. Sun (2005). Effect of skewness on fatigue life with mean stress correction. *Journal of Sound and Vibration 282*, 1231–1237.

Watson, P. and B. J. Dabell (1975). Cycle counting and fatigue damage. In *Symposium on Statistical Aspects of Fatigue Testing*, Warwick, UK. Warwick University.

Wen, Y.-K. (1990). *Structural Load Modeling and Combination for Performance and Safety Evaluation*. Amsterdam, The Netherlands: Elsevier Science Publishers B.V.

Wen, Y. K. and H. C. Chen (1987). On fast integration for time variant structural reliability. *Probabilistic Engineering Mechanics 2*(3), 156–162.

Wen, Y. K. and H. C. Chen (1989a). System reliability under time varying loads, I. *Journal of Engineering Mechanics, ASCE 115*(4), 808–823.

Wen, Y. K. and H. C. Chen (1989b). System reliability under time varying loads, II. *Journal of Engineering Mechanics, ASCE 115*(4), 824–839.

Wen, Y. K. and H. T. Pearce (1981). Recent developments in probabilistic load combinations. In *Proceedings on Probabilistic Methods in Structural Engineering*, St. Louis, Mo. ASCE.

Winterstein, S., T. Ude, C. A. Cornell, P. Bjerager, and S. Haver (1993). Environmental parameters for extreme response: Inverse FORM with omission factors. In *Proceedings*

6th International Conference on Structural Safety and Reliability (ICOSSAR'93). Innsbruck, Austria: Balkema.
Winterstein, S. R. (1985). Non-normal responses and fatigue damage. *Journal of Engineering Mechanics, ASCE 111*(10), 1291–1295.
Winterstein, S. R. (1988). Nonlinear vibration models for extremes and fatigue. *Journal of Engineering Mechanics, ASCE 114*, 1772–1790.
Wirsching, P. and M. C. Light (1980). Fatigue under wide band random loading. *Journal of the Structural Division, ASCE 106*(7), 1593–1607.
Wolf, J. P. (1994). *Foundation Vibration Analysis using Simple Physical Models*. Englewood Cliffs, NJ: PTR Prentice Hall.
Wood, W. L. (1984). A further look at Newmark, Houbolt, etc. time-stepping formulae. *International Journal for Numerical Methods in Engineering 20*, 1009–1017.
Wu, M. K. and T. Moan (1996). Linear and Nonlinear Hydroelastic Analysis of High Speed Vessels. *Journal of Ship Research 40*(2), 149–163.
Wu, M. K. and T. Moan (2005). Efficient calculation of wave-induced ship responses considering structural dynamic effects. *Applied Ocean Research 27*, 81–96.
Yago, K. and H. Endo (1996). Model experimental and numerical calculation of the hydroelastic behaviour of matlike VLFS. In *Proceedings of International Workshop on Very Large Floating Structures*. Hayama, Japan.
Yasuzawa, Y., K. Kagawa, D. Kawano, and K. Kitabayashi (1997). Dynamic response of a large flexible floating structure in regular waves. In *Proceedings of the 16th International Conference on Offshore Mechanics and Arctic Engineering*. New Orleans, USA: ASME.
Yeung, R. W. (1989). Added mass and damping of a vertical cylinder in finite-depth water. *Applied Ocean Research 11*(3), 119–133.
Yue, Q. J., Y. Qu, X. J. Bi, and T. Kärnä (2007). Ice Force Spectrum on Narrow Conical Structures. *Cold regions Science and Technology 49*(2), 161–169.
Zhang, B. and T. Moan (2005). Analysis of fatigue crack propagation in typical welded joints of FPSOs. In *Proceedings 24th International Conference on Offshore Mechanics and Arctic Engineering*, pp. OMAE2005–67058. New York: ASME.
Zhao, W. and M. J. Baker (1992). On the probability density function of rainflow stress range for stationary Gaussian processes. *International Journal of Fatigue 12*(2), 121–135.
Zienkiewicz, O. C. and R. L. Taylor (2006). *The Finite Element Method for Solid and Structural Mechanics* (Sixth Edition). Amsterdam, The Netherlands: Elsevier.
Zienkiewicz, O. C., R. L. Taylor, and J. Z. Zhu (2005). *The Finite Element Method: Its Basis and Fundamentals* (Sixth Edition). Amsterdam, The Netherlands: Elsevier.

Index

For EU product safety concerns, contact us at Calle de José Abascal, 56–1°, 28003 Madrid, Spain or eugpsr@cambridge.org.

www.ingramcontent.com/pod-product-compliance
Ingram Content Group UK Ltd.
Pitfield, Milton Keynes, MK11 3LW, UK
UKHW050030190126
467120UK00010B/451
* 9 7 8 0 5 2 1 8 8 1 5 5 5 *